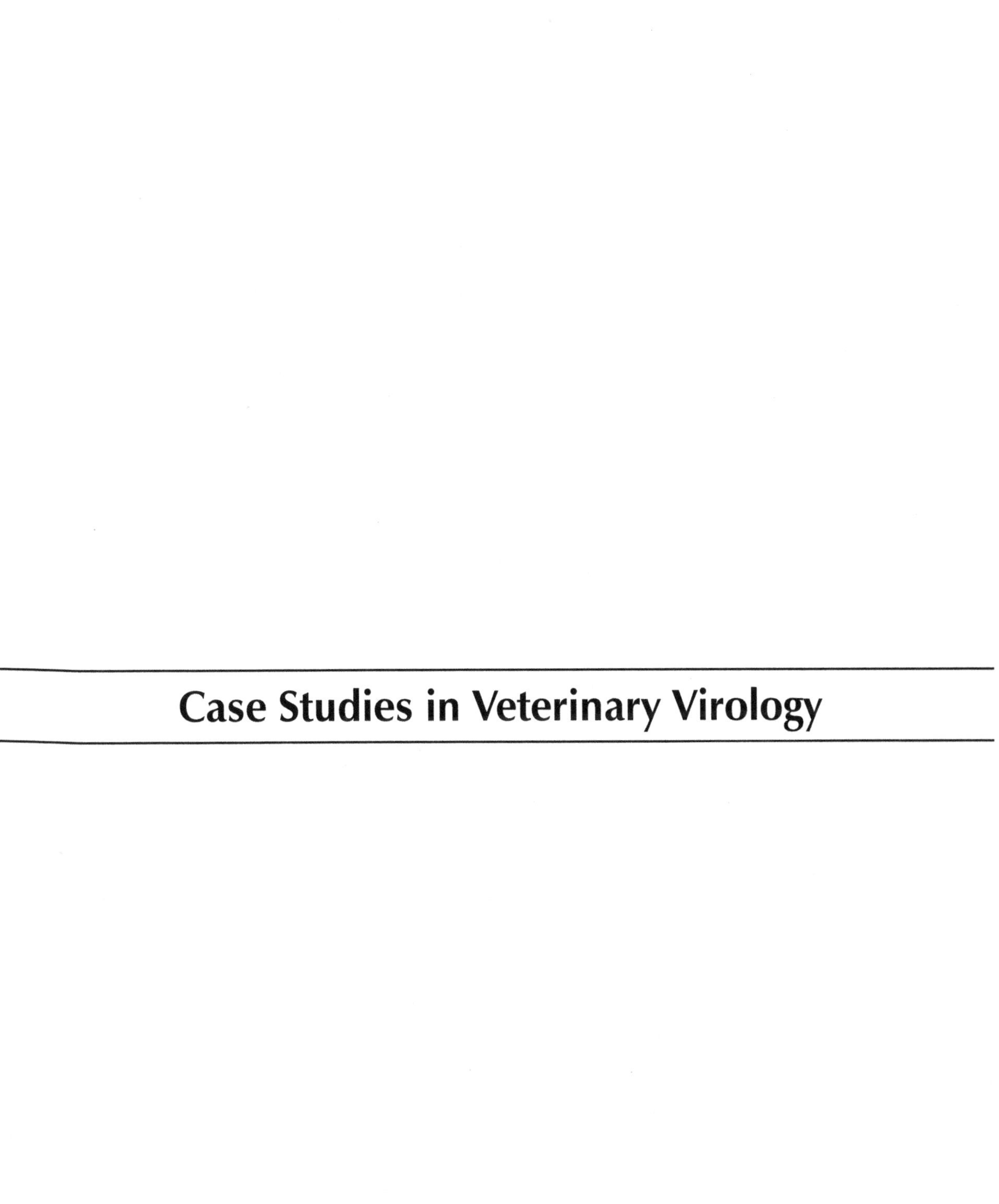

Case Studies in Veterinary Virology

Case Studies in Veterinary Virology

Editor: Travis Schroeder

Cataloging-in-Publication Data

Case studies in veterinary virology / edited by Travis Schroeder.
p. cm.
Includes bibliographical references and index.
ISBN 978-1-63927-519-9
1. Veterinary virology. 2. Veterinary virology--Case studies. 3. Veterinary medicine--Case studies.
4. Animals--Diseases. 5. Virus diseases. I. Schroeder, Travis.
SF780.4 .C37 2022
636.089 601 94--dc23

American Medical Publishers,
41 Flatbush Avenue,
1st Floor, New York,
NY 11217, USA

ISBN 978-1-63927-519-9 (Hardback)

Contents

Preface

Veterinary virology is a major branch of veterinary medicine that studies viruses in non-human animals. Some of the viruses studied under this discipline are rhabdoviruses, foot and mouth disease viruses, pestiviruses, parvoviruses, coronaviruses, toroviruses and influenza etc. Rhabdoviruses are single stranded, negative sense RNA viruses which can infect a wide variety of animals. A few examples of rhabdoviruses are rabies virus and vesicular stomatitis virus. Foot and mouth disease viruses are non-enveloped, positive strand, RNA viruses that cause foot and mouth diseases in animals such as cattle, sheep and pigs. Pestiviruses are made up of single stranded, positive-sense RNA genomes. Diseases like classic swine fever and bovine viral diarrhea are caused due to these viruses. Parvoviruses are one of the tiniest viruses. They cause diseases in the gastrointestinal tract and lymphatic system. This book explores all the important aspects of veterinary virology. It also elucidates some of the vital pieces of work conducted across the world, on various topics related to this field. The extensive content of this book provides the readers with a thorough understanding of the subject.

The researches compiled throughout the book are authentic and of high quality, combining several disciplines and from very diverse regions from around the world. Drawing on the contributions of many researchers from diverse countries, the book's objective is to provide the readers with the latest achievements in the area of research. This book will surely be a source of knowledge to all interested and researching the field.

In the end, I would like to express my deep sense of gratitude to all the authors for meeting the set deadlines in completing and submitting their research chapters. I would also like to thank the publisher for the support offered to us throughout the course of the book. Finally, I extend my sincere thanks to my family for being a constant source of inspiration and encouragement.

Editor

Canine distemper virus isolated from a monkey efficiently replicates on Vero cells expressing non-human primate SLAM receptors but not human SLAM receptor

Na Feng[1,2†], Yuxiu Liu[3†], Jianzhong Wang[1†], Weiwei Xu[2], Tiansong Li[4], Tiecheng Wang[2], Lei Wang[6], Yicong Yu[2], Hualei Wang[2], Yongkun Zhao[2], Songtao Yang[2], Yuwei Gao[2*], Guixue Hu[1*] and Xianzhu Xia[1,2,5*]

Abstract

Background: In 2008, an outbreak of canine distemper virus (CDV) infection in monkeys was reported in China. We isolated CDV strain (subsequently named Monkey-BJ01-DV) from lung tissue obtained from a rhesus monkey that died in this outbreak. We evaluated the ability of this virus on Vero cells expressing SLAM receptors from dog, monkey and human origin, and analyzed the H gene of Monkey-BJ01-DV with other strains.

Results: The Monkey-BJ01-DV isolate replicated to the highest titer on Vero cells expressing dog-origin SLAM ($10^{5.2\pm0.2}$ $TCID_{50}$/ml) and monkey-origin SLAM ($10^{5.4\pm0.1}$ $TCID_{50}$/ml), but achieved markedly lower titers on human-origin SLAM cells ($10^{3.3\pm0.3}$ $TCID_{50}$/ml). Phylogenetic analysis of the full-length H gene showed that Monkey-BJ01-DV was highly related to other CDV strains obtained during recent CDV epidemics among species of the Canidae family in China, and these Monkey strains CDV (Monkey-BJ01-DV, CYN07-dV, Monkey-KM-01) possessed a number of amino acid specific substitutions (E276V, Q392R, D435Y and I542F) compared to the H protein of CDV epidemic in other animals at the same period.

Conclusions: Our results suggested that the monkey origin-CDV-H protein could possess specific substitutions to adapt to the new host. Monkey-BJ01-DV can efficiently use monkey- and dog-origin SLAM to infect and replicate in host cells, but further adaptation may be required for efficient replication in host cells expressing the human SLAM receptor.

Keywords: Canine distemper virus (CDV), Monkey, SLAM, H protein

Background

Canine distemper virus (CDV) is a single-stranded, negative-sense, nonsegmented RNA virus of genus Morbillivirus within the family Paramyxoviridae. CDV is a highly contagious pathogen that can cause disease with high morbidity and mortality in immunologically naive hosts as a result of viral tropism for the cutaneous, respiratory, gastrointestinal, and central and peripheral nervous systems [1]. CDV has a broad host range and primarily affects animals belonging to the *Canidae* (e.g. dogs, wolves, and foxes) and *Mustelidae* (e.g. ferrets, badgers, and mink) families [2–4]. Previous studies had implicated CDV in the pathogenesis of Paget's disease [5], and natural CDV infection of non-human primates has been reported [6–8]. In 2006, a CDV outbreak occurred in rhesus monkeys (Macaca mulatta) at a breeding farm in Guangxi province, China, with a morbidity rate (60 %) and a mortality rate (≈30 %), unexpectedly [6]. Two additional CD occurred in monkeys were reported in 2008. One occurred in rhesus monkeys at a laboratory animal center in Beijing with a reported 60 %

* Correspondence:
gywtext@gmail.com; huguixue901103@163.com; xiaxzh@cae.cn
†Equal contributors
[2]Military Veterinary Research Institute of Academy of Military Medical Sciences, Key Laboratory of Jilin Province for Zoonosis Prevention and Control, Changchun 130122, China
[1]College of Animal Science and Technology, Jilin Agricultural University, Changchun 130118, China
Full list of author information is available at the end of the article

(12/20) mortality [7], another occurred in Japan following the importation of cynomolgus monkeys (Macaca fascicularis) from China and was associated with a 10.6 % (46/432) mortality [8].

Host cell infection initiates with viral binding to receptor proteins on the surface of cells. The CDV viral envelope contains two integral glycoproteins, the hemagglutinin (H) protein and fusion (F) protein. The H protein mediates the binding of the virus to the cell membrane, and the F protein serves to fuse viral and host membranes, thereby enabling release of the viral contents into the cytoplasm [2]. The CDV-H glycoprotein mediates viral attachment through specific interactions with signaling lymphocyte activation molecule (SLAM) [9] or nectin-4 cellular receptors [10, 11]. SLAM is expressed on a subset of immune cells, while nectin-4 is expressed on epithelial cells of various organs. Although SLAM is a main receptor for morbilliviruses, each morbillivirus preferentially uses the SLAM of its host animals, as the specific residues within SLAM responsible for mediating interactions with the CDV-H protein can vary by species [12]. The specificity of the CDV-H protein for the SLAM receptor imposes species barriers and is partly responsible for restricting CDV host range [12, 13]. Nectin-4 is highly conserved among different animals [10]. Unlike the SLAM receptor, the characteristics of binding between nectin-4 and CDV were unknown. In this study, we evaluated the replication capacity of a CDV isolate obtained from a naturally infected monkey on Vero cells expressing dog, monkey, or human SLAM receptor proteins to better understand how CDV-H protein receptor specificity affects host range restriction.

Results

Surface expression of dog, monkey, and human SLAM on Vero cells

Vero cells were transfected with expression plasmids to express the dog, monkey, or human version of the SLAM protein. The expression plasmids included an HA epitope tag to allow confirmation of the surface expression of the SLAM protein. After cells stably transfected with G418 had been selected, SLAM expression was examined by flow cytometry using an HA tag-specific monoclonal antibody. The results showed that dog, monkey, and human SLAM was expressed equally on the surface of the resulting cell lines (46.5 %, 45.7 % and 45 %, respectively) but not on the empty vector-transfected Vero cells (Fig. 1). These cells were named Vero/DogSLAM, Vero/MonkeySLAM and Vero/HumanSLAM, respectively. For virus isolation study, these cells were passaged fewer than five times.

Isolation of CDV from the lung tissue of a deceased rhesus monkey

We used the Vero/DogSLAM cells to isolate CDV from lung samples collected previously from a deceased rhesus monkey that had displayed signs of CD. Inoculation of the Vero/DogSLAM cells with supernatants from the homogenized tissue resulted in an obvious cytopathic effect (CPE), whereas no CPE was observed after four blind passages of the supernatants on the Vero cells (Fig. 2, upper panels). Nucleoprotein expression in the cells was analyzed by indirect immunofluorescence assay (IFA). The nucleoprotein antigen was detected in Vero/DogSLAM cells inoculated with homogenized tissue supernatants but was not detected the inoculated Vero cells (Fig. 2, lower panels). The viruses isolated from the inoculated Vero/DogSLAM cells were named Monkey-BJ01-DV.

Phylogenetic analysis of the H protein of Monkey-BJ01-DV

We sequenced the full-length H protein of the Monkey-BJ01-DV (GenBank accession number KM923900) and performed a phylogenetic analysis of this strain and others deposited in GenBank (Table 1). The Monkey-BJ01-DV was highly similar to the other CYN07-dV monkey strains (AB687720; 99.8 %; 1823/1824 nucleotides) and Monkey-KM-01 (FJ405224; 99.0 %; 1807/1824

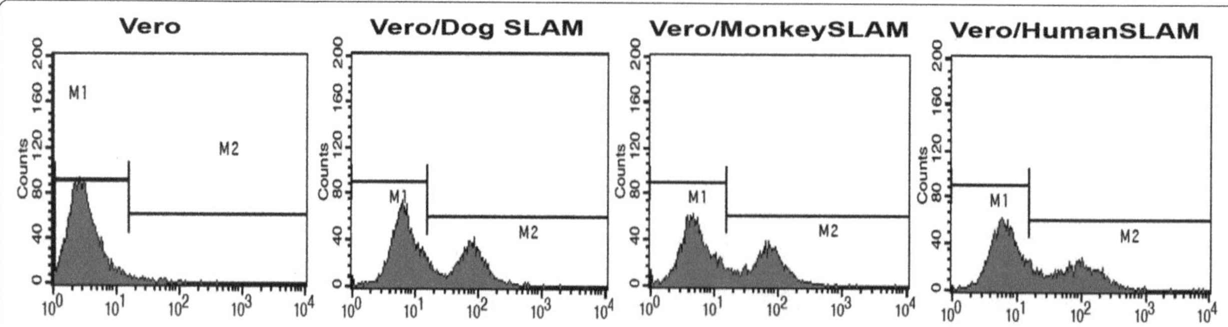

Fig. 1 Stable expression of dog, monkey and human SLAM protein on Vero cells. Vero cells were transfected with expression plasmids to mediate expression of HA epitope-tagged SLAM proteins from dog, monkey, and human and selected with G418. The control Vero cells were transfected by empty vector. The resulting cell lines were named Vero/DogSLAM,Vero/MonkeySLAM and Vero/HumanSLAM. Anti-influenza virus HA epitope MAb was used to confirm surface expression of each version of the SLAM protein by flow cytometry

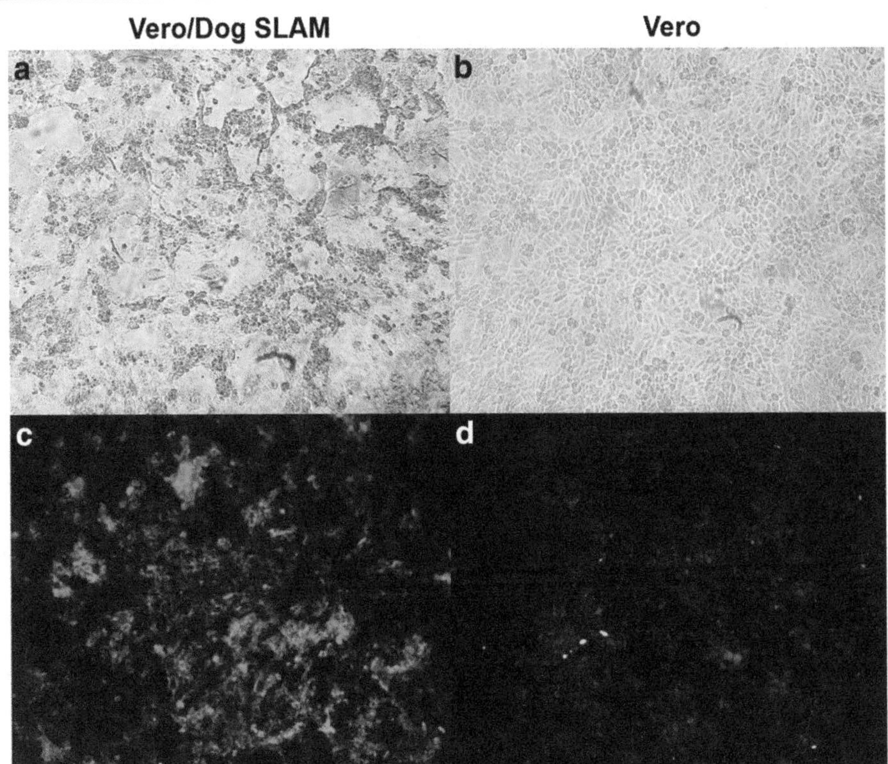

Fig. 2 Infection of Vero cells expressing dog SLAM protein with CDV from a deceased rhesus monkey. Microscopic evaluation of Vero/DogSLAM cells (**a**) and Vero cells (**b**) following inoculation with Monkey-BJ01-DV at a MOI of 0.01. Inoculation of Vero/DogSLAM cells resulted in obvious CPE. Detection of CDV antigens in Vero cells expressing the dog SLAM protein (**c**) and untransfected Vero cells (**d**) 64 h following inoculation with Monkey-BJ01-CDV at a MOI of 0.01. Cells were fixed and probed with a mouse monoclonal anti-CDV nucleoprotein antibody. A FITC-conjugated goat anti-mouse secondary antibody was used for detection

Table 1 Amino acid differences between the hemagglutinin of Monkey-BJ01-DV and other CDV isolates

Amino acid sequence identity[a]	Sequence no.	24	276	365	392	435	530	542	549	597
98.4 %	JN896331 (PS-dog)	S	E	V	Q	D	G	I	Y	R
97.5 %	AB286946 (MD231-dog)	b	b	A	b	b	b	b	b	b
98.4 %	JQ732173 (LDH (06)-fox)	b	b	b	b	b	b	b	b	b
98.4 %	FJ848530 (BJ080514-dog)	b	b	b	b	b	b	b	b	b
98.2 %	EU325720 (fox-Hebei07)	b	b	b	b	b	b	b	b	b
98.0 %	EU325721 (fox-HLJ07)	b	b	b	b	b	b	b	b	b
97.4 %	EU325724 (mink-LN)	b	b	b	b	b	b	b	b	b
97.9 %	EU325728 (raccoon-JL07)	b	b	b	b	b	b	b	b	b
97.5 %	GQ332530 (dog-Wuhan)	b	b	b	b	b	b	b	b	b
97.3 %	CDV-TM-CC	b	b	b	b	b	b	b	b	b
———	Monkey-BJ01-DV	F	V	A	R	Y	b	F	b	H
99.8 %	FJ405223 (monkey-BJ-01)	F	V	A	R	Y	b	F	b	b
99.8 %	AB687720 (mCNY07-dV)	F	V	A	R	Y	b	F	b	H
99.0 %	FJ405224 (monkey-KM-01)	b	V	A	R	Y	b	F	b	b

[a]Percent identity in hemagglutinin amino acid sequence when compared to Monkey-BJ01-DV
[b]Indicate the same residues as the ones of JN896331 (PS-dog)

nucleotides), which were isolated from a cynomolgus monkey and from a rhesus monkey, respectively [6, 8]. These three monkey strains are each associated with CDV outbreaks among monkeys in China and belong to the Asia type I lineage. The CDV H protein is responsible for host cell attachment, is the most variable protein described for all members of the genus Morbillivirus and is an important determinant of the CDV host range [3]. We found that the H protein from these three strains of monkey-CDV possessed a number of amino acid specific substitutions compared to the H protein of CDV epidemic in other animals at the same period, as shown in Table 1: E276V, Q392R, D435Y and I542F. The glycine (G) residue at position 530 and the tyrosine (Y) residue at position 549, which correspond to the partial SLAM-receptor binding region, were conserved among all three monkey CDV isolates and among CDV isolates from members of *Canidae* and *Mustelidae* (Table 1).

Replication of Monkey-BJ01-DV on Vero cells expressing SLAM from dog-, monkey-, or human- origin

We next evaluated the replication ability of the Monkey-BJ01-DV in engineered Vero/SLAM cell lines. Monkey-BJ01-DV replicated to the highest titers on Vero/DogSLAM ($10^{5.2\pm0.2}$ $TCID_{50}$/ml) and Vero/MonkeySLAM ($10^{5.4\pm0.1}$ $TCID_{50}$/ml) at 48 h post-infection, whereas replication on the Vero/HumanSLAM cells was reduced approximately 100-fold ($10^{3.3\pm0.3}$ $TCID_{50}$/ml; Fig. 3). The dog-CDV isolate from a Tibetan mastiff replicated to high titers on the Vero/DogSLAM cells and displayed reduced titers on the Vero cells expressing SLAM from monkey- or human- origin. Our results indicate that both the monkey- and dog-CDV isolates can efficiently replicate on Vero cells expressing either the monkey- or dog-origin SLAM receptor, but replicate less efficiently on cells expressing the SLAM receptor of human- origin.

Discussion

Morbilliviruses, including measles virus (MV), CDV, and rinderpest virus (RPV), are thought to have originated from a common ancestor several thousand years ago [14]. In general, however, each virus is able to naturally infect a relatively restricted number of host species. MV infection is limited to primates, CDV infection is limited to members of the *Canidae*, *Mustelidae* and *Procyonidae* families, and RPV infection is limited to ruminants [15]. Experimentally, monkeys are susceptible to MV infection and dogs to CDV infection [16]. Despite the apparent host range restriction of morbilliviruses, CDV has crossed species barriers, suggesting a potential ongoing expansion of host range. The first report of natural CDV infection of non-human primates was reported in 1989 among Japanese monkeys [15]. In 2006, a CDV epidemic affected a rhesus monkey (Macaca mulatta) breeding farm in China, inflicting high morbidity (60 %) and mortality (≈30 %) in young monkeys [6]. In 2008, Japan imported 432 cynomolgus monkeys (Macaca fascicularis) from China and 46 died from CDV while held in quarantine [8]. We isolated CDV from a deceased rhesus monkey (Monkey-BJ01-DV) following a 2008 outbreak in Beijing in which 12 of 20 rhesus monkeys died from CDV [7]. Collectively, these reports suggest significant changes in the epidemiology of CDV caused by an expanding CDV host range and/or viral virulence.

The complete genomes of three monkey-origin CDV isolates associated with outbreaks in China have been reported

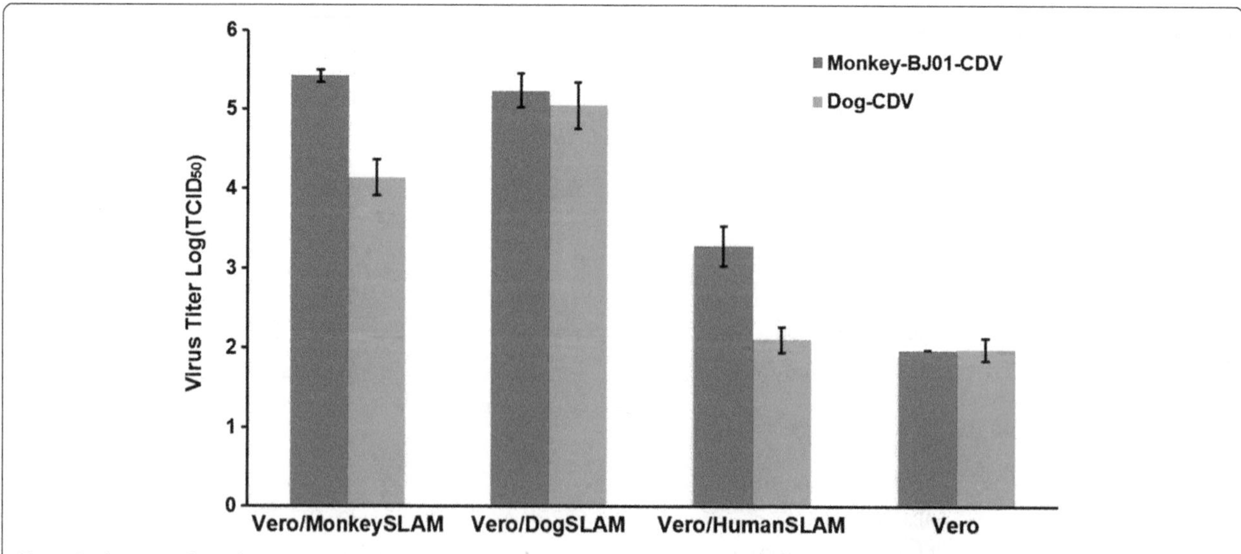

Fig. 3 Replication of Monkey-BJ01-DV and CDV-TM-CC in Vero/SLAM cells. Cells were infected at a MOI of 0.01. Viral titers were measured 48 h post-infection

(Monkey-BJ01-DV (KF856711), CYN07-dV (AB687720) and MKY-KM-08 (HM852904) and are highly related to other epidemic CDV strains affecting species of the *Canidae* and *Mustelidae* families in China [6, 8] (Table 1). The H protein mediates viral attachment by binding to one or more cellular receptors [9, 11]. Several studies have speculated about the impact of specific amino acid substitutions within the H protein on interactions with the SLAM receptor. In vitro receptor-binding studies showed that amino acid residues 527, 528, 529, and 552 of the H protein are conserved among all morbilliviruses and are crucial for CDV-H to SLAM dependent fusion [4]. In particular, residues 530 and 549 both fall into receptor-binding domains located on propeller β-sheet 5 of the H protein [17]. Sequence analysis of the H gene of all three monkey CDV strains revealed a glycine (G) and a tyrosine (Y) at amino acid positions 530 and 549 of the partial SLAM-receptor binding region. G530 and Y549 are typically found in viral strains obtained from dogs in China, whereas other amino acid residues are present in CDV isolates obtained from wildlife [18, 19]. This suggests that the CDV isolated from monkeys was transmitted from domestic dogs rather than from other wildlife. In this study, we found that amino acid Y549 was conserved within CDV lineages of the isolates analyzed, regardless of host species. CDV transmission in wild carnivore and non-canid species may most often occur between individuals within a species, and may also be influenced by a range of additional factors such as population size and ranging patterns. The three monkey CDV strains possessed E276V, Q392R, D435Y and I542F substitutions, which are unique changes when compared to the other Asia type I lineage strains. In particularly, the I542F substitution falls with the SLAM-binding regions of the H protein. The H protein amino acid substitutions identified among monkey CDV isolates may help explain recent changes in CDV host range.

The Monkey-BJ01-DV can efficiently grow on the Vero cells expressing SLAM from dog- and monkey-origin, but not the cells expressing SLAM receptor of human- origin. Interestingly, while the amino acid sequence identity of dog and monkey SLAM is only 63.6 %, the Monkey-BJ01-DV strain can replicate on the Vero cells expressing SLAM receptor of dog-origin as efficiently as the Vero cells expressing monkey-origin SLAM. Factors other than the receptor binding, such as intracellular replication of the viruses may also be important for the establishment of infection. Further studies are required to understand the mechanisms by which CDV can cross species barriers.

Conclusions

Canine distemper virus isolated from a deceased rhesus monkey efficiently replicates on the Vero cells expressing non-human primate SLAM receptors but not human SLAM receptor. The monkey origin-CDV-H protein could possess specific substitutions to adapt to the new host.

Methods

Plasmids and cell lines

Peripheral blood samples from a dog and rhesus monkey were collected from the Animal Hospital of Jilin University and the Animal Laboratory Center of the Academy of Military Medical Sciences, respectively. Human peripheral blood was kindly provided by the voluntary enrolled in this study. Peripheral blood mononuclear cells (PBMCs) of the dog, rhesus monkey and human were isolated from peripheral blood using dog and human lymphocyte separation media (TBD, Tianjin, China) according to the manufacturer protocols, respectively. Total RNA was extracted from the dog, rhesus monkey and human PBMCs after 2–4 h of stimulation with 2.5-3.0 mg of phytohemagglutinin (PHA) per milliliter and was used for reverse transcription with oligo (dT) primers. The presence and location of signal peptide cleavage sites in the amino acid sequences from the dog-SLAM (AF325357), rhesus monkey-SLAM (XM-001117605) and human-SLAM (NM-003037) were predicted by the Signal P 3.0 software. The SLAM gene of the dog, rhesus monkey and human without the signal peptide were encoded using various combinations of the primers designed on the basis of known SLAM sequences. The primer sets used were as follows: dogSLAM-F: 5′-GCCTCGAGACAGGTGAGAGCTTGATGAAT-3′ with the XhoI site underlined and dogSLAM-R: 5′-G CAGATCTTCAGCTCTCTGGGAACGTCAC-3′ with the BglII site underlined for dog; monkey/humanSLAM-F: 5′-GCCTCGAGGCAAGCTATGGAACAGGTGGG-3′; monkeySLAM-R: 5′-GCAGATCTTCAGCTCTCTG-GAAGTGTCACACT-3′; and humanSLAM-R: 5′-TCAGATCTCTGGRARYGTCACRCT-3′), respectively. After checking against the NCBI reference sequence of dog SLAM, rhesus monkey SLAM and human SLAM, the cDNA encoding the DogSLAM, MonkeySLAM, and HumanSLAM proteins were inserted into the pCAGGS (Neo) vector possessing the immunoglobulin Igk leader sequence (GAG ACAGACACACTCCTGCTATGGGTA CTGCTGCTCTGGGTTCCAGGTTCCACTGGTGAC) and the influenza virus hemagglutinin (HA) epitope (TATCCATATGATGTTCCAGATTATGCT) [20], generated pCAGDogSLAM, pCAGMonkeySLAM and pCA-GHumanSLAM. Vero cells constitutively expressing dog SLAM (Vero/DogSLAM), rhesus monkey SLAM (Vero/MonkeySLAM) and human SLAM (Vero/HumanSLAM) were generated by transfecting Vero cells with pCAGDog-SLAM, pCAGMonkeySLAM or pCAGHumanSLAM, respectively, and control Vero cells were transfected by empty vector. These cells were maintained in DMEM with 5 % FBS and 0.8 mg/ml Geneticin (G418; Invitrogen) in a humidified atmosphere at 37 °C and 5 % CO_2. The above -described cells were stained with anti-influenza virus HA epitope MAb 12CA5 (Boehringer Mannheim), and then stained with fluorescein isothiocyanate (FITC)-conjugated goat anti-mouse IgG (Abcam, Cambridge science Park, UK). These stained cells were analyzed on a FACScan machine (Becton Dickinson) [2].

Virus isolation

Tissue samples used for virus isolation were obtained from the lungs of a deceased rhesus monkey during the CDV epizootic in Beijing in 2008 [7]. Lung tissue was suspended in cold phosphate- buffered saline (PBS) with antibiotics and was grinded into a homogenate. Homogenized lung samples were centrifuged at 2500 rpm for 5 min; the supernatant was collected and centrifuged an additional 5 min at 5000 rpm. Supernatants were inoculated onto monolayers of Vero/DogSLAM cells, from which a CDV isolate was subsequently obtained and named Monkey-BJ01-DV. An additional CDV-TM-CC strain was isolated from a Tibetan mastiff in our laboratory [21]. Normal Vero cells and Vero/DogSLAM cells were plated in 24-well plates and infected with the Monkey-BJ01-DV. Productive CDV infection from the cultured cells was analyzed with a mouse monoclonal anti-CDV nucleoprotein antibody. A FITC-conjugated goat anti-mouse IgG was used as the secondary antibody, cells were analyzed with a fluorescence microscope (BX51FL; Olympus, Japan).

RT-PCR and sequencing analysis of the Monkey-BJ01/ Monkey-BJ01-DV

Total RNA was extracted from the lung of died monkey/ Monkey-BJ01-DV by using TRIzol reagent (Molecular Research Center Inc., USA) according to the manufacturer instructions. Reverse transcription (RT) was carried out using the Superscript II reverse transcriptase (Invitrogen, USA) according to the standard protocol. The CDV-H gene was cloned using the Ex-Taq DNA polymerase (TaKaRa) with the following primers: CDV-HF: GCGAATTCATGCTCTCCTACCAAGACAAGGTG with the EcoRI site underlined and according to the CDV-HR: GGCCCTCGAGTCAAGGTTTTGAACGATTAC with XhoI site underlined. The PCR products were cloned into a pMD 18-T vector (TaKaRa) and were sequenced. At least five clones of each PCR products were analyzed to acquirie the accurate sequence. The sequence determined in this study had been registered at the GenBank under accession numbers FJ405223 and KM923900.

Replication of CDV on Vero cells expressing the SLAM receptor from various animal species

Vero/DogSLAM, Vero/MonkeySLAM, and Vero/Human-SLAM cells and Vero cells (empty vector transfected cells) (1.8×10^5 cells/well) were cultured in 24-well plates and infected with Monkey-BJ01-DV or CDV-TM-CC at a multiplicity of infection (MOI) of 0.01. The cells and supernatants were harvested 48 h post-infection and were stored at −80 °C until the virus titers were measured using the Vero/DogSLAM cells by the limiting dilution method and expressed as $TCID_{50}$. Three independent experiments were performed to evaluate the CDV replication on the SLAM-expressing Vero cells.

Abbreviations

CDV, canine distemper virus; CPE, cytopathic effect; F, fusion; FITC, fluorescein isothiocyanate; G, glycine; H, hemagglutinin; IFA, immunofluorescence staining; MOI, multiplicity of infection; MV, measles virus; PBMCs, peripheral blood mononuclear cells; PBS, phosphate-buffered saline; PHA, phytohemagglutinin; RPV, rinderpest virus; RT, reverse transcription; SLAM, signaling lymphocyte activation molecule; Y, tyrosine

Acknowledgements

We thank Animal Hospital of Jilin University, Animal Laboratory Center of the Academy of Military Medical Sciences and the voluntary enrolled in this study for providing the peripheral bloods specimens. We also thank the Peter Wilker for editing the manuscript.

Funding

This work was supported by Chinese Special Fund for Agro-scientific Research in the Public Interest (201303042).

Authors' contributions

YG and XX designed and oversaw the experiments. NF, YL and JW wrote the manuscript. NF, TL, LW, XW and YY carried out the laboratory experiments. TW, YL, HW and YZ analysed the data, interpreted the results. YG, SY and GH were involved in experimental design and manuscript revision. All authors have read and approved the submitted manuscript.

Competing interests

The authors declare that they have no competing interests.

Consent for publication

Not applicable.

Author details

[1]College of Animal Science and Technology, Jilin Agricultural University, Changchun 130118, China. [2]Military Veterinary Research Institute of Academy of Military Medical Sciences, Key Laboratory of Jilin Province for Zoonosis Prevention and Control, Changchun 130122, China. [3]National Research Center for Veterinary Medicine, Luoyang, Henan 471000, China. [4]College of Chemistry and Biology, Beihua University, Jilin 132013, China. [5]Jiangsu Co-innovation Center for Prevention and Control of Important Animal Infectious Diseases and Zoonosis, Yangzhou 225009, China. [6]Department of Animal Science and Veterinary Medicine, Henan Institute of Science and Technology, Xinxiang 453003, China.

References

1. Osterhaus AD, de Swart RL, Vos HW, Ross PS, Kenter MJ, Barrett T. Morbillivirus infections of aquatic mammals: newly identified members of the genus. Vet Microbiol. 1995;44(2–4):219–27.
2. von Messling V, Zimmer G, Herrler G, Haas L, Cattaneo R. The hemagglutinin of canine distemper virus determines tropism and cytopathogenicity. J Virol. 2001;75(14):6418–27.
3. Pomeroy LW, Bjornstad ON, Holmes EC. The evolutionary and epidemiological dynamics of the paramyxoviridae. J Mol Evol. 2008;66(2):98–106.
4. von Messling V, Oezguen N, Zheng Q, Vongpunsawad S, Braun W, Cattaneo R. Nearby clusters of hemagglutinin residues sustain SLAM-dependent canine distemper virus entry in peripheral blood mononuclear cells. J Virol. 2005;79(9):5857–62.
5. Mee AP, Sharpe PT. Dogs, distemper and Paget's disease. BioEssays. 1993;15(12):783–9.
6. Qiu W, Zheng Y, Zhang S, Fan Q, Liu H, Zhang F, Wang W, Liao G, Hu R. Canine distemper outbreak in rhesus monkeys, China. Emerg Infect Dis. 2011;17(8):1541–3.
7. Sun Z, Li A, Ye H, Shi Y, Hu Z, Zeng L. Natural infection with canine distemper virus in hand-feeding Rhesus monkeys in China. Vet Microbiol. 2010; 141(3–4):374–8.
8. Sakai K, Nagata N, Ami Y, Seki F, Suzaki Y, Iwata-Yoshikawa N, Suzuki T, Fukushi S, Mizutani T, Yoshikawa T et al. Lethal canine distemper virus outbreak in cynomolgus monkeys in Japan in 2008. J Virol. 2013;87(2):1105–14.
9. Seki F, Ono N, Yamaguchi R, Yanagi Y. Efficient isolation of wild strains of canine distemper virus in Vero cells expressing canine SLAM (CD150) and their adaptability to marmoset B95a cells. J Virol. 2003;77(18):9943–50.
10. Noyce RS, Delpeut S, Richardson CD. Dog nectin-4 is an epithelial cell receptor for canine distemper virus that facilitates virus entry and syncytia formation. Virology. 2013;436(1):210–20.
11. Pratakpiriya W, Seki F, Otsuki N, Sakai K, Fukuhara H, Katamoto H, Hirai T, Maenaka K, Techangamsuwan S, Lan NT et al. Nectin4 is an epithelial cell receptor for canine distemper virus and involved in neurovirulence. J Virol. 2012;86(18):10207–10.
12. Ohishi K, Ando A, Suzuki R, Takishita K, Kawato M, Katsumata E, Ohtsu D, Okutsu K, Tokutake K, Miyahara H et al. Host-virus specificity of morbilliviruses predicted by structural modeling of the marine mammal SLAM, a receptor. Comp Immunol Microbiol Infect Dis. 2010;33(3):227–41.
13. Nikolin VM, Osterrieder K, von Messling V, Hofer H, Anderson D, Dubovi E, Brunner E, East ML. Antagonistic pleiotropy and fitness trade-offs reveal specialist and generalist traits in strains of canine distemper virus. PLoS One. 2012;7(12), e50955.
14. Norrby E, Sheshberadaran H, McCullough KC, Carpenter WC, Orvell C. Is rinderpest virus the archevirus of the Morbillivirus genus? Intervirology. 1985;23(4):228–32.
15. Yoshikawa Y, Ochikubo F, Matsubara Y, Tsuruoka H, Ishii M, Shirota K, Nomura Y, Sugiyama M, Yamanouchi K. Natural infection with canine distemper virus in a Japanese monkey (Macaca fuscata). Vet Microbiol. 1989;20(3):193–205.
16. DeLay PD, Stone SS, Karzon DT, Katz S, Enders J. Clinical and immune response of alien hosts to inoculation with measles, rinderpest, and canine distemper viruses. Am J Vet Res. 1965;26(115):1359–73.
17. McCarthy AJ, Shaw MA, Goodman SJ. Pathogen evolution and disease emergence in carnivores. Proc Biol Sci. 2007;274(1629):3165–74.
18. Muller A, Silva E, Santos N, Thompson G. Domestic dog origin of canine distemper virus in free-ranging wolves in Portugal as revealed by hemagglutinin gene characterization. J Wildl Dis. 2011;47(3):725–9.
19. Sekulin K, Hafner-Marx A, Kolodziejek J, Janik D, Schmidt P, Nowotny N. Emergence of canine distemper in Bavarian wildlife associated with a specific amino acid exchange in the haemagglutinin protein. Vet J (London, England : 1997). 2011;187(3):399–401.
20. Hara Y, Suzuki J, Noguchi K, Terada Y, Shimoda H, Mizuno T, Maeda K. Function of Feline Signaling Lymphocyte Activation Molecule as a Receptor of Canine Distemper Virus. J Vet Med Sci. 2013;75(8):1085–9.
21. Li W, Li T, Liu Y, Gao Y, Yang S, Feng N, Sun H, Wang S, Wang L, Bu Z et al. Genetic characterization of an isolate of canine distemper virus from a Tibetan Mastiff in China. Virus Genes. 2014;49(1):45–57.

Tween-20 transiently changes the surface morphology of PK-15 cells and improves PCV2 infection

Tao Hua[1,2,3,4], Xuehua Zhang[1,2,3,4], Bo Tang[1,2,3,4], Chen Chang[1,2,3,4], Guoyang Liu[1,2,3,4], Lei Feng[1,2,3,4], Yang Yu[1,2,3,4], Daohua Zhang[1,2,3,4*] and Jibo Hou[1,2,3,4*]

Abstract

Background: Low concentrations of nonionic surfactants can change the physical properties of cell membranes, and thus and in turn increase drug permeability. Porcine circovirus 2 (PCV2) is an extremely slow-growing virus, and PCV2 infection of PK-15 cells yields very low viral titers. The present study investigates the effect of various nonionic surfactants, namely, Tween-20, Tween-28, Tween-40, Tween-80, Brij-30, Brij-35, NP-40, and Triton X-100 on PCV2 infection and yield in PK-15 cells.

Result: Significantly increased PCV2 infection was observed in cells treated with Tween-20 compared to those treated with Tween-28, Tween-40, Brij-30, Brij-35, NP-40, and Triton X-100 ($p < 0.01$). Furthermore, 24 h incubation with 0.03% Tween-20 has shown to induce significant cellular morphologic changes (cell membrane underwent slight intumescence and bulged into a balloon, and the number of microvilli decreased), as well as to increase caspase-3 activity and to decrease cell viability in PCV2-infected PK-15 cells cmpared to control group; all these changes were restored to normal after Tween-20 has been washed out from the plate.

Conclusion: Our data demonstrate that Tween-20 transiently changes the surface morphology of PK-15 cells and improves PCV2 infection. The findings of the present study may be utilized in the development of a PCV2 vaccine.

Keywords: Nonionic surfactant, Tween-20, PCV2, Viral infection, Cellular morphologic change

Background

Porcine circovirus 2 (PCV2), which belongs to family Circoviridae, genus Circovirus, is the smallest non-enveloped, single-stranded, circular DNA virus that replicates autonomously. PCV2 was identified in the mid-1990s as the causative agent of post-weaning multisystemic wasting syndrome (PMWS), and is one of the most economically important viral pathogens among all major swine-producing countries [1]. Previous studies have shown that viral antigens, RNA transcripts, and progeny viruses all increase in a time-dependent manner during productive infection [1, 2]. The PK-15 cell line, which is widely used in PCV2 propagation, does not undergo efficient viral infection [3]. In China, several virus-inactivated vaccines derived from Chinese PCV2 strains have been extensively utilized in controlling PMWS and other porcine circovirus-associated disease [4]. Therefore, increasing the infection and replication of PCV2 in PK-15 cells may potentially facilitate in vaccine production, particularly in terms of efficiency and profitability. Several methods of increasing viral yield have been reported [5–11].

Previous studies have suggested that nonionic surfactants increase drug permeability through the cell membranes, thereby improving bioavailability [12–16]. When present at low concentrations, these surfactants are incorporated into the lipid bilayer, forming polar defects that alter the physical properties of cell membranes. In addition, nonionic surfactants promote membrane transport of various materials such as hydrocortisone and lidocaine across hairless mouse skin as mediated by Tween-80 [15], 5-flourouracil across hairless mouse skin by 6-fold using Tween-20 [16], and fluorescein in

* Correspondence: 782644144@qq.com; houjiboccvv@163.com
[1]Institute of Veterinary Immunology & Engineering, Jiangsu Academy of Agricultural Sciences, Nanjing 210014, China
Full list of author information is available at the end of the article

corneal tissues by Tween-20 and Brij-35 [13]. The aim of the present study was to investigate the effect of Tween-20, Tween-28, Tween-40, Tween-80, Brij-30, Brij-35, NP-40, and Triton X-100 on PCV2 infection and yield in PK-15 cells.

Methods

Virus, cells, and reagents

PCV2 strain DBN-SX07 was isolated from a piglet (Piglet was bought from a commercial pig farm in China's Shanxi province and was euthanized by an anesthetic overdose with the pentobarbital before collected the samples) in China (GenBank Accession No. FJ660968). PCV-free PK-15 cells, purchased from the China Institute of Veterinary Drug Control (Beijing, China), were maintained in minimum essential medium (MEM) (Gibco, Carlsbad, CA, USA) supplemented with 5% calf serum (CS) (Gibco Carlsbad, CA, USA), 100 U/mL penicillin (Sigma-Aldrich, St. Louis, MO, USA) and 0.1 mg/mL streptomycin (Sigma-Aldrich, St. Louis, MO, USA). Nonionic surfactants Tween-20, Tween-28, Brij-30, Brij-35 NP-40, and Triton X-100 were obtained from Sigma (St. Louis, MO, USA), while Tween-40 and Tween-80 were purchased from CRODA (Shanghai, China).

Effect of different nonionic surfactants on PCV2 infection in PK-15 cells

The highest concentrations of each nonionic surfactant that does not affect PK-15 cell viability 24 h after incubation period were used (Table1). PK-15 cells were seeded into the wells of a 96-well plate (Corning Incorporated, Shanghai, China) at a density of 2×10^5 cells/mL, with a volume of 100 μL for each well. After 24 h, Cell culture medium was then removed, and cells were consequently incubated for 23 h in a 5% CO2 incubator at 37 °C with or without different concentrations of nonionic surfactants (diluted in cell culture medium without CS) (Table 1), following the incubation with PCV2 at a multiplicity of infection (MOI) of 0.5 for 1 h at 37 °C and 5% CO_2 in the presence or absence of nonionic surfactants. 24 h post treatment, the viral inoculum and nonionic surfactants were washed off and PK-15 cells were further incubated in cell culture medium containing 2% CS, 100 U/mL penicillin and 0. 1 mg/mL streptomycin at 37 °C with 5% CO_2. 72 h later, the untreated and treated cells were fixed in cold 80% acetone (Nanjing Chemical Reagent CO., Nanjing, China) at 4 °C for 10 min. PCV2-infected PK-15 cells were identified using an indirect immunofluorescence assay (IFA) as described Section IFA. PCV2-infected PK-15 cells were counted and analyzed using fluorescence microscope Zeiss Axio Vert (Carl Zeiss AG, Oberkochen, Germany). The number of infected cells among the untreated cells was used as a reference, and all results were expressed as relative percentages to this reference. Data were expressed as the means of at least three independent experiments.

Table 1 Effect of nonionic surfactants on PCV2 infection in PK-15 cells

Agent	Concentration	Relative % of PCV2-infected cells (± S.D.)[a]
Tween-20	0.03%	880 ± 128
	0.02%	715 ± 152
	0.01%	380 ± 128
Tween-28	0.1%	140 ± 18
	0.05%	110 ± 20
	0.03%	90 ± 10
Tween-40	0.1%	175 ± 52
	0.05%	180 ± 37
	0.03%	145 ± 15
Tween-80	0.2%	430 ± 75
	0.1%	350 ± 60
	0.03%	270 ± 45
Brij-30	0.0005%	175 ± 35
	0.0003%	190 ± 43
	0.0001%	230 ± 45
Brij-35	0.0005%	250 ± 30
	0.0003%	458 ± 84
	0.0001%	469 ± 60
NP-40	0.02%	150 ± 45
	0.01%	400 ± 75
	0.005%	220 ± 58
Triton X-100	0.02%	232 ± 58
	0.01%	400 ± 13
	0.005%	460 ± 67

[a]The percentages of PCV2-infected PK-15 cells following treatment with different agents are expressed relative to the number of PCV2-infected cells in untreated PK-15 cells. The data are expressed as the mean ± standard deviation of three experiments

Immunofluorescence assay (IFA) analysis

PK-15 cells, which were inoculated with PCV2 in 96-well culture plates, were rinsed with phosphate buffered saline (PBS) (Wuhan Goodbio technology CO., Nanjing, China) nd fixed with cold 80% acetone for 10 min at 4 °C. The cells were washed, and then incubated for 1 h with anti-PCV2 antibody (VMRD, USA) diluted 1:200 in PBS with 0.05% Tween 20 (PBS-T) at 37 °C. After washing with PBS-T, cells were incubated with Staphylococcal protein A conjugated with fluorescein (1:50 diluted in PBS-T) as a secondary antibody (Boshide, Wuhan, China) for 45 min at 37 °C. After five rinses, cells were observed under a fluorescence microscope Zeiss Axio Vert (Carl Zeiss AG, Oberkochen, Germany).

Kinetics of PCV2 replication in PK-15 cells treated with nonionic surfactants

PK-15 cells were seeded into the wells of a 24-well plate (Corning Incorporated, Shanghai, China) at a density of 2×10^5 cells/mL, with a volume of 0.5 mL for each well. After 24 h, the culture medium was removed, and cells were washed and incubated for 23 h in a 5% CO2 incubator at 37 °C with or without different concentrations of nonionic surfactants diluted in cell culture medium without CS (Table 2). Subsequently, PK-15 cells were inoculated with PCV2 (MOI = 0.5) for 1 h at 37 °C and 5% CO_2 in the presence or absence of nonionic surfactants. After 24 h of treatment, the viral inoculum and nonionic surfactants were washed off and PK-15 cells were further incubated in cell culture medium containing 2% CS, 100 U/mL penicillin and 0. 1 mg/mL at 37 °C with 5% CO_2. The medium and cells from triplicate wells of each inoculation group were harvested every 24 h through 96 h post treatment (hpt) and stored at – 70 °C until virus titration.

Scanning Electron microscopy (SEM) analysis

PK-15 cells were grown on 18 × 18 mm coverslips (Sail Brand, Guangdong, China) in 6-well plates (Corning Incorporated, Shanghai, China) at a density of 2×10^5 cells/mL, with a volume of 2 mL for each well. After 24 h, the culture medium was removed. PK-15 cells were then washed and incubated for 23 h in a 5% CO_2 incubator at 37 °C with or without 0.03% Tween-20 diluted in cell culture medium without CS. After that, PK-15 cells were inoculated with PCV2 (MOI = 0.5) for 1 h at 37 °C and 5% CO_2 in the presence or absence of 0.03% Tween-20. 24 h post treatment, the viral inoculum and nonionic surfactants were washed off and PK-15 cells were further incubated in cell culture medium containing 2% CS, 100 U/mL penicillin and 0.1 mg/mL streptomycin at 37 °C with 5% CO_2. After 0, 24, 48, and 72 h post Tween-20 treatment, cells were prepared for SEM analysis. Briefly, the PK-15 cells were washed with PBS, followed by fixation in 3% glutaraldehyde (Sigma-Aldrich, St. Louis, MO, USA) for 2 h at room temperature. After three washes with distilled water for 30 min at room temperature, the cells were serially dehydrated in ethanol (Nanjing Chemical Reagent Co., Nanjing, China) (30%, 70%, 96%, 3 × 100%, 15 min for each step), critical point-dried, sputter-coated with gold particles, and stored in a desiccator until observation. Finally, all specimens were examined in a Zeiss EVO-LS10 SEM (Carl Zeiss AG, Oberkochen, Germany).

Table 2 Kinetics of PCV2 replication in PK-15 cells treated with or without different nonionic surfactants

Agent	Concentration	PCV2 titer $\log_{10}$ ($TCID_{50}$/mL)			
		24 hpt[a]	48 hpt	72 hpt	96 hpt
Control	0	1.3	2.7	3.2	3.3
Tween-20	0.03%	2.7	4.0	4.5	4.5
Tween-28	0.1%	1.5	2.7	3.2	3.3
Tween-40	0.1%	1.7	3	3.3	3.5
Tween-80	0.2%	2.2	3.5	3.8	3.7
Brij-30	0.0003%	1.7	2.8	3.5	3.5
Brij-35	0.0003%	2.3	3.3	3.8	3.8
NP-40	0.01%	2.2	3.3	3.5	3.5
Triton X-100	0.01%	2.2	3.3	3.8	3.7

[a]hours post-treatment

Measurement of caspase-3 activity

PK-15 cells were placed in the wells of a 6-well plate at a density of 2×10^5 cells/mL, with a volume of 2 mL for each well. After 24 h, the culture medium was removed. PK-15 cells were washed and incubated for 23 h in a 5% CO2 incubator at 37 °C with or without 0.03% Tween-20 diluted in cell culture medium without CS. Afterwards, PK-15 cells were inoculated with PCV2 (MOI = 0. 5) for 1 h at 37 °C and 5% CO_2 in the presence or absence of 0.03% Tween-20. After 24 h of treatment, the viral inoculum and nonionic surfactants were washed off and PK-15 cells were further incubated in cell culture medium containing 2% CS, 100 U/mL penicillin and 0. 1 mg/mL streptomycin at 37 °C with 5% CO_2. The PCV2-infected PK-15 cells were collected at 0, 24, 48, and 72 h post Tween-20 treatment. Caspase-3 activity was determined by a colorimetric assay, which was based on the ability of caspase-3 to convert acetyl-Asp-Glu-Val-Asp p-nitroanilide (Ac-DEVD-pNA) into a yellow formazan product (p-nitroanilide). An increase in the absorbance at a wavelength of 405 nm was indicative of caspase-3 activation. The culture medium and PK-15 cells were collected at indicated times. The cells were rinsed with cold PBS, and lysed with lysis buffer (100 μL/2×10^6 cells) for 15 min on ice. The cell lysates were centrifuged at 18,000 *g* for 10 min at 4 °C. Caspase-3 activity was determined using a caspase-3 activity kit (Beyotime Institute of Biotechnology, Nantong, China) following the manufacturer's protocol.

Cell viability measurement

The effect of Tween-20 on cell viability was determined by using the MTT [3-(4, 5-dimethylthiazol-2-yl)-2,5-diphenyl tetrazolium bromide, MTT] assay following the manufacturer's instructions (Merk Millipore, Shanghai, China). PK-15 cells were seeded into a 96-well plate at a density of 2×10^5 cells/mL, with a volume of 100 μL for each well.. After 24 h, PK-15 cells were washed and incubated for 23 h in a 5% CO_2 incubator at 37 °C with or without 0.03% Tween-20 diluted in cell culture medium without CS. Afterwards, PK-15 cells were inoculated with PCV2 (MOI = 0.5) for 1 h at 37 °C and 5% CO_2 in the presence or absence of 0.03% Tween-20. 24 h later, the viral inoculum and nonionic surfactants were washed off and PK-15 cells were further incubated in cell culture medium containing 2% CS, 100 U/mL penicillin and 0.1 mg/mL streptomycin at 37 °C with 5% CO_2. The PCV2-infected PK-15 cells were collected at 0, 24, 48, and 72 h post treatment with Tween-20. Approximately 10 μL of MTT (5 mg/mL) was added onto each well of the 96-well plate and then incubated for another 4 h at 37 °C. After incubation, the culture medium was removed, and 100 μL of acidified isopropanol (Sigma-Aldrich, St. Louis, MO, USA) was added to each well to dissolve the precipitate at room temperature. Absorbance was measured at a wavelength of 570 nm using a Stat Fax-2100 spectrophotometer (Awareness Technology, Inc., USA). Each treatment was performed in triplicate, and the viability of treated cells was expressed as the relative percentage of live cells relative to that of the untreated control cells.

Statistical analysis

Statistical analysis was performed using GraphPad PRISM software (version 5.02 for Windows; GraphPad Software, Inc.). The data were analyzed to establish their significance using one-way or two-way ANOVA followed by a least-significant difference test. The data were expressed as the mean ± SD. Differences were regarded as significant at $p < 0.01$.

Results

Effect of different nonionic surfactant on PCV2 infection

The PK-15 cells were treated with different concentrations of nonionic surfactants to investigate its effect on PCV2 infection (Table 1). The relative number of PCV2-infected cells in PK-15 cells were 880 ± 128%, 140 ± 18%, 180 ± 37%, 430 ± 75%, 230 ± 45%, 469 ± 60%, 400 ± 75%, and 460 ± 67% when PK-15 cells were treated with 0.03% Tween-20, 0.1% Tween-28, 0.05% Tween-40, 0.2% Tween-80, 0.0001% Brij-30, 0.0001% Brij-35, 0.01% NP-40, and 0.005% Triton X-100, respectively (Table 1). 0.03% Tween-20 treatment increased PCV2-infected PK-15 cells by up to 8.8 times compared to untreated PK-15 cells. The number of PCV2-infected cells from PK-15 cells treated with 0.03% Tween-20 was significantly higher compared to those treated with Tween-28, Tween-40, Brij-30, Brij-35, NP-40, Triton X-100, and untreated PK-15 cells ($p < 0.01$, Table 1 and Fig. 1). The number of PCV2-infected cells in PK-15 cells treated with Tween-80, Brij-35, NP-40 and Triton X-100 was significantly higher than the untreated PK-15 cells, but significantly lower than PK-15 cells treated with Tween-20 (Table 1). Furthermore, no significant changes were observed when treating cells with Tween-28 or Tween-40 compared to untreated PK-15 cells (Table 1). The relative number of PCV2-infected cells in PK-15 cells were 880 ± 128%, 715 ± 152% and 380 ± 128% when PK-15 cells were treated with 0.03%, 0.02% and 0.01% Tween-20, respectively (Table 1). After increasing the concentration of Tween-20 (> 0.03%) for 24 h, cell viability was significantly affected and the number of PCV2-infected cells decreased (data not shown). When nonionic surfactants exceeded the highest concentration in Table 1, cell viability would be significantly affected and the number of PCV2-infected cells decreased (data not shown). The highest concentrantion of Brij-35, NP-40 and Triton X-100 in Table 1 didn't show the highest effect on promoting the number of PCV2-infected

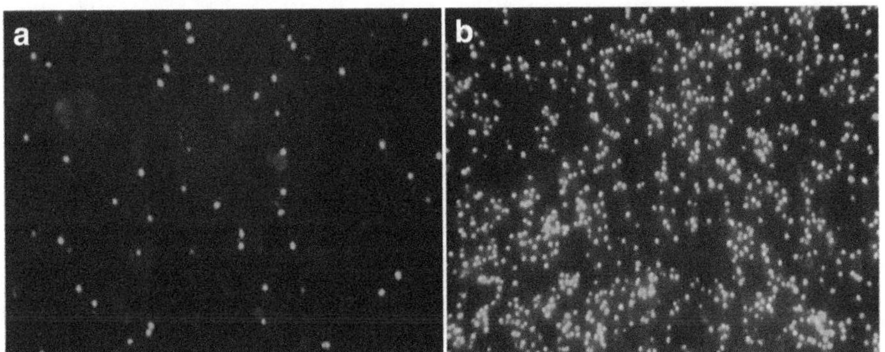

Fig. 1 Effect of Tween-20 on PCV2 infection in PK-15 cells. PK-15 cells were treated with or without 0.03% Tween-20 for 24 h, and simultaneously infected PCV2 (MOI = 0.5) for 1 h. After a 24-h treatment, the mixed solution of Tween-20 and PCV2 was washed off and the PK-15 cells were further incubated in cell culture medium containing 2% CS. After 72 h post treatment, the PCV2-infected cells were assessed using an immunofluorescence assay. The number of PCV2-infected cells from PK-15 cells treated with 0.03% Tween-20 was significantly higher compared to PCV2-infected PK-15 cells without Tween-20 treatment ($p < 0.01$, Table 1). **a** PCV2-infected PK-15 cells without Tween-20 treatment as control. **b** PCV2-infected PK-15 cells treated with Tween-20. Magnification: × 100

cells. Some function of cells may be affected at the highest concentration of Brij-35, NP-40 and Triton X-100 in Table 1.

Kinetics of PCV2 replication in PK-15 cells

The kinetics of PCV2 replication was determined in PK-15 cells treated with or without different nonionic surfactants (Table 2). After the initial infection, the replication levels of PCV2 were detected in PK-15 cells. The results showed that all viral stocks, originating from the infected cells, had low initial titers (Table 2). 72 h post-treatment, the PCV2 titers of the PK-15 cells treated with 0.03% Tween-20 rapidly increased and were higher ($10^{4.5}$ $TCID_{50}$/mL) compared to other treatments (Table 2).

Assessment of morphologic changes in PK-15 cells

PK-15 cellular morphologic changes at 0, 24, 48, and 72 h post 0.03% Tween-20 treatment were analyzed using SEM (Fig. 2). The surface of PCV2-infected PK-15 cells without Tween-20 treatment showed an abundance of microvilli and was rough in appearance. After 24 h treatment with Tween-20, the cells membrane of the PK-15 cells exhibited slight intumescence and bulged into a balloon, and the number of microvilli significantly decreased; while, all those changes were restored to normal after Tween-20 has been washed out from the plate. These findings indicated that the surface structure of PK-15 cells recovered after transient treatment with 0.03% Tween-20.

Tween-20 transiently promotes caspase-3 activation

PCV2 has been shown to induce apoptosis in cultured cells through activation of caspase-8, followed by activation of the caspase-3 pathway [17]. Tween-20 can induce membrane damage and initiate apoptosis [18]. To determine whether Tween-20 improves PCV2-induced apoptosis, cell

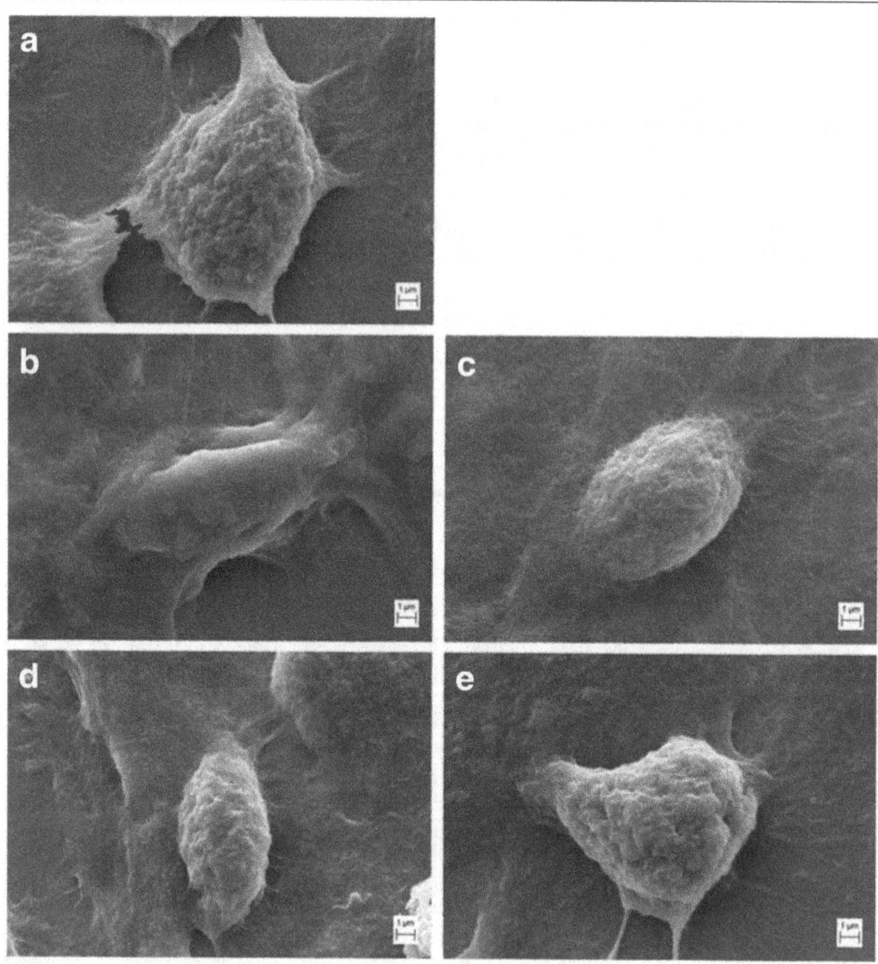

Fig. 2 The morphologic changes of PCV2-infected PK-15 cells were observed by SEM at the indicated times of 0.03% Tween-20 post treatment. **a** PCV2-infected PK-15 cells without Tween-20 treatment as control. **b** PCV2-infected PK-15 cells at 0 h post Tween-20 treatment. **c** PCV2-infected PK-15 cells at 24 h post Tween-20 treatment. **d** PCV2-infected PK-15 cells at 48 h post Tween-20 treatment. **e** PCV2-infected PK-15 cells at 72 h post Tween-20 treatment. Bar, 1 μm

lysates were harvested at various time points and assayed for caspase-3 activity. Following infection with PCV2 alone, a time-dependent increase in the cleavage of ρNA (a product of caspase-3 cleaving Ac-DEVD-ρNA) was observed throughout the course of post-infection. Caspase-3 activity in PK-15 cells were 0.53 ± 0.05, 1.09 ± 0.22, 1.21 ± 0.12, and 1.27 ± 0.19 U/mg protein (Pro.) when cells were infected with PCV2 alone at 0, 24, 48, and 72 h post PCV2 infection, respectively (Fig. 3). The percentage of increase caspase-3 activity were 110 ± 15%, 200 ± 13%, 234 ± 32%, and 239 ± 22% when PK-15 cells were infected with PCV2 alone compared to the control cells at 0, 24, 48, and 72 h post PCV2 infection, respectively. This indicated that caspase-3 was progressively activated by PCV2 infection (Fig. 3). Consequently, we examined the effect of 0.03% Tween-20 on caspase-3 activity in PCV2-infected cells. Caspase-3 activity in PK-15 cells were 2.46 ± 0.51, 1.64 ± 0. 22, 1.45 ± 0.19, and 1.45 ± 0.24 U/mg Pro. when PK-15 cells were treated 0.03% Tween-20 and simultaneously infected PCV2 at 0, 24, 48, and 72 h post post-treatment, respectively (Fig. 3). The percentage of increase caspase-3 activity was 503 ± 26%, 304 ± 22%, 279 ± 22%, and 273 ± 0.24% when PK-15 cells were treated 0.03% Tween-20 and simultaneously infected PCV2 compared to the control cells at 0, 24, 48, and 72 h post-treatment, respectively. The percentage of increase caspase-3 activity in PK-15 cells were 462 ± 84%, 152 ± 19%, 119 ± 7%, and 114 ± 1% when PK-15 cells were treated 0.03% Tween-20 and simultaneously infected PCV2 compared to untreated PCV2-infected cells at 0, 24, 48, and 72 h post-treatment, respectively. Caspase-3 activity in PCV2-infected cells treated 0.03% Tween-20 significantly increased compared to that in the PCV2-infected cells at 0 h post-treatment, while its activity returned to normal after removing Tween-20 from the plate (no significant difference in caspase-3 activity was observed between treated and untreated PCV2-infected cells at 24, 48, and 72 h post Tween-20 treatment). To sum up, these findings were indicative of a decrease in caspase-3 activation after removal of Tween-20.

Transient treatment with 0.03% Tween-20 does not significantly affect cell viability

MTT assay was used to examined whether 0.03% Tween-20 affects cell viability (Fig. 4). The PK-15 cells infected with PCV2 did not show an adverse change in cell viability compared to that in the control cells. Importantly, 0 h post- treatment with Tween-20 significantly decreased cell viability, which was then restored to normal after removing Tween-20 from the plate (no significant difference was observed between treated anduntreated PCV2-infected cells ($p > 0.01$)). These findings indicate that cell viability was increased when Tween-20 was washed off.

Discussion

Nonionic surfactants are a category of surfactants with uncharged hydrophilic and hydrophobic heads [19]. Nonionic surfactants can form structures in which hydrophilic heads are oriented opposite to the aqueous solutions, and hydrophobic heads opposite to the organic solutions. Based on this property, low concentrations of these surfactants can be incorporated into the lipid bilayer, forming polar defects that alter the physical properties of the cell membranes. When the lipid bilayer is saturated, mixed micelles begin to form, resulting in the removal of phospholipids from the cell membranes and membrane solubilization. Marsh and Maurice [13] have evaluated the effect of nonionic surfactants on corneal permeability and toxicity in humans, and found that Tween-20 and Brij-35 are the most effective in

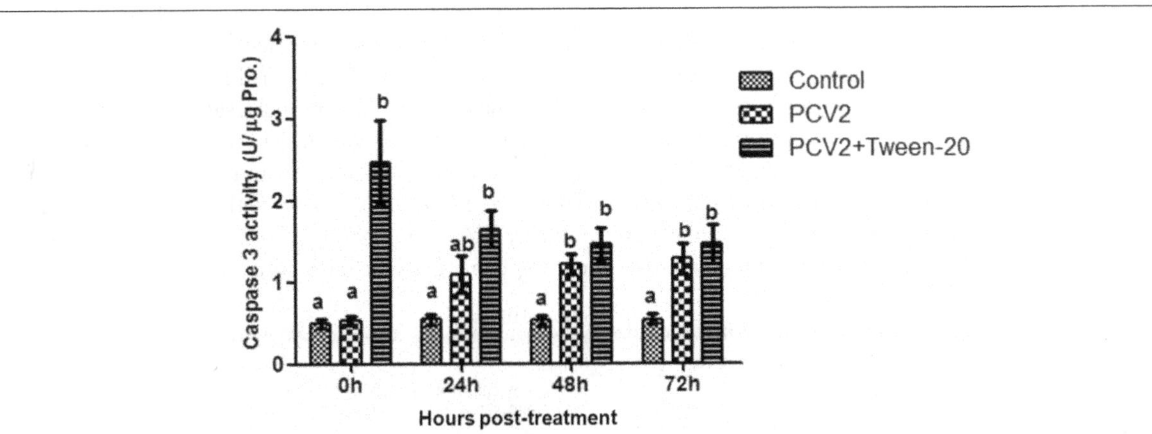

Fig. 3 Effect of 0.03% Tween-20 on caspase-3 activation of PCV2-infected PK-15 cells. Cell lysates were harvested at the indicated post treatment times. Caspase-3 activity was measured by using a colorimetric assay based on the ability of caspase-3 to convert Ac-DEVD-pNA into a yellow formazan product. The data are expressed as the mean ± SD ($n = 3$). Within each time point, means with different letters (a, b, c) are significantly different from each other ($p < 0.01$)

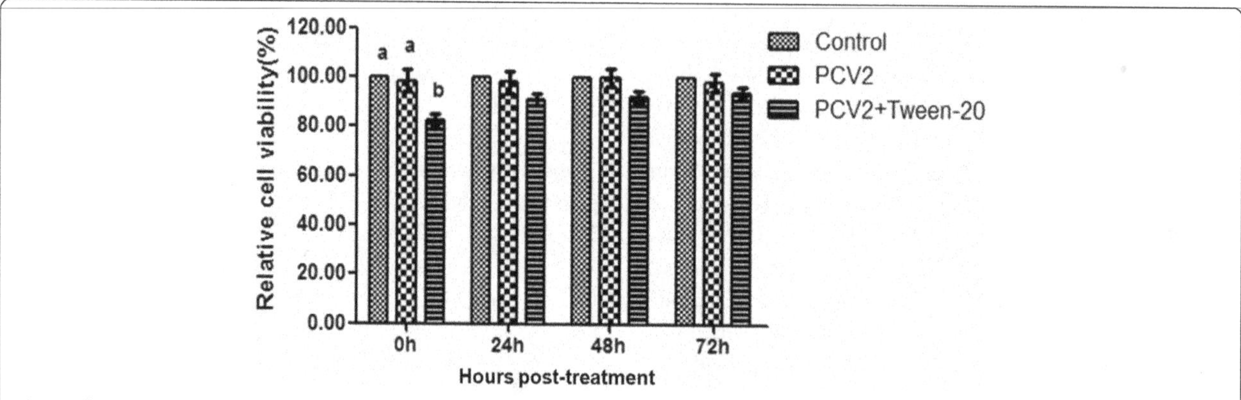

Fig. 4 Effect of 0.03% Tween-20 on the viability of PCV2-infected PK-15 cells. Cell viability was determined by using an MTT assay at the indicated post treatment times. The percentage of relative cell viability is expressed as themean ± SD (n = 3). Within each time point, values with different letters (a, b) are significantly different from each other ($p < 0.01$)

increasing corneal permeability. In the present study, the effects of nonionic surfactants Tween-20, Tween-28, Tween-40, Tween-80, Brij-30, Brij-35, NP-40, and Triton X-100 on PCV2 infection in PK-15 cells were investigated. Interestingly, Tween-20 treatment significantly increased the number of PCV2-infected cells compared to control and other nonionic surfactant treatments groups ($p < 0.01$, Table 1, and Fig. 1).

Cytotoxicity is an inherent property of various nonionic surfactants [12, 18, 20–22]. These nonionic surfactants can induce membrane damage and initiate apoptosis. However, a previous study has shown that their cytotoxicity could be reduced by using the appropriate type and/or number of side chains [20]. Tween-80 has the lowest cytotoxicity in normal human fibroblast cultures compared to Texapon N40, Tween-60, Texapon K1298, Triton X-100, and benzethonium chloride [22]. The cytotoxicity of nonionic surfactants can be further reduced using lower concentrations [18]. The application of Tween-20 concentration range of 0.013%–0.025% has shown to exert considerable cytotoxicity in both multidrug resistance cell lines and their parental cells after 48 h exposure. Tween-20 at concentrations < 0.01% is non-toxic to all cells, showing > 90% cell survival. In the present study, 24 h treatment with 0.03% Tween-20 induced cellular morphologic changes (cell membrane underwent slight intumescence and bulged into a balloon, and the number of microvilli decreased) (Fig. 2), increased caspase-3 activity (Fig. 3) and decreased cell viability (Fig. 4) in PCV2-infected PK-15 cells compared to control group; while all these changes were restored to normal after Tween-20 has been washed out from the plate.

Due to the low replication efficiency of PCV2, researchers have adopted various ways of improving virus titers [5–11]. PCV multiplication is inducible by treating infected cell cultures with D-glucosamine (D-G) [6]. Cholesterol removal enhances PCV2 replication in epithelial cells treated with methyl-β-cyclodextrin [9]. Some studies have shown that the number of PCV2-infected cells increases after treating PK-15 cells with either interferon-gamma, or inhibitors of endosomal-lysosomal system acidification such as ammonium chloride, chloroquine diphosphate, and monensin [5, 8]. The present study showed that PK-15 cells treated with Tween-20 significantly increased PCV2 infection compared to other nonionic surfactants, including Tween-28, Tween-40, Tween-80, Brij-30, Brij-35, NP-40, and Triton X-100 and untreated PK-15 cells ($p < 0.01$).

Conclusions

The present study examined the effects of nonionic surfactants on PCV2 infection. We demonstrated that PCV2-infected PK-15 cells treated with Tween-20 showed an increase in PCV2 infection and yield compared to other nonionic surfactants such as Tween-28, Tween-40, Tween-80, Brij-30, Brij-35, NP-40, and Triton X-100 and untreated PK-15 cells. Furthermore, SEM analysis showed that Tween-20 could transiently change the surface morphology and structure of PK-15 cells to improve PCV2 infection. After transient treatment with Tween-20, SEM and caspase-3, and MTT assays indicated a restoration of the surface structure and viability of PK-15 cells. Therefore, PK-15 cells treated with Tween-20 may be potentially used in increasing PCV2 infection, which in turn may facilitate the vaccine production.

Abbreviations

Ac-DEVD-pNA: acetyl-Asp-Glu-Val-Asp p-nitroanilide; CS: Calf serum; hpt: Hours post treatment; IFA: Immunofluorescence assay; MEM: Minimum essential medium; MOI: multiplicity of infection; PBS: Phosphate buffered saline; PBS-T: PBS with 0.05% Tween 20; PCV2: Porcine circovirus 2; Pro: Protein.; SEM: Scanning electron microscopy

Acknowledgments

We would like to thank Laigen Hu (Chengdu Tecbond Biological Products Co.) for kindly providing PCV2 strain DBN-SX07. We would also like to thank Dr. Yinghua Tang (Jiangsu Academy of Agricultural Science) for his positive advice on experimental design and Dr. Lei Feng (Jiangsu Academy of Agricultural Sciences) for their help throughout the study. We thank LetPub (www.letpub.com) for its linguistic assistance during the preparation of this manuscript.

Funding
This study was supported by the Independent Innovation of Agricultural Sciences Program of Jiangsu Province [No. CX (14)5044] and Agro-scientific Research in the Public Interest (201303046).

Authors' contributions
JH and DZ directed the research, reviewed the data and manuscript, and directed the manuscript revisions. TH and XZ conducted the research, compiled data, and wrote the paper. BT, CC and GL contributed to performance of Kinetics of PCV2 replication in PK-15 cells and participated in drafting the manuscript. LF put forward a lot of positive advices on experimental design, and analysed the data throughout the study. YY was involved in data analysis and participated in drafting the manuscript. All authors have read and approved the manuscript.

Competing interests
The authors declare that they have no conflicts of interest.

Author details
[1]Institute of Veterinary Immunology & Engineering, Jiangsu Academy of Agricultural Sciences, Nanjing 210014, China. [2]National Research Center of Engineering and Technology for Veterinary Biologicals, Jiangsu Academy of Agricultural Science, Nanjing 210014, China. [3]Key lab of Food Quality and Safety of Jiangsu Province—State Key laboratory Breeding Base, Jiangsu Academy of Agricultural Science, Nanjing 210014, China. [4]Jiangsu Co-innovation Center for Prevention and Control of Important Animal Infectious Diseases and Zoonoses, Yangzhou 225009, China.

References
1. Allan GM, McNeilly F, Kennedy S, Daft B, Clarke EG, Ellis JA, Haines DM, Meehan BM, Adair BM. Isolation of porcine circovirus-like viruses from pigs with a wasting disease in the USA and Europe. J Vet Diagn Investig. 1998; 10(1):3–10.
2. Cheung AK, Bolin SR. Kinetics of porcine circovirus type 2 replication. Arch Virol. 2002;147(1):43–58.
3. Zhu Y, Lau A, Lau J, Jia Q, Karuppannan AK, Kwang J. Enhanced replication of porcine circovirus type 2 (PCV2) in a homogeneous subpopulation of PK15 cell line. Virology. 2007;369(2):423–30.
4. Ge X, Wang F, Guo X, Yang H. Porcine circovirus type 2 and its associated diseases in China. Virus Res. 2012;164(1–2):100–6.
5. Meerts P, Misinzo G, Nauwynck HJ. Enhancement of porcine circovirus 2 replication in porcine cell lines by IFN-gamma before and after treatment and by IFN-alpha after treatment. J Interf Cytokine Res. 2005;25(11):684–93.
6. Tischer I, Peters D, Rasch R, Pociuli S. Replication of porcine circovirus: induction by glucosamine and cell cycle dependence. Arch Virol. 1987;96(1–2):39–57.
7. Misinzo G, Delputte PL, Lefebvre DJ, Nauwynck HJ. Increased yield of porcine circovirus-2 by a combined treatment of PK-15 cells with interferon-gamma and inhibitors of endosomal-lysosomal system acidification. Arch Virol. 2008;153(2):337–42.
8. Misinzo G, Delputte PL, Nauwynck HJ. Inhibition of endosome-lysosome system acidification enhances porcine circovirus 2 infection of porcine epithelial cells. J Virol. 2008;82(3):1128–35.
9. Misinzo G, Delputte PL, Lefebvre DJ, Nauwynck HJ. Porcine circovirus 2 infection of epithelial cells is clathrin-, caveolae- and dynamin-independent, actin and rho-GTPase-mediated, and enhanced by cholesterol depletion. Virus Res. 2009;139(1):1–9.
10. Beach NM, Juhan NM, Cordoba L, Meng XJ. Replacement of the replication factors of porcine circovirus (PCV) type 2 with those of PCV type 1 greatly enhances viral replication in vitro. J Virol. 2010;84(17):8986–9.
11. Yang X, Chen F, Cao Y, Pang D, Ouyang H, Ren L. Comparative analysis of different methods to enhance porcine circovirus 2 replication. J Virol Methods. 2013;187(2):368–71.
12. Sahoo RK, Biswas N, Guha A, Sahoo N, Kuotsu K. Nonionic surfactant vesicles in ocular delivery: innovative approaches and perspectives. Biomed Res Int. 2014;2014:263604.
13. Marsh RJ, Maurice DM. The influence of non-ionic detergents and other surfactants on human corneal permeability. Exp Eye Res. 1971;11(1):43–8.
14. Kaur IP, Smitha R. Penetration enhancers and ocular bioadhesives: two new avenues for ophthalmic drug delivery. Drug Dev Ind Pharm. 2002;28(4):353–69.
15. Sarpotdar PP, Zatz JL. Evaluation of penetration enhancement of lidocaine by nonionic surfactants through hairless mouse skin in vitro. J Pharm Sci. 1986;75(2):176–81.
16. Rigg PC, Barry BW. Shed snake skin and hairless mouse skin as model membranes for human skin during permeation studies. J Invest Dermatol. 1990;94(2):235–40.
17. Liu J, Chen I, Kwang J. Characterization of a previously unidentified viral protein in porcine circovirus type 2-infected cells and its role in virus-induced apoptosis. J Virol. 2005;79(13):8262–74.
18. Yang S, Liu J, Chen Y, Jiang J. Reversal effect of Tween-20 on multidrug resistance in tumor cells in vitro. Biomed Pharmacother. 2012;66(3):187–94.
19. Moghassemi S, Hadjizadeh A. Nano-niosomes as nanoscale drug delivery systems: an illustrated review. J Control Release. 2014;185:22–36.
20. Weiszhar Z, Czucz J, Revesz C, Rosivall L, Szebeni J, Rozsnyay Z. Complement activation by polyethoxylated pharmaceutical surfactants: Cremophor-EL, Tween-80 and Tween-20. Eur J Pharm Sci. 2012;45(4):492–8.
21. Li D, Wu X, Yu X, Huang Q, Tao L. Synergistic effect of non-ionic surfactants tween 80 and PEG6000 on cytotoxicity of insecticides. Environ Toxicol Pharmacol. 2015;39(2):677–82.
22. Arechabala B, Coiffard C, Rivalland P, Coiffard LJ, de Roeck-Holtzhauer Y. Comparison of cytotoxicity of various surfactants tested on normal human fibroblast cultures using the neutral red test, MTT assay and LDH release. J Appl Toxicol. 1999;19(3):163–5.

Evaluation of biosecurity measures to prevent indirect transmission of porcine epidemic diarrhea virus

Yonghyan Kim, My Yang, Sagar M. Goyal, Maxim C-J. Cheeran and Montserrat Torremorell*

Abstract

Background: The effectiveness of biosecurity methods to mitigate the transmission of porcine epidemic diarrhea virus (PEDV) via farm personnel or contaminated fomites is poorly understood. This study was undertaken to evaluate the effectiveness of biosecurity procedures directed at minimizing transmission via personnel following different biosecurity protocols using a controlled experimental setting.

Results: PEDV RNA was detected from rectal swabs of experimentally infected (INF) and sentinel pigs by real-time reverse transcription polymerase chain reaction (rRT-PCR). Virus shedding in INF pigs peaked at 1 day post infection (dpi) and viral RNA levels remained elevated through 19 dpi. Sentinel pigs in the low biosecurity group (LB) became PEDV positive after the first movement of study personnel from the INF group. However, rectal swabs from pigs in the medium biosecurity (MB) and high biosecurity (HB) groups were negative during the 10 consecutive days of movements and remained negative through 24 days post movement (dpm) when the first trial was terminated.

Viral RNA was detected at 1 dpm through 3 dpm from the personal protective equipment (PPE) of LB personnel. In addition, at 1 dpm, 2 hair/face swabs from MB personnel were positive; however, transmission of virus was not detected. All swabs of fomite from the HB study personnel were negative.

Conclusions: These results indicate that indirect PEDV transmission through contaminated PPE occurs rapidly (within 24 h) under modeled conditions. Biosecurity procedures such as changing PPE, washing exposed skin areas, or taking a shower are recommended for pig production systems and appear to be an effective option for lowering the risk of PEDV transmission between groups of pigs.

Keywords: Porcine epidemic diarrhea virus, Indirect transmission, Farm personnel, Animal movement, Biosecurity, Fomites

Background

Porcine epidemic diarrhea (PED) is a highly contagious viral disease that causes severe diarrhea in pigs [1]. In May 2013, the porcine epidemic diarrhea virus (PEDV) was first reported in the US, causing significant economic losses to the swine industry due to high mortality rates in piglets. Over 8 million pigs died due to PEDV, leading to an estimated total industry economic loss of more than 1.8 billion US dollars [2]. PEDV is an enveloped, single-stranded, positive-sense RNA virus belonging to the *Coronaviridae* family in the genus *Alphacoronavirus* [1, 3]. The virus is shed in feces of infected pigs and transmitted via the fecal-oral route. PEDV can be transmitted either by direct contact between infected and susceptible pigs or indirectly through contaminated fomites. Transmission via pig transportation has been reported as a major risk factor for the spread of PEDV [4, 5]. Five percent of PEDV negative trailers became contaminated during the unloading process at slaughterhouse facilities handling infected pigs [4]. Contaminated feed has also been implicated in the spread of PEDV, and both food ingredients (viz. dried spray plasma) and cross-contamination at the feed mill or from other sources have been implicated in the spread of PEDV [6–11]. PEDV has also been detected in air samples and

* Correspondence: torr0033@umn.edu
Department of Veterinary Population Medicine, College of Veterinary Medicine, University of Minnesota, 1988 Fitch Ave, St. Paul, MN 55108, USA

aerosol transmission has been suspected as a potential source of disease transmission in high pig dense areas [12].

In order to mitigate transmission of PEDV within and between farms, producers employ a range of biosecurity practices. These practices include disinfecting footwear and changing clothing of visitors or personnel prior to entering farm premises, washing and sanitizing delivery trucks or vehicles entering the farm, and controlling insects. Controlling transmission via feed can be done by using feed additives such as formaldehyde [13] and transmission via transport can be minimized by implementing proper cleaning and disinfection methods [14]. However, these implementations are not always practical or cost effective. The effectiveness of methods to mitigate transmission via farm personnel or contaminated fomites are less understood given that intervention strategies at the farm level have not been properly investigated. Furthermore, Stevenson et al. indicated that even shower-in/shower-out facilities with excellent biosecurity protocols also reported PEDV outbreaks [15]. Given the limited knowledge available on how biosecurity procedures may disrupt the transmission cycle of PEDV, the present study was undertaken to evaluate the effectiveness of biosecurity procedures directed at minimizing transmission via personnel following different biosecurity protocols using a controlled experimental setting.

Methods

Animals and animal housing

Forty-eight, 3-week-old crossbred pigs including both male and females were obtained from a farm with no history of PEDV infection and were housed at the St. Paul animal research isolation units at the University of Minnesota. After arrival (2 days before the start of the study), rectal swabs from all pigs were collected and tested by real-time reverse transcription polymerase chain reaction (rRT-PCR) for PEDV, transmissible gastroenteritis virus (TGEV) and porcine delta coronavirus (PDCoV) at the University of Minnesota Veterinary Diagnostic Laboratory (St. Paul, MN, USA).

The pigs were housed in 17 separate rooms that were independently operated from each other as described below. All individual rooms had anterooms with footbaths, a sink for hand and face washing, a storage area of 2.08 m^2, and an animal housing area of 7.28 m^2. Rooms were connected through a clean common hallway as shown in Fig. 1. The floor of the animal housing area was constructed with solid concrete and each animal housing area had a single water line with two water nipples as a source of drinking water. Prior to introducing the pigs into the rooms, environmental swabs were collected from the floors and confirmed PEDV negative by rRT-PCR. Ventilation for all rooms was kept under negative differential pressure to the main corridor, having one air inlet and one exhaust vent per room. The air supply was conditioned with a 3 ply panel filter (TRI-DEK® 15/40, TRI-DIM Filter Corp., Louisa, VA, USA) and 100% of exhaust air was filtered through a HEPA filter (XH Absolute HEPA filter, Camfil, Stockholm, Sweden).

The pigs were randomly distributed as described below and treated with a single intramuscular dose of enrofloxacin (0.5 mL/pig; Baytril®, Bayer HealthCare AG, Leverkusen, Germany) to control respiratory disease associated with Haemophilus parasuis 1 day prior to infection. In the first trial of the study, forty eight 3-week-old piglets were randomly assigned to 5 experimental groups: (i) 10 pigs were infected with PEDV and 2 contact sentinel pigs were housed together to serve as contact sentinels (INF group); (ii) 10 pigs (5 replicates of 2 each) were assigned to low biosecurity (LB) sentinel groups; (iii) 10 pigs (5 replicates of 2 each) were assigned to medium biosecurity (MB) sentinel groups; (iv) 10 pigs (5 replicates of 2 each) were assigned to high biosecurity (HB) sentinel groups and (v) six pigs were assigned to a negative control (NC) group, which were uninfected and handled separately. Trial 1 lasted a total of 27 days.

In the 2nd trial of the study, twenty-three 6-week-old pigs were assigned to 4 different groups: (i) 3 pigs were infected with PEDV and 1 contact sentinel pig were assigned to the INF group; (ii) 4 pigs (1 replicate) were assigned to LB sentinel group; (iii) 12 pigs (3 replicates of 4 each) were assigned to MB sentinel groups; and (iv) 3 pigs (1 replicate) were assigned to NC group. Trial 2 lasted a total of 13 days.

Study personnel

Study personnel were exclusively assigned to handle pigs in this study and had no direct contact with other pigs or with PEDV-infected pigs from another source for the entire duration of this study. Personnel entering LB, MB, and HB sentinel rooms had direct contact with infected pigs in the INF room only and performed all necessary procedures (e.g. pig fecal swab collection, pig blood collection, feeding of pigs, cleaning of room) in their designated rooms as assigned during the deputed sampling and movement days.

Clothing and personal protective equipment

All study personnel showered in the research animal facility prior to donning the facility-dedicated clothing and personal protective equipment (PPE). After showering, personnel put on clean scrubs, a pair of disposable plastic boots and entered the animal isolation corridor after stepping through an iodine footbath. In the animal isolation clean hallway, personnel put on disposable Tyvek® coverall (DuPont, Wilmington, DE, USA), nitrile gloves, and a bouffant cap for the first trial of the study. In the

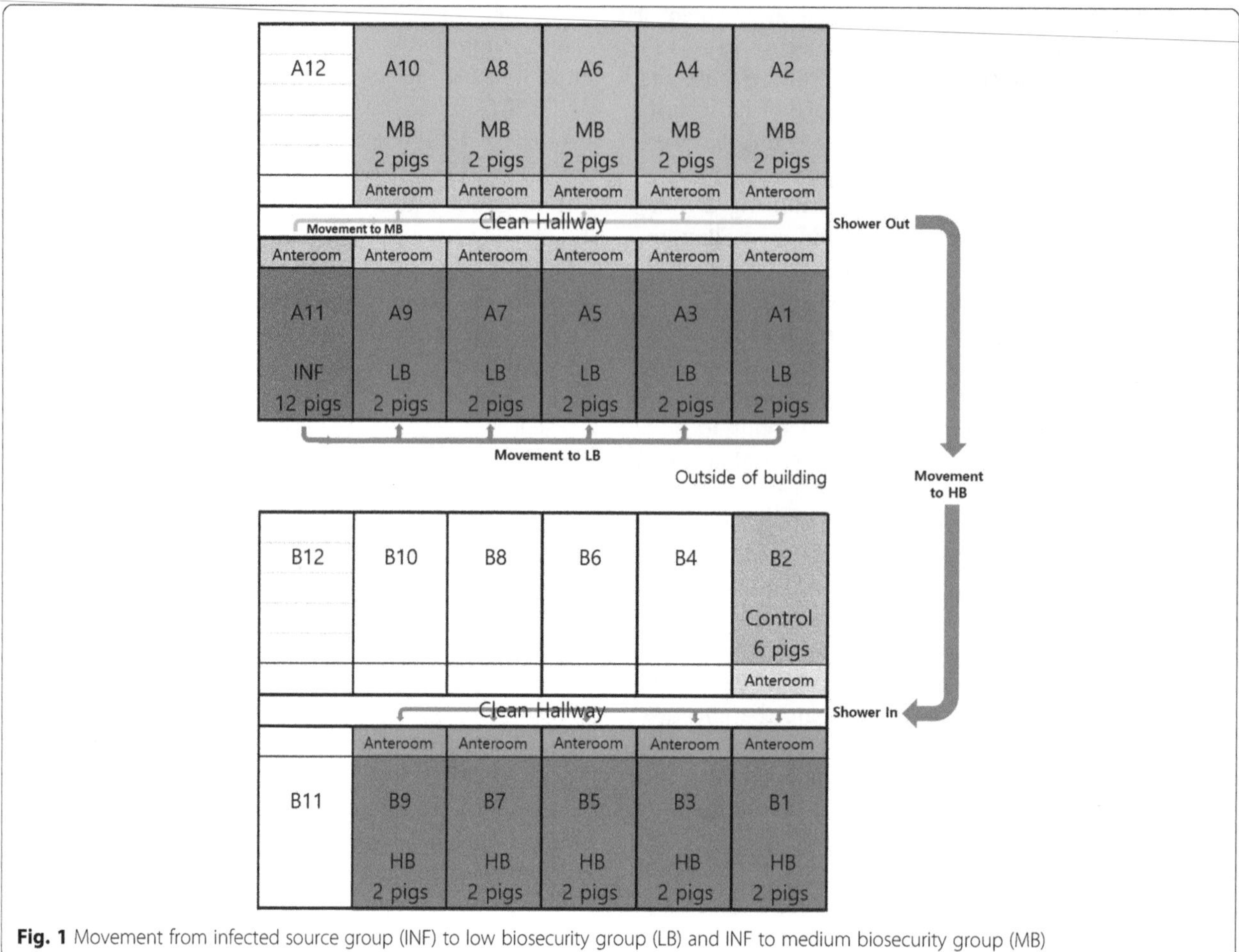

Fig. 1 Movement from infected source group (INF) to low biosecurity group (LB) and INF to medium biosecurity group (MB)

2nd trial, a cloth coverall was used instead of Tyvek® coverall. Upon entry into the anterooms through another iodine footbath, personnel put on face shield and room-specific rubber boots. Before entering the animal housing area, personnel put on another pair of disposable plastic boots over their rubber boots.

Experimental design

Infected group (INF)

For the first trial of the study, ten PEDV negative 3-week-old pigs were inoculated with PEDV (USA/Colorado/2013) strain of passage 16 via the intra-gastric route. Each pig was infected with 10 mL of the virus inoculum containing 3.6×10^4 50% tissue culture infective dose ($TCID_{50}$) per mL. Two uninfected PEDV negative 3-week-old pigs were housed with infected animals to serve as sentinels to assess transmission by direct contact. During the 2nd trial, which was performed after the 1st, three PEDV negative 6-week-old pigs were inoculated with gastrointestinal mucosal scrapings obtained from animals infected with the PEDV virulent strain, by intra-gastric route. An uninfected 6-week-old pig was added to this group as a direct contact sentinel. All study personnel interacted with infected pigs for the movements.

Movement between experimental groups

A movement was defined as the process when study personnel moved from the INF room to either the LB, MB, or HB sentinel rooms (Table 1). The first movement started approximately 44 h following the experimental inoculation of pigs in the INF group at a time when direct contact transmission was known to have occurred.

Exposure of study personnel to INF pigs

All study personnel who participated in movements between experimental groups were in contact first with pigs in the INF group for 45 min. Personnel interacted directly with the pigs by handling the pigs, collecting samples from them, and allowing pigs to come in contact with personnel clothes and PPE, e.g. biting, sniffing, and rubbing. Accordingly, potential infectious secretions and feces could be transferred to clothing and PPE worn by the study personnel.

Table 1 Summary of experimental design and biosecurity procedures followed prior to entry into the low, medium or high biosecurity rooms and the negative control room

From	To	Procedures	Trial 1	Trial 2
INFECTED (INF)	Low Biosecurity (LB)	Direct movement from INF to LB through soiled corridor	10 pigs	4 pigs
		No change of clothes or footwear between INF and LB	2 pigs/room	4 pigs/room
		No washing of hands or face	5 replicates	1 replicate
	Medium Biosecurity (MB)	Movement from INF to MB through clean corridor only after procedures were followed	10 pigs	12 pigs
		Wash hands and face	2 pigs/room	4 pigs/room
		Change clothes and footwear	5 replicates	3 replicates
	High Biosecurity (HB)	Movement from INF to HB through clean corridor only after procedures were followed	10 pigs	NA
		Shower	2 pigs/room	NA
		Change clothes and footwear	5 replicates	NA
	Negative control (NC)	No movement of people or fomities between INF. LB, MB or HB and negative control	6 pigs	3 pigs
		Dedicated study personnel different from personnel attending the other groups	6 pigs/room	3 pigs/room
		Shower, clean clothes and footwear each time entering the room	1 replicate	1 replicate

Movement from infected room to LB rooms

Following the interaction period with pigs in the INF group, study personnel who were designated to LB rooms placed their used nitrile gloves, disposable plastic boots, bouffant cap and coveralls into a clean plastic bag while in the storage area of the INF room. LB room study personnel exited INF room through a soiled (outside) corridor and entered directly into the LB sentinel holding room through an exit door in the soiled corridor, without stepping into iodine footbaths (Fig. 1). The LB room study personnel re-donned their used PPE, including nitrile gloves, disposable plastic boots, bouffant cap and coveralls, in the LB anteroom area. Prior to initiating contact with LB sentinel pigs, each person collected four separate swab samples from their (i) used coveralls, (ii) used disposable plastic boots, (iii) used nitrile gloves, and (iv) used bouffant cap and face/hair area in the LB storage area. After collecting the swab samples, personnel collected rectal swab samples from LB room sentinel pigs and interacted with LB room sentinel pigs for 45 min as previously described. All study personnel designated to LB rooms did not wash their hands and face prior to contact with LB sentinel pigs. These movements were scheduled once a day for 9 more consecutive days and terminated after LB sentinel pigs tested positive for PEDV.

Movement from infected room to MB rooms

Following interaction with INF pigs, study personnel collected four separate swabs from the surface of used (i) coveralls, (ii) disposable plastic boots, (iii) nitrile gloves, and (iv) bouffant caps and face/hair area in the INF room storage area. All MB room study personnel exited the INF room through the anteroom and removed their used coveralls, disposable plastic boots, latex gloves, and bouffant cap, and washed their hands and face with soap and water for approximately 20–40 s according to the Centers for Disease Control and Prevention (CDC) guidelines [18] prior to exiting the room into the clean corridor (Fig. 1). In the clean hallway, study personnel donned new coveralls and bouffant cap and collected four separate fomite swab samples from the new PPE, including (i) coveralls, (ii) disposable plastic boots, and from their (iii) hands, and (iv) bouffant cap and hair/face area prior to entering the anteroom of MB sentinel rooms. Here, study personnel washed again their hands and face and then, put on gloves, protective eyewear and room-specific rubber boots. Before entering the MB room animal housing area, personnel put on another pair of disposable plastic boots over their rubber boots. MB study personnel collected rectal swab samples from sentinel pigs and interacted with them, as described above. These movements were completed once a day over nine more consecutive days.

Movement from infected room to HB rooms

The HB animals were housed in a separate building located approximately 10 m away. After interacting with pigs in the LB or MB treatment group, study personnel showered with soap and shampoo for approximately 10 min before donning a new set of facility-dedicated scrubs and a pair of new disposable plastic boots. Study personnel donned new PPE, interacted again with pigs in the INF group, and took a full shower again before donning a new set of facility-dedicated scrubs and a pair of new disposable plastic boots before entering the isolation unit where the HB animals were housed. Study personnel entered the animal isolation hallway through an iodine footbath, then donned new coveralls and bouffant cap. Each of the study personnel collected four separate fomite swab samples from the new PPE, hands, and bouffant cap and hair/face area as described above. In the anteroom, study personnel washed their hands and face again, donned gloves, protective eyewear, and room-specific rubber boots with disposable plastic boots over them before entering the animal housing area. All study personnel collected rectal swab samples from HB room sentinel pigs and interacted with them as described above. These movements were completed twice a day over nine more consecutive days.

Collection of rectal and fomite swabs

Fomite and rectal swab samples were collected using a sterile rayon-tipped swab (BD CultureSwab™, liquid Stuart medium, single plastic applicator, Becton Dickinson and Co., Sparks, MD, USA). Fomite swabs were collected from coveralls, disposable plastic boots, hands or nitrile gloves, bouffant cap, face and hair areas using a zigzag pattern to cover maximum surface area prior to interacting with pigs in each biosecurity group. Rectal swabs were collected daily. Following collection, each swab was suspended in 2 mL transport media solution of Dulbecco's minimal essential medium (Gibco® DMEM, Thermo Fisher Scientific Inc., Waltham, MA, USA) containing 2% Bovine Albumin Fraction V 7.5% solution (Gibco® BSA, Thermo Fisher Scientific Inc., Waltham, MA, USA), 1% Antibiotic-Antimycotic, 100× solution (Gibco® Anti-Anti, Thermo Fisher Scientific Inc., Waltham, MA, USA), 0.15% Trypsin-TPCK, 1 mg/mL (Sigma-Aldrich, St. Louis, MO, USA) and 0.1% Gentamicin-Sulfate, 50 mg/mL (Lonza Inc., Walkersville, MD, USA). An aliquot (50 μL) of the swab suspension sample was used to extract RNA for rRT-PCR, and the remainder of the samples were stored at −80 °C. Swab samples were tested for the presence of PEDV Spike (S) gene by rRT-PCR. Briefly, RNA was extracted from eluent using the MagMAX™-96 Viral RNA Isolation Kit (ThermoFisher Scientific, Waltham, MA, USA), according to the manufacturer's instructions. A primer pair was designed to amplify a portion of the PEDV S gene with the following sequences: Forward 1910: ACGTCCCTTTACTTTCAATTCACA and Reverse 2012: TATACTTGGTACACACATCCAGAGTCA. PCR amplification was quantified using a FAM labeled probe 1939: FAM-TGAGTTGATTACTGGCACGCCTAAACCAC-BHQ. The primers and hydrolysis probe set were added to the AgPath-ID™ One-Step RT-PCR Reagents (Thermo-Fisher Scientific, Waltham, MA, USA) with 5 μl of extracted total RNA and amplified with the ABI 7500 Fast Real-Time PCR System (Thermo Fisher Scientific, Waltham, MA, USA) using the following condition: reverse transcription at 48 °C for 10 min; denaturation at 95 °C for 10 min; 45 cycles of denaturation at 95 °C for 15 s and annealing at 60 °C for 45 s.

Results

Fecal shedding

In both studies, pigs had limited clinical signs of diarrhea. Diarrhea was mild and transient in about half of the pigs. In both studies, PEDV RNA was detected by rRT-PCR from rectal swabs of pigs in the INF group at 1-day post infection (dpi), indicative of virus shedding from inoculated pigs. Rectal swabs of direct contact sentinel pigs, co-housed with the INF group in both trials, tested rRT-PCR positive at 2 dpi (Tables 2 and 3), 1 day after virus was detected in inoculated pigs. Virus shedding in pigs of the INF group, measured as viral RNA copies per rectal swab, peaked at 1 dpi and viral RNA levels remained elevated through 19 dpi (Fig. 2). During the 2nd trial, rectal swabs of INF pigs remained positive until 12 dpm when that experiment was terminated (Fig. 3).

Movements were started at 2 dpi of the INF group. Sentinel pigs in the LB group tested PEDV positive on rectal swabs 24 h after the first movement. Viral RNA was detected in 10 out of 10 sentinel pigs during the 1st trial and 4 out of 4 sentinel pigs in the 2nd trial (Tables 2 and 3). Viral shedding in the LB group of 1st trial was undetectable after 21 dpm (Fig. 2). Rectal swabs from pigs in the MB and HB groups tested rRT-PCR negative during the 10 consecutive days of movement and remained negative through 24 dpm, when the first trial

Table 2 Number of porcine epidemic diarrhea virus positive pigs (1st trial)

Days Post Infection	−1	0	1	2	3	4	5	6	7	8	9	10	11	12
Infection group	0/10	0/10	10/10	10/10	10/10	10/10	10/10	10/10	10/10	10/10	10/10	10/10	10/10	10/10
Infection group sentinel	0/2	0/2	0/2	2/2	2/2	2/2	2/2	2/2	2/2	2/2	2/2	2/2	2/2	2/2
Days After Movement				0	1	2	3	4	5	6	7	8	9	10
Low biosecurity				0/10	9/10	10/10	10/10	10/10	10/10	10/10	10/10	10/10	10/10	10/10
Medium biosecurity				0/10	0/10	0/10	0/10	0/10	0/10	0/10	0/10	0/10	0/10	0/10
High biosecurity				0/10	0/10	0/10	0/10	0/10	0/10	0/10	0/10	0/10	0/10	0/10

Table 3 Number of porcine epidemic diarrhea virus positive pigs (2nd trial)

Days Post Infection	−1	0	1	2	3	4	5	6	7	8	9	10	11	12
Infection group	0/3	0/3	3/3	3/3	3/3	3/3	3/3	3/3	3/3	3/3	3/3	3/3	3/3	3/3
Infection group sentinel	0/1	0/1	0/1	1/1	1/1	1/1	1/1	1/1	1/1	1/1	1/1	1/1	1/1	1/1
Days After Movement				0	1	2	3	4	5	6	7	8	9	10
Low biosecurity				0/4	4/4	4/4	4/4	4/4	4/4	4/4	4/4	4/4	4/4	4/4
Medium biosecurity				0/12	0/12	0/12	0/12	0/12	0/12	0/12	0/12	0/12	0/12	0/12

was terminated. Rectal swabs from pigs in the NC group remained negative for the entire duration of the study.

Fomite swabs

Fomite swab samples collected on 1, 2, and 3 dpm from hair/face, hands, coverall, and boots prior to contact with each group of sentinel pigs were tested by rRT-PCR to determine where PEDV was carried on each person that could potentially contribute to virus transmission (Table 4). Viral RNA was detected at 1 dpm through 3 dpm from all LB group PPE fomite swab samples during the first trial. PEDV was detected in one coverall swab in the 2nd study at 1 dpm and in all PPE at 2 and 3 dpm. In addition, at 1 dpm, 2 hair/face swabs from MB personnel were positive in the 1st study, even though transmission of virus was not detected. All fomite swabs from HB study personnel tested negative.

Discussion

Although several aspects of PEDV transmission have been examined, the efficiency by which biosecurity measures prevent indirect transmission of PEDV has been largely unexplored. In the current study, we sought to address this by modifying biosecurity measures in a controlled experimental design, using study personnel to simulate movements between rooms that reflect situations within swine farms around the country. Graded biosecurity stringency was designed into movements made by study personnel between a known infected room and sentinel rooms. As expected, direct-contact sentinel pigs showed signs of PEDV infection 24 h after viral shedding was detected in infected pigs supporting the view that PEDV is highly contagious [16]. Movements between INF and sentinel rooms were designed to begin when viral shedding peaked in the source group at 2 dpi. Movements to the LB rooms simulated indirect transmission in the absence of biosecurity protocols. Interestingly, transmission to the LB sentinel groups happened surprisingly rapidly. Virus shedding in the LB sentinels was detected 24 h after the first movement into the room again providing proof of the contagious nature of PEDV. Samples from PPE of all study personnel in contact with experimentally infected pigs were found to be contaminated with PEDV by rRT-PCR, and transmitted infection to the LB sentinel pigs even though virus infectivity on PPE was not tested. This information is relevant since it helps explain the rapid spread of PEDV within populations even in the absence of direct contact pig transmission.

Among the graded biosecurity measures designed to break the virus transmission cycle, movements into the MB sentinel groups showed no evidence of transmission even though swabs from MB study personnel's hair and face were PEDV rRT-PCR positive. Transmission of PEDV with MB protocols may have been limited by low dose of virus, presence of non-infectious virus, inadequate interaction of pigs with contaminated PPE/surfaces, or the decreased efficiency of fecal oral transmission route from these contaminated areas. Similar experiments using an influenza virus transmission model showed a breakdown of medium biosecurity measures after 10 consecutive movements [17]. Swabs from

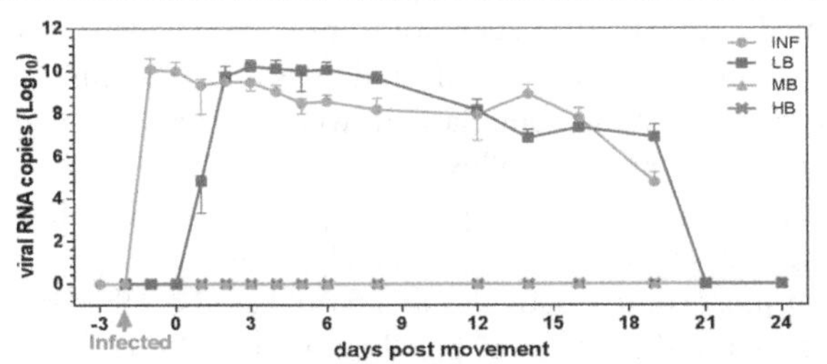

Fig. 2 Viral shedding of pigs (1st trial). Movements were terminated at 10 dpi. Data presented are average values of viral RNA copies (± SD) of infected source group (INF) (n = 12), low biosecurity group (LB) (n = 10), medium biosecurity group (MB) (n = 10), and high biosecurity group (HB) (n = 10) groups

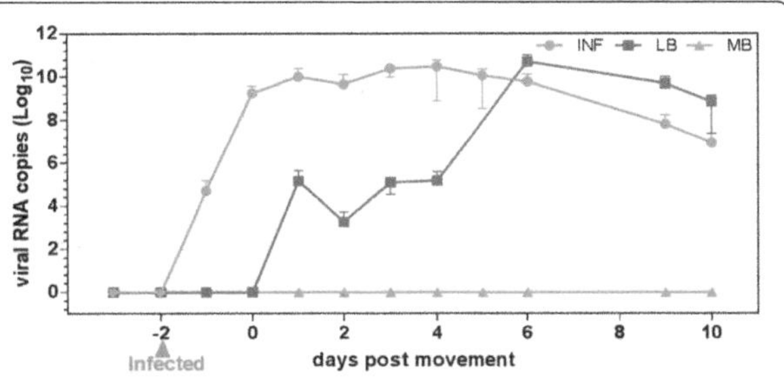

Fig. 3 Viral shedding of pigs (2nd trial). Movements were terminated at 10 dpi. Data presented are average values of viral RNA copies (± SD) of infected source group (INF) (n = 4), low biosecurity group (LB) (n = 4), and medium biosecurity group (MB) (n = 12)

Table 4 Number of porcine epidemic diarrhea virus positive fomite swabs prior to contact with pigs in the respective groups and mean (±SD) cycle threshold RT-PCR values for positive samples

Group	Swab	Movement day					
		1st study			2nd study		
		1	2	3	1	2	3
Negative	Bouffant cap, hair, face area	(0/5)[a]	(0/5)	(0/5)	(0/4)	(0/4)	(0/4)
	Coverall	(0/5)	(0/5)	(0/5)	(0/4)	(0/4)	(0/4)
	Hands	(0/5)	(0/5)	(0/5)	(0/4)	(0/4)	(0/4)
	Boots	(0/5)	(0/5)	(0/5)	(0/4)	(0/4)	(0/4)
LB	Bouffant cap, hair, face area	(3/5) (31.58 ± 1.03) [b]	(2/5) (33.62 ± 0.16)	(5/5) (32.66 ± 1.58)	(0/1)	(1/1) (31.62)	(1/1) (33.58)
	Coverall	(5/5) (26.16 ± 3.17)	(5/5) (29.28 ± 2.22)	(5/5) (27.96 ± 3.96)	(1/1) (33.40)	(1/1) (29.27)	(1/1) (24.60)
	Used gloves	(5/5) (28.81 ± 3.83)	(4/5) (28.01 ± 2.98)	(5/5) (28.76 ± 2.21)	(0/1)	(1/1) (28.03)	(1/1) (30.01)
	Boots	(5/5) (26.30 ± 4.44)	(5/5) (27.42 ± 6.22)	(5/5) (24.51 ± 3.94)	(0/1)	(1/1) (28.74)	(1/1) (28.54)
MB	Bouffant cap, hair, face area	(2/5) (30.75 ± 0.93)	(0/5)	(0/5)	(0/3)	(0/3)	(0/3)
	Coverall	(0/5)	(0/5)	(0/5)	(0/3)	(0/3)	(0/3)
	Hands	(0/5)	(0/5)	(0/5)	(0/3)	(0/3)	(0/3)
	Boots	(0/5)	(0/5)	(0/5)	(0/3)	(0/3)	(0/3)
HB	Bouffant cap, hair, face area	(0/5)	(0/5)	(0/5)			
	Coverall	(0/5)	(0/5)	(0/5)			
	Hands	(0/5)	(0/5)	(0/5)			
	Boots	(0/5)	(0/5)	(0/5)			

[a] Number of positive
[b] Ct value (avg. ± S.D)

HB group study personnel tested PEDV rRT-PCR negative even on the hair and face. Similarly, HB sentinel pigs were rRT-PCR negative for 10 consecutive days. These results together indicate that taking a shower and changing PPE before contacting pigs is an ideal way to completely prevent indirect viral transmission in conditions generally seen in farms. Although our results also support that only changing PPE and washing skin exposed areas is beneficial to decrease the risk of PEDV transmission, there may still be an inherent risk of PEDV transmission from contaminated body surfaces on personnel. Hence, only changing PPEs might not be the most effective way to protect against spread of PEDV.

Our results also indicate that breaches in biosecurity procedures can very rapidly transmit PEDV to naïve herds through indirect means (viz. contaminated fomites). These findings also suggest that contact with PEDV contaminated fomites for a sufficient time is an efficient source of infection and likely plays a role in the rapid transmission of PEDV when there is adequate contact with fomites. Previous studies have suggested that fomites may be an effective mode of PEDV transmission [6, 7, 9, 11–13, 16]. These previous studies rely on PCR detection of viral RNA particles or demonstration of infectious PEDV in cell culture assays to suggest the possibility of transmission by fomites. For example, airborne transmission [12], vehicles [4], feed [7, 10, 16], storage bags [6], personnel working with pigs [6] and other fomites have tested positive for PEDV indicating their possible role in viral transmission and should be considered as a source for virus spread. However, these studies lack a tangible demonstration of the ability of contaminated fomites to infect pigs, either in an experimental setting or in a farm, except for the role of contaminated feed. The present study provides evidence that personnel exposed to infected pigs can transmit the virus to a naïve population, when basic biosecurity procedures are not followed.

The experimental design in the present studies allowed a 45-min contact time with animals under the assumption that most routine activities in a farm, based on the size of the pen and number of pigs housed, may be completed within that time frame. However, one cannot rule out the possibility that transmission may occur with medium biosecurity, if longer interaction periods or larger infected source groups were used in the design. Contact with infected pigs for more than 45 min and/or 10 movements could have increased the probability of PEDV transmission to naïve sentinel pigs. However, we observed that with low biosecurity procedures, transmission and infection of PEDV was both efficient and rapid. This data provides evidence that spread of PEDV within farms may occur efficiently with failures in biosecurity procedures.

Results presented here should be considered carefully as many factors, including contact time, exposure time, viral dose, time after exposure to virus and other experimental conditions may influence the outcome of transmission studies. However, this experimental study highlights the main advantages of good biosecurity procedures in breaking the transmission cycle between rooms. The fact that PEDV transmission occurred under low biosecurity procedures indicated that the virus could spread easily through contaminated fomites worn by personnel. These results provide critical information to develop effective biosecurity procedures and will have potential applications for the development and implementation of transmission control policies in swine production systems. Our results are also relevant to design biosecurity measures to control the spread of other pathogens of similar characteristics and transmission routes than PEDV such as transmissible gastroenteritis virus and porcine deltacoronavirus.

Conclusions

In conclusion, these results indicate the indirect transmission of PEDV through contaminated personnel PPEs occurs rapidly under modeled conditions and to prevent transmission between groups of pigs, changing PPE and/or taking a shower is recommended as an effective option to lower the risk of virus spread.

Abbreviations

dpi: Days post infection; dpm: Days post movement; HB: High biosecurity; INF: Infection; LB: Low biosecurity; MB: Medium biosecurity; NC: Negative control; PDCoV: Porcine delta coronavirus; PED: Porcine epidemic diarrhea; PEDV: Porcine epidemic diarrhea virus; PPE: Personal protective equipment; rRT-PCR: Real-time reverse transcription polymerase chain reaction; TCID: Tissue culture infective dose; TGEV: Transmissible gastroenteritis virus

Acknowledgements

The authors would like to acknowledge Ang Su, Andres Diaz, Dane Goede, Fabian Chamba, Fabio A. Vannucci, Fernando L. Leite, Hunter Baldry, Jessica Johnson, Jisun Sun, Kruthikaben Patel, Luiza R. Roos, Michael Rahe, Nitipong Homwong and Venkatramana *D. Krishna* from the University of Minnesota College of Veterinary Medicine for their assistance with animal care, sample collection and study execution.

Funding

This study was funded in part by the state of Minnesota and by the University of Minnesota College of Veterinary Medicine Emerging and Zoonotic Diseases Signature program (MN-62-092).

Authors' contributions

YK participated in study design, implementation of the study, analysis, and writing of the manuscript. MY participated in study design and laboratory testing. SMG participated in study design and implementation of the study. MCC participated in study design, implementation of the study and revised the manuscript. MT conceived the study, participated in study design, implementation of the study and revised the manuscript. All authors have read and approved the final version of the manuscript.

Competing interests

The authors declare that they have no competing interests.

Consent for publication

Not applicable.

References

1. Saif LJ, Pensaert MB, Sestack K, Yeo SG, Jung K. Coronaviruses. In: Straw BE, Zimmerman JJ, Karriker LA, Ramirez A, Schwartz KJ, Stevenson GW, editors. Diseases of swine. Ames: Wiley-Blackwell; 2012. p. 501–24.
2. Paarlberg PL. Updated estimated economic welfare impacts of porcine epidemic diarrhea virus (PEDV); 2014. p. 1–38.
3. Hofmann M, Wyler R. Quantitation, biological and physicochemical properties of cell culture-adapted porcine epidemic diarrhea coronavirus (PEDV). Vet Microbiol. 1989;20(2):131–42.
4. Lowe J, Gauger P, Harmon K, Zhang J, Connor J, Yeske P, Loula T, Levis I, Dufresne L, Main R. Role of transportation in spread of porcine epidemic diarrhea virus infection, United States. Emerg Infect Dis. 2014;20(5):872–4.
5. O'Dea EB, Snelson H, Bansal S. Using heterogeneity in the population structure of U.S. swine farms to compare transmission models for porcine epidemic diarrhoea. Sci Rep. 2016;6:22248.
6. Scott A, McCluskey B, Brown-Reid M, Grear D, Pitcher P, Ramos G, Spencer D, Singrey A. Porcine epidemic diarrhea virus introduction into the United States: root cause investigation. Prev Vet Med. 2016;123:192–201.
7. Canadian Food Inspection A. Update: Canadian Food Inspection Agency Investigation into Feed as a Possible Source of Porcine Epidemic Diarrhea (PED). In.; 2014. Available from: http://www.inspection.gc.ca/animals/terrestrial-animals/diseases/other-diseases/ped/2014-03-03/eng/1393891410882/1393891411866.
8. Dee S, Clement T, Schelkopf A, Nerem J, Knudsen D, Christopher-Hennings J, Nelson E. An evaluation of contaminated complete feed as a vehicle for porcine epidemic diarrhea virus infection of naive pigs following consumption via natural feeding behavior: proof of concept. BMC Vet Res. 2014;10(1):176.
9. Bowman AS, Krogwold RA, Price T, Davis M, Moeller SJ. Investigating the introduction of porcine epidemic diarrhea virus into an Ohio swine operation. BMC Vet Res. 2015;11(3):38.
10. Pasick J, Berhane Y, Ojkic D, Maxie G, Embury-Hyatt C, Swekla K, Handel K, Fairles J, Alexandersen S. Investigation into the role of potentially contaminated feed as a source of the first-detected outbreaks of porcine epidemic diarrhea in Canada. Transbound Emerg Dis. 2014;61:397–410.
11. Gerber PF, Xiao CT, Chen Q, Zhang J, Halbur PG, Opriessnig T. The spray-drying process is sufficient to inactivate infectious porcine epidemic diarrhea virus in plasma. Vet Microbiol. 2014;174(1–2):86–92.
12. Alonso C, Goede DP, Morrison RB, Davies PR, Rovira A, Marthaler DG, Torremorell M. Evidence of infectivity of airborne porcine epidemic diarrhea virus and detection of airborne viral RNA at long distances from infected herds. Vet Res. 2014;45(1):73.
13. Dee S, Neill C, Singrey A, Clement T, Cochrane R, Jones C, Patterson G, Spronk G, Christopher-Hennings J, Nelson E. Modeling the transboundary risk of feed ingredients contaminated with porcine epidemic diarrhea virus. BMC Vet Res. 2016;12(1):51.
14. Bowman AS, Nolting JM, Nelson SW, Bliss N, Stull JW, Wang Q, Premanandan C. Effects of disinfection on the molecular detection of porcine epidemic diarrhea virus. Vet Microbiol. 2015;179(3–4):213–8.

15. Stevenson GW, Hoang H, Schwartz KJ, Burrough ER, Sun D, Madson D, Cooper VL, Pillatzki A, Gauger P, Schmitt BJ, et al. Emergence of porcine epidemic diarrhea virus in the United States: clinical signs, lesions, and viral genomic sequences. J Vet Diagn Investig. 2013;25(5):649–54.
16. Dee S, Neill C, Clement T, Singrey A, Christopher-Hennings J, Nelson E. An evaluation of porcine epidemic diarrhea virus survival in individual feed ingredients in the presence or absence of a liquid antimicrobial. Porcine Health Manage. 2015;1(1):1–10.
17. Allerson MW, Cardona CJ, Torremorell M. Indirect transmission of influenza a virus between pig populations under two different Biosecurity settings. PLoS One. 2013;8(6):e67293.

4

Phylogenetic assessment reveals continuous evolution and circulation of pigeon-derived virulent avian avulaviruses 1 in Eastern Europe, Asia, and Africa

Mahmoud Sabra[1,2], Kiril M. Dimitrov[2], Iryna V. Goraichuk[2,3], Abdul Wajid[4,5], Poonam Sharma[2], Dawn Williams-Coplin[2], Asma Basharat[4], Shafqat F. Rehmani[4], Denys V. Muzyka[3], Patti J. Miller[2] and Claudio L. Afonso[2*]

Abstract

Background: The remarkable diversity and mobility of Newcastle disease viruses (NDV) includes virulent viruses of genotype VI. These viruses are often referred to as pigeon paramyxoviruses 1 because they are normally isolated and cause clinical disease in birds from the *Columbidae* family. Genotype VI viruses occasionally infect, and may also cause clinical disease in poultry. Thus, the evolution, current spread and detection of these viruses are relevant to avian health.

Results: Here, we describe the isolation and genomic characterization of six Egyptian (2015), four Pakistani (2015), and two Ukrainian (2007, 2013) recent pigeon-derived NDV isolates of sub-genotype VIg. These viruses are closely related to isolates from Kazakhstan, Nigeria and Russia. In addition, eight genetically related NDV isolates from Pakistan (2014–2016) that define a new sub-genotype (VIm) are described. All of these viruses, and the ancestral Bulgarian ($n = 2$) and South Korean ($n = 2$) viruses described here, have predicted virulent cleavage sites of the fusion protein, and those selected for further characterization have intracerebral pathogenicity index assay values characteristic of NDV of genotype VI (1.31 to 1.48). A validated matrix gene real-time RT-PCR (rRT-PCR) NDV test detect all tested isolates. However, the validated rRT-PCR test that is normally used to identify the virulent fusion gene fails to detect the Egyptian and Ukrainian viruses due to mismatches in primers and probe. A new rapid rRT-PCR test to determine the presence of virulent cleavage sites for viruses from sub-genotypes VIg was developed and evaluated on these and other viruses.

Conclusions: We describe the almost simultaneous circulation and continuous evolution of genotype VI Newcastle disease viruses in distant locations, suggesting epidemiological connections among three continents. As pigeons are not migratory, this study suggests the need to understand the possible role of human activity in the dispersal of these viruses. Complete genomic characterization identified previously unrecognized genetic diversity that contributes to diagnostic failure and will facilitate future evolutionary studies. These results highlight the importance of conducting active surveillance on pigeons worldwide and the need to update existent rapid diagnostic protocols to detect emerging viral variants and help manage the disease in affected regions.

Keywords: Newcastle disease virus, NDV, Pigeons, Genotype VI, rRT-PCR, Mismatches, Evolution, Next-generation sequencing

* Correspondence: Claudio.Afonso@ars.usda.gov
[2]Exotic and Emerging Avian Viral Diseases Research Unit, Southeast Poultry Research Laboratory, US National Poultry Research Center, Agricultural Research Service, USDA, 934 College Station Road, Athens, GA 30605, USA
Full list of author information is available at the end of the article

Background

Virulent Newcastle disease viruses (NDV), synonymous with avian avulaviruses 1 (AAvV-1), are the causative agents of Newcastle disease (ND). Newcastle disease is one of the most important infectious diseases of poultry because of its worldwide distribution and devastating economic effects for the poultry industry [1–3]. The disease is highly contagious, presents high morbidity and mortality, and is classified as a notifiable disease by the World Organisation for Animal Health (OIE) [3]. AAvV-1 (along with another 12 serotypes, namely AAvV 2–13) belongs to genus *Avulavirus* of the family *Paramyxoviridae* [4, 5]. Several additional AAvV, have been recently proposed as potential new serotypes [6–9]. Avian avulaviruses 1 are enveloped, have a single stranded, non-segmented, negative sense RNA genome with helical capsid symmetry [10]. Three different genomic sizes (15,186 nucleotides [nt], 15,192 nt and 15,198 nt) have been identified so far, and all known AAvV-1 are divided into two major genetic groups, class I and class II [11]. Currently, AAvV-1 isolates are classified into 18 class II and one class I genotypes based on the complete coding sequences of the fusion protein [12, 13]. For the purpose of this work, the taxa name Newcastle disease virus will be used.

Pigeons may be infected with NDV of all genotypes, but are particularly susceptible to the genotype VI genetic variants, also referred to as pigeon paramyxovirus 1 (PPMV-1). These genotype VI viruses are mainly isolated from Rock Pigeons (*Columba livia*), but also have been isolated from other members of the family *Columbidae*, e.g. feral Eurasian Collared Doves (*Streptopelia decaocto*) [14]. Viruses of genotype VI are endemic in the pigeon population throughout the world, and can be distinguished as being variants of NDV by the patterns produced in a hemagglutination inhibition (HI) assay [15, 16] when tested with a panel of monoclonal antibodies. Genotype VI viruses have evolved rapidly since their emergence in the Middle East during the 1960s and are currently divided into at least eleven sub-genotypes (and one more putative for a group of viruses identified in Ethiopia), namely VIa-VIj (respectively VII) [13, 17].

Commonly, these viruses are an example of viruses that have a cleavage site motif that is generally associated with virulent viruses. However, most of them are considered to be of intermediate or low virulence for chickens, as assessed through the intracerebral pathogenicity index (ICPI) test [18, 19]. Nevertheless, their pathogenicity might be increased after passage in poultry species [20, 21].

Diagnostic testing and rapid detection of NDV are important steps in controlling an ND outbreak. Virus isolation in specific pathogen free (SPF) embryonated chicken eggs (ECE), followed by identification using hemagglutination (HA) and HI assays with a NDV-monospecific antiserum [3] is considered "gold standard" for NDV diagnostics. This approach is time-consuming and laborious and often requires up to ten days [22]. Molecular diagnostics assays are a viable alternative to classical diagnostic assays and are widely used. Several protocols for the detection of NDV by reverse-transcription PCR (RT-PCR) have been published in the last decade [23]. Real-time RT-PCR (rRT-PCR) is a rapid screening assay allowing for detection and pathotyping of NDV directly from diagnostic specimens and different protocols based on the use of hydrolysis probes, SybrGreen or LUX primers, have been published [24–26]. Real-time RT-PCR is highly dependent on genetic similarity between the primers/probe and the target genome, and protocols have to be updated as the genomes of pathogens accumulate mutations over time.

To improve our understanding of the distribution and evolution of NDV of genotype VI, viruses of this genotype circulating in five countries from 1982 to 2016 were isolated and studied. For this purpose, the following studies were done: i) isolation and biological and phylogenetic characterization of genotype VI viruses from different geographical locations; ii) analyses of the complete fusion protein gene coding sequences and complete genome sequences obtained from the studied viruses; iii) evaluation of their epidemiological relation to other circulating NDVs; and iv) optimization of rRT-PCR diagnostic test for detection of virulent variants of sub-genotype VIg as a result of identified failure of the current validated diagnostic protocol.

Methods

Sample collection and isolates background data

The samples and viruses studied here were collected from different species (pigeons, chickens and quail) in Egypt, Pakistan, South Korea, Ukraine and Bulgaria, representing three different continents (Africa, Asia and Europe). During April 2015, one hundred sixty-seven (n = 167) oropharyngeal and cloacal/fecal swabs were collected in Egypt from apparently healthy Rock Pigeons. Samples were collected from pigeon lofts in El Fayoum and Qena provinces, and a live bird market (LBM) in Cairo named Souq al-Goma'a (also named Souq Sayeda Aisha or Friday Market). The swabs were placed immediately into tubes with 3 ml of brain-heart-infusion broth (Difco, New Zealand) supplemented with penicillin G (10,000 IU /ml), amphotericin B (20 μg/ml), and gentamycin (1000 μg/ml). After collection, the samples were labelled, stored on ice, transported to the lab where kept frozen at −76 °C, and shipped on dry ice to the Southeast Poultry Research Laboratory (SEPRL) of the United States Department of Agriculture (USDA). Additionally, three more samples were collected from pigeons in Pakistan in 2015 and shipped to SEPRL, USDA. Six additional NDV from repositories in Bulgaria (n = 2), Ukraine (n = 2), and South Korea (n = 2), were also sent to SEPRL for further characterization. The Bulgarian and the Ukrainian viruses were passaged in eggs 3 times and 2 times, respectively. The Korean isolates were passaged 2 times at SEPRL and passage information was missing prior to their receiving. Nine more NDV, isolated between 2014 and 2016 from healthy and diseased pigeons (pet and zoo birds), were studied at the Quality Operations Laboratory (QOL) at the University of Veterinary and Animal Sciences (UVAS), Lahore, Pakistan.

Virus isolation, virus propagation and intracerebral pathogenicity index test

Initial screening of all Egyptian samples employing the NDV and avian influenza matrix gene rRT-PCR assays [24, 27], revealed that out of 167 oropharyngeal and cloacal swabs, 71 samples had cycle threshold (Ct) values ≤35 (40.3%) in the NDV assay and all samples were negative in the avian influenza assay. At SEPRL, thirty-one (selected to achieve representativeness of all locations) of the 71 rRT-PCR positive Egyptian samples and three Pakistani samples from 2015 (designated 21A, 22A and 25A) were selected for further studies and inoculated into 9-to-11-day-old SPF ECE, following standard procedures [28]. The SPF ECE and chickens used to characterize these viruses were from the SEPRL SPF White Leghorn flock. The allantoic fluids from both eggs with embryo mortality and embryos alive at the end of the incubation period were collected and tested by hemagglutination assay. All the hemagglutinating agents were confirmed as NDV using a HI assay with NDV specific antiserum [3]. Additionally, The NDV obtained from repositories in Ukraine, Bulgaria and South Korea were propagated into 9-to-11-day-old SPF ECE following the same procedures [28]. Assessment of the virulence in vivo of three selected viruses was done by the ICPI test using one-day-old SPF chickens following established procedures [3].

Complete fusion protein gene sequencing

RNA from the two South Korean isolates was extracted from infected allantoic fluid using TRIzol LS Reagent (Invitrogen, Carlsbad, CA, USA) following the manufacturer's instructions at SEPRL. RT-PCR was performed to amplify the complete F gene using the Superscript™ III One-step RT-PCR kit with Platinum Taq DNA polymerase (Invitrogen, Carlsbad, CA, USA), per manufacturer's instructions using a set of F-gene specific primers (4331F/5090R, MSF1/NDVR2, 4927F/5673R and 5491F/6341R) [29] (see Additional file 1: Table S1). The PCR amplicons were processed and sequenced as described previously [30]. RNA isolation and nucleotide sequencing of nine Pakistani viruses were performed in Quality operations laboratory at University of Veterinary and Animal Science in Pakistan as follows: RNA extraction and RT-PCR F-gene amplification were performed as described above using a set of F-gene specific primers (see Additional file 1: Table S1). The amplicons were electrophoresed using 1% agarose gel and purified using QIAquick® Gel Extraction Kit (Qiagen, Valencia, CA, USA). The purified products were sequenced using the ABI 3130 automated sequencer (Applied Biosystem Inc., Foster City, CA, USA), as described previously [30].

Complete genome sequencing using next-generation sequencing (NGS)

Total viral RNA of six Egyptian (positive in virus isolation assay), two Ukrainian, three Pakistani, and two Bulgarian viruses were extracted from the infected allantoic fluids using QIAmp® Viral RNA Mini Kit (Qiagen, USA) according to manufacturer's instructions. The recovered RNA was quantified using the Qubit® RNA HS Assay Kit (Life Technologies, USA) in the Qubit® Fluorometer instrument (Invitrogen, USA). Newcastle disease virus RNA was captured and enriched using Sera-Mag beads (GE Healthcare Life Sciences, USA), and three biotin-labeled oligonucleotide probes targeting three different positions in the NDV genome: 1) 8-AGA GAA TCT GTG AGG TAC GA/3Bio/, 2) 5905-TTC TCA AGT CAT CGT GAC AG/3Bio/, and 3) 12226-CCC TGC ATC TCT CTA CAG/3Bio/. Reverse transcription was performed using the Moloney Murine Leukemia Virus Reverse Transcriptase (M-MLV RT) kit (Thermo Scientific, USA) according to the manufacturer's instructions. The cDNA products were recovered and purified using the Agencourt® RNAClean® XP beads (Beckman Coulter, USA) according to the manufacturer's instructions and quantified using Qubit® ssDNA Assay Kit (Thermo Fisher Scientific, USA) in the Qubit instrument. The purified cDNA products were tagmented and amplified for NGS by using 1 ng/5 μl (0.25 ng/μl in water) of the cDNA product employing the Nextera XT DNA Library Preparation Kit (Illumina, USA) following manufacturer's protocol. The two Bulgarian samples were processed using the KAPA Stranded RNA-Seq Library Preparation Kit for Illumina platforms (Kapa Biosystems, USA) according to the manufacturer's instructions.

The distribution size and concentration of DNA in the prepared libraries were checked on a Bioanalyzer 2100 and Qubit instrument using Agilent High Sensitivity DNA Kit (Agilent Technologies, Germany) and Qubit® dsDNA HS Assay Kit (Life Technologies, USA), respectively. Paired-end sequencing (2 × 250 base pairs) of the generated libraries was performed on an Illumina MiSeq instrument using the 500 cycle MiSeq Reagent Kit version 2 (Illumina, USA). Raw sequence data were analyzed and assembled using MIRA version 3.4.1 [31] within a customized workflow on the Galaxy platform [32] as described previously [33]. Short internal gaps at the 3′ UTR of the nucleoprotein gene were closed using Sanger technology with primers designed using the sequences obtained from NGS (see Additional file 1: Table S1). The 5′ and 3′ ends of the genomes reported here were sequenced and confirmed as described previously [34].

Collection of sequences and phylogenetic analyses

All available NDV complete genome and complete fusion protein gene coding sequences were downloaded from GenBank and curated, resulting in two large datasets (n = 331 and n = 1406, respectively). Together with the sequences obtained in the current study, all sequences were aligned using Multiple Alignment with Fast Fourier Transformation (MAFFT v7.017) [35] as implemented in the Geneious software v8.1.4 [36]. The datasets were used for two preliminary maximum likelihood phylogenetic analyses using MEGA6 [37]. Based on evolutionary relatedness to the viruses sequenced here, two smaller datasets (including newly obtained sequences) of complete fusion protein gene coding sequences (n = 82) (see Additional file 1: Table S2) and complete genome sequences (n = 83) (see Additional file 1: Table S3) were parsed from the larger datasets. Sequences of selected representative isolates from other NDV genotypes were also included in each of the datasets. The coding regions of the complete genome and complete fusion gene were used to construct final phylogenetic trees using MEGA6. To select best-fit substitution model, the Bayesian Information Criterion (BIC) and corrected Akaike Information Criterion (AICc) values were estimated using MEGA6. The General Time Reversible (GTR) model as implemented in MEGA6 with a discrete gamma distribution (4 categories [+G, parameter = 0.6519 for the full fusion gene tree and 0.5984 for the complete genome tree]) with 1000 bootstrap replicates was used in all data analysis. The phylogenetic trees were visualized and edited using tree explorer implemented in MEGA6 and branch lengths are proportional to the differences between the isolates. The evolutionary distances were inferred using MEGA6 and showed as the average number of base substitutions per site. Analyses were conducted using the maximum composite likelihood model [38] with a gamma distribution (shape parameter = 1) of rate variation among sites. For all analyses, the codon positions included were the 1st, 2nd, 3rd, and noncoding and positions containing gaps and missing data were eliminated from the datasets. The classification criteria proposed by Diel et al. for naming sub-genotypes and genotypes were followed in the current study [12].

Real-time reverse transcription polymerase chain reaction (rRT-PCR)

Allantoic fluids from all of the samples studied at SEPRL were subjected to the USDA-validated matrix gene rRT-PCR (M-gene assay), as well as to the USDA-validated fusion gene rRT-PCR assay (F-gene assay) described previously by Wise et al. [24]. An additional test was performed using pigeon-specific fusion protein gene assay as described previously by Kim et al. [39]. AgPath one-Step rRT-PCR Reagents (Thermo Scientific, USA) was used and the reactions were carried out in the Cepheid Real-Time Thermal Cycler (Life Science, USA). The samples that had a Ct value ≤35 were considered positive in all assays. Due to the failure of the validated F-gene rRT-PCR assay [24] and pigeon-specific fusion protein gene assay [39] to detect some of the genotype VI viruses studied here, the probe and the forward primer described by Kim et al., [39] were optimized as follows: F-4876 5′–[6-FAM] AAG CGY TTC TGT CTC YTT CCT CCT [BHQ_1]–3′ and (F + 4837) 5′- TGA TTC CAT CCG CAG GAT ACA AG -3′. Additionally, the reverse primer for the pigeon-specific F-gene assay (F-4837) was replaced with a new primer (F-4943) 5′-GCT GCT GTT ATC TGT GCC GA-3′. The optimized probe and newly designed primers were analyzed by OligoAnalyzer 3.1. (Integrated DNA Technologies, USA, https://eu.idtdna.com/calc/analyzer) tool and checked for self-annealing, hairpin loops and heterodimers. The sequences of the optimized primers and probes along with the tested fifteen NDV and selected representatives from other genotypes were aligned and compared to the M-gene and F-gene primers and probes and the pigeon-specific fusion probe to identify variable sites that determined the different outcomes of the F-gene rRT-PCR assay [24, 39] (see Additional file 2: Figs. S1, S2, and S3).

Results

Virus isolation and biological properties

Twenty-four NDV isolates from pigeons, chickens and quail, isolated in five different countries were studied here (Table 1). Most samples were obtained from pigeons (pet pigeons living in lofts, n = 12; zoo pigeons, n = 2; live bird markets, n = 2; and pigeons of unknown habitats, n = 6). Out of 31 oropharyngeal and cloacal Egyptian swabs from healthy birds inoculated in eggs, six contained live NDV and were fully characterized. The three Pakistani clinical samples were found to be positive for NDV by virus isolation and HI. All viruses obtained from repositories were successfully re-isolated after propagation and confirmed as NDV by HI. The ICPI values of selected viruses (pigeon/Egypt/Giza/11/2015, pigeon/Ukraine/Doneck/3/2007 and pigeon/Pakistan/Lahore/25A/2015) were 1.31, 1.48 and 1.46,

Table 1 Background information data for Newcastle disease viruses isolated in Egypt, Ukraine, Pakistan, Bulgaria and South Korea analyzed in this study

Isolate name	Collection year	Country	Location	Host		Health status	Husbandry	Flock size	Age	Sequence coverage	Cleavage site motif (positions 113–117)	Genotype	GenBank acc.#
				Scientific name	Common name								
Giza/11	2015	Egypt	Giza	*Columba livia*	pigeon	apparently healthy	live bird market	NA	NA	complete genome	RQKR↓F	VI g	KY042129
Helwan/44	2015	Egypt	Helwan	Columba livia	pigeon	apparently healthy	live bird market	NA	NA	complete genome	RQKR↓F	VI g	KY042130
Qena/56	2015	Egypt	Qena	Columba livia	pigeon	apparently healthy	pet	200	2 years	complete genome	RQKR↓F	VI g	KY042131
El Fayom/73	2015	Egypt	El Fayom	Columba livia	pigeon	apparently healthy	pet	NA	NA	complete genome	RQKR↓F	VI g	KY042132
El Fayom /79	2015	Egypt	El Fayom	Columba livia	pigeon	apparently healthy	pet	NA	NA	complete genome	RQKR↓F	VI g	KY042133
El Fayom/84	2015	Egypt	El Fayom	Columba livia	pigeon	apparently healthy	pet	NA	NA	complete genome	RQKR↓F	VI g	KY042134
Ukraine/Kharkiv/2301	2013	Ukraine	Kharkiv	*Columba livia*	pigeon	dead/sick	pet	NA	2 years	complete genome	RQKR↓F	VI g	KY042127
Ukraine/Doneck/3	2007	Ukraine	Doneck	*Columba livia*	pigeon	dead/sick	pet	NA	2 years	complete genome	RQKR↓F	VI g	KY042128
Jhang/115[a]	2015	Pakistan	Jhang	*Columba livia*	pigeon	dead/sick	pet	110	2 year	full fusion	RKKR↓F	VI g	KY042137
Lahore/125[a]	2015	Pakistan	Lahore	*Columba livia*	pigeon	dead/sick	pet	200	1.5 years	full fusion	RKKR↓F	VI g	KY042136
Lahore/126[a]	2015	Pakistan	Lahore	*Columba livia*	pigeon	dead/sick	pet	200	1.5 years	full fusion	RKKR↓F	VI g	KY042138
Lahore/146[a]	2016	Pakistan	Lahore	*Columba livia*	pigeon	dead/sick	pet	55	1 year	full fusion	RKKR↓F	VI g	KY042139
Lahore/AW-1[a]	2014	Pakistan	Lahore	*Columba livia*	pigeon	dead	wildlife	39	2 years	full fusion	RQRR↓F	VI m	KU862297
Lahore/AW-2[a]	2015	Pakistan	Lahore	*Columba livia*	pigeon	dead/sick	pet	25	2 years	full fusion	RQRR↓F	VI m	KU862298
Lahore/AW-3[a]	2015	Pakistan	Lahore	*Columba livia*	pigeon	dead/sick	pet	200	3 months	full fusion	RQRR↓F	VI m	KU862299
Jallo-Lahore/221A[a]	2016	Pakistan	Lahore	*Columba livia*	pigeon	sick	zoo	20	2 years	full fusion	RQKR↓F	VI m	KY042140
Jallo-Lahore/221B[a]	2016	Pakistan	Lahore	*Columba livia*	pigeon	sick	zoo	20	2 years	full fusion	RQKR↓F	VI m	KY042141
Lahore/21A	2015	Pakistan	Lahore	*Columba livia*	pigeon	NA	NA	NA	NA	complete genome	RQKR↓F	VI m	KX236100
Lahore/22A	2015	Pakistan	Lahore	*Columba livia*	pigeon	dead/sick	pet	30	8 months	complete genome	RQKR↓F	VI m	KY042135
Lahore/25A	2015	Pakistan	Lahore	*Columba livia*	pigeon	NA	pet	20	1 year	complete genome	RQKR↓F	VI m	KX236101
Mokresh	1982	Bulgaria	Mokresh	*Gallus gallus*	chicken	NA	NA	NA	NA	complete genome	RQKR↓F	VI	KY042126
Dolnolinevo	1992	Bulgaria	Dolno Linevo	*Gallus gallus*	chicken	NA	NA	NA	NA	complete genome	RQKR↓F	VI c	KY042125
93-58GG	1993	South Korea	NA	*Gallus gallus*	chicken	NA	broiler farm	NA	14 days	full fusion	RRKR↓F	VI c	KY042142
88 M	1988	South Korea	NA	*Coturnix coturnix*	quail	NA	NA	NA	NA	full fusion	RRKR↓F	VI c	KY042143

[a] these viruses were studied in Pakistan at the Quality Operations Laboratory (QOL) of the University of Veterinary & Animal Sciences (UVAS)
NA = not available

respectively, and these indexes are typical for NDV with moderate virulence (mesogenic) in chickens [28].

Complete fusion protein gene and complete genome sequencing

The complete genome sequences of thirteen of the studied viruses were obtained at SEPRL, and the genome characteristics of these complete genomes are typical for NDV (See Additional file 1: Table S4). Additional eleven complete fusion gene coding sequences were also obtained and utilized for prediction of virulence and phylogenetic analysis. The fusion protein cleavage site of NDV is a major virulence determinant marker [40], and the deduced amino acid sequences of the fusion protein cleavage site revealed that all 24 isolates had multiple basic amino acids at positions 113–116, and a phenylalanine residue at position 117 (Table 1). The cleavage site motif of the Egyptian and Ukrainian viruses was ^{113}RQKR↓F^{117} (n = 8). The Pakistani viruses (n = 12) had the following three motifs, ^{113}RKKRF^{117}, ^{113}RQRR↓F^{117}, and ^{113}RQKR↓F^{117}, sharing the last motif with the Bulgarian viruses (n = 2). The South Korean (n = 2) viruses had a ^{13}RRKR↓F^{117} motif (Table 1). All of these motifs are characteristic of virulent NDV [3, 41].

Distance and phylogenetic analyses

In order to determine the phylogenetic relationship between the studied viruses and other NDV isolated worldwide, the complete fusion gene coding sequences obtained in the current work, along with sequences of highly related viruses, were used to construct a phylogenetic tree (n = 82) (Fig. 1 and Additional file 1: Table S2). The full fusion coding region of all available class II NDV and of genotype VI (n = 1430 and n = 281, respectively, both including the 24 sequences from the current study) was also used to estimate evolutionary distances (Tables 2 and 3). A second phylogenetic analysis was performed using 83 complete genome-concatenated coding sequences of viruses pertaining to class II (see Additional file 1: Table S3 and Additional file 2: Fig. S4). Both, the full fusion gene and the complete genome phylogenetic analyses displayed similar topology confirming the phylogenetic classification of the viruses studied here into different sub-genotypes related to viruses previously isolated in Asia, Africa and Eastern Europe. The phylogenetic analyses (Fig. 1 and Additional file 2: Fig. S4) demonstrated that these 24 NDV studied here the clustered with viruses of genotype VI. The isolates from pigeons grouped with previously described viruses from columbine birds, while the chicken isolates grouped with other viruses from chickens. The topology of the full fusion phylogenetic tree (Fig. 1) indicated the viruses from Egypt (n = 6) (0%–3.1% genetic distance among themselves) and Ukraine (n = 2) (1.9% genetic distance between themselves) isolated from pigeons grouped together and clustered within sub-genotype VIg. These viruses were closely related to viruses isolated from pigeons in Russia, Ukraine, Kazakhstan, and Nigeria during 2005–2014 (Fig. 1) [42–44]. These results were consistent with the results of the complete genome analysis (see Fig. S4). Four Pakistani viruses isolated during 2015–2016 clustered together in a separate monophyletic branch within sub-genotype VIg (Fig. 1). The genetic distance of these Pakistani viruses (4.7%) to the rest of the VIg viruses shows higher nucleotide diversity within the sub-genotype.

The remaining eight Pakistani viruses (0%–3.6% genetic distance among themselves, see Additional file 1: Table S5) isolated from pigeons during 2014–2016 did not cluster within any of the previously known sub-genotypes within genotype VI. These Pakistani isolates grouped with a Pakistani virus isolated from pigeon (KU885949/pigeon/Pakistan/MZS-UVAS/2014) creating a separate branch in the phylogenetic tree (Fig. 1) and were 8.5% to 12% distant from the rest of the sub-genotypes in genotype VI (Table 2). This new group of viruses fulfills all classification criteria set by Diel et al. [12] and was named as a novel, hitherto undescribed, sub-genotype of class II genotype VI, namely sub-genotype VIm. The viruses from the newly designated sub-genotype VIm were also more than 10% (11.2–24.2%) distant from all known NDV class II genotypes (Table 3).

The topology of the phylogenetic tree based on the full fusion gene sequences revealed that the Bulgarian virus (Chicken/Bulgaria/Mokresh/1982) grouped together with isolates from Ethiopia (during 2011–2012) [13] (Fig.1). However, the Bulgarian isolate was quite divergent from other members in this sub-genotype showing high genetic distance (10–11%). When this Bulgarian isolate was phylogenetically analyzed based on complete genome, it shared common ancestor with sub-genotypes VIg and VIm described above (see Additional file 2: Fig. S4). The other Bulgarian virus (chicken/Bulgaria/Dolno Linevo/1992), as well as both South Korean viruses obtained from a chicken and a quail, grouped within sub-genotype VIc with other chicken isolates from Sweden, Japan and China indicating that they belong to the older genotype VI sub-genotype (Fig. 1).

Rapid diagnostic and development of new primers and probe

While all tested viruses were positive with the M-gene rRT-PCR assay [24], the F-gene rRT-PCR assay used to identify virulent NDV [24] failed to detect eight viruses from Egypt (n = 6) and Ukraine (n = 2) (Table 4). The pigeon-specific F-gene assay [39] failed to detect all fifteen NDV (the viruses sequenced in Pakistan were not

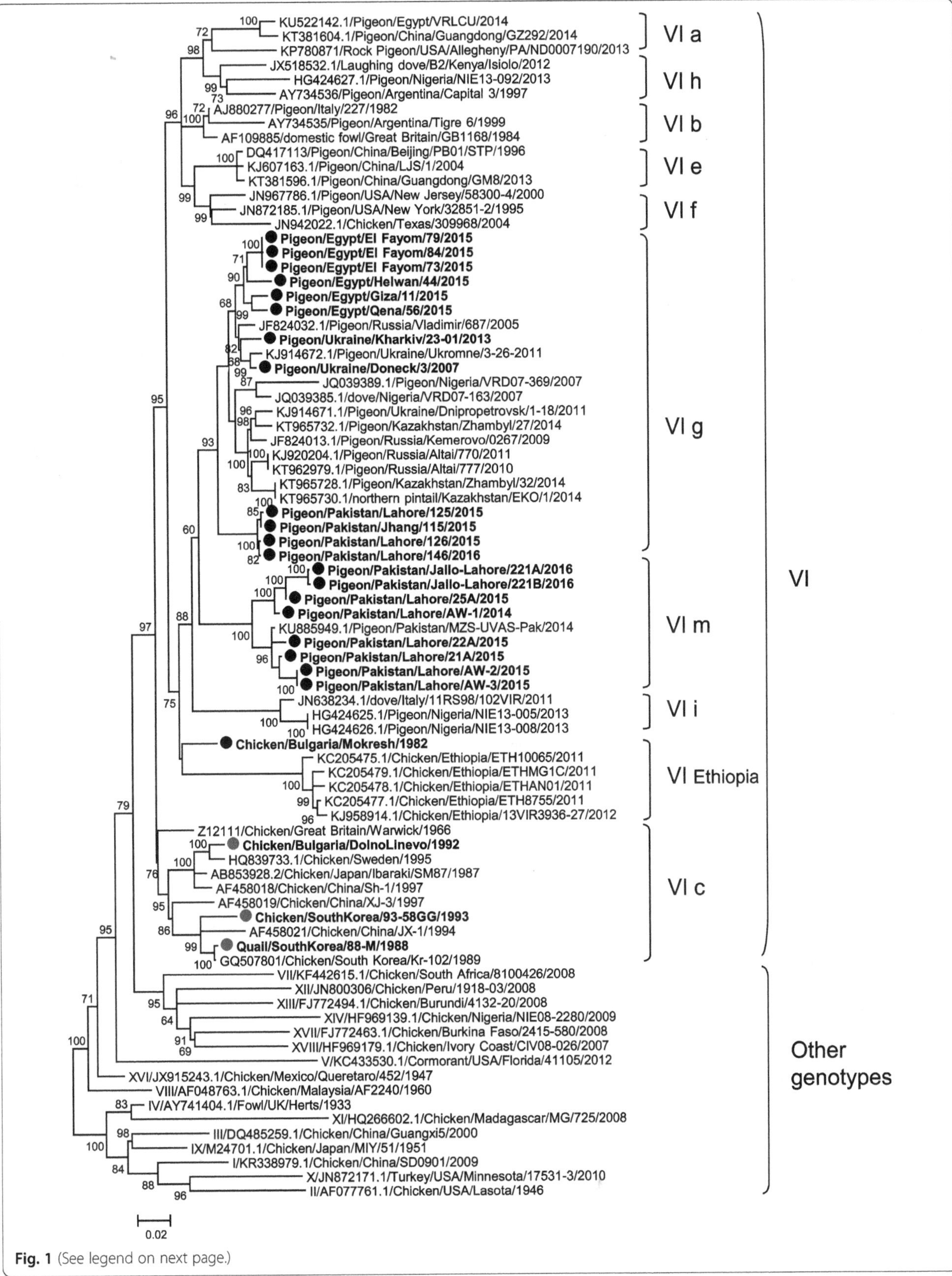

Fig. 1 (See legend on next page.)

(See figure on previous page.)
Fig. 1 Phylogenetic analysis based on the complete nucleotide sequence of the fusion gene of viruses representing Newcastle disease viruses of class II. Only bootstrap values greater or equal to 60% are visualized. There were a total of 1650 positions in the final dataset. The strains sequenced in this study are highlighted in bold font and have a circle symbol in front the taxa name. Provisional designation of genotypes is indicated on the right

tested, designated in Table 1). The genomic sequences of the viruses that failed detection were used to design new primers and probe based on a previous assay [39]. The new pigeon-specific F-gene test successfully detected all the genotype VI NDV (n = 15) that were not detected by the previously available protocol (Table 4). Upon further analysis, the new pigeon-specific fusion gene rRT-PCR assay was evaluated utilizing total RNA from allantoic fluids infected with virulent viruses of different genotypes and also side by side comparison of equal amount of total RNAs were performed with the M-gene and F-gene assays [24]. The results revealed that while the new fusion test recognized most effectively viruses of sub-genotype VIg, although with higher Ct values, it also recognized virulent viruses of seven different genotypes (II, Vb, VIa, VIIi, XII, XIVb, and XVIIa) (Table 4). While three Pakistani viruses (pigeon/Pakistan/Lahore/21A/2015, pigeon/Pakistan/Lahore/22A/2015 and pigeon/Pakistan/Lahore/25A/2015) were detected by the regular F-gene assay, using the new set of primers and probe resulted in four to nine lower Ct values (Table 4).

Discussion

Here, we present the relationship between NDV of genotype VI isolated from Egypt, Ukraine, Pakistan, Bulgaria and South Korea based on their full fusion and complete genome characterization. Our data confirm the concurrent evolution and mobility of viruses of two sub-genotypes of genotype VI NDV across 3 continents. Viruses of genotype VI were first isolated from pigeons in the Middle East in 1960s and spread rapidly throughout Northern Africa to Europe and the rest of the world [45–47]. In Egypt, NDV was first identified in 1947 [48] on the basis of virus isolation into ECE and serologically by HI tests. Clinical signs in pigeons consistent with ND have been seen in Egypt since early 1981, and infection with NDV was serologically confirmed in diseased pigeons in the delta area of the Nile in 1984 [49, 50]. In Bulgaria, ND was first detected in 1943 [51], however the viruses of genotype VI were first found in Bulgaria in the mid-1970s [45]. To the best of our knowledge, the first identification of genotype VI NDV in Ukraine, Pakistan and South Korea has not been previously documented. The phylogenetic relationship among Eastern European, African and Asian viruses suggest the circulation of related viruses in pigeons across three continents.

Pigeons are not migratory and the circulation of these closely related viruses isolated from pigeons in six distant countries (Fig. 1) within eight years (2007–2015) is epidemiologically important. In many countries, pigeons (*Columba livia*) live freely as synantropic birds, or are bred for a variety of different purposes, such as a source of meat, pet companion birds, or for laboratory experiments in biology and cognitive science. Viruses of genotype VI have been reported previously to circulate in apparently healthy pigeons [52–54], and in the current work it was confirmed that at least sub-genotype VIg is seemingly maintained in healthy pigeons kept in captivity in Egypt (all samples from Egypt were collected from healthy pigeons). However, the mechanism of spread of these viruses at long distances remains unknown. A possible explanation for the spread of genotype VI NDV is the contact between columbid birds during competition flights, exhibitions, or due to the intensive international

Table 2 Evolutionary distances[a] between class II Newcastle disease virus of genotype VI estimated using the complete fusion gene coding sequences

Sub-genotype (number of analyzed sequences)	VI a	VI b	VI c	VI e	VI f	VI g	VI h	VI i	VI Ethiopia
VI a (n = 165)									
VI b (n = 10)	0.071								
VI c (n = 19)	0.091	0.069							
VI e (n = 16)	0.080	0.063	0.083						
VI f (n = 16)	0.075	0.058	0.082	0.059					
VI g (n = 25)	0.096	0.080	0.088	0.097	0.091				
VI h (n = 9)	0.071	0.071	0.098	0.084	0.081	0.103			
VI i (n = 4)	0.113	0.100	0.104	0.108	0.106	0.093	0.120		
VI Ethiopia (n = 8)	0.125	0.099	0.106	0.119	0.119	0.107	0.123	0.125	
VI m (n = 9)	0.111	0.091	0.099	0.112	0.101	0.085	0.112	0.107	0.120

[a] The numbers of base substitutions per site from averaging over all sequence pairs between groups within genotype VI are shown. The analysis involved 281 nucleotide sequences. There were a total of 1647 positions in the final dataset

Table 3 Evolutionary distances[a] between class II Newcastle disease virus genotypes and sub-genotype VIm estimated using the complete fusion gene coding sequences

Genotype (number of analyzed sequences)	I	II	III	IV	V	VI	VII	VIII	IX	X	XI	XII	XIII	XIV	XV	XVI	XVII	XVIII
I (*n* = 139)																		
II (*n* = 171)	0.125																	
III (n = 10)	0.117	0.143																
IV (*n* = 5)	0.104	0.130	0.084															
V (*n* = 90)	0.191	0.206	0.178	0.148														
VI (*n* = 272)	0.185	0.204	0.177	0.136	0.159													
VII (*n* = 476)	0.180	0.213	0.168	0.139	0.159	0.132												
VIII (*n* = 5)	0.144	0.165	0.133	0.100	0.129	0.119	0.122											
IX (*n* = 35)	0.107	0.127	0.093	0.079	0.169	0.170	0.164	0.125										
X (*n* = 11)	0.115	0.114	0.141	0.129	0.207	0.201	0.199	0.164	0.125									
XI (*n* = 14)	0.201	0.224	0.191	0.130	0.229	0.234	0.239	0.197	0.172	0.224								
XII (*n* = 9)	0.192	0.224	0.179	0.152	0.169	0.129	0.118	0.126	0.177	0.205	0.248							
XIII (*n* = 44)	0.186	0.216	0.178	0.146	0.165	0.140	0.117	0.126	0.167	0.204	0.233	0.112						
XIV (*n* = 56)	0.221	0.260	0.221	0.184	0.191	0.167	0.143	0.155	0.216	0.235	0.283	0.134	0.139					
XV (*n* = 6)	0.145	0.132	0.135	0.109	0.163	0.145	0.112	0.127	0.108	0.159	0.210	0.149	0.144	0.178				
XVI (*n* = 4)	0.173	0.199	0.169	0.129	0.168	0.165	0.169	0.129	0.161	0.189	0.233	0.170	0.166	0.199	0.167			
XVII (*n* = 56)	0.182	0.222	0.187	0.152	0.171	0.151	0.132	0.138	0.174	0.212	0.235	0.124	0.120	0.132	0.155	0.184		
XVIII (*n* = 18)	0.192	0.217	0.185	0.153	0.172	0.139	0.125	0.136	0.176	0.209	0.236	0.119	0.115	0.134	0.148	0.179	0.109	
VI m (*n* = 9)	0.193	0.218	0.185	0.158	0.176	0.112	0.143	0.134	0.182	0.222	0.242	0.148	0.155	0.182	0.155	0.184	0.173	0.157

[a] The numbers of base substitutions per site from averaging over all sequence pairs between genotypes are shown. The analysis involved 1430 nucleotide sequences. There were a total of 1596 positions in the final dataset

trade of such birds [55]. Other possibility is the international trade of live birds from other species or avian products between countries, either by legal or illegal routes of importation and exportation. In addition, NDV of genotype VI have also been isolated from birds from non-columbidae species kept in captivity and from wild birds, including partridges, pheasants, swans, falcons, blackbirds, cockatoos, budgerigars, raptors, partridges, crested ibises, waterfowl, starlings, pintails, gannets, and buzzards [44, 56–59]. However, while other NDV (e.g. genotype VII) are known to infect wild birds and this possibility can't be excluded, no evidence exists that wild birds, especially migratory, play a role in the spread of genotype VI viruses. The presence of genotype VI viruses in fecal and oral swabs suggests that viral replication, which could result in virus transmission and possible outbreaks in poultry, as seen previously, is occurring in pigeons. Although sporadic, ND outbreaks in poultry caused by viruses of genotype VI have been reported [60–63], and the potential of the virus to cause clinical disease in poultry must not be under estimated.

The phylogenetic analyses revealed the complexity of NDV genotype VI and the challenges in the classification of its sub-genotypes. As shown in Table 2, some sub-genotypes have more than 10% distance compared to the rest of the sub-genotypes within genotype VI. In addition, the newly designated sub-genotype VIm is more than 10% distant from the rest of the genotypes in NDV class II (Table 3). These viruses topologically fall into a lower-order group (see Fig. 1 and Additional file 2: Fig. S4) compared to the existing genotype VI. Some of these groups of viruses meet, or will eventually meet the classification criteria for consideration as new genotypes. However, as naming new genotypes that originated from existing sub-genotypes may create confusion in the field of NDV classification, here, the most diverse group of viruses in genotype VI is named as sub-genotype VIm. In the future it may be appropriate to utilize rules similar to those put forth the WHO/OIE/FAO H5N1 Evolution Working Group for the nomenclature of highly pathogenic H5N1 avian influenza viruses [64] as already proposed by Susta et al. [65]. According to those rules, when a group is split into subgroups of higher order, the newly named groups remain part of the existing original lower-order group (e.g., VIa.1 or VIi.1 and so on). We believe that the criteria for naming new genotypes need to be updated; however, this has to be done based on international consensus rather than by individual scientific teams [65].

The failure of the validated F-gene rRT-PCT assay [24] designed to specifically detect viruses from the outbreak that occurred in southern California in 2002–2003 to detect the sub-genotype VIg Egyptian and the Ukrainian

Table 4 Results of testing selected[a] Newcastle disease viruses analyzed in this study by real-time RT-PCR using different sets of primers and probes

Isolate name	sub/genotype	Cycle threshold (Ct) values[b]		
		Matrix gene test[c]	Fusion-specific gene test[d]	New pigeon-specific fusion gene primers and probe (annealing temperature 56 °C)
pigeon/Egypt/Giza/11/2015	VI g	16.12	0	13.55
pigeon/Egypt/Helwan/44/2015	VI g	12.07	0	20.27
pigeon/Egypt/Qena/56/2015	VI g	12.27	0	15.92
pigeon/Egypt/El Fayom/73/2015	VI g	12.11	0	19.97
pigeon/Egypt/El Fayom /79/2015	VI g	14.72	0	15.57
pigeon/Egypt/El Fayom/84/2015	VI g	12.42	0	14.34
pigeon/Ukraine/Kharkiv/2301/2013	VI g	15.53	0	18.4
pigeon/Ukraine/Doneck/3/2007	VI g	12.11	0	21.62
pigeon/Pakistan/ Lahore/21A/2015	VI m	12.03	23.35	19.66
pigeon/Pakistan/Lahore/22A/2015	VI m	12.99	26.92	22.45
pigeon/Pakistan/Lahore/25A/2015	VI m	16.38	26.49	17.81
chicken/Bulgaria/Mokresh/1982	VI	13.72	24.25	21.83
chicken/Bulgaria/Dolnolinevo/1992	VI c	12.94	18.3	15.54
chicken/South Korea/9358GG/1993	VI c	12.31	26.61	24.84
quail/South Korea /88 M/1988	VI c	12.08	16.51	24.52
hawk/Mexico 663-ZM03/2008 (KC808489.1)	II	18.52	0	38.85
chicken/Belize/4224-3/08 (JN872163.1)	V b	16.23	18.95	28.02
parrot/Israel/2012/841 (KF792020.1)	VI a	16.63	26.41	32.42
chicken/KY-Israel/2013/50 (KF792019.1)	VII i	13.35	26.29	33.29
poultry/Peru/1918-03/2008 (JN800306.1)	XII	13.16	20.31	35.68
NG-707/GM.GMM.17-18 T (KC568207.1)	XIV b	12.29	17.07	32.05
chicken/DominicanRepublic499-31/2008 (JX119193.1)	XVI	17.44	17.92	0
NG-694/YB.GSH1.9-10C (KC568215.1)	XVII a	13.67	17.18	29.33

[a] some Pakistani viruses were characterized only in Pakistan and were not submitted to SEPRL for further studies (designated in Table 1)
[b] the pigeon-specific F-gene assay [39] failed to detect all fifteen NDV (the viruses sequenced in Pakistan were not tested)
[c,d] primers and probes previously described by Wise et al. [24]

isolates (Table 4) was attributed to the mismatches in the probe and/or the primers. Here, several new variable sites that resulted in nucleotide mismatches were identified, suggesting continuous variation at the site used for the design of the test (see Additional file 2: Figsure S1, S2, and S3). The pigeon-specific probe, designed to detect dove/Italy/2736/2000 and US pigeon viruses [39] also had significant mismatches to the viruses tested here. The increasing number of mismatches of different probes and primers indicates that genotype VI is composed of a highly diverse group of viruses that is not covered by a single test. The newly developed pigeon test successfully identified the genotype VI isolates tested here (including the sub-genotype VIg); however, the new fusion test is not genotype specific, and positive detection of viruses of other genotypes is also possible (although with decreased sensitivity). Albeit also dependent on specific primers and not routinely performed in all laboratories, sequencing of the fusion gene will provide data allowing definitive diagnostics and characterization.

Conclusion

In summary, genotype VI NDV continue to evolve in Africa, Asia and Europe suggesting the need for a constant surveillance and characterization of these viruses. The described epidemiological connections among viruses isolated from non-migratory birds on three continents creates uncertainty regarding the mechanism of spread of these viruses. Further studies are needed to elucidate this issue and understand the possible role of human activity in the dispersal of these viruses. Complete genomic characterization identified previously unrecognized genetic diversity that contributes to diagnostic failure and will facilitate future evolutionary studies. These results highlight the importance of updating existent rapid diagnostic protocols to detect emerging viral variants and help manage the disease in affected regions. The obtained rRT-PCR results suggest that the developed pigeon-specific F-gene assay, run in conjunction with the USDA-validated M-gene rRT-PCR assay, should be effective in detecting viruses from sub-genotype VIg, and those from the newly classified sub-genotype VIm.

Additional files

Additional file 1: Table S1. Nucleotide sequences of primers used in PCR amplification, and sequencing of the NDV isolates used in this study. **Table S2.** List of the NDV used for construction of full fusion phylogenetic tree presented in Fig. 1. Highlighted in bold font are the viruses studied in the current work. **Table S3.** List of the NDV used for construction of complete genome phylogenetic tree presented in Additional file 2: Fig. S4. Highlighted in bold font are the viruses studied in the current work. **Table S4.** Characteristics of the thirteen complete genomes of Newcastle disease viruses of genotype VI sequenced in this study. **Table S5.** Estimated pairwise evolutionary distances among viruses of the new sub-genotype VIm. (DOCX 60 kb)

Additional file 2: Figure S1 A, B and C. Mismatches between the tested viruses and: **A)** previous fusion probe designed by Wise et al. [24]; **B)** pigeon-specific fusion probe designed by Kim et al. [66]; and **C)** the optimized pigeon specific probe in this study, respectively. Sequences are in order of 5′ to 3′. **Figure S2. A and B** Mismatches between the tested viruses and: **A)** previous fusion forward primer designed by Wise et al. [24]; and **B)** the optimized fusion forward primer in this study, respectively. Sequences are in order of 5′ to 3′. **Figure S3. A and B** Mismatches between the tested viruses and: **A)** previous fusion reverse primers designed by Wise et al. [24]; **B)** and the new fusion reverse in this study, respectively. Sequences are in order of 5′ to 3′. **Figure S4.** Phylogenetic analysis based on the complete genome concatenated coding sequence of viruses representing NDV class II. Only bootstrap values greater or equal to 60% are visualized. There were a total of 13,697 positions in the final dataset. The strains sequenced in this study are highlighted in bold font and have a circle symbol in front the taxa name. Provisional designation of genotypes is indicated on the right. (DOCX 1990 kb)

Acknowledgements
The authors gratefully acknowledge Tim Olivier for his technical assistance and Dr. Hassan El-Sayed Helal for his help in collecting Egyptian clinical samples. The mention of trade names or commercial products in this publication is solely for the purpose of providing specific information and does not imply recommendation or endorsement by the U.S. Department of Agriculture. The USDA is an equal opportunity provider and employer.

Funding
This work was supported by the U.S. Department of Agriculture ARS CRIS 6040-32,000-072, the U.S. Department of State BEP/CRDF NDV 31063, and by the Egyptian Cultural and Education Bureau.

Author's contributions
CLA, PJM, MS, and KMD conceived this project. MS, AW, AB, SFR, DWC and DVM coordinated field sampling efforts and isolated viruses. MS, PS, IVG, DWC and KMD prepared and sequenced the viral isolates. MS, DWC, KMD, and CLA optimized the diagnostic tests. MS, KMD, PJM and CLA conducted analyses of the data. MS, KMD, PJM, and CLA wrote and edited the manuscript. All authors read and approved the final manuscript.

Consent for publication
Not applicable.

Competing interests
The authors declare that they have no competing interests.

Author details
[1]Department of Poultry Diseases, Faculty of Veterinary Medicine, South Valley University, Qena 83523, Egypt. [2]Exotic and Emerging Avian Viral Diseases Research Unit, Southeast Poultry Research Laboratory, US National Poultry Research Center, Agricultural Research Service, USDA, 934 College Station Road, Athens, GA 30605, USA. [3]National Scientific Center Institute of Experimental and Clinical Veterinary Medicine, 83 Pushkinskaya Street, Kharkiv 61023, Ukraine. [4]Quality Operations Laboratory (QOL), University of Veterinary and Animal Sciences, Syed Abdul Qadir Jilani Road, Lahore 54000, Pakistan. [5]Institute of Biochemistry and Biotechnology, University of Veterinary and Animal Sciences, Syed Abdul Qadir Jilani Road, Lahore 54000, Pakistan.

References

1. Aldous EW, Alexander DJ. Detection and differentiation of Newcastle disease virus avian paramyxovirus type 1. Avian Pathol. 2001;30:117–28. doi: 10.1080/03079450120044515
2. Miller PJ, Decanini EL, Afonso CL. Newcastle disease: Evolution of genotypes and the related diagnostic challenges. Infect Genet Evol. 2010;10:26–35. doi: 10.1016/j.meegid.2009.09.012
3. OIE. Newcastle disease. In *Biological Standards Commission, Manual of diagnostic tests and vaccines for terrestrial animals: mammals, birds and bees*, vol. Volume 1, Part 2, Chapter 2.3.14, 7th edition. pp. 555–74. Paris, France: World Organisation for Animal Health; 2012:555-574.
4. Afonso CL, Amarasinghe GK, Banyai K, Bao Y, Basler CF, Bavari S, Bejerman N, Blasdell KR, Briand FX, Briese T, et al. Taxonomy of the order Mononegavirales: update 2016. Arch Virol. 2016;161:2351–60. doi: 10.1007/s00705-016-2880-1
5. Goraichuk I, Sharma P, Stegniy B, Muzyka D, Pantin-Jackwood MJ, Gerilovych A, Solodiankin O, Bolotin V, Miller PJ, Dimitrov KM, Afonso CL. Complete Genome Sequence of an Avian Paramyxovirus Representative of Putative New Serotype 13. Genome Announc. 2016;4:e00729–16. doi: 10.1128/genomeA.00729-16
6. Thampaisarn R, Bui VN, Trinh DQ, Nagai M, Mizutani T, Omatsu T, Katayama Y, Gronsang D, Le DH, Ogawa H, Imai K. Characterization of avian paramyxovirus serotype 14, a novel serotype, isolated from a duck fecal sample in Japan. Virus Res. 2016;228:46–57. doi: 10.1016/j.virusres.2016.11.018
7. Thomazelli LM, Araujo J, Oliveira DB, Sanfilippo L, Ferreira CS, Brentano L, Pelizari VH, Nakayama C, Duarte R, Hurtado R, et al. Newcastle disease virus in penguins from King George Island on the Antarctic region. Vet Microbiol. 2010; doi: 10.1016/j.vetmic.2010.05.006
8. Lee HJ, Kim JY, Lee YJ, Lee EK, Song BM, Lee HS, Choi KS. A Novel Avian Paramyxovirus (Putative Serotype 15) Isolated from Wild Birds. Front Microbiol. 2017;8:786. doi: 10.3389/fmicb.2017.00786
9. Neira V, Tapia R, Verdugo C, Barriga G, Mor S, Ng TFF, Garcia V, Del Rio J, Rodrigues P, Briceno C, et al. Novel Avulaviruses in Penguins. Antarctica Emerg Infectious Diseases. 2017;23:1212–4. doi: 10.3201/eid2307.170054
10. Seal BS, King DJ, Sellers HS. The Avian Response to Newcastle Disease Virus. Dev Comp Immunol. 2000;24:257–68. doi: 10.1016/S0145-305X(99)00077-4
11. Czeglédi A, Ujvári D, Somogyi E, Wehmann E, Werner O, Lomniczi B. Third genome size category of avian paramyxovirus serotype 1 (Newcastle disease virus) and evolutionary implications. Virus Res. 2006;120:36–48. doi: 10.1016/j.virusres.2005.11.009

12. Diel DG, da Silva LH, Liu H, Wang Z, Miller PJ, Afonso CL. Genetic diversity of avian paramyxovirus type 1: proposal for a unified nomenclature and classification system of Newcastle disease virus genotypes. Infect Genet Evol. 2012;12:1770–9. doi: 10.1016/j.meegid.2012.07.012
13. Dimitrov KM, Ramey AM, Qiu X, Bahl J, Afonso CL. Temporal, geographic, and host distribution of avian paramyxovirus 1 Newcastle disease virus. Infect Genet Evol. 2016;39:22–34. doi: 10.1016/j.meegid.2016.01.008
14. Terregino C, Cattoli G, Grossele B, Bertoli E, Tisato E, Capua I. Characterization of Newcastle disease virus isolates obtained from Eurasian collared doves (Streptopelia decaocto) in Italy. Avian Pathol. 2003;32:63–8. doi: 10.1080/0307945021000070732
15. Collins MS, Alexander DJ, Brockman S, Kemp PA, Manvell RJ. Evaluation of mouse monoclonal antibodies raised against an isolate of the variant avian paramyxovirus type 1 responsible for the current panzootic in pigeons. Arch Virol. 1989;104:53–62. doi: 10.1007/BF01313807
16. Ujvari D, Wehmann E, Kaleta EF, Werner O, Savic V, Nagy E, Czifra G, Lomniczi B. Phylogenetic analysis reveals extensive evolution of avian paramyxovirus type 1 strains of pigeons (Columba livia) and suggests multiple species transmission. Virus Res. 2003;96:63–73. doi: 10.1016/S0168-1702(03)00173-4
17. Xue C, Xu X, Yin R, Qian J, Sun Y, Wang C, Ding C, Yu S, Hu S, Liu X, et al. Identification and pathotypical analysis of a novel VIk sub-genotype Newcastle disease virus obtained from pigeon in China. Virus Res. 2017;238: 1–7. doi: 10.1016/j.virusres.2017.05.011
18. Dortmans JC, Fuller CM, Aldous EW, Rottier PJ, Peeters BP. Two genetically closely related pigeon paramyxovirus type 1 (PPMV-1) variants with identical velogenic fusion protein cleavage sites but with strongly contrasting virulence. Vet Microbiol. 2009; doi: 10.1016/j.vetmic.2009.11.021.
19. Meulemans G, van den Berg TP, Decaesstecker M, Boschmans M. Evolution of pigeon Newcastle disease virus strains. Avian Pathol. 2002;31:515–9. doi: 10.1080/0307945021000005897
20. Collins MS, Strong I, Alexander DJ. Evaluation of the molecular basis of pathogenicity of the variant Newcastle disease viruses termed "pigeon PMV-1 viruses". Arch Virol. 1994;134:403–11. doi: 10.1007/BF01310577
21. Fuller CM, Collins MS, Easton AJ, Alexander DJ. Partial characterisation of five cloned viruses differing in pathogenicity, obtained from a single isolate of pigeon paramyxovirus type 1 (PPMV-1) following passage in fowls' eggs. Arch Virol. 2007;152:1575–182. doi: 10.1007/s00705-007-0963-8
22. Dimitrov KM, Clavijo A, Sneed L. RNA extraction for molecular detection of Newcastle disease virus – comparative study of three methods. Rev Méd Vét. 2014;165:172–5.
23. Cattoli G, Monne I. Molecular Diagnosis of Newcastle Disease Virus. In: Capua I, Alexander DJ, editors. . Springer Milan: Avian Influenza and Newcastle Disease: A Field and Laboratory Manual; 2009. p. 127–32.
24. Wise MG, Suarez DL, Seal BS, Pedersen JC, Senne DA, King DJ, Kapczynski DR, Spackman E. Development of a real-time reverse-transcription PCR for detection of newcastle disease virus RNA in clinical samples. J Clin Microbiol. 2004;42:329–38. doi: 10.1128/JCM.42.1.329-338.2004
25. Pham HM, Konnai S, Usui T, Chang KS, Murata S, Mase M, Ohashi K, Onuma M. Rapid detection and differentiation of Newcastle disease virus by real-time pcr with melting-curve analysis. Arch Virol. 2005;150:2429–38. doi: 10. 1007/s00705-005-0603-0
26. Antal M, Farkas T, German P, Belak S, Kiss I. Real-time reverse transcription-polymerase chain reaction detection of Newcastle disease virus using light upon extension fluorogenic primers. J Vet Diagn Investig. 2007;19:400–4. 17609351.
27. Spackman E, Senne DA, Myers TJ, Bulaga LL, Garber LP, Perdue ML, Lohman K, Daum LT, Suarez DL. Development of a real-time reverse transcriptase PCR assay for type A influenza virus and the avian H5 and H7 hemagglutinin subtypes. J Clin Microbiol. 2002;40:3256–60. doi: 10.1128/ JCM.40.9.3256-3260.2002
28. Alexander DJ, Swayne DE. Newcastle disease virus and other avian paramyxoviruses. In: Swayne DE, Glisson JR, Jackwood MW, Pearson JE, Reed WM, editors. A Laboratory Manual for the Isolation and Identification of Avian Pathogens. 4th ed. Kennett Square, PA: The American Association of Avian Pathologists; 1998. p. 156–63.
29. Miller PJ, Dimitrov KM, Williams-Coplin D, Peterson MP, Pantin-Jackwood MJ, Swayne DE, Suarez DL, Afonso CL. International biological engagement programs facilitate Newcastle disease epidemiological studies. Front Public Health. 2015;3:235. doi: 10.3389/fpubh.2015.00235
30. Miller PJ, Haddas R, Simanov L, Lublin A, Rehmani SF, Wajid A, Bibi T, Khan TA, Yaqub T, Setiyaningsih S, Afonso CL. Identification of new sub-genotypes of virulent Newcastle disease virus with potential panzootic features. Infect Genet Evol. 2015;29:216–29. doi: 10.1016/j.meegid.2014.10.032
31. Chevreux B, Wetter T, Suhai S. Genome sequence assembly using trace signals and additional sequence information. In: Computer Science and Biology, vol. 99. Hanover, Germany; 1999. p. 45–56.
32. Afgan E, Baker D, Van den Beek M, Blankenberg D, Bouvier D, Čech M, Chilton J, Clements D, Coraor N, Eberhard C. The Galaxy platform for accessible, reproducible and collaborative biomedical analyses: 2016 update. Nucleic Acids Res. 2016:gkw343. doi: 10.1093/nar/gkw343
33. Dimitrov KM, Sharma P, Volkening JD, Goraichuk IV, Wajid A, Rehmani SF, Basharat A, Shittu I, Joannis TM, Miller PJ, Afonso CL. A robust and cost-effective approach to sequence and analyze complete genomes of small RNA viruses. Virol J. 2017;14:72. doi: 10.1186/s12985-017-0741-5
34. Brown PA, Briand F-X, Guionie O, Lemaitre E, Courtillon C, Henry A, Jestin V, Eterradossi N. An alternative method to determine the 5′ extremities of non-segmented, negative sense RNA viral genomes using positive replication intermediate 3′ tailing: Application to two members of the Paramyxoviridae family. J Virol Methods. 2013;193:121–7. doi: 10.1016/j. jviromet.2013.05.007
35. Katoh K, Misawa K, Kuma K, Miyata T. MAFFT: a novel method for rapid multiple sequence alignment based on fast Fourier transform. Nucleic Acids Res. 2002;30:3059–66. doi: 10.1093/nar/gkf436
36. Kearse M, Moir R, Wilson A, Stones-Havas S, Cheung M, Sturrock S, Buxton S, Cooper A, Markowitz S, Duran C, et al. Geneious Basic: an integrated and extendable desktop software platform for the organization and analysis of sequence data. Bioinformatics. 2012;28:1647–9. doi: 10.1093/bioinformatics/bts199
37. Tamura K, Stecher G, Peterson D, Filipski A, Kumar S. MEGA6: molecular evolutionary genetics analysis version 6.0. Mol Biol Evol. 2013;30:2725–9. doi: 10.1093/molbev/mst197
38. Tamura K, Nei M, Kumar S. Prospects for inferring very large phylogenies by using the neighbor-joining method. Proc Natl Acad Sci U S A. 2004;101: 11030–5. doi: 10.1073/pnas.0404206101
39. Kim LM, Afonso CL, Suarez DL. Effect of probe-site mismatches on detection of virulent Newcastle disease viruses using a fusion-gene real-time reverse transcription polymerase chain reaction test. J Vet Diagn Investig. 2006;18: 519–28. 17121078.
40. Glickman RL, Syddall RJ, Iorio RM, Sheehan JP, Bratt MA. Quantitative Basic Residue Requirements in the Cleavage-Activation Site of the Fusion Glycoprotein as a Determinant of Virulence for Newcastle Disease Virus. J Virol. 1988;62:354–6. 3275436.
41. Miller PJ, Koch G. Newcastle disease. In: Swayne DE, Glisson JR, McDougald LR, Nolan LK, Suarez DL, Nair V, editors. Diseases of Poultry. 13th ed. Hoboken, New Jersey: Wiley-Blackwell; 2013. p. 89–138.
42. Pchelkina IP, Manin TB, Kolosov SN, Starov SK, Andriyasov AV, Chvala IA, Drygin VV, Yu Q, Miller PJ, Suarez DL. Characteristics of pigeon paramyxovirus serotype-1 isolates (PPMV-1) from the Russian Federation from 2001 to 2009. Avian Dis. 2013;57:2–7. doi: 10.1637/10246-051112-Reg.1
43. Yurchenko KS, Sivay MV, Glushchenko AV, Alkhovsky SV, Shchetinin AM, Shchelkanov MY, Shestopalov AM. Complete Genome Sequence of a Newcastle Disease Virus Isolated from a Rock Dove (Columba livia) in the Russian Federation. Genome Announc. 2015:3e01514–4. doi: 10.1128/ genomeA.01514-14
44. Benson DA, Clark K, Karsch-Mizrachi I, Lipman DJ, Ostell J, Sayers EW. GenBank. Nucleic Acids Res. 2015;43:D30. doi: 10.1093/nar/gku1216
45. Czeglédi A, Herczeg J, Hadjiev G, Doumanova L, Wehmann E, Lomniczi B. The occurrence of five major Newcastle disease virus genotypes (II, IV, V, VI and VIIb) in Bulgaria between 1959 and 1996. Epidemiol Infect. 2002;129: 679–88. doi: 10.1017/S0950268802007732
46. Lomniczi B, Wehmann E, Herczeg J, Ballagi-Pordany A, Kaleta EF, Werner O, Meulemans G, Jorgensen PH, Mante AP, Gielkens AL, et al. Newcastle disease outbreaks in recent years in western Europe were caused by an old (VI) and a novel genotype (VII). Arch Virol. 1998;143:49–64. doi: 10.1007/ s007050050267
47. Alexander DJ, Russell PH, Parsons G, Elzein EMEA, Ballouh A, Cernik K, Engstrom B, Fevereiro M, Fleury HJA, Guittet M, et al. Antigenic and biological characterisation of avian paramyxovirus type I isolates from pigeons an international collaborative study. Avian Pathol. 1985;14:365–76. doi: 10.1080/03079458508436238
48. Daubney R, Mansy W. The occurrence of Newcastle disease in Egypt. J Comp Pathol Ther. 1948;58:189–200. doi: 10.1016/S0368-1742(48)80019-6
49. Kaleta E, Alexander D, Russell P. The first isolation of the avian PMV-1 virus

responsible for the current panzootic in pigeons. Avian Pathol. 1985;14:553–7. doi: 10.1080/03079458508436258
50. Vindevogel H, Duchatel JP. Panzootic Newcastle disease virus in pigeons. In: Alexander DJ, editor. Newcastle disease. Boston: Kluwer Academic Publishers; 1988. p. 184–96.
51. Semerdzhiev B. Fowl plague in Bulgaria. Vet Sbirka. 1946;49:105–17.
52. Wang J, Liu H, Liu W, Zheng D, Zhao Y, Li Y, Wang Y, Ge S, Lv Y, Zuo Y, et al. Genomic Characterizations of Six Pigeon Paramyxovirus Type 1 Viruses Isolated from Live Bird Markets in China during 2011 to 2013. PLoS One. 2015;10:e0124261. doi: 10.1371/journal.pone.0124261
53. Byarugaba DK, Mugimba KK, Omony JB, Okitwi M, Wanyana A, Otim MO, Kirunda H, Nakavuma JL, Teillaud A, Paul MC, Ducatez MF. High pathogenicity and low genetic evolution of avian paramyxovirus type I (Newcastle disease virus) isolated from live bird markets in Uganda. Virol J. 2014;11:173. doi: 10.1186/1743-422X-11-173
54. Teske L, Ryll M, Rautenschlein S. Epidemiological investigations on the role of clinically healthy racing pigeons as a reservoir for avian paramyxovirus-1 and avian influenza virus. Avian Pathol. 2013;42:557–65. doi: 10.1080/03079457.2013.852157
55. Aldous EW, Fuller CM, Ridgeon JH, Irvine RM, Alexander DJ, Brown IH. The evolution of pigeon paramyxovirus type 1 (PPMV-1) in Great Britain: a molecular epidemiological study. Transbound Emerg Dis. 2014;61:134–9. doi: 10.1111/tbed.12006
56. Alexander DJ. Newcastle Disease and other Avian Paramyxoviridae Infections. In: Calnek BW, Barnes HJ, Beard CW, McDougald LR, Saif YM, editors. Diseases of Poultry. 10th ed. Ames, Iowa: Iowa State University Press; 1997. p. 541–69.
57. Aldous EW, Fuller CM, Mynn JK, Alexander DJ. A molecular epidemiological investigation of isolates of the variant avian paramyxovirus type 1 virus (PPMV-1) responsible for the 1978 to present panzootic in pigeons. Avian Pathol. 2004;33:258–69. https://doi.org/10.1080/0307945042000195768
58. Irvine RM, Aldous EW, Manvell RJ, Cox WJ, Ceeraz V, Fuller CM, Wood AM, Milne JC, Wilson M, Hepple RG, et al. Outbreak of Newcastle disease due to pigeon paramyxovirus type 1 in grey partridges (Perdix perdix) in Scotland in October 2006. Vet Rec. 2009;165:531–5. doi: 10.1136/vr.165.18.531
59. Krapez U, Steyer AF, Slavec B, Barlic-Maganja D, Dovc A, Racnik J, Rojs OZ. Molecular characterization of avian paramyxovirus type 1 (Newcastle disease) viruses isolated from pigeons between 2000 and 2008 in Slovenia. Avian Dis. 2010;54:1075–80. doi: 10.1637/9161-111709-ResNote.1
60. Alexander DJ, Parsons G, Marshall R. Infection of fowls with Newcastle disease virus by food contaminated with pigeon faeces. Vet Rec. 1984;115: 601–2. 6523693.
61. Alexander DJ. Newcastle disease in the European Union 2000 to 2009. Avian Pathol. 2011;40:547–58. https://doi.org/10.1080/03079457.2011.618823
62. Alexander DJ, Wilson GW, Russell PH, Lister SA, Parsons G. Newcastle disease outbreaks in fowl in Great Britain during 1984. Vet Rec. 1985;117:429–34. doi: 10.1136/vr.117.17.429
63. Abolnik C, Gerdes GH, Kitching J, Swanepoel S, Romito M, Bisschop SP. Characterization of pigeon paramyxoviruses (Newcastle disease virus) isolated in South Africa from 2001 to 2006. Onderstepoort J Vet Res. 2008; 75:147–52.
64. WHO. Continued evolution of highly pathogenic avian influenza A (H5N1): updated nomenclature. Influenza Other Respir Viruses. 2012;6:1–5. doi: 10. 1111/j.1750-2659.2011.00298.x
65. Susta L, Dimitrov KM, Miller PJ, Brown CC, Afonso CL. Reply to "May Newly Defined Subgenotypes Va and Vb of Newcastle Disease Virus in Poultry Be Considered Two Different Genotypes?". J Clin Microbiol. 2016;54:2205–6. doi: 10.1128/jcm.00914-16
66. Kim LM, King DJ, Suarez DL, Wong CW, Afonso CL. Characterization of class I Newcastle disease virus isolates from Hong Kong live bird markets and detection using real-time reverse transcription-PCR. J Clin Microbiol. 2007;45: 1310–4. doi: 10.1128/JCM.02594-06

No evidence of enteric viral involvement in the new neonatal porcine diarrhoea syndrome in Danish pigs

N. B. Goecke[1*], C. K. Hjulsager[1], H. Kongsted[4], M. Boye[1,6], S. Rasmussen[1], F. Granberg[2], T. K. Fischer[3], S. E. Midgley[3], L. D. Rasmussen[3,5], Ø. Angen[1,3], J. P. Nielsen[6], S. E. Jorsal[1] and L. E. Larsen[1]

Abstract

Background: The aim of this study was to investigate whether the syndrome New Neonatal Porcine Diarrhoea Syndrome (NNPDS) is associated with a viral aetiology. Four well-managed herds experiencing neonatal diarrhoea and suspected to be affected by NNPDS were included in a case-control set up. A total of 989 piglets were clinically examined on a daily basis. Samples from diarrhoeic and non-diarrhoeic piglets at the age of three to seven days were selected for extensive virological examination using specific real time polymerase chain reactions (qPCRs) and general virus detection methods.

Results: A total of 91.7% of the animals tested positive by reverse transcription qPCR (RT-qPCR) for porcine kobuvirus 1 (PKV-1) while 9% and 3% were found to be positive for rotavirus A and porcine teschovirus (PTV), respectively. The overall prevalence of porcine astrovirus (PAstV) was 75% with 69.8% of the PAstV positive pigs infected with PAstV type 3. No animals tested positive for rotavirus C, coronavirus (TGEV, PEDV and PRCV), sapovirus, enterovirus, parechovirus, saffoldvirus, cosavirus, klassevirus or porcine circovirus type 2 (PCV2). Microarray analyses performed on a total of 18 animals were all negative, as were eight animals examined by Transmission Electron Microscopy (TEM). Using Next Generation de novo sequencing (de novo NGS) on pools of samples from case animals within all herds, PKV-1 was detected in four herds and rotavirus A, rotavirus C and PTV were detected in one herd each.

Conclusions: Our detailed analyses of piglets from NNPDS-affected herds demonstrated that viruses did not pose a significant contribution to NNPDS. However, further investigations are needed to investigate if a systemic virus infection plays a role in the pathogenesis of NNPDS.

Keywords: Diarrhoea, Neonatal piglets, NNPDS, Virus

Background

Since 2008, field experiences on a new diarrhoeic syndrome in neonatal piglets referred to as New Neonatal Porcine Diarrhoea Syndrome (NNPDS) have been reported in Denmark and elsewhere [1–4]. The prevalence of well-known enteric pathogens as well as gross- and histological findings in age-matched diarrhoeic- and non-diarrhoeic piglets from four Danish herds have recently been reported [5]. In that study, no association between the presence of diarrhoea and the detection of enterotoxigenic *Escherichia coli, Clostridium perfringens* type A or C, rotavirus A, coronavirus, *Clostridium difficile, Cryptosporidium spp., Giardia spp., Cystoisospora suis* or *Strongyloides ransomi* was revealed. The conclusion of these detailed examinations was that no known single causative pathogen could be related to the presence of neither clinical disease nor pathological lesions [5, 6].

The aim of the present study was to perform a detailed investigation on possible viral involvement in NNPDS. Selected samples from the previously examined herds were tested for the presence of a range of specific viruses previously found related to enteric conditions in pigs

* Correspondence: nicbg@vet.dtu.dk
[1]National Veterinary Institute, Technical University of Denmark, Kemitorvet, Lyngby DK-2800, Denmark
Full list of author information is available at the end of the article

and other species. Viruses that have been shown to cause diarrhoea in pigs are rotavirus, coronavirus, norovirus and sapovirus [7–11], while viruses such as porcine kobuvirus 1 (PKV-1) and the five porcine astrovirus types (PAstV1–5) so far only have been associated with diarrhoea in pigs in a few studies [12–14]. Viruses like enterovirus, parechovirus, saffoldvirus, cosavirus and klassevirus, all belonging to the family *Picornaviridae,* are known human enteric pathogens, but so far there have not been reports on the presence of these viruses in pigs. Porcine teschovirus (PTV) and porcine circovirus type 2 (PCV2) are on the other hand enzootic in pig herds in most countries, but have not been proven to be a primary cause of diarrhoea in pigs [15–18]. Systemic PCV2 may, however, indirectly contribute to enteric diseases due to its immunosuppressive effect [19]. In the present study, samples were examined for the above mentioned viruses by ELISA, conventional reverse transcription PCR (RT-PCR) or real time RT-PCR (RT-qPCR). In addition, selected samples were investigated for the presence of viruses in general by Transmission Electron Microscopy (TEM), pan-viral microarray and de novo sequencing using Next Generation Sequencing (NGS).

Methods

Herds and animals

Four well-managed herds affected by severe neonatal diarrhoea for at least one year were selected for the study. Approximately 15 case (diarrhoeic) and 15 control (non-diarrhoeic) piglets per herd were selected for euthanasia. All selected piglets were at the age of three to seven days. In addition, tissue samples from the euthanized animals and a selected number of rectal swabs from the day of euthanasia were included. For details on inclusion criteria and definitions on case and control animals see Kongsted et al. (2013) [5].

Selection and storing of samples

Live piglets were transported to the laboratory and euthanized within six hours after selection in the herds. Immediately after euthanasia, samples from ileum were snap-frozen on dry ice and stored at −80 °C until further use. Rectal swabs taken in the herds were frozen immediately and kept in a freezer with dry-ice in the herds until transportation to the laboratory, where they were stored at −80 °C until further analyses.

Nucleic acid extraction

For extraction of RNA for coronavirus (TGEV, PEDV and PRCV) analyses up to 40 mg ileum tissue with content was homogenized in 300 µl chilled 1-Thioglycerol Homogenization Solution (Promega, Nacka, Sweden) in a TissueLyserII (QIAGEN, Copenhagen, Denmark) at 20 Hz for 2 min, vortexed and heated at 70 °C for 2 min and then stored on ice. 300 µl of lysis buffer were mixed into the sample and 10 µl DNase added afterwards. RNA was extracted from all of the homogenate on a Maxwell® automated purification robot with the Maxwell® 16 LEV SimplyRNA Tissue Kit (Promega) according to instructions from the supplier and eluted in 70 µl nuclease free water. The samples were stored at −80 °C until analysis.

Extraction of nucleic acid for PAstV1–5, PKV-1, PTV and PCV2 analyses was performed on samples consisting of ileum with content and on ileum content. For samples of ileum (with content) a 10% homogenate was prepared in RLT plus buffer (QIAGEN). One 5 mm stainless steel bead (QIAGEN) was added to each sample and the samples were homogenized in a TissueLyser II (QIAGEN) for 2 min at 30 Hz. The homogenate was centrifuged for 3 min at 12.000 rpm and the supernatant (400 µL) was used for nucleic acid extraction. For samples of ileum content a 10% homogenate in ATL buffer (QIAGEN) was prepared. One stainless steel bead, 5 mm, was added to each sample and the samples were homogenized in a Tissuelyser II and the supernatant (at least 350 µL) was used for nucleic acid extraction. Extraction of nucleic acid from ileum with content and ileum content samples was automated on the QIAsymphony SP system (QIAGEN) using QIAsymphony RNA kit (QIAGEN) protocol RNA_CT_400_V5 with an elution volume of 100 µL and QIAsymphony DSP virus/pathogen kit (QIAGEN) protocol complex200_V5_DSP without addition of carrier RNA and with an elution volume of 110 µL. The samples were stored at −80 °C until analysis.

Extraction of RNA for rotavirus A and C, norovirus, sapovirus, enterovirus, parechovirus, saffoldvirus, cosavirus and klassevirus was performed on samples consisting of ileum with content. The samples were prepared as a 10% suspension in minimal essential medium and centrifuged at 3500×g for 30 min. Nucleic acids were extracted from 200 µl sample material using the MagNa Pure LC Total Nucleic Acid Isolation Kit (Roche Diagnostics, Hvidovre, Denmark) on the MagNa Pure LC or MagNa Pure 96 (Roche Diagnostics) instruments according to the manufacturer's specifications. The samples were stored at −80 °C until analysis.

Detection of viral RNA by RT-PCR or RT-qPCR

In order to detect viral pathogens, samples were tested in different PCR assays. In the RT-qPCR and qPCR assays samples were analysed in duplicates, while samples were analysed as single reactions in the RT-PCR assays. For each separate PCR run positive and negative (nuclease-free water, Amresco) PCR controls were included.

The samples were initially tested by a conventional pan-corona RT-PCR assay designed to detect a wide range of coronaviruses [20]. In house RT-qPCR assays were used to test samples for porcine epidemic diarrhoea virus (PEDV), porcine respiratory corona virus (PRCV) and transmissible gastroenteritis virus (TGEV). TGEV and PRCV were detected simultaneously in a duplex RT-qPCR by combining two published assays [21, 22]. The RT-qPCR was performed as a one-step RT-PCR reaction using the RNA UltraSense ™ One-Step Quantitative RT-PCR System (Invitrogen, Carlsbad, California, USA). Primer and probe concentrations and the PCR assay conditions were as described using the MX3005p qPCR system (Stratagene, Santa Clara, USA). One minor modification was made since the TGEV probe had the fluorescent dye Cy5 at the 5′ end to distinguish its signal from the FAM marked TGEV/PRCV probe. Detection of PEDV was performed under the conditions described by Kim et al., (2007) with the modifications that the RT-qPCR was performed as a one-step RT-PCR reaction using RNA UltraSense ™ One-Step Quantitative RT-PCR System (Invitrogen) and the amplification was performed on the MX3005p qPCR system (Stratagene) [21].

In house RT-qPCR assays were used to test samples for PAstV1–5 and PKV-1 (primer and probe sequences are available from the authors upon request). For detection of PTV, previously published primer (PTV-F: 5CTCCTGACTGGGYAATGGG-3′, PTV-R: 5′-TGTCAGGCAGCACAAGTCCA-3′) and probe (PTV-FAM: 5′-FAM- CACCAGCGTGGAGTTCCTGTAT GGG-BHQ1–3′) sequences were modified and used [23]. PCR amplifications were performed with AgPath-ID™ One-Step RT-PCR Kit (Life Technologies, Carlsbad, California, USA). The RT-qPCR assays were performed in a total volume of 15 μL containing 2 μL template, forward and reverse primer (400 nM each), probe (120 nM), 1× RT-PCR Buffer, 1× RT-PCR Enzyme Mix and nuclease free water. PCR amplification was carried out in RotorGeneQ (QIAGEN) with following thermal cycling conditions: 45 °C for 10 min, 95 °C for 10 min, 45 cycles of 95 °C for 15 s and 60 °C for 45 s.

For each of the RT-qPCR assays specific for PAstV1–5 and PKV-1 a positive standard was used to test the sensitivity of the PCR assays and to quantify the viral load in the samples. The viral load was calculated as log10 to the genome copy number per reaction for the virus.

The samples were furthermore tested for a range of viruses. A previously published RT-qPCR assay targeting the NSP3 gene and designed to detect all rotavirus A genotypes from humans and animals, was used [24]. A published RT-qPCR assay was used to test the samples for rotavirus C virus by targeting the VP7 gene [25]. The presence of norovirus was tested using a previously described multiplex RT-qPCR for Norwalk virus (GI & GII) [26]. The samples were tested for sapovirus by a previously described RT-qPCR assay [27]. Previously described RT-qPCR assays were used to test for enterovirus, parechovirus, saffoldvirus, cosavirus and klassevirus [28, 29].

Detection of viral DNA by qPCR

Purified DNA was quantified for PCV2 against a standard curve using the Primer Probe Energy Transfer qPCR (PriProEt-RT-PCR) assay as previously described [30].

ELISA for rotavirus A

Contents of jejunum from all samples were examined for rotavirus group A by an enzyme immunoassay ProSpecT® Rotavirus (Thermo Fisher Scientific, Massachusetts, USA) according to the manufacturer's instructions.

Analysis for unknown viruses

For investigation of the samples for the presence of viruses in general, three different methods were used; TEM, pan-viral microarray and de novo NGS.

The TEM analyses were performed on frozen ileum including content from animals with defined and severe villus atrophy previously described by a commercial provider (Bio-imaging unit at Animal and Plant Health Agency (APHA), Weybridge, UK) using standard methods at a magnification of 34,000× [6]. Confirmation of the presence of virus in a sample was based on size, shape, fine structure and surface morphological differentiation.

The microarray analyses were performed on frozen ileum, including content from selected animals with defined and severe villus atrophy as previously described [6]. The microarray consisted of a pan-viral microarray containing 47,000 probes covering all the virus entries in GenBank at the time of its design and was developed in cooperation with APHA in the UK. The protocol for sample preparation and test of samples was as previously described [31]. The estimated sensitivity of the assay was 3–6 log10 copies/reaction.

The de novo NGS analyses were performed on frozen ileum including content from selected animals with defined and severe villus atrophy [6]. The samples were tested in pools of five animals from each of the four herds. Sample preparation and nucleic acid isolation was performed as previously described [32]. By making two parallel extractions of RNA and DNA from each sample, generating cDNA, and pooling the material from all samples in each group, enough material was generated to avoid pre-amplification. Each pooled sample was sequenced on an Ion Torrent PGM system using the 200-bp read chemistry and an Ion 316 chip. This was performed as described earlier [33] at the Uppsala Genome Center, SciLifeLab, Sweden. The resulting reads were assembled using MIRA [34] with the standard settings for de novo assembly of Ion Torrent data. Taxonomic classification of assembled contigs was enabled by Blastn and Blastx searches against local copies of NCBI's nucleotide and protein databases using the Blast + package [35] with default settings. Evaluating the taxonomic data for potential viruses, candidate reference genomes were identified and retrieved from GenBank in FASTA format. Alignments of contigs against the nucleotide sequences of the reference genomes were performed using the CodonCode Aligner software (CodonCode Corporation).

Statistical analysis

Logistic regression was used for the comparison of the viral load (copies per reaction) between case and control pigs overall and in the four herds. The outcome in this study was diarrhoea and no diarrhoea. For comparison of the viral load (copies per reaction) between the four herds (herd 1–4), the one-way analysis of variance (ANOVA) test was used. The analyses were performed using R version 3.2.3 [36].

Results

Prevalence of enteric viruses in case and control pigs

Samples from a total of 46 case animals and 46 control animals were tested for coronavirus (Table 1). Ileum tissue, ileum contents and rectal swabs from animals in herds 1, 2 and 3 and rectal swabs from animals in herd 4 were tested in the conventional pan-corona assay. All samples tested negative. Some of the samples generated a band at the agarose gel at about the expected size, but subsequent sequencing revealed that the band represented unspecific amplification of porcine DNA (data not shown). Samples from all animals in herds 1–3 (Table 1) were tested in RT-qPCR assay specific for PEDV, PRCV and TGEV with negative results. Ileum samples were omitted from herd 4 in these analyses due to lack of material.

Table 1 The number of case and control samples tested in specific and nonspecific virus tests

Herd	Day	Case animals (*n*)	Control animals (*n*)
1	3	7[a]	6
	5	6	7
2	4	2	2
	5	6	6
	7	3	4
3	4	8	7
	6	6	6
4	3	8	6
	5	0	2
Summary		46	46

[a]: One sample was not tested by pan-corona PCR

By ELISA, only one case animal tested positive for rotavirus A. Using RT-qPCR, test of 76 ileum (with content) samples from herds 1–3 (Table 1) revealed a total of seven positive samples, of which six samples were collected from case animals and one from a control animal. The numbers of positive samples in each herd were two, three and two for herds 1, 2 and 3, respectively. The sample that tested positive for rotavirus A in ELISA was the most positive sample in the RT-qPCR assay (quantification cycle (Cq) value of 19). All 76 samples tested for rotavirus C by RT-qPCR yielded negative results. A total of 13 rectal swab samples from case pigs representing all four herds were tested for enterovirus, parechovirus, saffoldvirus, cosavirus, aichivirus and klassevirus by virus specific RT-qPCR assays. All samples yielded negative results. Faecal swab samples from the 76 animals in herds 1–3 (Table 1) were tested in RT-qPCR specific for sapovirus and norovirus with negative results.

Prevalence of PKV-1, PAstV1–5, PTV and PCV2 in case and control pigs

Ileum samples from a total of 47 case pigs and 49 control pigs were tested for PKV-1 in a specific RT-qPCR assay. The overall prevalence of PKV-1 was found to be 91.7% (88/96) and the virus was detected in all four herds with prevalence ranging between 80 and 100%. The overall prevalence for the case pigs was 93.6% compared to 89.8% for the control pigs (Table 2). PKV-1 had a similar mean value (±SD) of viral load for the case (4.60 ± 1.76) and control (4.79 ± 1.72) pigs (Table 3). The log-transformed copy numbers for the positive samples are shown in Fig. 1 for the four herds. No statistically difference in viral load between the case and control pigs was observed for PKV-1 ($p > 0.05$) overall or in the four herds (Table 3).

Ileum samples from a total of 47 case pigs and 49 control pigs were tested in five RT-qPCR assays specific for each of the five PAstV types. In total, 75% (72/96) of the animals were positive for at least one of the PAstVs. Among the PAstV types, PAstV3 had the highest overall prevalence of 69.8%, followed by PAstV5 (12.5%) and PAstV4 (11.5%) while none of the samples tested positive for PAstV1 or PAstV2. PAstV3 was detected in all four herds, while PAstV4 and PAstV5 only were found in herds 1–3 (Table 2).

The overall prevalence of PAstVs in the case pigs was 70.2% compared to 79.6% for the control pigs. PAstV3 was found in 63.8% of the case pigs and in 75.5% of the control pigs. For PAstV4 and PAstV5, 19.2% and 12.8% of the case pigs were positive, respectively. In comparison, 4.1% and 12.2% of the control pigs were positive for these two viruses (Table 2).

The mean value (±SD) of viral load for PAstV3 was 7.50 ± 1.78 for the case pigs and 7.89 ± 2.15 for the

Table 2 Prevalence of PKV-1 and PAstV1–5. The number of positive samples, followed by percentage in parentheses

Herd	Status	No. of samples	PKV-1	PAstV1	PastV2	PastV3	PastV4	PAstV5
1	Case	12	10/12 (83.3%)	0/12 (0%)	0/12 (0%)	5/12 (41.7%)	2/12 (16.7%)	1/12 (8.3%)
	Control	13	10/13 (76.9%)	0/13 (0%)	0/13 (0%)	13/13 (100%)	0/13 (0%)	0/13 (0%)
2	Case	11	11/11 (100%)	0/11 (0%)	0/11 (0%)	11/11 (100%)	4/11 (36.4%)	2/11 (18.2%)
	Control	11	11/11 (100%)	0/11 (0%)	0/11 (0%)	11/11 (100%)	1/11 (9.1%)	4/11 (36.4%)
3	Case	12	12/12 (100%)	0/12 (0%)	0/12 (0%)	7/12 (58.3%)	3/12 (25%)	3/12 (25%)
	Control	13	13/13 (100%)	0/13 (0%)	0/13 (0%)	6/13 (46.2%)	1/13 (7.7%)	2/13 (15.4%)
4	Case	12	11/12 (91.7%)	0/12 (0%)	0/12 (0%)	7/12 (58.3%)	0/12 (0%)	0/12 (0%)
	Control	12	10/12 (83.3%)	0/12 (0%)	0/12 (0%)	7/12 (58.3%)	0/12 (0%)	0/12 (0%)
Summary	Case	47	44/47 (93.6%)	0/47 (0%)	0/47 (0%)	30/47 (63.8%)	9/47 (19.2%)	6/47 (12.8%)
	Control	49	44/49 (89.8%)	0/49 (0%)	0/49 (0%)	37/49 (75.5%)	2/49 (4.1%)	6/49 (12.2%)
Total		96	88/96 (91.7%)	0/96 (0%)	0/96 (0%)	67/96 (69.8%)	11/96 (11.5%)	12/96 (12.5%)

control pigs (Table 3). The log-transformed copy numbers for the positive samples are shown in Fig. 2. No statistically difference was observed between the viral load in the case and control pigs for PAstV3 ($p > 0.05$) overall. However, examining of the individual herds using logistic regression showed that there was a statistical significant difference between the viral load for PAstV3 for the positive case and control pigs in herd 2 ($p = 0.032$), while no significant difference was observed in herd 1 ($p = 0.17$), 3 ($p = 0.87$), or 4 ($p = 0.59$). The mean value (±SD) of viral load for PAstV4 was 5.62 ± 0.68 and 5.11 ± 0.40 for the case and control pigs (Table 3), respectively. For PAstV5 the values were 5.08 ± 0.61 and 5.01 ± 0.32 for the case and control pigs (Table 3). No statistically difference was observed between the viral load in the case and control pigs for either PAstV4 or PAstV5.

Furthermore, the 47 case and 49 control samples were tested for PTV and PCV2 with a specific RT-qPCR and qPCR assay, respectively. Three out of 96 animals tested positive for PTV, whereas none of the pigs tested positive for PCV2 (data not shown).

Table 3 Viral load ($\log_{10}$ copies per reaction) of PKV-1 and PAstV3–5. The values represent Mean ± SD

Herd	Status	PKV-1	PAstV3	PAstV4	PAstV5
1	Case	4.52 ± 1.66	6.45 ± 1.20	5.09 ± 0.01	5.54
	Control	4.13 ± 1.36	8.36 ± 2.23	–	–
2	Case	5.58 ± 1.52	9.07 ± 1.04[a]	5.72 ± 0.52	4.53 ± 0.32
	Control	4.49 ± 1.79	7.15 ± 2.40[a]	4.82	5.05 ± 0.40
3	Case	4.70 ± 1.93	6.77 ± 1.35	5.83 ± 1.03	5.29 ± 0.63
	Control	5.54 ± 1.88	7.85 ± 1.38	5.39	4.93 ± 0.01
4	Case	3.59 ± 1.50	6.52 ± 1.90	–	–
	Control	4.82 ± 1.63	8.18 ± 2.21	–	–
Summery	Case	4.60 ± 1.76	7.50 ± 1.78	5.62 ± 0.68	5.08 ± 0.61
	Control	4.79 ± 1.72	7.89 ± 2.15	5.11 ± 0.40	5.01 ± 0.32

[a] A statistical significant difference ($p = 0.032$) between the viral load for the case and control pigs

Co-infection with different viruses

Viral co-infection with two or more viruses was observed in all herds. Different combinations of PAstVs were detected in 15 of the 96 pigs (15.6%) (Table 4), while co-infection with other viruses was found in seven case pigs and one control pig. These pigs were infected with different combinations of rotavirus A, PTV, PKV-1 and PAstVs.

Nonspecific virus tests

For investigation of the presence of virus in general, selected samples were analysed with TEM, pan-viral microarray and de novo NGS. Eight samples from case animals with histological lesions typical of viral infections (villus atrophy) representing all four herds were analysed by TEM with negative results. A total of 18 samples from 13 case and five control animals were tested by microarray. None of the samples tested conclusively positive for any of the viruses present on the pan-viral microarray (data not shown). Samples from five case animals from each herd were pooled and tested by de novo NGS. Endogenous retrovirus was present in all four pools (data not shown). PKV-1 was also present in all four pools as evident by several reads (Table 5). In addition, the pools of samples from herd 1 were positive for rotavirus C, herd 2 were positive for rotavirus A and herd 3 were positive for PTV. One of the samples included in the pool from herd 2 was from the animal that also tested positive for rotavirus A by RT-qPCR and by ELISA.

Discussion

The present study is part of a series of studies focusing on identifying the cause of NNPDS in Danish pigs,

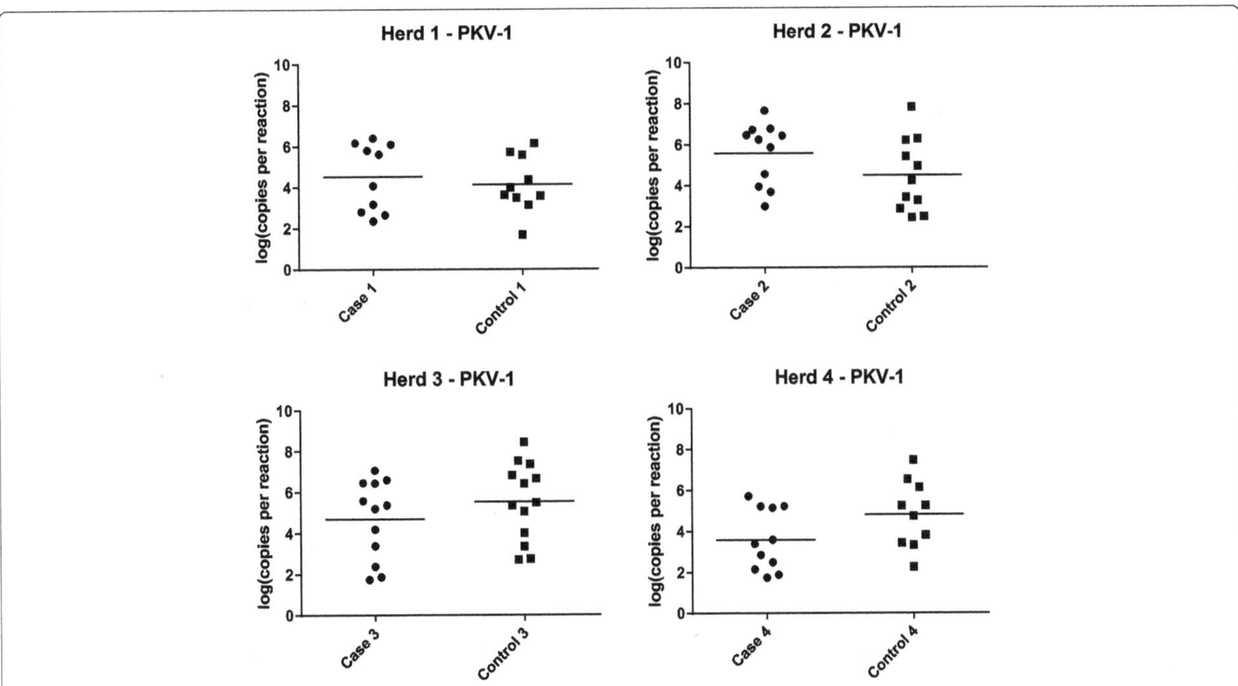

Fig. 1 Viral load (log10 copies per reaction) of PKV-1. The copy number for the positive PKV-1 samples in herd 1–4 for case and control pigs. The straight line shows the mean value

which has previously been shown not to be clearly associated with known bacterial or parasitic pathogens [5, 6]. In the present study the tests for viruses were broadened by testing selected samples for a range of specific viruses linked to enteric disorders in pigs or humans.

The initial test of the samples for rotavirus A virus by a commercial available ELISA generated only one positive sample [5]. Totally seven of 76 (9.2%) samples tested positive when retested in a RT-qPCR assay specific for rotavirus A. The difference between the outcome of tests

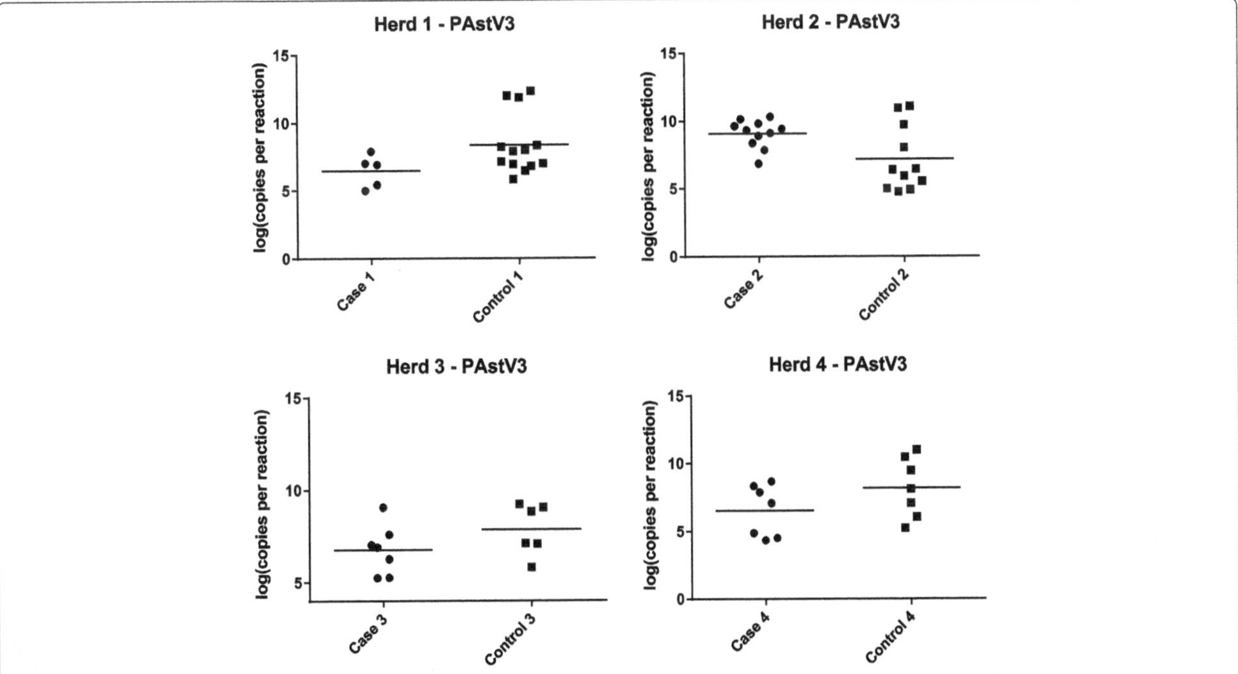

Fig. 2 Viral load (log10 copies per reaction) of PAstV3. The copy number for the positive PAstV3 samples in herd 1–4 for case and control pigs. The straight line shows the mean value

Table 4 Co-infection with two or three PAstV types in the same pig

Herd	No. of samples	PAstV3 + V4	PAstV3 + V5	PAstV4 + V5	PAstV3 + V4 + V5
1	25	1/25 (4%)	1/25 (4%)	0/25 (0%)	0/25 (0%)
2	22	3/22 (13.6%)	4/22 (18.2%)	0/22 (0%)	2/22 (9.1%)
3	25	2/25 (8%)	0/25 (0%)	1/25 (4%)	1/25 (4%)
4	24	0/24 (0%)	0/24 (0%)	0/24 (0%)	0/24 (0%)
Total	96	6/96 (6.3%)	5/96 (5.2%)	1/96 (1.0%)	3/96 (3.1%)

by ELISA and RT-qPCR most likely reflects the higher sensitivity of the PCR, which was also underlined by the fact that the ELISA positive sample was the sample with the lowest Cq value in the PCR. The low number of positive samples strongly indicated that rotavirus A was not a significant problem in very young animals in the NNPDS herds of the present study. None of the samples from pigs younger than one week of age were positive for rotavirus C by RT-qPCR, but one pool of samples, from the pigs in herd 1, investigated by de novo NGS generated 26 reads matching rotavirus C. The differences in the outcome of the two tests can be due to the specificity of the primers and probe used in the PCR. Since the assay was designed based on sequences specific for human rotavirus C it is unclear if the negative results in the PCR test reflected a low level of rotavirus C among Danish pigs or if the assay used did not detect porcine rotavirus C. Group C rotavirus was first identified in swine in 1980 [9], and diarrheal outbreaks associated with rotavirus C have been documented in nursing, weaning and post-weaning pigs, either alone or in mixed infection with other enteric pathogens [37]. In a recent study in the USA, 19.5% of the tested samples were found positive with a higher rotavirus C frequency in case (28.4%) piglets compared to non-case (6.6%) piglets [38]. In contrast, the combined results of the RT-qPCR and the NGS of the present study indicated that this virus did not contribute to NNPDS.

Testing of samples for the two porcine coronaviruses TGEV and PEDV generated the expected negative results. Denmark is considered free of these viruses, but PEDV has recently re-emerged as a significant pathogen in Asia and North America [39] and has also been detected in several European countries including Germany [40].

Table 5 Results of Next Generation Sequencing

Herd	Case animals (*n*)	Result of test of pools of 5 samples
1	5	Porcine rotavirus (group C) - [1 Contig, 26 reads]
		Porcine kobuvirus - [6 Contigs, 206 reads]
2	5	Rotavirus (group A) - [1 Contig, 10 reads]
		Porcine kobuvirus - [1 Contig, 1282 reads]
3	5	Porcine teschovirus - [3 Contigs, 288 reads]
		Porcine kobuvirus - [16 Contigs, 1479 reads]
4	5	Porcine kobuvirus - [11 Contigs, 262 reads]

The testing of samples for sapovirus gave negative results, indicating that sapovirus did not contribute to NNPDS. Porcine sapovirus has been shown experimentally to induce mild to moderate diarrhoea in pigs [41, 42], but epidemiological evidence for a causative role of sapovirus in natural cases of suckling pig diarrhoea is scarce. In a European survey from 2010, sapovirus was detected in faecal samples in 1.6%, 12.2% and 43.4% of suckling pigs from Hungary, Spain and Slovenia, respectively, but there was no association between diarrhoea and detection of sapovirus [43]. In the same study, faecal samples from 57 two-eight weeks old pigs with diarrhoea from 31 Danish herds were tested for sapovirus, with positive results in 68% of the herds, covering 44% of the pigs. The test used both in the European study and the present Danish study was identical [27], however, the pigs included in the present study were younger than the pigs tested in the previous study, which may explain the difference in test results.

Norovirus was not detected in any of the samples. Norovirus has been shown to be prevalent in slaughter pigs in USA [44] and in Belgium [45] whereas samples from pigs aged 1–16 weeks tested negative in Spain [46]. In 2007, a screening of 56 routine laboratory submissions from 31 Danish swine herds revealed that 16% of the herds were positive (unpublished results). The age of the pigs included in the screening was not known, but the combined results indicate that porcine norovirus is indeed prevalent in Danish herds, and the negative outcome of the test described in the present study is probably due to the young age of the animals tested.

Thirteen of the samples from case pigs were screened in a multiplex qPCR, developed for diagnostic use in humans, for the suspected emerging human enteric pathogens enterovirus, parechovirus, saffoldvirus, cosavirus and klassevirus. These samples were negative, and since there have been no reports on these virus being present in pigs it was decided to omit testing of the remaining samples.

The viruses PAstVs and PKV-1 were found with high prevalence in the present study using RT-qPCRs. These viruses have been detected both in diarrhoeic and healthy pigs worldwide, and their role as causative agents of diarrhoea in pigs is not established [10, 47–52], however, some studies have found an association between detection of either PAstVs or PKV-1 and diarrhoea in pigs [12–14]. In the present study, a high overall prevalence of PAstVs was detected, with PAstV3 as the most prevalent type in all four herds. PAstV4 and PAstV5 were present in three herds, while PAstV1 and PAstV2 were not detected in any of the herds. No statistical significant difference between the viral load in the case pigs and control pigs was observed when looking at the data overall. So based on this analysis, there was no evidence for the case pigs having a higher viral load than the control pigs or that detection of PAstV led to diarrhoea in the pigs. In herd 2, however, there was a significant higher viral load of PAstV3 in the case pigs compared to the control pigs. In this herd all pigs were found to be positive for PAstV3, so in this herd it was not the presence of the virus but merely the amount of virus that could be related to diarrhoea. Nevertheless, the results did not support that PAstVs are generally involved in NNPDS.

The presence of two or three PAstV types in the same pig was found in a few of the samples, which is in accordance with previous findings [49]. An overall high prevalence of PAstV in pigs has been observed elsewhere, but to our knowledge this study is the first to detect PAstV3 as the dominant type. PAstV3 has been found in USA, Canada, Croatia and East Africa, but only in few of the tested pigs [47, 49, 53–55] and some of these studies found PAstV3 to be most prevalent in young piglets. Xiao and co-workers found the highest prevalence (4.72–5.3%) of PAstV3 in suckling pigs (0–20 days), and in the study by Luo et al., 2011, PAstV3 was only detected in this age group [47, 49]. Thus, the results in this study also suggested an association between PAstV3 and the early growing stage of the pigs. However, further investigations are needed to confirm if this is the case in all herds. A high overall prevalence of PAstVs has been detected in Croatia (89%), Canada (79.2%), USA (64%), Czech Republic (34.2%) and Germany (20.8%), where the most prevalent PAstV type differs between the countries [47, 49, 54, 56, 57].

In the present study, 91.7% of the animals tested positive for PKV-1 by RT-qPCR. The virus was found in all four herds, but no statistical significant difference between the viral load in the case and control pigs was observed either overall or at herd level. The PCR result was consistent with the findings by the de novo NGS, where PKV-1 also was detected in all herds. A high prevalence in young piglets has also been observed by others, and it has been indicated that there is a higher rate of infection with PKV-1 in young pigs than in older pigs [50, 51, 55]. A study conducted in 40 Korean herds detected a higher prevalence of PKV-1 in diarrhoeic vs. non-diarrhoeic piglets [13]. However, only 4% of the diarrhoeic pigs were not co-infected with other pathogens. Recent studies have found the prevalence of PKV-1 to be similar in diarrhoeic and healthy pigs [58–61], although, two of the studies showed that diarrheic suckling piglets was the most frequently infected group but the results were not statistically significantly different from the healthy piglets [58, 59]. Altogether, these finding indicated that PKV-1 is not a significant primary pathogen in natural cases of diarrhoea in young piglets.

Only few samples were positive for PTV. This result is consistent with the de novo NGS, where only case samples were included, which also found herd 3 to be PTV positive. To date, 13 serotypes of PTV have been associated with a variety of clinical diseases. PTV-1 strains were associated with highly fatal, non-suppurative encephalomyelitis of pigs (Teschen disease) in the 1930–1950s. Today, less virulent talfan strains of PTV-1 are more widespread, and PTVs are detected in swine herds worldwide often together with a variety of other common swine pathogens. In a Chinese study, PTV was found in 96.7% of 30 culled four-eight weeks old post-weanling piglets by nested RT-PCR, but this study concluded that PTV was only marginally related to non-suppurative encephalitis [62]. However, other PTV types may cause other disease symptoms [63].

None of the samples were positive for PCV2. This virus is regarding the primary causative agent of post-weaning multisystemic wasting syndrome (PMWS), which has had a huge influence on the pig production worldwide during the last 10 years [64, 65]. So far, PCV2 has not been shown to be a primary causative agent of diarrhoea, and the clinical description of PMWS related diarrhoea is considered to be a result of the lymphoid depletion, which is seen during PCV2 infections. This can lead to immunosuppression which may predispose for co-infections with other pathogens leading to diarrhoea [18, 66]. PCV2 related diseases have not been described in very young piglets so the negative findings in the present study were not surprising.

During recent years, newer techniques using a metagenomic approach, such as microarrays and de novo NGS, have been developed and used to detect emerging and re-emerging viruses in humans and in animals [33, 67]. The advantages of these techniques are that they can detect all viruses irrespective of prior knowledge of the virus genetic sequence. The only prerequisite is that the new virus has some level of identity to previously sequenced viruses in order to be picked up by the downstream bioinformatic filtering of the data [68]. The general virological analyses employed in the present study were so expensive that it was not economically possible to test all samples. The pan viral microarray chip used in the present study included more than

40.000 probes covering all the viruses present in the GenBank [31]. The array failed to detect any virus in the relatively few (n = 18) samples tested, indicating that no, or small amounts of, virus particles were present. The detection level of the chip is low compared to qPCR and de novo NGS [69]. In accordance, eight of eight samples examined by TEM, which also has a relatively poor level of detection, also gave negative results.

Compared to other studies on pig faeces, very few viruses were detected by de novo NGS in the present study. A previous study conducted on a US farm detected PKV-1 (23% of all reads), PAstV (22%), enterovirus (14%), sapoviruses (5.7%), sapeloviruses (1.5%), coronaviruses (0.69%), bocaviruses (0.22%) and PTV in 0.03% of the reads [70]. In the present study only PKV-1, PAstV and PTV were detected. Interestingly, the US study found that, except for PKV-1, the prevalence of most viruses was not greater in animals with diarrhoea than in control animals. The difference in viral prevalence between the studies may be explained by difference in age of the animals tested, but the most likely explanation is that the US study used random PCR amplification prior to library building, which greatly improves the sensitivity but also introduces a bias, which can prevent a meaningful quantitative analysis [71]. The sensitivity of the NGS protocol used for each of the pathogens has not been determined so it is possible that a low copy virus has been missed. Experimental infections of animals with filtered faeces from affected animals would be a way to investigate if viral components contribute to the NNPD syndrome, but the pilot studies performed by the project group gave inconclusive results (unpublished observations).

Conclusion

In conclusion, the analyses performed on samples from four Danish pig herds in this study did not find a significant contribution of enteric viruses to NNPDS. Despite the fact that a high prevalence and viral load of PAstV3 and PKV-1 were detected, these viruses did not seem to be the causative agents of NNPDS. However, since only enteric samples were analysed, further studies are needed to investigate whether a systemic virus infection may play a role in the pathogenesis of NNPDS.

Acknowledgements

The authors would like to thank herd-owners and their staff for help and cooperation. Furthermore, we would like to thank the technical personnel at the National Veterinary Institute, the Pig Research Centre, Statens Serum Institut, the Swedish University of Agricultural Sciences and Department of Veterinary and Animal Sciences, University of Copenhagen.

Funding

The study was funded by the Innovation Foundation of the Ministry of Food, Agriculture and Fisheries, Denmark (project no. 3412–09-02519).

Authors' contributions

All author contributed to the design of the study. Selection of samples was done by HK. PCR analyses were conducted and interpreted by NBG, CKH, SR, TKF, SEM and LDR. NGS analyses were performed by FG. NBG and LEL participated in drafting the manuscript, while all authors participated in proofreading of the manuscript. All authors read and approved the final manuscript.

Consent for publication

Not applicable.

Competing interests

The authors declare that they have no competing interests.

Author details

[1]National Veterinary Institute, Technical University of Denmark, Kemitorvet, Lyngby DK-2800, Denmark. [2]Department of Biomedical Sciences and Veterinary Public Health (BVF), Swedish University of Agricultural Sciences (SLU), Uppsala, Sweden. [3]Statens Serum Institut (SSI), Artillerivej 5, Copenhagen S DK-2300, Denmark. [4]Pig Research Centre, Danish Agriculture and Food Council, Vinkelvej 13, DK-8620 Kjellerup, Denmark. [5]National Veterinary Institute, Technical University of Denmark, Lindholm, Kalvehave DK-4771, Denmark. [6]Department of Veterinary and Animal Sciences, Faculty of Health and Medical Sciences, University of Copenhagen, Gronnegaardsvej 15, DK-1870 Frederiksberg, Denmark.

References

1. Gin T, Guennec J, Le MH, Martineau G. Clinical and laboratory investigations in 10 French pig herds dealing with enzootic neonatal diarrhoea. Proc 21st IPVS. 2010:758.
2. Melin L, Wallgren P, Mattson S, Stampe M, Löfstedt M. Neonatal diarrhoea in piglets from E coli vaccinated sows in Sweden Proc 21st IPVS 2010;290.
3. Svensmark S. New neonatal diarrhoea syndrome in Denmark. Proc 1st ESPHM. 2009:27.
4. Wallgren P, Mattsson S, Merza M. New neonatal porcine diarrhoea: Aspects on etiology. Proc. 22nd IPVS Congr. Jeju, Korea. 2012;76.
5. Kongsted H, Jonach B, Haugegaard S, Angen O, Jorsal SE, Kokotovic B, et al. Microbiological, pathological and histological findings in four Danish pig herds affected by a new neonatal diarrhoea syndrome. BMC Vet Res. 2013;9:206.
6. Jonach B, Boye M, Stockmarr A, Jensen TK. Fluorescence in situ hybridization investigation of potentially pathogenic bacteria involved in neonatal porcine diarrhea. BMC Vet Res. 2014;10:68.
7. Pensaert MB, de Bouck PA. New coronavirus-like particle associated with diarrhea in swine. Arch Virol. 1978;58:243–7.
8. Bohl EH, Kohler EM, Saif LJ, Cross RF, Agnes AG, Theil KW. Rotavirus as a cause of diarrhea in pigs. J Am Vet Med Assoc. 1978;172:458–63.
9. Saif LJ, Bohl EH, Theil KW. Rotavirus-like, calcivirus-like, and 23nm virus-like particles associated with diarrhea in young pigs. J Clin Microbiol. 1980:105–11.
10. Sisay Z, Wang Q, Oka T, Saif L. Prevalence and molecular characterization of porcine enteric caliciviruses and first detection of porcine kobuviruses in US swine. Arch Virol. 2013;158:1583–8.
11. Wang QH, Souza M, Funk JA, Zhang W, Saif LJ. Prevalence of noroviruses and sapoviruses in swine of various ages determined by reverse transcription-PCR and microwell hybridization assays. J Clin Microbiol. 2006; 44:2057–62.
12. Mor SK, Chander Y, Marthaler D, Patnayak DP, Goyal SM. Detection and molecular characterization of porcine astrovirus strains associated with swine diarrhea. J Vet Diagnostic Investig. 2012;24:1064–7.
13. Park SJ, Kim HK, Moon HJ, Song DS, Rho SM, Han JY, et al. Molecular detection of porcine kobuviruses in pigs in Korea and their association with diarrhea. Arch Virol. 2010;155:1803–11.

14. Shimizu M, Shirai J, Narita M, Yamane T. Cytopathic astrovirus isolated from porcine acute gastroenteritis in an established cell line derived from porcine embryonic kidney. J Clin Microbiol. 1990;28:201–6.
15. Buitrago D, Cano-Gómez C, Agüero M, Fernandez-Pacheco P, Gómez-Tejedor C, Jiménez-Clavero MAA. Survey of porcine picornaviruses and adenoviruses in fecal samples in Spain. J Vet Diagn Investig. 2010;22:763–6.
16. Cano-Gómez C, Palero F, Buitrago MD, García-Casado MA, Fernández-Pinero J, Fernández-Pacheco P, et al. Analyzing the genetic diversity of teschoviruses in Spanish pig populations using complete VP1 sequences. Infect Genet Evol. 2011;11:2144–50.
17. La Rosa G, Muscillo M, Di Grazia A, Fontana S, Iaconelli M, Tollis M. Validation of RT-PCR assays for molecular characterization of porcine teschoviruses and enteroviruses. J Vet Med Ser B Infect Dis Vet Public Heal 2006;53:257–265.
18. Johansen M, Nielsen M, Dahl J, Svensmark B, Bækbo P, Kristensen CS, et al. Investigation of the association of growth rate in grower-finishing pigs with the quantification of Lawsonia Intracellularis and porcine circovirus type 2. Prev Vet Med. 2013;108:63–72.
19. Segalés J. Porcine circovirus type 2 (PCV2) infections: clinical signs, pathology and laboratory diagnosis. Virus Res. 2012:10–9.
20. Vijgen L, Moës E, Keyaerts E, Li S, Van Ranst MA. Pancoronavirus RT-PCR assay for detection of all known coronaviruses. Methods Mol Biol. 2008;454:3–12.
21. Kim SH, Kim IJ, Pyo HM, Tark DS, Song JY, Hyun BH. Multiplex real-time RT-PCR for the simultaneous detection and quantification of transmissible gastroenteritis virus and porcine epidemic diarrhea virus. J Virol Methods. 2007;146:172–7.
22. Vemulapalli R, Gulani J, Santrich CA. Real-time TaqMan?? RT-PCR assay with an internal amplification control for rapid detection of transmissible gastroenteritis virus in swine fecal samples. J Virol Methods. 2009;162:231–5.
23. Zhang C, Wang Z, Hu F, Liu Y, Qiu Z, Zhou S, et al. The survey of porcine teschoviruses in field samples in China with a universal rapid probe real-time RT-PCR assay. Trop Anim Health Prod. 2013;45:1057–61.
24. Pang XL, Lee B, Boroumand N, Leblanc B, Preiksaitis JK, Ip CCY. Increased detection of rotavirus using a real time reverse transcription-polymerase chain reaction (RT-PCR) assay in stool specimens from children with diarrhea. J Med Virol. 2004;72:496–501.
25. Logan C, O'Leary JJ, O'Sullivan N. Real-time reverse transcription-PCR for detection of rotavirus and adenovirus as causative agents of acute viral gastroenteritis in children. J Clin Microbiol. 2006;44:3189–95.
26. Kageyama T, Kojima S, Shinohara M, Uchida K, Fukushi S, Hoshino FB, et al. Broadly reactive and highly sensitive assay for Norwalk-like viruses based on real-time quantitative reverse transcription-PCR. J Clin Microbiol. 2003;41:1548–57.
27. Oka T, Katayama K, Hansman GS, Kageyama T, Ogawa S, FT W, et al. Detection of human sapovirus by real-time reverse transcription-polymerase chain reaction. J Med Virol. 2006;78:1347–53.
28. Nielsen ACY, Böttiger B, Midgley SE, Nielsen LPA. Novel enterovirus and parechovirus multiplex one-step real-time PCR-validation and clinical experience. J Virol Methods. 2013;193:359–63.
29. Nielsen ACY, Gyhrs ML, Nielsen LP, Pedersen C, Böttiger B. Gastroenteritis and the novel picornaviruses aichi virus, cosavirus, saffold virus, and salivirus in young children. J Clin Virol. 2013;57:239–42.
30. Hjulsager CK, Grau-Roma L, Sibila M, Enøe C, Larsen L, Segalés J. Inter-laboratory and inter-assay comparison on two real-time PCR techniques for quantification of PCV2 nucleic acid extracted from field samples. Vet Microbiol. 2009;133:172–8.
31. Gurrala R, Dastjerdi A, Johnson N, Nunez-Garcia J, Grierson S, Steinbach F, et al. Development of a DNA microarray for simultaneous detection and genotyping of lyssaviruses. Virus Res. 2009;144:202–8.
32. Granberg F, Vicente-Rubiano M, Rubio-Guerri C, Karlsson OE, Kukielka D, Belák S, et al. Metagenomic detection of viral pathogens in Spanish honeybees: co-infection by aphid lethal paralysis, Israel acute paralysis and Lake Sinai viruses. PLoS One. 2013;8:e57459.
33. Belák S, Karlsson OE, Blomström AL, Berg M, Granberg F. New viruses in veterinary medicine, detected by metagenomic approaches. Vet Microbiol. 2013:95–101.
34. Chevreux B, Pfisterer T, Drescher B, Driesel AJ, Müller WEG, Wetter T, et al. Using the miraEST assembler for reliable and automated mRNA transcript assembly and SNP detection in sequenced ESTs. Genome Res. 2004;14: 1147–59.
35. Camacho C, Coulouris G, Avagyan V, Ma N, Papadopoulos J, Bealer K, et al. BLAST+: architecture and applications. BMC Bioinformatics. 2009;10:421.
36. R Core Team. R: A language and environment for statistical [Internet]. 2013. Available from: http://www.r-project.org/.
37. Saif LJ, Jiang B, Nongroup A. Rotaviruses of humans and animals. Curr Top Microbiol Immunol. 1994;185:339–71.
38. Amimo JO, Vlasova AN, Saif LJ. Prevalence and genetic heterogeneity of porcine group C rotaviruses in nursing and weaned piglets in Ohio, USA and identification of a potential new VP4 genotype. Vet Microbiol. 2013;164: 27–38.
39. Huang YW, Dickerman AW, Piñeyro P, Li L, Fang L, Kiehne R, et al. Origin, evolution, and genotyping of emergent porcine epidemic diarrhea virus strains in the united states. MBio. 2013;4:e00737–13.
40. Stadler J, Zoels S, Fux R, Hanke D, Pohlmann A, Blome S, et al. Emergence of porcine epidemic diarrhea virus in southern Germany. BMC Vet Res. 2015; 11:142.
41. Flynn WT, Saif LJ, Moorhead PD. Pathogenesis of porcine enteric calicivirus-like virus in four-day-old gnotobiotic pigs. Am J Vet Res. 1988;49:819–25.
42. Guo M, Hayes J, Cho KO, Parwani AV, Lucas LM, Saif LJ. Comparative pathogenesis of tissue culture-adapted and wild-type Cowden porcine enteric calicivirus (PEC) in gnotobiotic pigs and induction of diarrhea by intravenous inoculation of wild-type PEC. J Virol. 2001;75:9239–51.
43. Reuter G, Zimšek-Mijovski J, Poljšak-Prijatelj M, Di Bartolo I, Ruggeri FM, Kantala T, et al. Incidence, diversity, and molecular epidemiology of sapoviruses in swine across Europe. J Clin Microbiol. 2010;48:363–8.
44. Scheuer KA, Oka T, Hoet AE, Gebreyes WA, Molla BZ, Saif LJ, et al. Prevalence of porcine noroviruses, molecular characterization of emerging porcine sapoviruses from finisher swine in the United States, and unified classification scheme for sapoviruses. J Clin Microbiol. 2013;51:2344–53.
45. Mauroy A, Scipioni A, Mathijs E, Miry C, Ziant D, Thys C, et al. Noroviruses and sapoviruses in pigs in Belgium. Arch Virol. 2008;153:1927–31.
46. Halaihel N, Masía RM, Fernández-Jiménez M, Ribes JM, Montava R, De Blas I, et al. Enteric calicivirus and rotavirus infections in domestic pigs. Epidemiol Infect. 2010;138:542–8.
47. Luo Z, Roi S, Dastor M, Gallice E, Laurin MA, L'homme Y. Multiple novel and prevalent astroviruses in pigs. Vet Microbiol 2011;149:316–323.
48. Reuter G, Pankovics P, Boros Á. Identification of a novel astrovirus in a domestic pig in Hungary. Arch Virol. 2011;156:125–8.
49. Xiao CT, Giménez-Lirola LG, Gerber PF, Jiang YH, Halbur PG, Opriessnig T. Identification and characterization of novel porcine astroviruses (PAstVs) with high prevalence and frequent co-infection of individual pigs with multiple PAstV types. J Gen Virol. 2013;94:570–82.
50. Barry AF, Ribeiro J, Alfieri AF, van der Poel WHM, Alfieri AA. First detection of kobuvirus in farm animals in Brazil and the Netherlands. Infect Genet Evol. 2011;11:1811–4.
51. Khamrin P, Maneekarn N, Kongkaew A, Kongkaew S, Okitsu S, Ushijima H. Porcine Kobuvirus in piglets, Thailand. Emerg Infect Dis. 2009;15:2075–6.
52. Di Bartolo I, Angeloni G, Tofani S, Monini M, Ruggeri FM. Infection of farmed pigs with porcine kobuviruses in Italy. Arch Virol. 2015;160:1533–6.
53. Reuter G, Boldizsár Á, Pankovics P. Complete nucleotide and amino acid sequences and genetic organization of porcine kobuvirus, a member of a new species in the genus Kobuvirus, family Picornaviridae. Arch Virol. 2009; 154:101–8.
54. Brnić D, Jemeršić L, Keros T, Prpić J. High prevalence and genetic heterogeneity of porcine astroviruses in domestic pigs. Vet J. 2014;202: 390–2.
55. Amimo JO, Okoth E, Junga JO, Ogara WO, Njahira MN, Wang Q, et al. Molecular detection and genetic characterization of kobuviruses and astroviruses in asymptomatic local pigs in East Africa. Arch Virol. 2014;159: 1313–9.
56. Dufkova L, Scigalkova I, Moutelikova R, Malenovska H, Prodelalova J. Genetic diversity of porcine sapoviruses, kobuviruses, and astroviruses in asymptomatic pigs: an emerging new sapovirus GIII genotype. Arch Virol. 2013;158:549–58.
57. Machnowska P, Ellerbroek L, Johne R. Detection and characterization of potentially zoonotic viruses in faeces of pigs at slaughter in Germany. Vet Microbiol. 2014;168:60–8.
58. Zhou W, Ullman K, Chowdry V, Reining M, Benyeda Z, Baule C, et al. Molecular investigations on the prevalence and viral load of enteric viruses in pigs from five European countries. Vet Microbiol. 2016;182:75–81.
59. Jackova A, Sliz I, Mandelik R, Salamunova S, Novotny J, Kolesarova M, et al. Porcine kobuvirus 1 in healthy and diarrheic pigs: genetic detection and

characterization of virus and co-infection with rotavirus a. Infect Genet Evol. 2017;49:73–7.
60. Verma H, Mor SK, Abdel-Glil MY, Goyal SM. Identification and molecular characterization of porcine kobuvirus in U. S. Swine. Virus Genes. 2013;46: 551–3.
61. Chuchaona W, Khamrin P, Yodmeeklin A, Kongkaew A, Vachirachewin R, Kumthip K, et al. Detection and molecular characterization of porcine kobuvirus in piglets in 2009–2013 in northern Thailand. Trop Anim Health Prod Springer Netherlands. 2017;49:1077–80.
62. Chiu SC, SC H, Chang CC, Chang CY, Huang CC, Pang VF, et al. The role of porcine teschovirus in causing diseases in endemically infected pigs. Vet Microbiol. 2012;161:88–95.
63. Feng L, Shi HY, Liu SW, BP W, Chen JF, Sun DB, et al. Isolation and molecular characterization of a porcine teschovirus 1 isolate from China. Acta Virol. 2007;51:7–11.
64. Harding JCS, Clark EG. Recognizing and diagnosing postweaning multisystemic wasting syndrome (PMWS). J Swine Heal Prod. 1997;5:201–3.
65. Segalés J, Allan GM, Domingo M. Porcine circovirus diseases. Anim Health Res Rev. 2005;6:119–42.
66. Allan GM, Ellis JA. Porcine circoviruses: a review. J Vet Diagn Investig. 2000; 12:3–14.
67. Chen EC, Miller SA, DeRisi JL, Chiu CY. Using a pan-viral microarray assay (Virochip) to screen clinical samples for viral pathogens. J Vis Exp. 2011;50: e2536.
68. Hoffmann B, Scheuch M, Höper D, Jungblut R, Holsteg M, Schirrmeier H, et al. Novel orthobunyavirus in cattle, Europe, 2011. Emerg Infect Dis. 2012; 18:469–72.
69. Frey KG, Herrera-Galeano JE, Redden CL, Luu TV, Servetas SL, Mateczun AJ, et al. Comparison of three next-generation sequencing platforms for metagenomic sequencing and identification of pathogens in blood. BMC Genomics. 2014;15:96.
70. Shan T, Li L, Simmonds P, Wang C, Moeser A, Delwart E. The fecal Virome of pigs on a high-density farm. J Virol. 2011;85:11697–708.
71. Karlsson OE, Belák S, Granberg F. The effect of preprocessing by sequence-independent, single-primer amplification (SISPA) on metagenomic detection of viruses. Biosecur Bioterror. 2013;11(Suppl 1):S227–34.

Spatio-temporal trends and risk factors affecting West Nile virus and related flavivirus exposure in Spanish wild ruminants

Ignacio García-Bocanegra[1], Jorge Paniagua[1], Ana V. Gutiérrez-Guzmán[2], Sylvie Lecollinet[3], Mariana Boadella[4], Antonio Arenas-Montes[1], David Cano-Terriza[1], Steeve Lowenski[3], Christian Gortázar[2] and Ursula Höfle[2*]

Abstract

Background: During the last decade, the spread of many flaviviruses (Genus *Flavivirus*) has been reported, representing an emerging threat for both animal and human health. To further study utility of wild ruminant samples in West Nile virus (WNV) surveillance, we assessed spatio–temporal trends and factors associated with WNV and cross-reacting flaviviruses exposure, particularly Usutu virus (USUV) and Meaban virus (MBV), in wild ruminants in Spain. Serum samples from 4693 wild ruminants, including 3073 free-living red deer (*Cervus elaphus*), 201 fallow deer (*Dama dama*), 125 mouflon (*Ovis aries musimon*), 32 roe deer (*Capreolus capreolus*) and 1262 farmed red deer collected in 2003–2014, were screened for WNV and antigenically-related flavivirus antibodies using a blocking ELISA (bELISA). Positive samples were tested for neutralizing antibodies against WNV, USUV and MBV by virus micro-neutralization tests.

Results: Mean flavivirus seroprevalence according to bELISA was 3.4 ± 0.5 % in red deer, 1.0 ± 1.4 % in fallow deer, 2.4 ± 2.7 % in mouflon and 0 % in roe deer. A multivariate logistic regression model revealed as main risk factors for seropositivity in red deer; year (2011), the specific south-coastal bioregion (bioregion 5) and presence of wetlands. Red deer had neutralizing antibodies against WNV, USUV and MBV.

Conclusions: The results indicate endemic circulation of WNV, USUV and MBV in Spanish red deer, even in areas without known flavivirus outbreaks. WNV antibodies detected in a free-living red deer yearling sampled in 2010, confirmed circulation this year. Co-circulation of WNV and USUV was detected in bioregions 3 and 5, and of WNV and MBV in bioregion 3. Sampling of hunted and farmed wild ruminants, specifically of red deer yearlings, could be a complementary way to national surveillance programs to monitor the activity of emerging flaviviruses.

Keywords: West Nile virus, Usutu virus, Meaban virus, Red deer *Cervus elaphus*, Risk factors, Wild ruminants, Spain

Background

The distribution of vector-borne flaviviruses (family *Flavi viridae*) in the world has substantially increased over the last decades. During this period, many flavivirus infections have become a major public health concern due to continuous and growing reporting of outbreaks in humans [1]. Flaviviruses are mainly transmitted within an enzootic cycle involving ornithophilic mosquitoes or ticks as competent vectors, as well as wild birds as the main amplifying hosts in the wild. Most mammalian species including humans are considered dead-end or incidental hosts, because they can get infected but are not thought to be able to transmit the viruses.

During the last few years, six flaviviruses, including West Nile virus (WNV), Usutu virus (USUV), tick-borne encephalitis virus (TBEV), Bagaza virus (BAGV), Meaban virus (MBV) and louping-ill virus (LIV), have been detected in Europe [2, 3]. Five of them have circulated in Spain in the last decade. WNV exposure has been documented in mosquitoes [4], wild birds [5, 6] and different mammalian

* Correspondence: ursula.hofle@uclm.es
[2]Instituto de Investigación en Recursos Cinegéticos IREC, (CSIC-UCLM-JCCM), Ciudad Real, Spain
Full list of author information is available at the end of the article

species, including humans [7], horses [8], dromedary camels (*Camelus dromedarios*) [9], wild boar (*Sus scrofa*) and Iberian pigs and red foxes (*Vulpes vulpes*) [10]. Clinical disease and mortality associated with WNV infection has also been detected in wild birds, horses and humans in this country [11, 12]. USUV has been detected in mosquitoes [4, 13] and in both migratory and resident birds in Spain [14, 15]. Mortality associated with Louping-ill virus (LIV) infection was also detected in sheep and goats in northern Spain [16] and suspected in chamois (*Rupicapra pyrenaica*) [17]. An unusually high mortality due to Bagaza virus (BAGV) infection was also confirmed in free-living game birds in south-western Spain in 2010 [18]. Finally, Meaban virus (MBV) has been found in both yellow-legged gulls (*Larus michaelis*) and ticks (*Ornithodoros maritimus*) in north-eastern Spain [3].

Wild and domestic artiodactyls can be useful sentinel species for monitoring flavivirus activity [19–21]. Antibodies against St. Louis encephalitis virus (SLEV) and WNV have been found in white-tailed deer (*Odocoileus virginianus*) from the United States [22, 23], and against TBEV and WNV in different game species from the Czech Republic [24]. In addition, fatal cases of WN fever were reported in white-tailed deer and reindeer (*Rangifer tarandus*) in North America [25, 26]. Serosurveillance on WNV and related flaviviruses has also been performed in wild and domestic Spanish animals such as wild boar, red deer (*Cervus elaphus*), Iberian pigs, red foxes, and others [8, 10]. Accordingly, 0.2 % yearling red deer from south-western Spain had antibodies against WNV or cross-reacting flaviviruses [27].

As red deer is an important game species in Spain [28] and is also frequently farm raised, it could be an easily accessible, cost-effective species to use as a complementary tool to the national surveillance programs to monitor the activity of mosquito-borne flaviviruses [27]. While the analysis of samples from yearlings is useful for continuous surveillance and early warning systems, the (retrospective) analysis of serum samples of individuals of all age classes is a useful tool to explore temporal and spatial trends of flavivirus activity in a given region [23].

The aim of this study was to monitor seroprevalence of WNV and antigenically-related mosquito and tick-borne flaviviruses, particularly USUV and MBV, in wild ruminants in Spain, and, using red deer as the most distributed, abundant and readily available species, to assess the spatial–temporal trends and risk factors associated with the exposure to these flaviviruses in this species.

Methods

Study area

Samples of 4693 wild ruminants, including 3073 free-living and 1262 farmed red deer, 201 fallow deer (*Dama dama*), 125 mouflon (*Ovis aries musimon*) and 32 roe deer (*Capreolus capreolus*) were collected between 2003 and 2014 from hunting estates across Spain. Red deer were sampled in 130 hunting areas located throughout the five geographical bioregions of Spain previously defined for the national wildlife disease monitoring program [29, 30]: (1) Atlantic coast, characterized by a wet, temperate climate and abundant rainy seasons (32 sampling areas); (2) cereal plains, in which agriculture with cereal crops are dominant (7 sampling areas); (3) Continental Mediterranean ecosystem, with cold winters, hot dry summers, and rainy seasons in autumn and spring (70 sampling areas); (4) Interior mountains, consisting of small mountain chains with a continental Mediterranean climate (9 sampling areas); (5) and south coast, with a humid coastal Mediterranean climate, warm humid winters and hot dry summers (12 sampling areas). Samples of roe deer, fallow deer, and mouflon, were obtained only from bioregions 3 and 5 (south-central Spain).

Sampling

Blood samples were taken from the thoracic cavity of freshly killed hunted animals or during health inspection in the slaughterhouse. Samples of farmed red deer were obtained by puncture of the jugular vein using a sterile collection system during health inspections. All samples of free-living ruminants were grouped by season and habitat type (sites associated with large permanent wetlands i.e., rivers, lagoons or lakes, and areas without wetlands or large water bodies). Samples were classified by age according to dentition patterns (yearlings, sub-adults, adults) [31]. The animals sampled were classified according to gender and status (farmed or free-living). When kidney samples were available from red deer, the right kidney fat index (KFI) was obtained as an indicator of body condition [32]. Upon arrival at the laboratory, blood samples were centrifuged for 15 min at 1,800X *g* for serum separation and serum was stored at −20 ° C until analysis.

Laboratory analysis

All ruminants were tested for antibodies against an epitope of the WNV pre-membrane-envelope (prM-E) protein shared with other viruses of the Japanese encephalitis serocomplex. A commercially available blocking enzyme-linked immunosorbent assay (bELISA 10.WNV.K3 INGEZIM West Nile COMPAC®, Ingenasa, Madrid, Spain) was used in accordance with the manufacturer's recommendations. The bELISA was used as a serological screening tool and bELISA-positive and doubtful sera were confirmed by micro virus neutralization test (VNT) for the detection of specific neutralizing antibodies against WNV (Is98 strain, lineage 1) and USUV (It12 strain). Additionally, given the possibility of cross-reaction with antigenically-related flaviviruses not included within the Japanese encephalitis serocomplex such as Meaban virus (MBV), bELISA-positive and doubtful samples were also tested by VNT against this flavivirus (MBV; Brest ART707 strain). VNTs were performed as previously described [3, 33]. Samples that showed neutralization and absence of cytopathic effect at dilutions ≥ 10 for WNV and USUV and ≥ 20 for MBV were considered positive. Interpretation of results was based on comparison of VNT titers obtained in parallel against the three flaviviruses. The neutralizing immune response observed was considered specific when VNT titers for a given virus was ≥ 4-fold higher than titers obtained for the other viruses. Samples showing VNT titers differences ≤ 2-fold between the viruses examined were considered positive for flaviviruses but not conclusive for any specific virus.

Statistical analysis

The prevalence of antibodies against antigenically-related flaviviruses was estimated from the ratio of positives to the total number of samples, with the exact binomial confidence intervals of 95 %. Differences between species were analyzed using a Pearson's Chi-square test and a Fisher's exact test (when observations/category were <6). Due to the absence of seropositivity in roe deer and to the limited number of seropositive fallow deer and mouflon, the associated risk factors were only analyzed for red deer, the most widely distributed and abundant wild ruminant species in Spain. A Chi-square or Fisher's exact test were used to test the relevance of the explanatory variables (age class, gender, season, year, status, bioregion and wetland area) to the risk of an red deer being exposed to flaviviruses. Covariates correlated with a P-value < 0.20 in the bivariate analysis were included for further analysis. Biologically plausible confounding factors were assessed using Mantel-Haenszel analysis and confounding was considered to be potentially significant if odd ratios (ORs) shifted appreciably. Variables that altered the coefficients for the independent variables of interest by 30 % or more were removed from the model and were classified as confounding factors. Finally, a multiple logistic regression analysis [34] was performed including risk factors potentially associated with related flavivirus exposure (likelihood-ratio Wald's test, $P < 0.05$). The goodness of fit was assessed using the Hosmer–Lemeshow goodness-of-fit test. SPSS 22.0 software (IBM Corp., Armonk, NY, USA) was used for statistical analyses.

Results

A total of 153 out of the 4693 (3.3 %, $CI_{95\%}$: 2.7–3.8 %) wild ruminants were seropostive against antigenically-related flaviviruses using bELISA. A mean seroprevalence of 3.4 ± 0.5 % (148/4335) was obtained in red deer, 1.0 ± 1.4 % (2/201) in fallow deer, and 2.4 ± 2.7 % (3/125) in mouflon. No seropositivity was detected in the 32 roe deer tested. Statistically significant differences in seroprevalence among species were not observed.

Five (year, bioregion, wetland area, season and status) out of seven variables were selected ($P < 0.20$) from the bivariate analysis (Table 1). At least one seropositive red deer was detected in every year analyzed, with a significantly higher seroprevalence detected in 2011 (17.6 %; 44/250; $P < 0.001$). Forty-one out of the 130 (31.5 %) sampling areas presented at least one seropositive animal. The seropositivity by sampling areas was 3.1 % (1/32) in bioregion 1, 42.9 % (3/7) in bioregion 2, 44.3 % (31/70) in bioregion 3, 11.1 % (1/9) in bioregion 4 and 41.7 % (5/12) in bioregion 5. Even though seropositivity was detected in all bioregions, considering farmed and free-living red deer together, a significantly

Table 1 Seroprevalence to WNV and antigenically-related flaviviruses in wild ruminants in Spain

Variable	Categories	No. examined[a]/positive (%)	P-value
Species	Red deer	4335/148 (3.4)	0.174
	Fallow deer	201/2 (1.0)	
	Mouflon	125/3 (2.4)	
	Roe deer	32/0 (0.0)	
Age class	Juveniles	56/1 (1.8)	0.628
	Sub-adults	221/8 (3.6)	
	Adults	607/26 (4.3)	
Sex	Females	949/24 (2.5)	0.217
	Males	893/29 (3.2)	
Bioregion	1	437/1 (0.2)	<0.001
	2	506/3 (0.6)	
	3	2164/61 (2.8)	
	4	230/4 (1.7)	
	5	998/79 (7.9)	
Year	2003	502/9 (1.8)	<0.001
	2004	179/1 (0.6)	
	2005	338/5 (1.5)	
	2006	446/12(2.7)	
	2007	664/13 (2.0)	
	2008	551/15 (2.7)	
	2009	555/21 (3.8)	
	2010	576/20 (3.5)	
	2011	250/44 (17.6)	
	2012	74/1 (1.4)	
	2013	140/5 (3.6)	
	2014	60/2 (3.3)	
Season	Autumn	1754/52 (3.0)	0.017
	Spring	377/10 (2.7)	
	Summer	503/29 (5.8)	
	Winter	1701/57 (3.4)	
Wetland area	Presence	1634/109 (6.7)	<0.001
	Absence	2701/39 (1.4)	
Status	Free-living	3073/74 (2.4)	<0.001
	Farmed	1262/74 (5.9)	

[a]Missing values excluded

higher seroprevalence was found in bioregion 5 (7.9 %; 79/998) compared to the other bioregions ($P < 0.001$), and in bioregion 3 (2.8 %; 61/2164) compared to bioregion 1 (0.2 %; 1/437) ($P < 0.001$). Significantly higher seropositivity was also detected in bioregion 5 (3.0 %; 8/269) ($P = 0.003$), bioregion 4 (3.0 %; 4/134) ($P = 0.012$) and bioregion 3(2.9 %; 59/2017) ($P < 0.001$) compared to bioregion 1 (0.2 %; 1/437) when seroprevalence was compared only in free-living red deer. A significantly higher seroprevalence was obtained in summer (5.8 %; 29/503) as compared to autumn (3.0 %; 52/1754) ($P = 0.003$) and winter (3.4 %; 57/1701) ($P = 0.012$). Seroprevalence was significantly higher in farmed 5.9 % (74/1262) than in free-living 2.4 ± 0.5 % (74/3073) red deer ($P < 0.001$). Seropositivity was confirmed in three out of the five red deer farms tested, with the mean seroprevalence ranging from 0.4 % (1/243) to 9.7 % (71/729). In the red deer farm with the highest seropositivity, a significantly higher mean seroprevalence was found in autumn (18.5 %; 20/108) compared to spring (4.6 %; 6/131; $P = 0.002$) and summer (8.4 %; 26/309; $P = 0.012$), but not in winter (10.5 %; 19/181; $P = 0.113$). The final multivariate logistic regression model showed that the main risk factors associated with exposure to related flaviviruses in red deer were: year (2011), bioregion (5) and wetland areas (presence) (Table 2).

Table 2 Logistic regression model of potential risk factors associated with seroprevalence to WNV and antigenically-related flaviviruses in red deer in Spain

Variable	Categories	B	P-value	OR	95 % CI	
Year	2003	a	a	a	a	a
	2004	−1.174	0.271	0.309	0.038	2.496
	2005	−0.028	0.962	0.972	0.313	3.020
	2006	0.294	0.525	1.342	0.542	3.322
	2007	0.223	0.627	1.250	0.509	3.072
	2008	0.430	0.332	1.538	0.645	3.669
	2009	0.509	0.219	1.663	0.739	3.743
	2010	0.386	0.355	1.471	0.649	3.335
	2011	1.991	<0.001	7.320	3.404	15.739
	2012	−0.741	0.487	0.477	0.059	3.851
	2013	0.077	0.892	1.081	0.353	3.306
	2014	0.608	0.447	1.837	0.383	8.822
Bioregion	1	a	a	a	a	a
	2	0.602	0.605	1.826	0.186	17.946
	3	1.913	0.060	6.777	0.920	49.903
	4	1.784	0.115	5.954	0.648	54.709
	5	2.192	0.034	8.957	1.180	67.990
Wetland area	Absence	a	a	a	a	a
	Presence	1.097	<0.001	2.995	1.888	4.751

[a]Reference category

Testing by VNT was possible in 140 out of the 153 bELISA seropositive wild ruminants, while 13 samples (11 from red deer and 2 from fallow deer) could not be analyzed by VNT due to serum cytotoxicity. Of these, 25 (22 from red deer and 3 from mouflon) showed negative results against the three flaviviruses tested using VNT. Specific antibodies against WNV, USUV and MBV were confirmed in 103 (69.7 %; bioregions 2, 3, 4 and 5), 4 (2.7 %; bioregions 1, 3 and 5) and 2 (1.4 %; bioregion 3) red deer, respectively (Table 3, Fig. 1). Six red deer (five from bioregion 3 and one from bioregion 5) showed ≤ 2-fold VNT titer differences between WNV and USUV and were therefore considered positive for other related flaviviruses but not conclusive for any of these viruses. Taking into account possible co-infections, the overall antibody prevalence ranged between 2.2 and 2.4 % for WNV, 0.1 and 0.2 % for USUV and 0.04 % for MBV. Specific WNV, USUV and MBV antibodies were confirmed in 31, four and one sampling areas, respectively. WNV and USUV and WNV and MBV co-circulation (different individuals from the same sampling area with specific antibodies against WNV, USUV or MBV) were detected in two (bioregions 3 and 5) and one (bioregion 3) sampling areas, respectively (Fig. 1).

Discussion

The overall prevalence of antibodies against WNV and antigenically-related flaviviruses detected in wild ruminants (3.3 %) in the present study was consistent with that recently observed in captive zoo artiodactyls in Spain [15]. Similar seroprevalences were also found in white-tailed deer in the United States [22, 23]. In contrast, a limited number of wild ruminants from south-western Spain tested during surveillance of WNV outbreaks were negative for flavivirus antibodies using the same bELISA [8]. In our study, no antibodies were detected in roe deer, potentially due to the small sample size, while a similar study in the Czech Republic reported flavivirus exposure in roe deer, mouflons, fallow deer and red deer [24]. Remarkably, wild ruminants and specifically red deer showed lower seroprevalence to

Table 3 Virus neutralization test (VNT) titers obtained in parallel against West Nile virus (WNV), Usutu virus (USUV) and Meaban virus (MBV) from 115 sera positive by bELISA

VNT Titers	WNV	USUV	MBV	WNV and/or USUV
10	28	3		4 (WNV); 3 (USUV)
20	31		2	1 (WNV); 2 (USUV)
40	18	1		
80	12			
160	8			
320	3			1 (USUV)
≥640	3			1 (WNV)

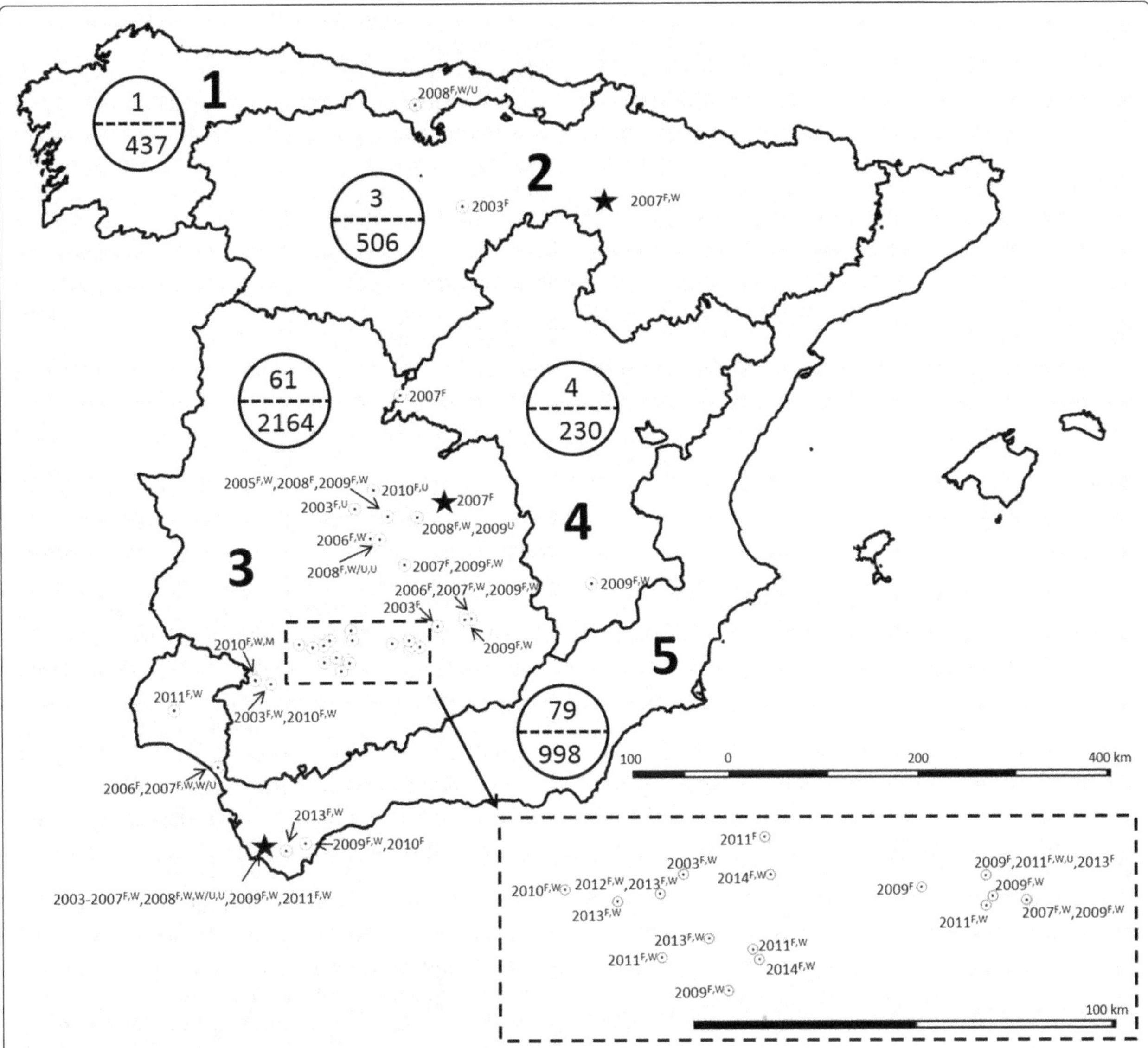

Fig. 1 Detection of antibodies to WNV and antigenically-related flaviviruses in red deer in the five different bioregions of Spain (large circles positives/n tested) and spatio-temporal distribution of exposure to flaviviruses, WNV, USUV and Meaban virus. *Open circles* represent hunting estates (only the 41 positive hunting estates are included), *black stars* the red-deer farms. Years in which antibody positive animals were detected are listed with superscript letters that list the pathogens against which antibodies were present: F = Flavivirus ELISA positive, W = WNV, W/U = WNV/USUV, U = USUV, M = MBV

WNV and related flaviviruses than other artiodactyl species such as wild boar from the same regions and periods [10, 27]. In North America, *Culex* family mosquitoes (flavivirus vector) have shown a preference of feeding on deer [35]. Similarly, *Culex* species sampled in Canary Islands (Spain) frequently fed on ruminant species [36], while a study in wetlands of bioregion 5 did show feeding on mammals in addition to birds and reptiles but without identifying the mammal hosts beyond humans and horses [37]. Thus, our results would suggest a reduced susceptibility of red deer to flaviviruses of the encephalitis antigenic complex and other cross reacting flaviviruses.

Multivariate logistic regression identified the year 2011, bioregion 5 and the presence of wetland areas as individual risk factors for antigenically-related flavivirus exposure in red deer. The seroprevalence data obtained indicates widespread but not homogeneous distribution of these flaviviruses in wild ruminants in Spain. Consistently with other studies, seroprevalence of flaviviruses in both farmed and free-living wild ruminants, mainly WNV, was higher in southern Spain, namely in bioregions 3, 4 and 5 [10, 38, 39]. In fact, bioregion 5 includes the provinces with the highest number of WNV outbreaks reported in horses and humans and the area where BAGV infection caused high mortality in wild game birds in 2010 [18, 40]. The results indicate that southern regions present the highest risk of flavivirus circulation in Spain. Even though a study conducted in 1980 described the presence of antibodies against flaviviruses in rodents in northern Spain [41], the extremely low prevalence (0.2 %) of flavivirus antibodies in north Spain (bioregion 1), which has a colder mean temperature [42], suggests lower activity of flaviviruses in this area.

Animals sampled in wetland areas were three times more likely to be exposed to related flaviviruses, which may be due to the larger populations and greater diversity of competent vectors. In addition, in late summer, the number of wetlands and waterholes in southern Spain becomes greatly reduced, aggregating animals in these hotspots [43], which may also act as breeding areas for mosquito species. Our results are consistent with those previously reported in France [44], and indicate that wetland areas may help to target risk-based surveillance in Spain.

Seropositivity of at least one animal in all sample years supports a more-or-less continuous circulation of antigenically-related flaviviruses in the study period. However, a significantly higher seroprevalence was found in 2011, 1 year after the highest number of WNV and Bagaza outbreaks were reported in Spain [18, 40]. In 2011, seroprevalence was detected in adult and sub-adult deer but not in yearling animals suggesting exposure to related flaviviruses of the wild ruminant populations at the end of 2010 and persistence of antibodies in infected animals. Accordingly, a seropositive yearling free-living red deer sampled in 2010 was positive to WNV by VNT, confirming WNV circulation this year. Similar results were obtained in juvenile red deer from a farm in the 2010 outbreak area [27]. Our results are in agreement with previous studies in which an increase in WNV seroprevalence was observed in red deer and white-tailed deer 2 years after the first outbreak occurred in the area [23, 27].

Intrinsic host factors such as age can influence exposure and susceptibility to flavivirus infection. In the present study, seroprevalence increased with age but no significant differences among age classes were observed. The results may be explained by the limited number of samples from yearlings analyzed, due to the fact that yearlings are not usually hunted. A low seroprevalence in yearling red deer from Spain was also previously reported by [27]. Moreover, lower seroprevalences were detected in yearling while-tailed deer but no association with age was observed [22, 23]. Our findings are also consistent with the increase in WNV and related flavivirus seroprevalence with age found in wild boar [10]. Although flavivirus antibody persistence in artiodactyl species is unknown, a longer time span of possible exposure and antibody persistence could be associated with the higher seroprevalence found in adult individuals. Further research with larger sample sizes for the different age classes is needed to have a better understanding about the association between age and flavivirus infection.

Significantly higher seroprevalence to antigenically-related flavivirus was found in farmed red deer as compared to free-living ones, although this variable was not retained in the multivariate analysis. In contrast, free-living red deer sampled in open and fenced sampling areas did not differ in seroprevalence. Our results indicate that samples from farmed red deer, which are accessible during all year, may be a valuable surveillance tool. Variations in the seroprevalence among seasons were also detected in the bivariate analysis, with significantly higher seroprevalence in summer. Moreover, significantly higher mean seroprevalence was found in autumn in farmed red deer sampled in bioregion 5. Most of the WNV outbreaks in horses in Spain have been reported in late summer and autumn [40], confirming the higher risk of flavivirus infection during these seasons. In addition, the higher seropositivity observed in autumn in farmed deer is also in agreement with what has been reported in wild birds in Spain [5, 6].

To the authors' knowledge, this is the first report on WNV, USUV and MBV antibodies in red deer, which confirms local circulation of these flaviviruses in Spain. Twenty two sera were positive by bELISA but negative to USUV, WNV and MBV in VNT. The results could be due to the circulation of other related flaviviruses [45]. Mortality due to BAGV and LIV infections have recently been detected in wild game birds and ruminant species respectively in Spain [16, 18]. Further studies are needed to determine the possible impact of these flaviviruses on wild ruminant populations. Unfortunately, VNT against BAGV and LIV could not be carried out in the present study. Six red deer presented positive results to WNV and USUV by VNT with titer differences ≤ 2-fold. This could be due to VNT cross reactions, however co-infections cannot be ruled out. In fact, both WNV and USUV as well as WNV and MBV co-circulation were confirmed (in two and one sampling areas, respectively). These results are of interest because a previous infection with a closely related flavivirus might confer cross-reactive immunity in the animal host, reducing the amplification of new strains with higher virulence introduced into the same region [46]. In fact, co-circulation of different flaviviruses, and the induction of cross-protective antibodies, has been suggested as an explanation for the high rate of seropositivity to WNV in birds and the limited number of cases reported in humans [46]. In Spain, both WNV lineage 1 and the putative new lineage, as well as BAGV, USUV and MB virus have been confirmed [47]. In contrast, the understanding of flavivirus co-infections both in host and vector species is still very limited.

Specific antibodies against WNV were confirmed in four out of the five bioregions. The higher WNV seroprevalence in bioregion 3 and 5 is in accordance with WNV distribution reported in horses in Spain [40]. Interestingly,

WNV antibodies were found in animals in areas where WNV outbreaks have not been reported to date. Our results constitute the first report of WNV circulation in bioregion 4, where habitat conditions do not favour WNV activity. Yearly re-emergence of WNV in Mediterranean countries is thought to be due to WNV overwintering and becoming endemic in local bird populations [48, 49]. The presence of animals seropositive to WNV throughout the sampling period (2003–2014) suggest an endemic circulation of WNV in southern Spain. In fact, after WNV was first reported in horses from southern Spain in 2010, outbreaks in this species have been reported every year [40].

During the last decade, USUV has rapidly expanded in different European countries, and has been detected in mosquitoes and vertebrate species [50]. Even though human infections are believed to be clinically mild or asymptomatic, neuroinvasive illness associated with USUV infection has been reported in Italy and Croatia [51, 52]. In our study, USUV infection was confirmed in red deer from bioregions 1, 3 and 5 in 2007 and 2008. The seroprevalence found in this species, as well as the USUV infections previously reported in mosquitoes, resident and migratory wild birds [4, 14, 15, 53], indicate circulation of this emerging flavivirus during the last decade in Spain. Further epidemiological and molecular research is required to determine the role of wild ruminants in the epidemiology of USUV and to determine USUV strains and lineages circulating in Spain.

Infection with MBV or a Meaban-like virus has been recently detected in yellow-legged and herring (*Larus argenta tus*) gulls from different breeding colonies in France and Spain [3], and in a great frigatebird (*Fregata minor*) and a sooty tern (*Onychoprion fuscatus*) in the Western Indian Ocean [54], although the zoonotic potential of this emerging flavivirus is still unknown. Two adult free-living red deer sampled in Seville province (bioregion 3) in 2010 showed antibodies against this flavivirus, which indicates that MBV is also circulating among wild ruminants in Spain. Interestingly, seropositivity against MBV has been detected precisely in the same province and year in a recent study carried out in waterfowl used as decoys in Spain (unpublished data), which suggests limited spread of this flavivirus in the study area. This result is consistent with a previous report of MBV infection in only two specific locations within the Mediterranean Basin [3].

Conclusions

The results obtained indicate local circulation of antigenically-related flaviviruses in wild ruminants in Spain. The significance of WNV, USUV and MBV infection in this species is unknown at this time and further research on aspects such as the genetic diversity, infection rates and competent vectors of these emerging flaviviruses is needed. Surveillance of hunted wild ruminants, and especially on farmed red deer, may be a useful, cost-effective, rapid and complementary tool to the national surveillance programs to monitor the activity of mosquito and tick-borne flaviviruses in Europe.

Abbreviations

BAGV: Bagaza virus; ELISA: Enzyme-linked immunosorbent assay; KFI: Kidney fat index; LIV: Louping-ill virus; MBV: Meaban virus; ORs: Odd ratios; SLEV: Saint Louis encephalitis virus; TBEV: Tick borne encephalitis virus; USUV: Usutu virus; VNT: Virus neutralization test; WNV: West Nile virus

Acknowledgements

We acknowledge the help of fellow students with sample collection in hunting drives. We are indebted to Dr. Pelayo Acevedo for his help with Fig. 1.

Funding

This study has been supported by projects PAC08-0296-7771 (JCCM), AG2008-02504GAN, FAU2008-00019-C03-01 and AGL2013-49159-C2-2-R. Gutierrez-Guzman AV, was a JCCM fellow (PAC08-0296-7771).

Authors' contributions

Conceived and designed the experiment: IGB, JP, AVGG, MB, AAM, DCT, CG, UH. Performed the experiment: IGB, JP, AVGG, MB, AAM, DCT, CG, UH. Contributed to/carried out analyses: IGB, JP, AVGG, SL, MB, AAM, DCT, SL, CG, UH. Performed statistical analysis: IGB, UH. Drafted and amended the manuscript: IGB, JP, AVGG, MB, AAM, DCT, CG, UH. All authors read and approved the final manuscript.

Competing interests

The authors declare that they have no competing interests.

Consent for publication

Not applicable.

Ethics approval and consent to participate

This study did not involve purposeful killing of animals. Samples from dead animals were collected from individuals legally hunted during the hunting seasons. These animals were legally hunted under Spanish and EU legislation and all hunters had hunting licenses. Blood from farmed live individuals was obtained from archive samples previously collected during routine handling procedures. None of the blood samples was collected specifically for this study, therefore, no ethical approval was deemed necessary. The collection of all samples was performed following routine procedures before the design of this study in compliance with the Ethical Principles in Animal Research. Protocols, amendments and other resources were done according to the guidelines approved by each Autonomous government following the R.D.1201/2005 of the Ministry of Presidency of Spain (10th October 2005, BOE 21st October 2005) (http://www.umh.es/_web_rw/ceie/docs/animales/1201_05%20proteccion%20animales%20experimentacion.pdf).

Author details

[1]Departamento de Sanidad Animal, Facultad de Veterinaria, Universidad de Córdoba-Agrifood Excellence International Campus (ceiA3), Rabanales, 14071 Córdoba, Spain. [2]Instituto de Investigación en Recursos Cinegéticos IREC, (CSIC-UCLM-JCCM), Ciudad Real, Spain. [3]ANSES, Laboratoire de Santé Animale de Maisons-Alfort, UMR 1161 Virologie, INRA, ANSES, ENVA, Maisons-Alfort F-94703, France. [4]Sabiotec, Camino de Moledores s.n., Ed. Polivalente UCLM, 13005 Ciudad Real, Spain.

References

1. Weissenböck H, Hubálek Z, Bakonyi T, Nowotny N. Zoonotic mosquito-borne flavivirus: worldwide presence of agents with proven pathogenicity and potencial candidates of future emerging diseases. Vet Microbiol. 2010;140:271–80.
2. Beck C, Jiménez-Clavero MA, Leblond A, Durand B, Nowotny N, Leparc-Goffart I, Zientara S, Jourdain E, Lecollinet S. Flaviviruses in Europe: complex circulation patterns and their consequences for the diagnosis and control of West Nile disease. Int J Environ Res Public Health. 2013;10:6049–83.
3. Arnal A, Gómez-Díaz E, Cerdà-Cuéllar M, Lecollinet S, Pearce-Duvet J, Busquets N, García-Bocanegra I, Pagès N, Vittecoq M, Hammouda A, Samraoui B, Garnier R, Ramos R, Selmi S, González-Solís J, Jourdain E, Boulinier T. Circulation of a meaban-like virus in yellow-legged gulls and seabird ticks in the Western Mediterranean basin. PLoS ONE. 2014;9, e89601.
4. Vázquez A, Jiménez-Clavero MA, Franco L, Donoso-Mantke O, Sambri V, Niedrig M, Zeller H, Tenorio A. Usutu virus - potential risk of human disease in Europe. Euro Surveill. 2011;16:22–6.
5. Figuerola J, Jiménez-Clavero MA, López G, Rubio C, Soriguer R, Gomez-Tejedor C, Tenorio A. Size matters: West Nile virus neutralizing antibodies in resident and migratory birds in Spain. Vet Microbiol. 2008;132:39–46.
6. García-Bocanegra I, Busquets N, Napp S, Alba A, Zorrilla I, Villalba R, Arenas A. Serosurvey of West Nile virus and other flaviviruses of the Japanese encephalitis antigenic complex in birds from Andalusia, Southern Spain. Vector Borne Zoonotic Dis. 2011;11:1107–13.
7. Bernabeu-Wittel M, Ruiz-Pérez M, del Toro MD, Aznar J, Muniain A, de Ory F. West Nile virus past infections in the general population of Southern Spain. Enferm Infecc Microbiol Clin. 2007;25:561–5.
8. García-Bocanegra I, Jaén-Téllez JA, Napp S, Arenas-Montes A, Fernández-Morente M, Fernández-Molera V, Arenas A. Monitoring of the West Nile Virus epidemic in Spain between 2010 and 2011. Transbound Emerg Dis. 2012;59:448–55.
9. Mentaberre G, Gutiérrez C, Rodríguez NF, Joseph S, González-Barrio D, Cabezón O, de la Fuente J, Gortazar C, Boadella M. A transversal study on antibodies against selected pathogens in dromedary camels in the Canary Islands, Spain. Vet Microbiol. 2013;167:468–73.
10. Gutiérrez-Guzmán A, Vicente J, Sobrino R, Perez-Ramírez E, Llorente F, Höfle U. Antibodies to West Nile virus and related flaviviruses in wild boar, red foxes and other mesomammals from Spain. Vet Microbiol. 2012;159:291–7.
11. Jiménez-Clavero MA, Sotelo E, Fernandez-Pinero J, Llorente F, Blanco JM, Rodriguez-Ramos J, Perez-Ramirez E, Höfle U. West Nile virus in golden eagles, Spain, 2007. Emerg Infect Dis. 2008;14:1489–91.
12. García-Bocanegra I, Jaén-Téllez JA, Napp S, Arenas-Montes A, Fernández-Morente M, Fernández-Molera V, Arenas A. West Nile outbreak in horses and humans, Spain, 2010. Emerg Infect Dis. 2011;17:2397–9.
13. Busquets N, Alba A, Allepuz A, Aranda C, Núñez JE. Usutu virus sequences in *Culex pipiens* (Diptera: *Culicidae*), Spain. Emerg Infect Dis. 2008;14:861–3.
14. Höfle U, Gamino V, Fernández de Mera IG, Mangold AJ, Ortíz JA, de la Fuente J. Usutu virus in migratory song thrushes, Spain. Emerg Infect Dis. 2013;19:1173–5.
15. Cano-Terriza D, Guerra R, Lecollinet S, Cerdà-Cuéllar M, Cabezón O, Almería S, García-Bocanegra I. Epidemiological survey of zoonotic pathogens in feral pigeons (*Columba livia* var. *domestica*) and sympatric zoo species in Southern Spain. Comp Immunol Microbiol Infect Dis. 2015;43:22–7.
16. Balseiro A, Royo LJ, Pérez-Martínez C, Fernández de Mera IG, Höfle U, Polledo L, Marreros N, Casais R, García-Marín JF. Louping Ill in goats, Spain, 2011. Emerg Infect Dis. 2012;18:976–8.
17. Ruiz-Fons F, Balseiro A, Willoughby K, Oleaga Á, Dagleish MP, Pérez-Ramírez E, Havlíková S, Klempa B, Llorente F, Martín-Hernando MP. Clinical infection of Cantabrian chamois (*Rupicapra pyrenaica parva*) by louping ill virus: New concern for mountain ungulate conservation? Eur J Wildlife Res. 2014;60:691–4.
18. García-Bocanegra I, Zorrilla I, Rodríguez E, Rayas E, Camacho L, Redondo I, Gómez-Guillamón F. Monitoring of the Bagaza virus epidemic in wild bird species in Spain, 2010. Transbound Emerg Dis. 2013;60:120–6.
19. Puerto FI, Lorono-Pino MA, Farfan-Ale JA, García-Rejón JE, Cetina-Trejo RC, Hidalgo-Martinez AC, Ramos C, Rosado-Paredes EP, Flores-Flores LF. Captive animals as sentinels for West Nile virus transmission in zoos from Yucatan and Tabasco states of Mexico. Am J Trop Med Hyg. 2007;77:134.
20. Nitatpattana N, Le Flohic G, Thongchai P, Nakgoi K, Palaboodeewat S, Khin M, Barbazan P, Yoksan S, González JP. Elevated Japanese encephalitis virus activity monitored by domestic sentinel piglets in Thailand. Vector Borne Zoonotic Dis. 2011;11:391–4.
21. Phoutrides E, Jusino-Mendez T, Pérez-Medina T, Seda-Lozada R, Garcia-Negron M, Davila-Toro F, Hunsperger E. The utility of animal surveillance in the detection of West Nile Virus activity in Puerto Rico. Vector Borne Zoonotic Dis. 2011;11:447–50.
22. Farajollahi A, Gates R, Crans W, Komar N. Serologic evidence of West Nile virus and St. Louis encephalitis virus infections in while-tailed deer (*Odocoileus virginianus*) from New Jersey, 2001. Vector Borne Zoonotic Dis. 2004;4:379–83.
23. Santaella J, Mclean R, Hall JS, Gill JS, Bowen RA, Hadow HH, Clark L. West Nile virus serosurvaillance in Iowa White-tailde deer (199–2003). Am J Trop Med Hyg. 2005;73:1038–42.
24. Juricova Z, Hubalek Z. Serological surveys for arboviruses in the game animals of southern Moravia (Czech Republic). Folia Zoo. 1999;48:185–9.
25. Palmer MV, Stoffregen WC, Rogers DG, Hamir AN, Richt JA, Pedersen DD. West Nile virus in reindeer (*Rangifer tarandus*). J Vet Diagn Invest. 2004;16:219–22.
26. Miller DL, Radi ZA, Baldwin C, Ingram D. Fatal West Nile virus infection in a white-tailed deer (*Odocoileus virginianus*). J Wildlife Dis. 2005;41:246–9.
27. Boadella M, Díez-Delgado I, Gutiérrez-Guzmán AV, Höfle U, Gortázar C. Do wild ungulates allow improved monitoring of flavivirus circulation in Spain? Vector Borne Zoonotic Dis. 2012;12:490–5.
28. Acevedo P, Ruiz-Fons F, Vicente J, Reyes-García AR, Alzaga V, Gortázar C. Estimating red deer abundance in a wide range of management situations in Mediterranean habitats. J Zool. 2008;276:37–47.
29. Muñoz P, Boadella M, Arnal M, de Miguel M, Revilla M, Martinez D, Vicente J, Acevedo P, Oleaga A, Ruiz-Fons F, Marín CM, Prieto JM, de la Fuente J, Barral M, Barberán M, Fernández de Luco D, Blasco JM, Gortázar C. Spatial distribution and risk factors of Brucellosis in Iberian wild ungulates. BMC Infect Dis. 2010;10:46.
30. MMARM, (Ministerio de Medio Ambiente y Medio Rural y Marino de España), Plan Nacional de Vigilancia Sanitaria en Fauna Silvestre. 2016. http://rasve.magrama.es/publica/programas/NORMATIVA%20Y%20PROGRAMAS%5CPROGRAMAS%5CFAUNA%20SILVESTRE%5CPLAN%20NACIONAL%20DE%20VIGILANCIA%20EN%20FAUNA%20SILVESTRE.PDF. Accessed 15 Apr 2016.
31. Sáenz de Buruaga M, Lucio-Calero A, Purroy-Iraizoz FJ. Reconocimiento de Sexo y Edad en Especies Cinegéticas. 1st ed. Spain: Edilesa; 2001.
32. Santos JPV, Fernández-de-Mera IG, Acevedo P, Boadella M, Fierro Y, Vicente J, Gortázar C. Optimizing the sampling effort to evaluate body condition in ungulates: a case study on red deer. Ecol Ind. 2013;30:65–71.
33. Chaintoutis SC, Dovas CI, Papanastassopoulou M, Gewehr S, Danis K, Beck C. Evaluation of a West Nile virus surveillance and early warning system in Greece, based on domestic pigeons. Comp Immunol Microbiol Infect Dis. 2014;37:131–41.
34. Hosmer DW, Lemeshow S. Applied Logistic Regression. 2nd ed. New York: Wiley; 2000.
35. Watts SL, Fitzpatrick DM, Maruniak JE. Blood meal identification from Florida mosquitoes (Diptera: *Culicidae*). Florida Entomol. 2009;92:619–22.
36. de la Martínez- Puente J, Martínez J, Ferraguti M, de la Morales- Nuez A, Castro N, Figuerola J. Genetic characterization and molecular identification of the bloodmeal sources of the potential bluetongue vector *Culicoides obsoletus* in the Canary islands, Spain. Parasit Vector. 2012;5:147.
37. Muñoz J, Ruiz S, Soriguer R, Alcaide M, Viana DS, David R, Vázquez A, Figuerola J. Feeding patterns of potential West Nile virus vectors in South-West Spain. PLoS ONE. 2012;7, e39549.
38. Figuerola J, Jiménez-Clavero MA, Rojo G, Gómez C, Soriguer R. Prevalence of West Nile virus neutralizing antibodies in colonial aquatic birds in southern Spain. Avian Pathol. 2007;36:209–12.
39. López G, Jiménez-Clavero MA, Vázquez A, Soriguer R, Gómez-Tejedor C, Tenorio A, Figuerola J. Incidence of West Nile virus in birds arriving in wildlife rehabilitation centers in Southern Spain. Vector Borne Zoonotic Dis. 2010;11:285–90.
40. RASVE (Red de Alerta Sanitaria Veterinaria). 2016. http://rasve.magrama.es/RASVE_2008/Publica/Focos/Consulta.aspx. Accessed 15 May 2016.
41. Chastel C, Launay H, Rogues G, Beaucornu JC. Infections à arbovirus en Espagne; enquête sérologique chez les petits mammiferes. Bull Soc Path Exot. 1980;73:384–90.
42. Ninyerola M, Pons X, Roure JM. Atlas climático digital de la Península Ibérica. Metodología y aplicaciones en Bioclimatología y geobotánica. 2005. http://opengis.uab.es/wms/iberia/pdf/acdpi.pdf. Accessed 3 May 2016.
43. Barasona JA, Latham MC, Acevedo P, Armenteros JA, Latham ADM, Gortázar C, Carro F, Soriguer RC, Vicente J. Spatiotemporal interactions between wild boar and cattle: implications for cross-species disease transmission. Vet Res. 2014;45:122.

44. Pradier S, Sandoz A, Paul MC, Lefebvre G, Tran A, Maingault J, Lecollinet S, Leblond A. Importance of wetlands management for West Nile Virus circulation risk, Camargue, Southern France. Int J Environ Res Public Health. 2014;11:7740–54.
45. Blitvich BJ, Fernandez-Salas I, Contreras-Cordero JF, Marlenee NL, González-Rojas JI, Komar N. Serologic evidence of West Nile virus infection in horses, Coahuila State, Mexico. Emerg Infect Dis. 2003;9:853–6.
46. Rizzoli A, Jiménez-Clavero MA, Barzon L, Cordioli P, Figuerola J, Koraka P, Martina B, Moreno A, Nowotny N, Pardigon N, Sanders N, Ulbert S, Tenorio A. The challenge of West Nile virus in Europe: knowledge gaps and research priorities. Euro Surveill. 2015;20(20). Available online: http://www.eurosurveillance.org/ViewArticle.aspx?ArticleId=21135.
47. Vázquez A, Sanchez-Seco MP, Ruiz S, Molero F, Hernandez L, Moreno J, Magallanes A, Tejedor CG, Tenorio A. Putative new lineage of West Nile virus, Spain. Emerg Infect Dis. 2010;16:549–52.
48. Monaco F, Savini G, Calistri P, Polci A, Pinoni C, Bruno R, Lelli R. 2009 West Nile disease epidemic in Italy: first evidence of overwintering in Western Europe? Res Vet Sci. 2011;91:321–6.
49. Di Sabatino D, Bruno R, Sauro F, Danzetta ML, Cito F, Iannetti S, Narcisi V, De Massis F, Calistri P. Epidemiology of West Nile disease in Europe and in the Mediterranean Basin from 2009 to 2013. Biomed Res Int. 2014;2014:907852.
50. Ashraf U, Ye J, Ruan X, Wan S, Zhu B, Cao S. Usutu virus: an emerging flavivirus in Europe. Viruses. 2015;7:219–38.
51. Pecorari M, Longo G, Gennari W, Grottola A, Sabbatini A, Tagliazucchi S, Savini G, Monaco F, Simone M, Lelli R, Rumpianesi F. First human case of Usutu virus neuroinvasive infection, Italy, August-September 2009. Euro Surveill. 2009;14. pii: 19446.
52. Santini M, Vilibic-Cavlek T, Barsic B, Barbic L, Savic V, Stevanovic V, Listes E, Di Gennaro A, Savini G. First cases of human Usutu virus neuroinvasive infection in Croatia, August-September 2013: clinical and laboratory features. J Neurovirol. 2015;21:92–7.
53. Llorente F, Pérez-Ramírez E, Fernández-Pinero J, Soriguer R, Figuerola J, Jiménez-Clavero MA. Flaviviruses in Game Birds, Southern Spain, 2011–2012. Emerg Infect Dis. 2013;19:1023–5.
54. Jaeger A, Lecollinet S, Beck C, Bastien M, Corre MLE, Dellagi K, Pascalis H, Boulinier T, Lebarbenchon C. Serological evidence for the circulation of flavivirus in seabird populations of the western Indian Ocean. Epidemiol Infect. 2015;21:1–9.

Development of pooled testing system for porcine epidemic diarrhoea using real-time fluorescent reverse-transcription loop-mediated isothermal amplification assay

Thi Ngan Mai[1,2†], Van Diep Nguyen[1,2†], Wataru Yamazaki[1,3], Tamaki Okabayashi[1,3], Shuya Mitoma[1], Kosuke Notsu[1], Yuta Sakai[1], Ryoji Yamaguchi[1], Junzo Norimine[1,3] and Satoshi Sekiguchi[1,3*]

Abstract

Background: Porcine epidemic diarrhoea (PED) is an emerging disease in pigs that causes massive economic losses in the swine industry, with high mortality in suckling piglets. Early identification of PED virus (PEDV)-infected herd through surveillance or monitoring strategies is necessary for mass control of PED. However, a common working diagnosis system involves identifying PEDV-infected animals individually, which is a costly and time-consuming approach. Given the above information, the thrusts of this study were to develop a real-time fluorescent reverse transcription loop-mediated isothermal amplification (RtF-RT-LAMP) assay and establish a pooled testing system using faecal sample to identify PEDV-infected herd.

Results: In this study, we developed an accurate, rapid, cost-effective, and simple RtF- RT-LAMP assay for detecting the PEDV genome targeting M gene. The pooled testing system using the RtF-RT-LAMP assay was optimized such that a pool of at least 15 individual faecal samples could be analysed.

Conclusions: The developed RtF-RT-LAMP assay in our study could support the design and implementation of large-scaled epidemiological surveys as well as active surveillance and monitoring programs for effective control of PED.

Keywords: PEDV, RtF-RT-LAMP, One-step RT-PCR, Pooled stool samples

Background

Porcine epidemic diarrhoea (PED) is caused by PED virus (PEDV), which is characterized by enteritis, vomiting, and watery diarrhoea. This leads to massive economic losses in the swine industry with high mortality in suckling pigs [1]. PED was first observed in England in 1971 and identified in Belgium in 1978 [2, 3]. The disease quickly spread to other European countries such as Belgium, England, Germany, France, and Switzerland in the 1980s, and later to Asian countries including Korea, China, Thailand, and Vietnam [4, 5]. Recently, several epidemics were reported in important swine-producing countries such as USA, Canada, and Japan [6–8].

Control PED programs require effective and rapid surveillance protocols, linked to prompt control procedures, to ensure that epidemics are brought under control quickly. Currently, the identification of infected herd is done by passive surveillance with required reporting of infected herds from veterinarians. However, veterinarians rely on herd demonstrating clinical signs of infection, which can lead to failure to accurately identify PED status and transmission of PEDV to healthy animals. Moreover, surveillance or monitoring is applied to individuals, which is associated with important financial and time obstacles. Therefore, well-designed surveys as well as sensitive, specific, rapid, and simple detection methods are necessary for the identification of infected herd to control PED. Loop-mediated isothermal amplification (LAMP)

* Correspondence: sekiguchi@cc.miyazaki-u.ac.jp
†Thi Ngan Mai and Van Diep Nguyen contributed equally to this work.
[1]Animal Infectious Disease and Prevention, Department of Veterinary Sciences, Faculty of Agriculture, University of Miyazaki, Miyazaki, Japan
[3]Center for Animal Disease Control, University of Miyazaki, Miyazaki, Japan
Full list of author information is available at the end of the article

combines rapidity, simplicity, and high specificity under isothermal conditions [9, 10]. We developed an accurate, timely, and simple real-time fluorescent reverse transcription LAMP (RtF-RT-LAMP) assay (from M gene) using pooled stool samples for PEDV detection. This assay can be applied to strategies for the control, monitoring, and surveillance of PED.

Methods

Viruses

PEDV NK94P6 and Fukuoka-1 Tr(−) strains which belong to classical clade (G1) were propagated in Vero-KY5 (Vero) cells. Trypsin was not used to culture PEDV in Vero cells. The NK94P6 and Fukuoka-1 strains were kindly provided by the National Institute of Animal Health, Japan, and the Fukuoka Chuo Livestock Hygiene Service Center, Fukuoka, Japan, respectively. The Vero cells were also provided by the National Institute of Animal Health, Japan. Briefly, Vero cells were cultured in Eagle's minimal essential medium (EMEM) (Sigma-Aldrich, Tokyo, Japan) supplemented with 10% (*v*/v) foetal bovine serum (FBS) (Funakoshi, Tokyo, Japan), 0.3% (*w*/*v*) tryptose phosphate broth (TPB) (Sigma-Aldrich), and 100 U/ml penicillin-streptomycin (Wako, Tokyo, Japan) at 37 °C in a humidified atmosphere containing 5% CO_2. Viruses were propagated in Vero cells cultured in EMEM with 2% FBS and 0.3% TPB at 37 °C. The titter of the PEDV NK94P6 and Fukuoka-1 Tr(−) strains were 2.8×10^6 $TCID_{50}$/ml and 2×10^4 $TCID_{50}$/ml, respectively.

Transmissible gastroenteritis coronavirus - TGEV (vaccine strain h-5; Nisseiken, Tokyo, Japan) was propagated in Vero cells; porcine reproductive and respiratory syndrome virus - PRRSV (live PRRS vaccine - Ingelvac PRRS® MLV- Boehringer Ingelheim company); Japanese Encephalitis virus – JEV and Getal virus - GV (live vaccine – Kyoto Biken company, Kyoto, Japan).

Primers

All primers for RtF-RT-LAMP were designed from the highly conserved M gene sequence of porcine epidemic diarrhoea virus CV777 strain (GenBank accession number: KT323979) using the Primer Explorer 4 (https://primerexplorer.jp/lamp4.0.0/index.html) (Additional file 1). They were synthesized using sequence-grade purification by Hokkaido System Science Co., Ltd. (Sapporo, Japan), which included an outer pair (F3, B3), inner primers (FIP, FIP1, BIP), and a loop pair (loopF, loopB) (Table 1). Nucleotide sequences specific for PEDV were detected by multiple alignments of 997 M gene sequences, available from the DDBJ/EMBL/GenBank database. The one-step RT-PCR used a previously published primer pair on the S gene [11].

RNA extraction

The total RNAs were extracted from 250 μl cell culture supernatants of PEDV, TGEV, and PRRSV, JEV, GV using a RNA extraction kit (ReliaPrep™ RNA Cell Miniprep System, Promega, USA), according to the manufacturer's instructions.

One-step RT-PCR

One-step RT-PCR was performed using AccessQuick™ RT-PCR System kits (Promega Corporation, WI, USA) as previously reported [12]. RT-PCR parameters included a reverse transcription step of 45 °C for 45 min and an incubation step of 94 °C for 2 min, 35 cycles at 94 °C for 30 s, 53 °C for 30 s, and 72 °C for 1 min, followed by the final extension at 72 °C for 10 min. The RT-PCR products were visualized by electrophoresis on a 1.5% agarose gel with ethidium bromide.

RtF-RT-LAMP

The RtF-RT-LAMP reaction was conducted in a final reaction volume of 25 μl consisting of 2 μl RNA template, FIP and BIP primers (1.6 μM each), Loop F and Loop B primers (0.8 μM each), F3 and B3 primers (0.2 μM each), Isothermal Mastermixes (OptiGene, UK), 0.15 u of AMV reverse transcriptase (15 u/μl; Invitrogen, USA). Amplification reactions were performed at 63 °C for 40 min (with fluorescence detection followed by melt curve analysis from 90 to 70 °C at 0.05 °C/s), and then

Table 1 Primers used for RtF-RT-LAMP in this study

Primer	ID	Sequence (5′ to 3′)	Gene location
F3	PED_F3_ID1	TCCTTATGGCTTGCATCAC	25,846–25,864
B3	PED_B3_ID1	CCGTAGACAATTGTTGTAGTGG	26,143–26,122
FIP	PED_FIP_ID1, PED_FIP_ID1modified	GTMGGCCCATCACAGAAGTAGTTTT GGTTGTGGCGCAGGACA	25,983–25,963 (TTTT) 25,903–25,919
BIP	PED_BIP_ID1	CCAACTGGTGTAACGCTAACACTTTTT TACCTGTACGCCAGTAGC	26,010–26,032 (TTTT) 26,087–26,070
LF	PED_LF_ID37	TTTCAGGATTGAAAGACCACCAAG	25,947–25,924
LB	PED_LB_ID6	GGTACATTGCTTGTAGAGGGCTATAA	26,040–26,065

M: A or C

heated at a start temperature of 98 °C and end temperature of 80 °C for 10 min with a ramp rate of 0.05 °C/sec to terminate the reactions using Genie® III (OptiGene, UK). The fluorescence of the reaction was measured in real time, verifying the start of the amplification.

Specificity of RtF-RT-LAMP

PEDV (adapted strain NK94P6), PRRSV (vaccine strain MLV), TEGV (vaccine strain h5), JEV (vaccine strain HmLu-SC) and GV (vaccine strain HAL-KB) were used as templates for RtF-RT-LAMP to analyse the specificity of RtF-RT-LAMP. Sterile ddH_2O was used as the negative control.

Sensitivity analysis of the RtF-RT-LAMP

To evaluate the sensitivity of the RtF-RT-LAMP, PEDV-infected Vero cell cultures of two strains (NK94P6 and Fukuoka-1 Tr(−)) with defined median tissue culture infective dose ($TCID_{50}$) was tenfold serial diluted with the supernatant of negative faecal samples. RNA was then extracted from 250 µl media of each dilution and used as a template for RtF-RT-LAMP and one-step RT-PCR as mentioned above.

Real-time RT-PCR

The quantitative One-Step PrimeScript RT-PCR kit (Takara Bio, Japan) was used for the real-time RT-PCR to quantitate two PEDV field strains (PEDV S INDEL and Non-S INDEL field strains). A 198 bp DNA fragment of the N gene was amplified with the primer sets of forward primer (qN306-F) 5'-CGCAAAGACTGAAC CCACTAAC-3′ and reverse primer (56R) 5'-TTGC CTCTGTTGTTACTTGGAGAT-3′. A TaqMan probe (ProbeN466–469) with the sequence of 5'-GCAG GAGTCGTGGTAATGGCAACA-3′ was labeled with the 5′-reporter dye 6-carboxyfluorescein (FAM) and the 3′-quencher BHQ3. Real-time RT-PCR was carried out in a 20 µl reaction containing 2 µl of RNA template, 10 µl of 2X One Step RT-PCR Buffer III, 0.4 µl of TaKaRa Ex Taq HS, 0.4 µl of both forward and reverse primer, 0.8 µl of Probe, 0.4 µl of ROX Reference Dye and 5.6 µl of RNase free water. The reactions were performed using a StepOnePlus™ Real-Time PCR System (Amplified Biosystems, USA) under the following conditions: initial reverse transcription at 42 °C for 5 min, followed by initial denaturation at 95 °C for 10 s, 40 cycles of denaturation at 95 °C for 5 s, and annealing and extension at 60 °C for 30 s. The results of amplification were analyzed by StepOne Software v2.3 (Amplified Biosystems). Tenfold serial dilutions of the transcripts were prepared at concentrations of 8.97×10^7 to 8.97×10^2 copies of PEDV per 1 µl volume that were used for obtaining the standard curves.

Detection of PEDV in clinical samples

A total of 99 faecal pig samples were collected from pig farms in Japan including 50 PED positive samples (from Kagoshima, Miyazaki, Aomori, Aichi prefectures) that were collected from December 2013 to August 2017 and 49 negative samples from a PED negative farm (Sumiyoshi farm, Miyazaki prefecture). These positive samples are classical, emerging Non-S INDEL, S INDEL, and S1 NTD-del PEDV variants and some positive samples are mixed infection of emerging non-S INDEL and S1 NTD-del PEDV variants. Faecal samples were prepared as a 10% (*w*/*v*) suspension in PBS (pH 7.2) and centrifuged at 2300 x g at 4 °C for 10 min. A 250-µl aliquot of supernatant was used for RNA extraction (ReliaPrep™ RNA Cell Miniprep System, Promega, USA). RNA was used as a template for detection of PEDV by one-step RT-PCR and RtF-RT-LAMP.

Pooled samples

The RtF-RT-LAMP assay was used to determine efficiency in pooled stool samples for future application in large-scaled epidemiological surveys. We determined the sensitivity of the RT-LAMP assay in pooled faecal samples to calculate the possible sample sizes that can be applied to the pooled technique. Each PEDV positive sample was pooled with PEDV negative samples in different pooling ratios including 1:4, 1:9, 1:14, 1:19, 1:24, 1:29, 1:34, 1:39, 1:44, and 1:49. A 50-µl aliquot from individual positive or negative samples was transferred to a new tube and carefully votexed. Then, 250 µl in each pooling ratio was used for RNA extraction (ReliaPrep™ RNA Cell Miniprep System, Promega, USA). RNA was used as a template for RtF-RT-LAMP.

Results

Specificity of the RtF-RT-LAMP assay

The PEDV NK94P6 strain and other related porcine viruses (PRRSV, TGEV, JEV, GV) were tested using the RtF-RT-LAMP assay to evaluate the specificity. Only PEDV was positive, and no LAMP products were detected in the reactions from other relevant porcine viruses or negative control used in this study (Fig. 1). The results indicated that the RtF-RT-LAMP assay was specific for PEDV and can be applied for distinguishing PEDV from other porcine viruses.

Sensitivity of the RtF-RT-LAMP assay

To evaluate the sensitivity of the RtF-RT-LAMP assay, the detection limit was compared to the conventional one-step RT-PCR by amplifying ten-fold serial dilutions from the cell culture of two PEDV strains (NK94P6 and Fukuoka-1 Tr(−)). The detection limit of the one-step RT-PCR of the NK94P6 strain and Fukuoka-1 Tr(−) strain were 2.8×10^3 $TCID_{50}$/ml and 2×10^2 $TCID_{50}$/ml,

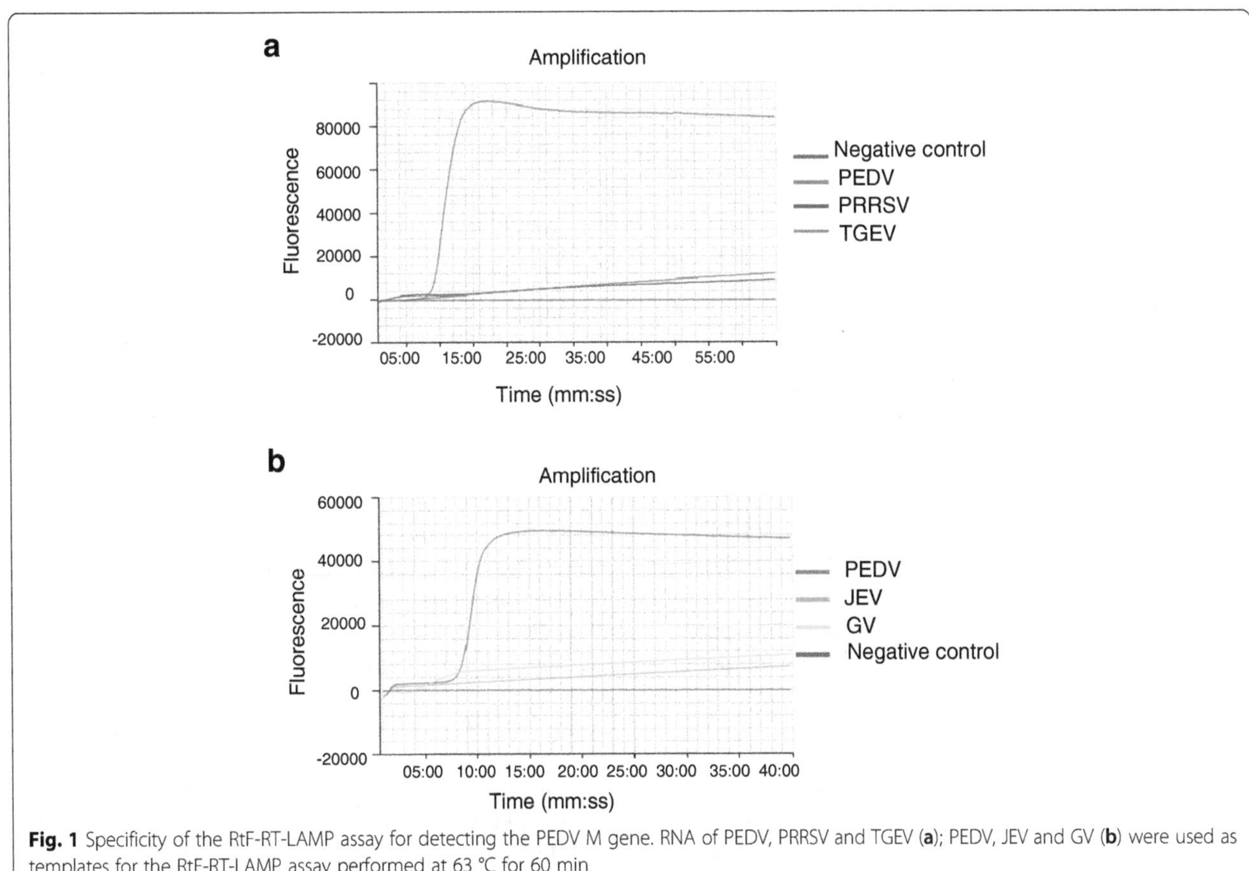

Fig. 1 Specificity of the RtF-RT-LAMP assay for detecting the PEDV M gene. RNA of PEDV, PRRSV and TGEV (**a**); PEDV, JEV and GV (**b**) were used as templates for the RtF-RT-LAMP assay performed at 63 °C for 60 min

while, the detection limit of the RtF-RT-LAMP assay was 2.8×10^1 $TCID_{50}$/ml and 2×10^0 $TCID_{50}$/ml, respectively (Tables 2 and 3). In addition, we used other two field strains that were PEDV S INDEL and Non-S INDEL strains to confirm the sensitivity of RtF-RT-LAMP (Additional files 2 and 3). This was much higher than that of the one-step RT-PCR. The sensitivity of the RtF-RT-LAMP assay was 100 times higher than that of one-step RT-PCR. Moreover, the real-time DNA fluorescence intensity from the reactions at all concentrations evaluated was high when the reactions were performed within 40 min. Therefore, the optimal reaction condition of the current RtF-RT-LAMP assay for PEDV was optimized for 40 min.

Detection of PEDV in clinical samples

To evaluate the sensitivity and specificity of the RtF-RT-LAMP assay to detect PEDV from clinical samples, one-step RT-PCR was used as the gold standard. A total of 99 clinical samples were tested by one-step RT-PCR that included 50 PED positive samples and 49 PED negative samples. All samples were tested by RtF-RT-LAMP assay. As shown in Tables 4, 49 PED negative samples were detected as negative and 50 PED positive samples were detected as positive by the RtF-RT-LAMP assay. No false negative or positive results were observed. Therefore, using one-step RT-PCR as the gold standard, the sensitivity and specificity of the RtF-RT-LAMP assay were 100%.

Table 2 Detection limits of one-step RT-PCR and RtF-RT-LAMP for the NK94P6 strain

TCID50	2.8×10^5	2.8×10^4	2.8×10^3	2.8×10^2	2.8×10^1	2.8×10^0	2.8×10^{-1}
One-step RT-PCR	+	+	+	–	–	–	
RtF-RT-LAMP (amplification time mm:ss)	+ (9:30)	+ (11:45)	+ (12:30)	+ (15:45)	+ (34:45)	–	–

+ Positive in duplicate
- Negative in duplicate
From 2.8×10^5 to 2.8×10^{-1}: tenfold serial dilution of 2.8×10^6 TCID50 PEDV NK94P6 strain

Table 3 Detection limits of one-step RT-PCR and RtF-RT-LAMP for the Fukuoka-1 Tr(−) strain

TCID50	2×10^3	2×10^2	2×10^1	2×10^0	2×10^{-1}	2×10^{-2}	2×10^{-3}
One-step RT-PCR	+	+	–	–	–	–	–
RtF-RT-LAMP (amplification time mm:ss)	+ 16:00	+ 18:15	+ 23:45	+ 35:45	±	–	–

+ Positive in duplicate
- Negative in duplicate
± One positive and one negative in duplicate
From 2×10^3 to 2×10^{-3}: tenfold serial dilution of 2×10^4 TCID50 PEDV Fukuoka-1 Tr(−) strain

Pooled sample

To estimate sample sizes for pooled faecal samples using the RtF-RT-LAMP assay, four positive samples were chosen from 50 positive samples based on amplification time in RT-LAMP and intensity of the electrophoresis band of RT-PCR products at different levels from the weakest positive to the strongest positive (Fig. 2). Sample 331 had the weakest electrophoresis band in RT-PCR. It was positive at the 25:45 min mark for amplification time, for which RT-LAMP can be positive until a pooling size of 15 samples. However, three other positive samples can be positive until a pooling size of 45 pooling or 50 samples (Table 5).

Discussion

In this study, we successfully developed an RtF-RT-LAMP assay for detection of PEDV in pooled faecal samples as an economical protocol for detection of infected herd in surveillance or monitoring strategies of PED. A sensitive, specific, rapid, and simple RtF-RT-LAMP assay including loop primers from the M gene for PEDV detection was developed. The reaction condition of the RtF-RT-LAMP was optimized by selecting a primer set and simple incubation at 63 °C for 40 min. The sensitivity of the RtF-RT-LAMP assay for PEDV detection was at least 100 times higher than that of one-step RT-PCR. Particularly, by semi-quantitative analysis, the RtF-RT-LAMP assay was applied to identifying the size for pooled stool samples. Using the RtF-RT-LAMP assay, at least a pool of 15 individual faecal samples could be applied instead of testing individual samples for cost saving in PED surveillance or monitoring programmes.

The PEDV M protein is a highly conserved trans-membrane protein that is the most abundant envelope component [13, 14]. Two reports have shown that the developed RT-LAMP method for the PEDV M gene has a higher sensitivity than the RT-LAMP method developed for the N gene [15, 16]. The use of LAMP for detecting PEDV has been reported [15–17]. However, the previously described RT-LAMP assays for detecting PEDV were not monitored by real-time florescent devices. Furthermore, they only used four primers for the LAMP assay, and only the N gene was used for designing primers. Moreover, some mismatches were found between primers and template in the 3′-end of some primers that were used in the Gou et al. study [17]. Mismatches, especially within the 3′-end primer region, affect both the stability of the primer-template duplex and the efficiency with which the polymerase extends the primer, potentially leading to biased results or even failure [18, 19]. In this study, all primers, including loop primers, were designed from the highly conserved M gene of PEDV. To achieve maximum sensitivity of detection for PEDV, the primer set used in this study included both FIP primers (PED_FIP_ID1 and PED_FIP_ID1modified). PED_FIP_ID1 and PED_FIP_ID1modified shared nucleotide identity with approximately 95 and 5% available PEDV sequences in the GenBank, respectively. Importantly, the entire procedure of current RT-LAMP could be completed in a simple process within 50 min. Using RT-PCR as the gold standard, the sensitivity and specificity of the RtF-RT-LAMP assay reached 100%. Interestingly, the sensitivity and specificity of the RT-LAMP assay were not highlighted in the previously described RT-LAMP methods for PEDV detection [15–17]. Our results also indicate that the sensitivity of the RtF-RT-LAMP assay was much higher than that of the one-step RT-PCR. LAMP is a simple, rapid, specific and cost-effective nucleic acid amplification method because it provides high amplification efficiency with DNA being amplified 10^9–10^{10} times in 15–60 min and use of 4 to 6 different primers to recognize 6 to 8 distinct regions on the target gene [9, 10]. In addition, LAMP is also applicable to RNA upon use of reverse transcriptase (RTase)

Table 4 Sensitivity and specificity of the RtF-RT-LAMP assay

		One-step RT-PCR		
		Number of positive samples	Number of negative samples	Total
RtF-RT-LAMP	Number of positive samples	50	0	50
	Number of negative samples	0	49	49
Total		50	49	99

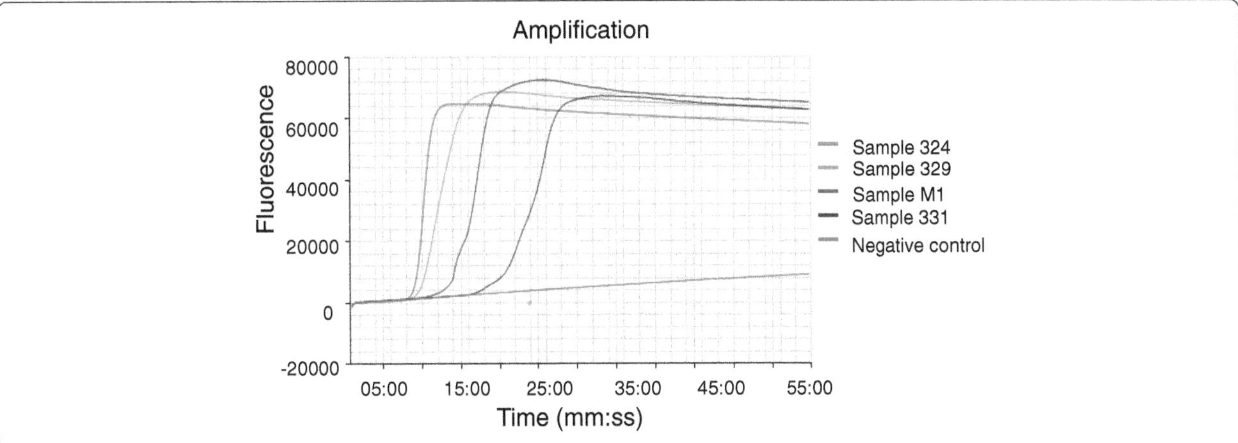

Fig. 2 Positive faecal samples were used for estimating the pooled size. Four faecal samples were chosen from strongest positive to weakest positive that based on amplification time in the RtF-RT-LAMP assay and the intensity of electrophoresis bands of RT-PCR products

together with DNA polymerase [20]. One study demonstrated that for PEDV detection, the sensitivity of qPCR was higher than that of RT-PCR [21]. RT-PCR and real-time RT-PCR techniques also demonstrate high specificity and sensitivity. However, these techniques require sophisticated and high-precision instruments (such as PCR and quantitative fluorescence PCR machines). Furthermore, the RT-PCR procedure is a time-consuming and complicated process. The RtF-RT-LAMP method showed distinct advantages with regards to detection time, as well as a simple process for rapid detection of PEDV.

Early identification of the infected herd through surveillance and monitoring strategies to enhance biosecurity measures is necessary to control PED. However, passive surveillance and individual testing could lead to important problems such being less effective as well as resulting in high cost and time commitments. Recently, a pooled sample technique has been developed and applied as a cost-efficient approach to surveillance or monitoring programs for pathogens such as Salmonella spp. in pigs and bovine viral diarrhoea virus in cattle [22, 23]. In this study, two pathogenically different PEDV strains and other two field strains were used to evaluate the sensitivity of the RtF-RT-LAMP assay. The RtF-RT-LAMP assay was much more sensitive than one-step RT-PCR even with different strains in TCID50 or copies. The optimal size for pooled stool samples was evaluated using a semi-quantitative method based on the amplification time in the RtF-RT-LAMP assay and intensity of the electrophoresis band for RT-PCR products. Even with the weakest positive electrophoresis band in one-step RT-PCR, in total 50 positive faecal samples was still positive in the RtF-RT-LAMP assay at a pooling size of 15, which was pooled from 14 individual negative faecal samples and one weak positive faecal sample. Furthermore, of the 50 PED positive samples, only two samples showed weak positive electrophoresis bands. Our pooled results indicate that a pool of at least 15 individual faecal samples can be applied using the RtF-RT-LAMP assay. In addition, the cost of one RT-PCR test was estimated based on the reagent cost, which was about 5 times more expensive than the RtF-RT-LAMP assay. In our study, testing pooled stool samples by RtF-RT-LAMP assay holds promise for surveillance and monitoring strategies. To our knowledge, this is the first study to provide evidence of the estimation of sample sizes for pooled stool samples for detecting PEDV using an accurate, simple, and timely RtF-RT-LAMP method. Our results will support the design and implementation of large-scaled epidemiological surveys as well as active surveillance or monitoring systems for effective control of PED. Further research will be required on a larger scale to confirm the effectiveness of the pooled sample protocol.

Table 5 Detection PEDV in pooled faecal samples by RtF-RT-LAMP

Pooled size	5	10	15	20	25	30	35	40	45	50
331	+	+	+	-	-	-	-	-	-	-
M1	+	+	+	+	+	+	+	+	+	-
329	+	+	+	+	+	+	+	+	+	+
324	+	+	+	+	+	+	+	+	+	+

Pooled size: Each PEDV positive sample was pooled with PEDV negative samples in different pooling ratios including 1:4, 1:9, 1:14, 1:19, 1:24, 1:29, 1:34, 1:39, 1:44, and 1:49
331, M1, 329, 324: Four positive samples were selected for pooling with a different number of negative samples
+: Positive by RtF-RT-LAMP
-: Negative by RtF-RT-LAMP

Conclusions

In this study, the highly sensitive, specific, rapid, and simple RtF-RT-LAMP assay based on the M gene, using a mobile device for detection of PEDV in pooled stool of at least 15 samples was shown to be an economical diagnosis test for PEDV detection. Use of these methods will not only be feasible, but also serve as effective surveillance and monitoring strategies to control PED.

Additional files

Additional file 1: RT-LAMP primers design for PEDV nucleotide detection. Nucleotide sequence alignments of M gene of seven PEDV strains. Representative M gene sequences in each strain are aligned with clustalW. Sequence data of designing primers for RT-LAMP in this study (KT323979.1), the sequence used for RT-PCR (JX435310.1 and JN089738.1), the sequence of G1b S INDEL strain (KY619833.1), the sequence of G2b/Non S INDEL/North America strain (KY619838.1), the sequence of G2a/Non S INDEL/Asian strain (KJ960178.1), the sequence of NK96P4C6 G1a classical strain (KY619828). Primer recognition sites are indicated with primer names. (DOCX 28 kb)

Additional file 2: Detection limits of one-step RT-PCR and RtF-RT-LAMP for PEDV S INDEL field strain. From 5.0×10^5 to 5.0×10^0: tenfold serial dilution of 5.0×10^6 copies PEDV S INDEL field strain. (DOCX 14 kb)

Additional file 3: Detection limits of one-step RT-PCR and RtF-RT-LAMP for the PEDV Non-S INDEL field strain. From 1.5×10^6 to 1.5×10^0: tenfold serial dilution of 1.5×10^7 copies PEDV Non-S INDEL field strain. (DOCX 14 kb)

Abbreviations

EMEM: Eagle's minimal essential medium; FBS: Foetal bovine serum; GV: Getal virus; JEV: Japanese Encephalitis virus; PED: Porcine epidemic diarrhoea; PEDV: Porcine epidemic diarrhoea virus; PRRSV: Porcine reproductive and respiratory syndrome virus; RtF-RT-LAMP: Real-time fluorescent reverse transcription loop-mediated isothermal amplification; RT-PCR: Reverse transcription polymerase chain reaction; TCID: Tissue culture infective dose; TGEV: Transmissible gastroenteritis coronavirus; TPB: Tryptose phosphate broth

Acknowledgments

We would like to thank the National Institute of Animal Health, Japan and Fukuoka Chuo Livestock Hygiene Service Center, Fukuoka, Japan for providing PEDV strains and Vero cells.

Funding

This work was supported by JSPS KAKENHI (Grant Number 15 K18786) and the Ito Foundation. The funders had no role in study design, or the collection, analysis, and interpretation of data. In addition, they were not involved in the writing of the report or the decision to submit the article for publication.

Authors' contributions

TNM collected samples and performed the RtF-RT-LAMP, analyzed the data and drafted the manuscript. VDN collected samples and performed the one-step RT-PCR. WY assembled the sequence data, designed primers and contributed to experimental design. TO, RY and JN contributed to the study design. SM, KN and YS helped in laboratory analysis. SS conceptualized and supervised the study. SS also analyzed and revised the manuscript. All authors read, commented on and approved the final version of the manuscript.

Competing interests

The authors declare that they have no competing interests.

Author details

[1]Animal Infectious Disease and Prevention, Department of Veterinary Sciences, Faculty of Agriculture, University of Miyazaki, Miyazaki, Japan. [2]Faculty of Veterinary Medicine, Vietnam National University of Agriculture, Hanoi, Vietnam. [3]Center for Animal Disease Control, University of Miyazaki, Miyazaki, Japan.

References

1. Sun RQ, Cai RJ, Chen YQ, Liang PS, Chen DK, Song CX. Outbreak of porcine epidemic diarrhea in suckling piglets, China. Emerging Infect Dis. 2012;18:161–3.
2. Pensaert M, De Bouck P. A new coronavirus-like particle associated with diarrhea in swine. Arch Virol. 1978;58:243–7.
3. Woode GN, Bridger J, Hall GA, Jones JM, Jackson G. The isolation of reovirus-like agents (rotaviruses) from acute gastroenteritis of piglets (plates XVI). J Med Microbiol. 1976;9:203–9.
4. Song D, Park B. Porcine epidemic diarrhoea virus: a comprehensive review of molecular epidemiology, diagnosis, and vaccines. Virus Genes. 2012;44:167–75.
5. Duy DT, Toan NT, Puranaveja S, Thanawongnuwech R. Genetic characterization of porcine epidemic diarrhea virus (PEDV) isolates from southern Vietnam during 2009–2010 outbreaks. Thai J Vet Med. 2011;41:55–64.
6. Alvarez J, Goede D, Morrison R, Perez A. Spatial and temporal epidemiology of porcine epidemic diarrhea (PED) in the Midwest and southeast regions of the United States. Prev Vet Med. 2016;123:155–60.
7. Vlasova AN, Marthaler D, Wang Q, Culhane MR, Rossow K, Rovira A, James Collins JK. Distinct characteristics and complex evolution of PEDV strains, North America, may 2013–February 2014. Emerging Infect Dis. 2014;20(10):1620–8.
8. Sasaki Y, Alvarez J, Sekiguchi S, Sueyoshi M, Otake S, Perez A. Epidemiological factors associated to spread of porcine epidemic diarrhea in Japan. Prev Vet Med. 2016;123:161–7.
9. Notomi T, Okayama H, Masubuchi H, Yonekawa T, Watanabe K, Amino N, Hase T. Loop-mediated isothermal amplification of DNA. Nucleic Acids Res. 2000;28:e63.
10. Nagamine K, Hase T, Notomi T. Accelerated reaction by loop-mediated isothermal amplification using loop primers. Mol Cell Probes. 2002;16:223–9.
11. Kim SY, Song DS, Park BK. Differential detection of transmissible gastroenteritis virus and porcine epidemic diarrhea virus by duplex RT-PCR. J Vet Diagn Investig. 2001;13:516–20.
12. Van Diep N, Norimine J, Sueyoshi M, Lan NT, Hirai T, Yamaguchi R. US-like isolates of porcine epidemic diarrhea virus from Japanese outbreaks between 2013 and 2014. Springerplus. 2015;4:756.
13. Utiger A, Tobler K, Bridgen A, Suter M, Singh M, Ackermann M. Identification of proteins specified by porcine epidemic diarrhoea virus. Adv Exp Med Biol. 1995;380:287–90.
14. Zhang Z, Chen J, Shi H, Chen X, Shi D, Feng L, Yang B. Identification of a conserved linear B-cell epitope in the M protein of porcine epidemic diarrhea virus. Virol J. 2012;9:225.
15. Ren X, Li P. Development of reverse transcription loop-mediated isothermal amplification for rapid detection of porcine epidemic diarrhea virus. Virus Genes. 2011;42:229–35.
16. Yu X, Shi L, Lv X, Yao W, Cao M, Yu H, Wang X, Zheng S. Development of a real-time reverse transcription loop-mediated isothermal amplification method for the rapid detection of porcine epidemic diarrhea virus. Virol J. 2015;12:76.
17. Gou H, Deng J, Wang J, Pei J, Liu W, Zhao M, Chen J. Rapid and sensitive detection of porcine epidemic diarrhea virus by reverse transcription loop-mediated isothermal amplification combined with a vertical flow visualization strip. Mol Cell Probes. 2015;29:48–53.
18. Lefever S, Pattyn F, Hellemans J, Vandesompele J. Single-nucleotide polymorphisms and other mismatches reduce performance of quantitative PCR assays. Clin Chem. 2013;59:1470–80.
19. Stadhouders R, Pas SD, Anber J, Voermans J, Mes TH, Schutten M. The effect of primer-template mismatches on the detection and quantification of nucleic acids using the 5′ nuclease assay. J Mol Diagn. 2010;12:109–17.

20. Whiting SH, Champoux JJ. Properties of strand displacement synthesis by Moloney murine leukemia virus reverse transcriptase: mechanistic implications. J Mol Biol. 1998;278:559–77.
21. Masuda T, Tsuchiaka S, Ashiba T, Yamasato H, Fukunari K, Omatsu T, Furuya T, Shirai J, Mizutani T, Nagai M. Development of one-step real-time reverse transcriptase-PCR-based assays for the rapid and simultaneous detection of four viruses causing porcine diarrhea. Jpn J Vet Res. 2016;64:5–14.
22. Arnold ME, Cook AJ. Estimation of sample sizes for pooled faecal sampling for detection of Salmonella in pigs. Epidemiol Infect. 2009;137:1734–41.
23. Munoz-Zanzi CA, Johnson WO, Thurmond MC, Hietala SK. Pooled-sample testing as a herd-screening tool for detection of bovine viral diarrhea virus persistently infected cattle. J Vet Diagn Investig. 2000;12:195–203.

Development of real-time recombinase polymerase amplification assay for rapid and sensitive detection of canine parvovirus 2

Yunyun Geng[1†], Jianchang Wang[2,3†], Libing Liu[2,3], Yan Lu[1], Ke Tan[1*] and Yan-Zhong Chang[1*]

Abstract

Background: Canine parvovirus 2, a linear single-stranded DNA virus belonging to the genus *Parvovirus* within the family *Parvoviridae*, is a highly contagious pathogen of domestic dogs and several wild canidae species. Early detection of canine parvovirus (CPV-2) is crucial to initiating appropriate outbreak control strategies. Recombinase polymerase amplification (RPA), a novel isothermal gene amplification technique, has been developed for the molecular detection of diverse pathogens. In this study, a real-time RPA assay was developed for the detection of CPV-2 using primers and an exo probe targeting the CPV-2 nucleocapsid protein gene.

Results: The real-time RPA assay was performed successfully at 38 °C, and the results were obtained within 4–12 min for 10^5–10^1 molecules of template DNA. The assay only detected CPV-2, and did not show cross-detection of other viral pathogens, demonstrating a high level of specificity. The analytical sensitivity of the real-time RPA was 10^1 copies/reaction of a standard DNA template, which was 10 times more sensitive than the common RPA method. The clinical sensitivity of the real-time RPA assay matched 100% (n = 91) to the real-time PCR results.

Conclusion: The real-time RPA assay is a simple, rapid, reliable and affordable method that can potentially be applied for the detection of CPV-2 in the research laboratory and point-of-care diagnosis.

Keywords: Canine parvovirus, Exo probe, Recombinase polymerase amplification

Background

Canine parvovirus diseases, caused by canine parvovirus type 2 (CPV-2), is highly contagious and prevalent worldwide in domestic and wild canids. CPV-2 emerged as a novel pathogen in 1978 and spread rapidly worldwide [1, 2]. Within a few years, the original type 2 virus underwent a rapid and extensive evolution, and was replaced by two antigenic types, termed CPV-2a and CPV-2b. Recently, a new antigenic variant, CPV-2c, was reported in dogs [3]. CPV-2 is a small non-enveloped, linear single-stranded DNA virus belonging to the family *Parvoviridae*, genus *Parvovirus*, and is considered as a major cause for large numbers of animal deaths worldwide [4]. CPV2-infected dogs are characterized by a gastroenteritis disorder with clinical signs of anorexia, lethargy, vomiting, fever and diarrhea (from mucoid to hemorrhagic) [5, 6]. Because of infection and damage to the bone marrow, parvovirus can also cause acute leukopenia with lymphopenia and neutropenia. When CPV-2 shed into the feces, it can be spread by direct oral or nasal contact. Furthermore, CPV-2 is extremely stable in the environment and can survive for several months. Therefore, early, rapid and accurate diagnosis of CPV-2 infection would help veterinarians to implement appropriate strategies in time to improve disease management and prevent outbreaks, particularly within a shelter environment.

* Correspondence: angel2005123@126.com; angeltanke@yahoo.co.jp; chang7676@163.com
†Equal contributors
[1]College of Life Sciences, Hebei Normal University, No.20, Road E. 2nd Ring South, Yuhua District, Shijiazhuang, Hebei Province 050024, People's Republic of China
Full list of author information is available at the end of the article

Several traditional diagnosis methods exist for CPV-2, including direct observation by electron microscopy, virus isolation in a suitable cell culture system, serological tests such as the latex agglutination test (LAT), hemaghaemagglutination (HA) test, ELISA and so on. These methods are often time-consuming, laborious, and have low sensitivity [7, 8]. With the advances in molecular detection techniques, a substantial number of gene amplification-based assays have been described for CPV-2 diagnosis such as polymerase chain reaction (PCR), nested PCR, real-time PCR, reverse-transcription loop-mediated isothermal amplification (RT-LAMP) and insulated isothermal PCR (iiPCR) [9–15]. Among these methods, nested PCR, real-time PCR and RT-LAMP have shown high sensitivity. However, due to the requirement of an expensive thermocycler and well-experienced technicians, implementation of these assays is limited in the field and at the point- of -care (POC). The iiPCR method was reported for its sensitive detection of CPV-2, but the reaction time was about 60 min [9]. In addition, the SNAP test based on ELISA protocols for detection of viral antigens is commonly used for CPV-2 diagnosis and can be completed in about 8 min by employing the commercial kit, whereas PCR seems to be more sensitive than SNAP [8, 13, 16, 17]. Recently we reported on recombinase polymerase amplification (RPA) as a rapid, specific, sensitive and cost-effective molecular method for POC diagnosis of CPV-2 infection [18].

RPA is an isothermal gene amplification technique [19]. Similar to conventional PCR, the use of two opposing primers allows exponential amplification of the target sequence in RPA, but the latter is tolerant to 5–9 mismatches in the primer and probe without influencing the performance of the assay. RPA possesses superiority in speed, portability and accessibility compared to PCR, and it has been used to replace the PCR method for the molecular detection of diverse pathogens, such as fungi, parasites, bacteria and viruses [20–22]. Based on our previous study, we developed here a real-time RPA assay for simple, rapid, convenient and POC detection of CPV-2, which utilizes an exo probe and a portable, user-friendly POC tube scanner.

Methods

DNA/RNA extraction and RNA reverse transcription

CPV-2a, CPV-2b, CDV, CCoV and CPIV (shown in Table 1) were propagated in Madin–Darby canine kidney (MDCK, CRL-2936™) cells and pseudorabies virus (PRV) in Syrian baby hamster kidney (BHK-21, CCL-10™) cells (ATCC, Manassas, USA). Two hundred microliters of cell culture medium, after brief centrifugation to remove cell debris, was used for viral DNA and RNA extraction using the TIANamp Virus DNA kit (Tiangen Biotech Co., Ltd., Beijing, China) and Trizol reagent (Invitrogen, Beijing, China), respectively. Viral DNA and RNA were quantified using a ND-2000c spectrophotometer (NanoDrop, Wilmington, DE, USA). One hundred nanograms of extracted viral RNA was reverse transcribed to cDNA using the Primescript™ II 1st strand cDNA Synthesis kit (Takara) according to the manufacturer's instructions. The synthesized cDNA was then purified using the cDNA purification kit (Takara, Dalian, China) and quantified using a ND-2000c spectrophotometer. For viral DNA extraction from the clinical samples, the starting material consisting of 10 mg of feces for each sample was emulsified in 1 mL sterile phosphate-buffered saline (PBS) and centrifuged at 10000 rpm for 10 min at 4 °C. The supernatant was collected and used for viral DNA extraction using the TIANamp Virus DNA kit. Viral DNA extracted from each fecal sample was finally eluted in 20 μl of nuclease-free water. All DNA and cDNA templates were stored at −20 °C until assayed.

Table 1 List of Virus strains and clinical samples

Virus/Clinical samples	Strain	Reference
Canine parvovirus type 2a	CPV-b114	
Canine parvovirus type 2b	SJZ101	
Canine distemper virus	CDV-FOX-TA	[8]
Canine coronavirus	VR-809	
Canine parainfluenza virus	CPIV/A-20/8	
Pseudorabies virus	Barth-K61	
91 fecal swab samples[a]		No

[a]Ninety-one fecal swab samples were collected from the dogs sent to our laboratory from 2012 to 2016 and snap-frozen for storage at −80 °C. Seventy-six of the above clinical samples were detected as CPV-2 positive, and fifteen of them were CPV-2 negative by real-time PCR

Generation of standard DNA

To generate a CPV-2 standard DNA for the real-time RPA, a PCR product containing 1755 bp covering the region of interest of VP2 was amplified from the CPV-2a DNA using VP2-Forward and VP2-Reverse as primers (Table 2) and cloned into the pMD19-T vector using the pMD19-T Vector Cloning Kit (Takara) according to manufacturer's instructions. The resulting plasmid, pCPV-VP2, was transformed into *Escherichia coli* DH5α cells, and the positive clones were confirmed by sequencing using M13 primers (Invitrogen®, Carlsbad, CA, USA). pCPV-VP2 was purified with the SanPrep Plasmid MiniPrep Kit (Sangon Biotech Co., Ltd., Shanghai, China) and quantified using a ND-2000c spectrophotometer. The copy number of DNA molecules was calculated by the following formula: amount (copies/μL) = [DNA concentration (g/μL) / (plasmid length in base pairs × 660)] × 6.02 × 10^{23}.

RPA primers and exo probe

The primers used in this study have been described previously [18]. Nucleotide sequences of CPV-2a (GenBank:

Table 2 Sequences of primers and probes for CPV-2 PCR, real-time PCR and real-time RPA assay

Name	Sequence 5'-3'Amplication size (bp)	
VP2-FP	ATGAGTGATGGAGCAGTTCAACCAGAC	1775
VP2-RP	TTAATATAATTTTCTAGGTGCTAGTTGA	
CPV-FP	AAACAGGAATTAACTATACTAATATATTTA	93
CPV-RP	AAATTTGACCATTTGGATAAACT	
CPV-P	FAM-TGGTCCTTTAACTGCATTAAATAATG TACC-BHQ1	
CPV-RPA-FP	CACTTACTAAGAACAGGTGATGAATTTGCT ACAG	214
CPV-RPA-RP	AGTTTGTATTTCCCATTTGAGTTACACCACG TCT	
CPV-RPA-P	CCTCAAGCTGAAGGAGGTACTAACTTTGGT/ BHQ1-dT//THF //FAM-dT/ATAGGAGTTCAAC AAG-C3 spacer	

M24003, AB054215, KF803642), –2b (GenBank: M38245, AY869724, KF803611) and -2c (GenBank: FJ005196, KM236569) were aligned to identify conserved regions of the VP2 gene. Three exo probes were designed based on the conserved region of the VP2 gene. Both primer and probe sequences were 100% identical to their target sequences in the CPV-2a, 2b and 2c genomic DNA. The real-time RPA primers and probes were selected by testing the combination to yield the highest sensitivity (Table 2). Primers and exo probes were synthesized by Sangon Biotech Co., Ltd.

RPA assay

The real-time RPA reactions were performed in a 50 μL volume using a TwistAmp™ exo kit (TwistDX, Cambridge, UK). Other components included 420 nM each RPA primer, 120 nM exo probe, 14 mM magnesium acetate, and 1 μL of viral or sample DNA. All reagents except for the viral template and magnesium acetate were prepared in a master mix, which was distributed into each 0.2 mL freeze-dried reaction tube containing a dried enzyme pellet. One microliter of viral DNA was added to the tubes. Subsequently, magnesium acetate was pipetted into the tube lids, and then the lids were closed carefully. The magnesium acetate was centrifuged into the rehydrated material using a mini spin centrifuge. After briefly vortexing and centrifuging the reaction tubes once again, they were immediately placed in the Genie III scanner device to start the reaction at 38 °C for 20 min. The fluorescence signal was collected in real-time and increased markedly upon successful amplification.

Real-time PCR for CPV-2

Real-time PCR specific for CPV-2 was performed on the ABI 7500 instruments described previously with some modifications [18]. Premix Ex Taq™ (Takara Co., Ltd., Dalian, China) was applied in the real-time PCR and the reaction was performed as follows: 95 °C for 3 min, followed by 40 cycles of 95 °C for 10s and 60 °C for 32 s.

Specificity and analytical sensitivity analysis

Ten nanograms of viral DNA or cDNA was used as the template for the specificity analysis of the real-time RPA assay. The assay was evaluated against a panel of pathogens considered important in dogs, i.e. CPV-2a, CPV-2b, CDV, CCoV, CPIV and PRV.

The recombinant plasmid, pCPV-VP2, was 10-fold serially diluted to achieve DNA concentrations ranging from 10^5 to 10^0 copies/μL, which were used as the standard DNA for the CPV-2 RPA sensitivity assay. The real-time RPA was tested using the standard DNA in eight replicates. The threshold time was plotted against the molecules detected.

Validation with clinical samples

Viral DNA extracted from 91 clinical swab samples were detected by real-time RPA, and the results were compared with those obtained using real-time PCR as previously described [10].

Statistical methods

For the determination of analytical sensitivity of the real-time PRA assay by the molecular DNA standard, a semi-log regression was calculated using Prism software 5.0 (Graphpad Software Inc., SanDiego, CA). For exact determination, a probit regression was performed using the Statistical Product and Service Solutions software (IBM, Armonk, NY, USA). The evaluation of exo RPA with clinical samples data is presented as R^2 value.

Results

The specificity and analytical sensitivity of the real-time RPA assay were analyzed (Figs.1 and 2). Using 10 ng of viral DNA, cDNA or canine genome as template, the results showed that only CPV-2a and CPV-2b were detected by the real-time RPA assay, while the other four viruses, including CDV, CCoV, CPIV and PRV, and canine genome, were not (Fig. 1, n = 5). Thus, the real-time RPA assay results demonstrated good specificity for the detection of CPV-2.

To evaluate the sensitivity of the real-time RPA method, a dilution range of 10^5–10^0 copies/μL of standard DNA was used as templates, and the real-time RPA and real-time PCR were performed simultaneously. The limit of detection of the real-time RPA was 10^1 copies (Fig. 2a and b). A probit regression analysis using the results of eight complete molecular standard runs calculated that the limit of detection (LOD) of the real-time RPA was 10^1 copies per reaction in 95% of cases (Fig. 2c), which was the same as that of the real-time PCR applied in the study (data not

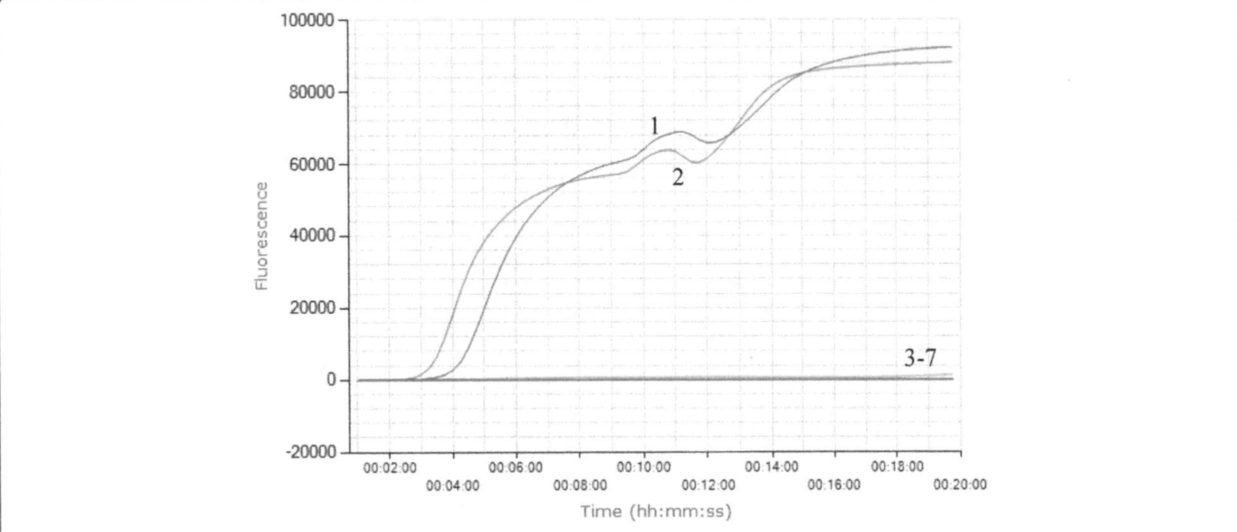

Fig. 1 Specificity of the real-time RPA assay for CPV-2 detection. Real-time RPA was carried out at 38 °C for 20 min using 10 ng of viral DNA or cDNA as template. The results showed real-time RPA only amplified the CPV-2a and CPV-2b DNA, but not other viruses tested (n = 5). Lane 1, CPV-2a; lane 2, CPV-2b; lane 3, CDV; lane 4, CCoV; lane 5, CPIV; lane 6, PRV; and lane 7, canine genomic DNA

shown). With the data from eight runs on the quantitative DNA standards, a semi-log regression analysis for the real-time RPA and real-time PCR was made. Run times of the real-time RPA assay were about 4–12 min for 10^5–10^1 copies, respectively (Fig. 2b), while the real-time PCR with Ct values between 21 and 36 (data not shown) required about 32–54 min to obtain the final results.

Results of the evaluation of exo RPA with clinical samples are shown in Table 2 and Fig. 3. The diagnostic performance of the real-time RPA assay to detect CPV-2 in the 91 clinical swab samples was compared to that of the real-time PCR. These two assays showed the same results (76 positive and 15 negative cases), and further analysis demonstrated that the real-time RPA had a diagnostic agreement of 100% with the real-time PCR (Table 3). No discrepancy was found in samples (14/76) containing low levels of CPV-2 DNA (Ct > 35, real-time PCR), indicating that the established real-time RPA reliably detected low amounts of CPV-2 in clinical samples. Positive samples had real-time RPA Ct values ranging from 17.52 to 36.29, indicating that the method was able to detect CPV-2 DNA across the entire range of the assay. Twenty-three positive samples were selected randomly, and the threshold time (TT) and cycle threshold (Ct) values of real-time RPA and real-time PCR, respectively, were well correlated with an R^2 value of 0.846 (Fig. 3). The relative sensitivity of the real-time RPA assay was further evaluated by comparing with the SNAP test. A total of 30 fecal swab samples were collected for this experiment. Twenty-four of the above tested clinical samples displayed CPV-2 positive, six of them were CPV-2 negative by real-time PCR assay. The result showed that the SNAP test was able to detect CPV antigen in 16/30 (53.3%) of analyzed samples. A higher detection rate (24/30, 80%) was obtained using real-time RPA for CPV-2 DNA. The relative sensitivity of the SNAP test was 66.7% (16/24), when compared to real-time RPA. The best correlation was observed again between conventional real-time PCR and real-time RPA analysis (shown in the Additional file 1: Table S1). The detection results of the swab samples showed that the real-time RPA method we developed was effective in detecting CPV-2.

Discussion

In this study, a real-time RPA method was developed based on an exo probe for the rapid and sensitive detection of CPV-2. With real-time RPA, only CPV-2a and CPV-2b among multiple viruses infectious to canines tested could be amplified, demonstrating the high specificity of this assay (Fig. 1). Using the CPV-2 plasmid DNA as a template, it was detected between 4 and 12 min for 10^5–10^1 copies of DNA (Fig. 2a). The limit of detection of the real-time assay was 10^1 copies/reaction, which was 10 times higher than that of the conventional RPA [18]. Further validation of this method was performed using clinical samples. The detection rate of the real-time RPA assay was comparable of that of the real-time PCR, although the former assay was much faster than the latter. The real-time RPA assay detected the positive samples within 4–12 min. While the CPV-2 detection results by real-time RPA could be obtained in about 30 min including the time for nucleic acid extraction, the reaction time for positive samples reached up to 1 h with real-time PCR. Additionally, the cost per reaction performed in the real-time RPA, real-time PCR assays and commercial SNAP tests including the cost of

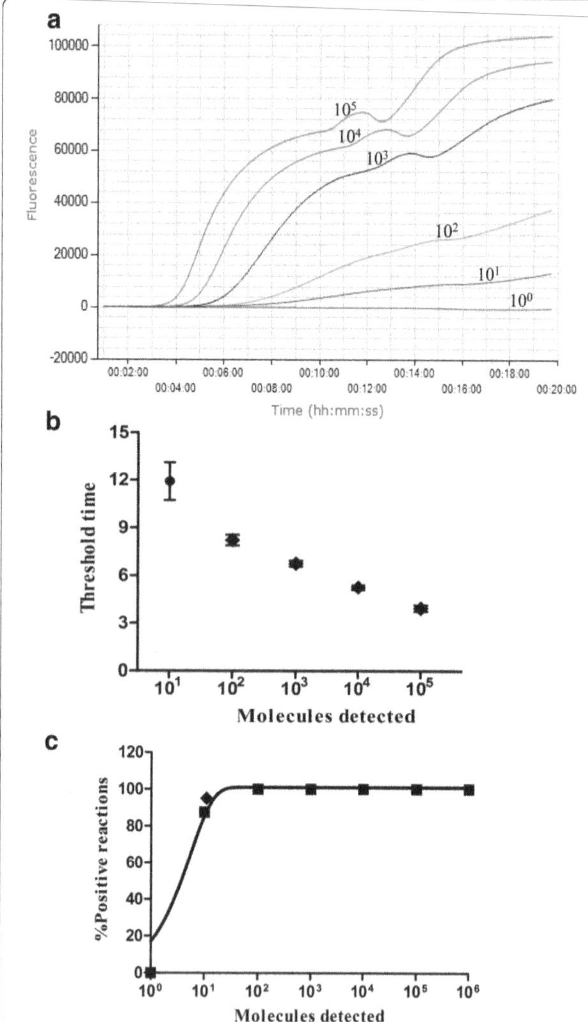

Fig. 2 Performance of the CPV-2 real-time RPA. **a** Fluorescence development over time using a dilution range of 10^5–10^0 copies of the CPV-2 standard DNA. **b** Semi-logarithmic regression of the data collected from eight CPV-2 real-time RPA test runs on the standard DNA using Prism Software 5.0. Run times of the real-time RPA were 4–12 min to detect 10^5 and 10^1 copies, respectively. **c** Probit regression analysis using SPSS software on data from eight runs. The limit of detection of 95% probability (11 copies) is indicated by a rhomboid

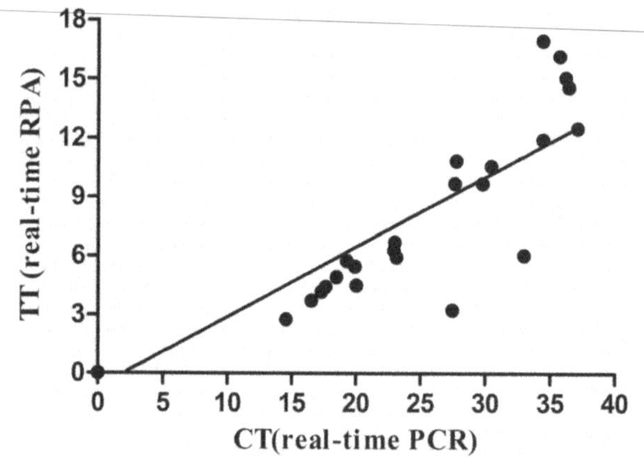

Fig. 3 Comparisons between results of real-time RPA and real-time PCR on clinical samples. DNA extracts from 34 clinical samples were screened. Linear regression analysis of real-time RPA threshold time (TT) values (y axis) and real-time PCR cycle threshold (CT) values (x axis) were determined using Prism software. R^2 value = 0.846

the reagents and the samples preparation was about 4.3, 6.0 and 5.0 US dollars, respectively. The cost of real-time RPA decreased 28.33%, when the method was compared to real-time PCR, and 14% respectively, when compared to SNAP tests. RPA has been widely explored for the molecular detection of diverse pathogens, including H7N9 virus and Dengue virus infection, and real field testing also has been achieved [23, 24]. Moreover, the portable POC tube scanner (Genie III, OptiGene Limited, West Sussex, United Kingdom) used in the study, weighing only 1.75 kg with the dimensions of 25 cm × 16.5 cm × 8.5 cm, is much more simple and less expensive than a real-time PCR machine and can be run on battery power for use in the field. For CPV-2, it is often sufficient to simply boil the fecal samples before running diagnostic PCR. In our subsequent experiments we obtained similar results using boiling only to extract CPV-2 from clinical samples. In recent years, a number of isothermal DNA amplification methods have been developed as a simple, rapid technique alternative to PCR-based amplification. In the LAMP assay for the rapid and sensitive detection of CPV-2, a set of four primers was needed, and the optimum time and temperature were 60 min and 65 °C, respectively [12]. The developed iiPCR could detect as low as 13 copies of CPV-2 DNA in about 60 min [9]. For the real-time RPA assay described in this work, 10^1 copies of CPV-2 DNA could be detected in 4 min, which was much more rapid than the above assays, including the common RPA assay. Compared to other isothermal amplification techniques, RPA does not require initial heating for DNA denaturation, and the results can be obtained in less than 12 min. The major advantages of RPA compared with other isothermal DNA amplification methods are that it (1) shows a certain degree of tolerance to common PCR inhibitors, (2) can tolerate a wide range of biological samples [19] and (3) utilizes reagents stored in lyophilized pellets,

Table 3 Detection of CPV-2 in clinical samples by real-time RPA and real-time PCR

		Real-time PCR		
		Positive	Negative	Total
Real-time RPA	Positive	76	0	76
	Negative	0	15	15
	Total	76	15	91

which are very stable and can react satisfactory at 25 °C for up to 12 weeks and at 45 °C for up to 3 weeks [25]. The SNAP test based on ELISA protocols for rapid detection of viral antigen is commonly used for CPV-2 clinical diagnosis, but it is less sensitive than RCR-based methods, especially RAP assay we developed in this study. The lower sensitivity of the SNAP test is generally associated with the host immune response, such as the very small amounts of virus shed in the feces during the late stage of infection and/or the presence of high CPV antibody titres in the gut lumen and so on. In our RPA assay, the detecting target was the nucleic acid of CPV-2. The virus-specific gene was amplified first, and its amplification products were tested. In this process, the detection signal is magnified hundreds of thousands of times. More importantly, it is not affected by the host immune response. One limitation of our assay is that it fails to distinguish CPV-2 vaccine from wild type stains. If vaccine strain is shed post vaccination it will be detected by this assay. All in all, these distinguishing features of the RPA assay and hands-on experience of real field testing using RPA from other research groups make us believe that RPA could work well in a clinical diagnostic setting. It may also be the most applicable approach for the field and POC diagnosis of infectious diseases.

Conclusion

In conclusion, the real-time RPA method based on an exo probe was successfully developed for the detection of CPV-2. With high sensitivity and specificity, the assay could be completed within 12 min. More importantly, the portability of the real-time RPA assay renders it applicable at quarantine stations, ports or sites of outbreaks. The effective and rapid real-time RPA assay developed in this study would be highly useful in the control of CPV-2, especially in resource-limited settings.

Abbreviations

BHK-21: Baby hamster kidney cells; CCoV: Canine coronavirus; CDV: Canine distemper virus; CPIV: Canine parainfluenza virus; CPV-2: Canine parvovirus 2; Ct: Cycle threshold; FP: Forward primer; iiPCR: Insulated isothermal PCR; LAMP: Loop-mediated isothermal amplification; LOD: Limit of detection; MDCK: Madin–darby canine kidney cells; n: Number; PCR: Polymerase chain reaction; POC: Point-of-care; PRV: Pseudorabies virus; RP: Reverse primer; RPA: Recombinase polymerase amplification; RT: Real-time reverse transcription; TT: Threshold time

Acknowledgements

We thank Dr. Rui Zhang at the China Agricultural University for valuable comments, and we also thank Dr. An-Wen Shao for language review and editing.

Funding

This work was supported by doctorate program funding of Hebei Normal University, Hebei Province, China (grant number 130401), Biology postdoctoral Science Foundation of Heibei Normal University (grant number 183717) and Natural Science Foundation Youth Project of Hebei province (C2017325001). The funders had no role in study design, in the collection, analysis and interpretation of data, in the writing the manuscript, or in the decision to submit the article for publication.

Authors' contributions

YYG, JCW, KT and YZC designed and conducted the experiment. YYG, JCW, LBL and YL performed the experiments and analyzed the data. YYG drafted the manuscript. All authors read, revised, and approved the final manuscript.

Consent for publication

Not applicable.

Competing interests

The authors declare that they have no competing interests.

Author details

[1]College of Life Sciences, Hebei Normal University, No.20, Road E. 2nd Ring South, Yuhua District, Shijiazhuang, Hebei Province 050024, People's Republic of China. [2]Center of Inspection and Quarantine, Hebei Entry-Exit Inspection and Quarantine Bureau, No.318 Hepingxilu Road, Shijiazhuang, Hebei Province 050051, People's Republic of China. [3]Hebei Academy of inspection and quarantine science and technology, No.318 Hepingxilu Road, Shijiazhuang, Hebei Province 050051, People's Republic of China.

References

1. Zhao Y, Lin Y, Zeng X, Lu C, Hou J. Genotyping and pathobiologic characterization of canine parvovirus circulating in Nanjing, China. Virol J. 2013;10:272.
2. Hoelzer K, Shackelton LA, Parrish CR, Holmes EC. Phylogenetic analysis reveals the emergence, evolution and dispersal of carnivore parvoviruses. The Journal of general virology. 2008;89(Pt 9):2280–9.
3. Perez R, Calleros L, Marandino A, Sarute N, Iraola G, Grecco S, Blanc H, Vignuzzi M, Isakov O, Shomron N, et al. Phylogenetic and genome-wide deep-sequencing analyses of canine parvovirus reveal co-infection with field variants and emergence of a recent recombinant strain. PLoS One. 2014;9(11):e111779.
4. Castanheira P, Duarte A, Gil S, Cartaxeiro C, Malta M, Vieira S, Tavares L. Molecular and serological surveillance of canine enteric viruses in stray dogs from Vila do Maio, Cape Verde. BMC Vet Res. 2014;10:91.
5. Timurkan M, Oguzoglu T. Molecular characterization of canine parvovirus (CPV) infection in dogs in Turkey. Vet Ital. 2015;51(1):39–44.
6. Decaro N, Desario C, Campolo M, Elia G, Martella V, Ricci D, Lorusso E, Buonavoglia C. Clinical and virological findings in pups naturally infected by canine parvovirus type 2 Glu-426 mutant. Journal of veterinary diagnostic investigation : official publication of the American Association of Veterinary Laboratory Diagnosticians, Inc. 2005;17(2):133–8.
7. Decaro N, Buonavoglia C. Canine parvovirus–a review of epidemiological and diagnostic aspects, with emphasis on type 2c. Vet Microbiol. 2012; 155(1):1–12.
8. Desario C, Decaro N, Campolo M, Cavalli A, Cirone F, Elia G, Martella V, Lorusso E, Camero M, Buonavoglia C. Canine parvovirus infection: which diagnostic test for virus? J Virol Methods. 2005;126(1–2):179–85.
9. Wilkes RP, Lee PY, Tsai YL, Tsai CF, Chang HH, Chang HF, Wang HT. An insulated isothermal PCR method on a field-deployable device for rapid and sensitive detection of canine parvovirus type 2 at points of need. J Virol Methods. 2015;220:35–8.
10. Decaro N, Elia G, Martella V, Desario C, Campolo M, Trani LD, Tarsitano E, Tempesta M, Buonavoglia C. A real-time PCR assay for rapid detection and

quantitation of canine parvovirus type 2 in the feces of dogs. Vet Microbiol. 2005;105(1):19–28.
11. Gizzi AB, Oliveira ST, Leutenegger CM, Estrada M, Kozemjakin DA, Stedile R, Marcondes M, Biondo AW. Presence of infectious agents and co-infections in diarrheic dogs determined with a real-time polymerase chain reaction-based panel. BMC Vet Res. 2014;10:23.
12. Sun Y-L, Yen C-H, C-F T. Visual detection of canine parvovirus based on loop-mediated isothermal amplification combined with enzyme-linked immunosorbent assay and with lateral flow dipstick. J Vet Med Sci. 2014; 76(4):509–16.
13. Schmitz S, Coenen C, Konig M, Thiel HJ, Neiger R. Comparison of three rapid commercial canine parvovirus antigen detection tests with electron microscopy and polymerase chain reaction. Journal of veterinary diagnostic investigation : official publication of the American Association of Veterinary Laboratory Diagnosticians, Inc. 2009;21(3):344–5.
14. Senda M, Parrish CR, Harasawa R, Gamoh K, Muramatsu M, Hirayama N, Itoh O. Detection by PCR of wild-type canine parvovirus which contaminates dog vaccines. J Clin Microbiol. 1995;33(1):110–3.
15. Wilkes RP, Tsai YL, Lee PY, Lee FC, Chang HF, Wang HT. Rapid and sensitive detection of canine distemper virus by one-tube reverse transcription-insulated isothermal polymerase chain reaction. BMC Vet Res. 2014;10:213.
16. Sun YL, Yen CH, CF T. Visual detection of canine parvovirus based on loop-mediated isothermal amplification combined with enzyme-linked immunosorbent assay and with lateral flow dipstick. J Vet Med Sci. 2014; 76(4):509–16.
17. Decaro N, Desario C, Billi M, Lorusso E, Colaianni ML, Colao V, Elia G, Ventrella G, Kusi I, Bo S, et al. Evaluation of an in-clinic assay for the diagnosis of canine parvovirus. Vet J. 2013;198(2):504–7.
18. Wang J, Liu L, Li R, Wang J, Fu Q, Yuan W. Rapid and sensitive detection of canine parvovirus type 2 by recombinase polymerase amplification. Archives Virol. 2016;161(4):1015–18.
19. Daher RK, Stewart G, Boissinot M, Bergeron MG. Recombinase polymerase amplification for diagnostic applications. Clin Chem. 2016;62(7):947–58.
20. Sakai K, Trabasso P, Moretti ML, Mikami Y, Kamei K, Gonoi T. Identification of fungal pathogens by visible microarray system in combination with isothermal gene amplification. Mycopathologia. 2014;178(1–2):11–26.
21. Boyle DS, McNerney R, Teng Low H, Leader BT, Perez-Osorio AC, Meyer JC, O'Sullivan DM, Brooks DG, Piepenburg O, Forrest MS. Rapid detection of mycobacterium tuberculosis by recombinase polymerase amplification. PLoS One. 2014;9(8):e103091.
22. Crannell ZA, Cabada MM, Castellanos-Gonzalez A, Irani A, White AC, Richards-Kortum R. Recombinase polymerase amplification-based assay to diagnose giardia in stool samples. The American journal of tropical medicine and hygiene. 2015;92(3):583–7.
23. Abd El Wahed A, Patel P, Faye O, Thaloengsok S, Heidenreich D, Matangkasombut P, Manopwisedjaroen K, Sakuntabhai A, Sall AA, Hufert FT, et al. Recombinase polymerase amplification assay for rapid diagnostics of dengue infection. PLoS One. 2015;10(6):e0129682.
24. Abd El Wahed A, Weidmann M, Hufert FT. Diagnostics-in-a-suitcase: development of a portable and rapid assay for the detection of the emerging avian influenza a (H7N9) virus. Journal of clinical virology : the official publication of the Pan American Society for Clinical Virology. 2015;69:16–21.
25. Lillis L, Siverson J, Lee A, Cantera J, Parker M, Piepenburg O, Lehman DA, Boyle DS. Factors influencing recombinase polymerase amplification (RPA) assay outcomes at point of care. Mol Cell Probes. 2016;30(2):74–8.

Co-localization of and interaction between duck enteritis virus glycoprotein H and L

Daishen Feng[1,2†], Min Cui[1,2†], Renyong Jia[1,2,3*], Siyang Liu[1,2], Mingshu Wang[1,2,3], Dekang Zhu[1,3], Shun Chen[1,2,3], Mafeng Liu[1,2,3], Xinxin Zhao[1,2,3], Yin Wu[1,2,3], Qiao Yang[1,2,3], Zhongqiong Yin[3] and Anchun Cheng[1,2,3*]

Abstract

Background: Duck Enteritis Virus (DEV), belonging to the α-herpesvirus subfamily, is a linear double-stranded DNA virus. Glycoprotein H and L (gH and gL), encoded by UL22 and UL1, are conserved in the family of herpesviruses. They play important roles as gH/gL dimers during viral entry into host cells through cell-cell fusion. The interaction between gH and gL has been confirmed in several human herpesviruses, such as Herpes Simplex Virus (HSV), Epstein-Barr virus (EBV) and Human Cytomegalovirus (HCMV). In this paper, we studied the interaction between DEV gH and gL.

Results: Recombinant plasmids pEGFP-N-gH and pDsRED-N-gL were constructed successfully. Expressions of both DEV gH and gL were observed after incubation of COS-7 cells transfected with pEGFP-N-gH and pDsRED-N-gL plasmids after 12 h, respectively. Also, the co-localization of a proportion of the gH and gL was detected in the cytoplasm of COS-7 cells after co-transfection for 24 h. Then, pCMV-Flag-gL and pCMV-Myc-gH recombinant plasmids were constructed and co-transfected into COS-7 cells. It was showed that both gH and gL were tested with positive results through co-immunoprecipitation and Western-blotting.

Conclusions: Our results demonstrated not only the co-localization of DEV gH and gL in COS-7 cells, but also the interaction between them. It will provide an insight for the further studies in terms of protein-protein interaction in DEV.

Keywords: DEV, Glycoprotein H , Glycoprotein L, Interaction, Co-localization, Co-immunoprecipitation

Background

Duck Enteritis Virus (DEV) is a linear double-stranded DNA virus, of the α-herpesvirus subfamily [34]. Herpesviruses, consisting of core, capsid, tegument and envelope, are a large family of the linear double-stranded DNA group of viruses [27, 28]. It has three subfamilies including α-, β- and γ- herpesvirus. The α-herpesvirus contains: Herpes Simplex Virus 1 (HSV-1), Herpes Simplex Virus 2 (HSV-2), Varicella-zoster Virus (VZV), and Pseudorabies Virus (PRV); β-herpesvirus: Human Cytomegalovirus (HCMV), Human Herpesvirus 6 and 7 (HHV-6 and HHV-7); and γ-herpesvirus: Epstein-Barr virus (EBV) and Kaposi's sarcoma-associated herpesvirus (KSHV, also known as Human Herpesvirus 8 (HHV-8)) [5, 19, 32]. DEV, also known as Duck Plague Virus (DPV) and Anatid Herpesvirus 1 (AnHV-1), mainly causes acute and contagious infection to waterfowl. Until now, the duck enteritis virus is one of the most severe diseases that seriously affect economy in the worldwide duck industry [23, 39].

Glycoprotein H is highly conserved in the herpesviruses. Heterodimer gH/gL, commonly formed through non-covalent linkages, is of great importance to membrane fusion of the enveloped viruses and host cells [6]. It has been studied that in the process of HSV-1 and HCMV entry into host cells, gB, gH and gL are three essential glycoproteins for the cell-cell fusion [1, 2, 24, 26]. Firstly, glycoproteins of herpesviruse bind with their specifically cellular receptors, in order to attaching to the membrane of host cells [10]. Then, gB, as a fusogen, is activated by gH/gL through interaction, therefore the fusogenic function of gB is promoted [23, 24, 30]. The

* Correspondence: jiary@sicau.edu.cn; chenganchun@vip.163.com
†Daishen Feng and Min Cui contributed equally to this work.
[1]Research Center of Avian Disease, College of Veterinary Medicine, Sichuan Agricultural University, Chengdu 611130, Sichuan, China
Full list of author information is available at the end of the article

function of connection between gB and gH/gL, also known as "core fusion machinery," is essential for herpesviruses to initiate membrane fusion [15]. For example, the first step of EBV entry into Epithelial cells is tethering gH/gL to its cellular membrane receptors, including integrin αγβ5, αγβ6 and αγβ8, which will change the conformation of gH/gL delicately due to receptor-binding. Then, the heterodimer will interact with gB and trigger its fusogenic function [5, 10, 11]. Therefore, gH/gL is more likely a regulator in the process of cellular membrane fusion.

Recently, the three-dimensional crystal structures and interaction domains of gH/gL of HSV-2 and EBV gH/gL have been solved [8, 26], which helped immensely in understanding the structure and function of gH/gL. However, the functional mechanisms of gH/gL are complicated and diverse in different herpesviruses. For instance, although the structural conformation of gH/gL is generally homologous among HSV-2, EBV and HCMV, some differences still exist, including divided domains, interdomain packing angles and fragment antigen-binding regions of the neutralizing antibody [8, 10]. When gH binds gL at the N-terminal domain (H1) of HSV-2, gH/gL therefore forms a boot-like heterodimer with a ~ 60° kink between domain H2 and H3 at the C terminus [8]. Similarly, comprising a kink between domains III and IV, the structure of HCMV gH/gL has the semblable boot-like heterodimer as HSV-2 [8–10, 26]. Furthermore, in most human herpesviruses, gL requires correct enfoldment to bind with gH. HSV-1 gL is more likely a scaffolding protein than a chaperone to traffic gH and promotes its surface expression, then gH anchors gL to the cell surface [12, 20, 33]. During membrane fusion, the fusogenic potential of HSV-1 gB is activated through interaction with gH/gL, before the alteration of gH/gL is initiated by gD [1, 4]. In short, the main function of HSV gH/gL is activating gB through direct interaction during entry of host cells.

Although the structure and function of gH/gL in some human herpesviruses have been studied to some extent, for example the interaction regions of gH and gL of HSV-2 and EBV have been studied and published already [8, 26], there is insignificant research data in the study of DEV gH and gL until now. According to the existence of gH/gL dimers in other herpesviruses, the hypothesis that DEV gH would interact with gL was proposed in this paper. In this study, the two expression plasmids, pEGFP-N-gH and pDsRED-N-gL, were co-transfected into COS-7 cells to analyze the co-localization of DEV gH and gL. Then, the co-immunoprecipitation (Co-IP) was conducted in order to detect the interactions between these two proteins. As a result, our study provided an evidence of interaction between DEV gH and gL, which will be helpful for the further study about the protein-protein interaction in DEV as well as other herpesviruses.

Methods

Virus and DNA extraction

The DEV CHv strain (GenBank No. JQ647509.1) was precured from Avian Disease Research Center of Sichuan Agricultural University. Monolayer cultures of duck embryo fibroblasts (DEFs) were infected with DEV CHv strain at a multiplicity of infection (MOI) of 0.1, which was incubated in Dulbecco's modified Eagle medium (DMEM) containing 10% fetal bovine serum (FBS). The use of duck embryos was approved by the Animal Ethics Committee of Sichuan Agricultural University (approval No. XF2014–18). Then, the cells were incubated at 37 °C until a cytopathic effect (CPE) appeared in about 80% of the cells. DNA of DEV was extracted from the infected cell lysis using DNA Extraction Kit for Virus (TianGen Biotech Co., LTD.).

Primer design and PCR amplification

A bioinformatics analysis of DEV CHv genome sequence (GenBank No. JQ647509.1) was performed and a pair of primers was designed to amplify the truncate UL1 gene (UL1t). The forward primer UL1t-1F and reverse primer UL1t-1R were showed in Table 1 (the underlined sequences was restriction enzyme *Hind*III site).

Two pairs of primer were designed to amplify the UL1 and UL22 gene respectively, which encode the gL and gH of DEV respectively. The two pairs of primer for amplifying UL1 gene are seen in Table 1 as UL1-1F/1R and UL1-2F/2R respectively. The other two pairs of primer for amplifying UL22 gene are also shown in Table 1 as UL22-1F/1R and UL22-2F/2R respectively. The DEV CHv DNA was used as template. The PCR was performed with 10 μL mixture as reactions containing 5 μL PrimeSTAR (premix) DNA polymerase, 0.2 μL each of primer at the concentration of around 5 nmol/OD, 0.4 μL DNA template at the concentration of 280 ng/μL, and 4.2 μL H_2O. Then, the PCR amplification was performed in pre-denaturation at 98 °C for 2 min, denaturation at 98 °C for 10s, annealing at 55 °C for 30s, extension at 72 °C for 30s, and final extension at 72 °C for 10 min after 30 cycle repeats.

Plasmid construction

Expression vectors pET-32a(+), pEGFP-C, pDsRED-N, pCMV-Flag and pCMV-Myc were available in the laboratory. After digested with both *EcoR*I and *Hind*III restriction enzymes respectively, the truncate DEV UL1 gene and vector pET-32a(+) were both linked to the recombinant expression plasmid pET-32a(+)-UL1t with solution I ligase. Plasmids pEGFP-N-gH and pDsRED-N-gL were constructed in the same way as above by digestion with *EcoR*I and *Kpn*I, while the plasmid pCMV-Flag-gL was constructed by digestion with *EcoR*I and *Sal*I, and pCMV-Myc-gH with *EcoR*I and

Table 1 The primers of amplification for both DEV UL1 and UL22

Name of Primers	Primers	Restriction Enzyme
UL1t-F	5'- CCGGAATTCATGTACCCGGTTATCGAAGGAG – 3'	*EcoR*I
UL1t-R	5'- CCCAAGCTTTCGTCGGCCATGGTGTCCGGG – 3	*Hind*III
UL1-1F	5'-CGGAATTCATGGGTTCCAGATGGAAGGC-3'	*EcoR*I
UL1-1R	5'- GGGGTACCCCTCGTCGGCCATGGTGTCCGG – 3'	*Kpn*I
UL1-2F	5'- CCGGAATTCATGGGTTCCAGATGGAAGGCC -3'	*EcoR*I
UL1-2R	5'-ACGCGTCGACCTATCGTCGGCCATGGTGTCC -3'	*Sal*I
UL22-1F	5'- CGGAATTCATGTCGCAGCTTACGGTGC -3'	*EcoR*I
UL22-1R	5'- GGGGTACCCCTTCTTCATTGCTTAACAGTTC -3'	*Kpn*I
UL22-2F	5'- CGGAATTCATGTCGCAGCTTACGGTG –3'	*EcoR*I
UL22-2R	5'- CCCTCGAGTCATTCTTCATTGCTTAACAG – 3'	*Xho*I

*Xho*I. Once these plasmids were constructed successfully, they were identified by restriction enzyme digestion and nucleotide sequencing.

Antibodies

Mouse polyclonal antibody against DEV CHv gL was prepared with purified UL1t protein, and identified by both indirect immunofluorescence and western-blotting respectively, which was detailed in the next section. The anti-Flag mouse and anti-Myc rabbit polyclonal antibodies (Beyotime Institute of Biotechnology, China) were available in the laboratory.

Cells culture and transfection

COS-7 cell lines, available in the laboratory, were cultured in DMEM supplement with 10% FBS at 37 °C in 5% CO_2 incubator. Mixture of transfection reagent containing recombinant plasmids, DMEM, Lipofectainine 3000 and P3000 (Invitrogen, USA) were used to transfect COS-7 cells with 6-well plates. 250 μL of the transfection reagent was seeded into each well of 6-well plates containing COS-7 cells at densities of 78–80%. DMEM was supplemented in each well up to 2 mL with 2% FBS. Then, the cells were cultured at 37 °C in 5% CO_2 incubator.

Preparation of DEV CHv anti-gL mouse polyclonal antibody

Recombinant plasmid pET-32a(+)-UL1t was transformed into *E. coli* Rosetta and induced to expression for 2 h, 3 h, 4 h and 5 h with 1 mM IPTG at 37 °C to ensure the best induced time, and induced to expression with 0.2, 0.4, 0.6, 0.8, 1.0 mM IPTG at 37 °C for 4 h to ensure the best concentration of IPTG. Then, UL1t protein was induced with 0.6 mL IPTG at 37 °C for 4 h following purified with nickel agarose to bind with the His-tag of the pET-32a(+)-UL1t plasmid. Finally, the purified UL1 protein was confirmed by SDS-PAGE analysis (data not shown). Four male Kunming mice (Chengdu Dashuo Laboratory Animal Technology Co., Ltd) around 18-22 g of 6 weeks old were immunized with purified UL1t protein for four times at 4r week intervals. After immunization, the serum containing anti-gL polyclonal antibody was collected, and the euthanasia via cervical dislocation of mice was performed according to the Canadian Council on Animal Care (CCAC) guidelines and the American Veterinary Medical Association (AVMA) guidelines.

Indirect immunofluorescence assay

After plasmids transfection for 48 h, the cells were processed as follows: fixation with 4% paraformaldehyde, permeabilization of cells with 0.25% Triton-X100, and sealing with 5% BSA for 1 h separately. The cells were incubated with one of the following antibodies for 1 h: anti-Myc rabbit or anti-gL mouse polyclonal antibody (1:1000 dilution), and incubated with goat anti-rabbit IgG or goat anti-mouse IgG secondary antibodies (1: 1000 dilution) for 1 h. Also, the cell nucleus was stained with 4′,6-diamidino-2-phenylindole (DAPI). Finally, cells were observed by fluorescence microscopy (Nikon, Japan) [31].

Co-immunoprecipitation

Based on the protocol of Co-immunoprecipitation [14], the COS-7 cells were harvested in phosphate buffered saline (PBS) after co-transfection of experimental group (pCMV-Flag-gL and pCMV-Myc-gH recombinant plasmids), and control group (Control 1: pCMV-Flag-gL and pCMV-Myc or Control 2: pCMV-Flag and pCMV-Myc-gH plasmids) into COS-7 cells respectively for 48 h, followed by incubation in lysis buffer (containing 1% PMSF) for 30 min on ice. Then, the cells were centrifuged for 30 min in a 4 °C (Thermo Fisher Scientific, USA) and 14,000×g speed to pellet debris. SureBeads Magnectic Beads System (Bio-Red) was introduced in the Co-IP experiment. SureBeads Protein A was washed thrice in PBST and added 100 μL into every

200 μL IP antibody (anti-Myc rabbit or anti-gL mouse polyclonal antibody were used at dilutions of 1:1000). After incubation for 30 min at room temperature, the beads were pulled to the side of the tube by the Magnetic Racks, followed by discarding of the supernatant. After being washed three times with PBST, the beads were added into the lysates, and incubated at room temperature for 1 h. Finally, the beads were pulled to the side by the Racks, and washed three times with PBST, followed by mixing evenly with 1 × SDS loading buffer for analysis through Western blotting [2].

Western blot

The extracts or lysates with loading buffer were electrophoresed by SDS-PAGE, and then electro-transferred to polyvinylidene fluoride (PVDF) membranes through trans-blot SD (Bio-Rad, USA). Membranes were incubated with one of the antibodies as follows: anti-gL mouse polyclonal antibody (1: 1000 dilution), anti-Flag mouse or anti-Myc rabbit polyclonal antibody (1: 1000 dilution). Then, either the HRP-conjugated goat anti-rabbit IgG or goat anti-mouse IgG (1: 3000 dilution) was incubated as secondary antibody. The membrane was washed three times by tris-buffered saline and tween 20 (TBST, containing 8 g NaCl, 0.2 g KCl, 3 g Tris-base, 0.05% Tween 20 in each 1 L, pH 7.4) after incubation with antibodies in each of the step above, and finally detected through ECL reagent [38].

Results

Construction of the expression plasmids

Sequencings of amplified DEV CHv UL1 and UL22 were identified through alignment with known sequences published in NCBI. The structure of recombinant plasmid pET-32a(+)-UL1t was confirmed by digestion with the restriction enzymes *EcoR*I and *Hind*III (Fig. 1a) and sequencing. The plasmids pEGFP-N-gH and pDsRED-N-gL were also identified by digestion with *EcoR*I and *Kpn*I (Fig. 1b), while the plasmid pCMV-Flag-gL was confirmed by digestion with *EcoR*I and *Sal*I (Fig. 1c), and pCMV-Myc-gH with *EcoR*I and *Xho*I (Fig. 1d). These recombinant plasmids were further confirmed by sequencing analysis (data not shown).

Intracellular localization of DEV gL and gH in COS-7 cells respectively

After transfection of COS-7 cells with pDsRED-N-gL and pEGFP-N-gH expression plasmids respectively, the cells were observed by fluorescence microscopy in 12 h, 24 h, 36 h, 48 h and 60 h. A red low fluorescence of plasmid pDsRED-N-gL was observed starting 12 h post transfection, with gradual increase in intensity from 24 h to 60 h, suggesting that the expression of gL started at 12 h and increased from 24 h to 60 h after transfection. The observation of red fluorescence also indicated the localization of expressed gL in COS-7. Compared with the even distribution in the cells of pDsRED (negative control), pDsRED-N-gL was randomly distributed as dots around the nucleus which was stained with DAPI

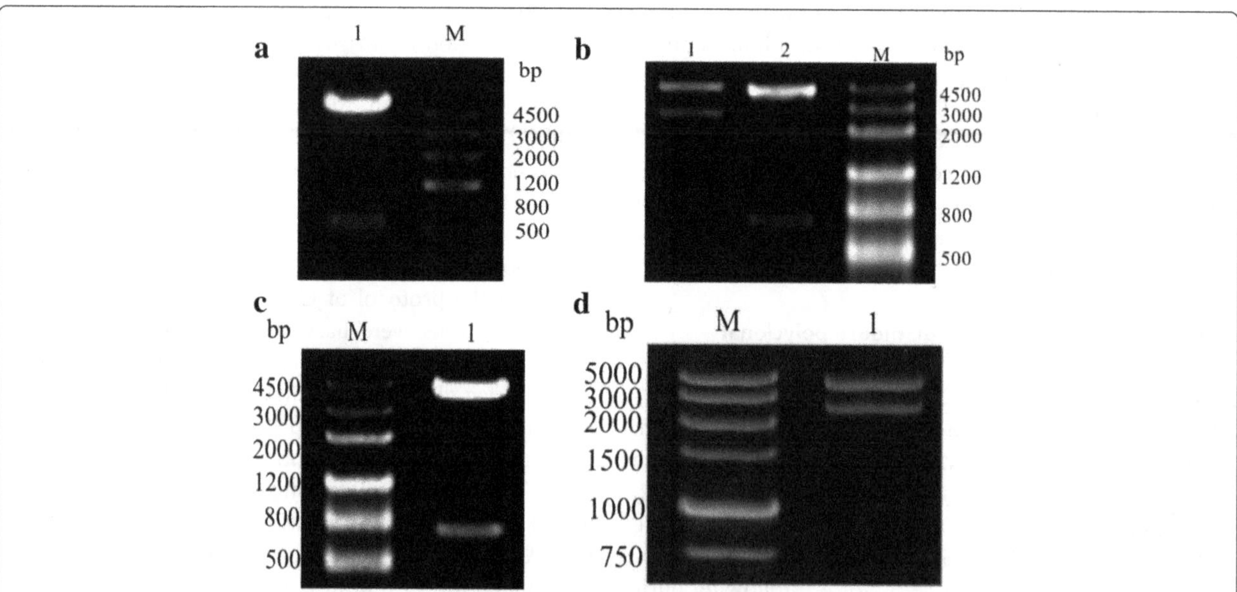

Fig. 1 The identification of recombinant plasmids with enzyme digestionThe expression plasmids were identified through enzyme digestion. **a** Lane 1: pET32a (+)-UL1t digested with two target enzymes; M: DNA Marker. **b** Lane 1: pEGFP-N-gH digested with two target enzymes; Lane 2: pDsRED-N-gL digested with two target enzymes; M: DNA Marker. **c** Lane 1: pCMV-Flag-gL digested with two target enzymes; M: DNA Marker. **d** Lane 1: pCMV-Myc-gH plasmid digested with two target enzymes; M: DNA Marker.

in blue (Fig. 2a). Similarly, the observation of green fluorescence suggested the localization of expressed gH in COS-7. The green fluorescence of plasmid pEGFP-N-gH started at 12 h and increased from 24 h to 60 h after transfection. It also showed that pEGFP-N-gH was randomly distributed in the cytoplasm as well as the nucleus, while the EGFP was distributed in the entire cells at 12 h, 24 h, 36 h, 48 h and 60 h (Fig. 2b). Thus, the expression of both DEV gL and gH appears to have started at 12 h, and increased from 24 h to 60 h after transfection. Finally, gL localized in the cytoplasm while the gH localized in both of the cytoplasm and the nucleus in COS-7 cells.

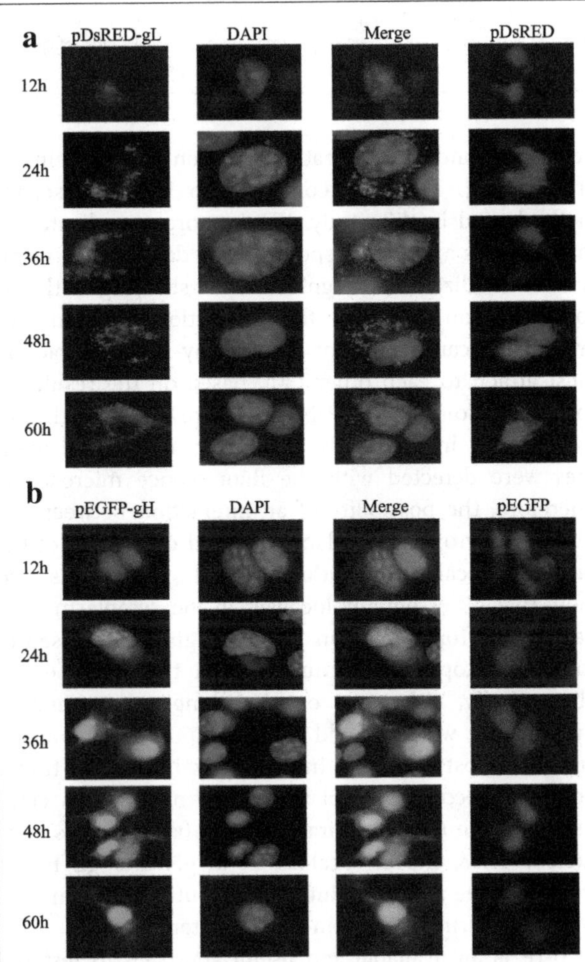

Fig. 2 Intracellular localization and expression phase of DEV gH and gL respectively in COS-7 cells. **a** In the first column, red shows the expression and intracellular localization of pDsRED-N-gL. In the second column, blue shows the nuclear stained with DAPI. The third column represents the merged images. The vector pDsRED-N as a negative control was showed in the fourth column. **b** Similarly, green shows the expression and intracellular localization of pEGFP-N-gH; blue shows the nuclear stained with DAPI. The vector pEGFP-N as a negative control was showed in the fourth column

Co-localization of gH and gL in COS-7 cells

In order to ascertain the co-localization of gH and gL, the two plasmids pDsRED-N-gL and pEGFP-N-gH were co-transfected into COS-7 cells. The vectors pDsRED-N and pEGFP-N were co-transfected into COS-7 cells as negative controls. After the nucleus was stained with DAPI, the cells with co-transfection plasmids were observed through fluorescence microscope (Fig. 3). As seen in Fig. 2, pDsRED-N-gL expressed as red fluorescence while pEGFP-N-gH expressed as green fluorescence. The expression was detected at 24 h, and gradually increased until 60 h. Interestingly, when the expression signals of pDsRED-N-gL, pEGFP-N-gH and nucleus merged together, optically an orange fluorescence appeared at points throughout the cytoplasm. If merged results suggest that the orange dots represented the co-localization sites of gH and gL (Fig. 3), then the fluorescence microscope observation as described above indicates the partial co-localization of gH and gL in COS-7 cells.

Co-immunoprecipitation analysis of the interaction of gH and gL

- On the basis of the co-transfection protocols above, recombinant plasmids pCMV-Myc-gH and pCMV-Flag-gL were co-transfected into COS-7 cells. The expressions of gH and gL after co-transfection was identified through IFA (Fig. 4). Compared to the negative control, green and red fluorescence was detected after incubation with either anti-gL or anti-Myc specific antibody (Fig. 4a, b), indicating the expression of gL and gH in COS-7 cells respectively. After co-transfection of experimental and control group 1 or 2 into COS-7 cells for 48 h, the glycoproteins in the experiment group pulled down by anti-gL antibody and detected by anti-Myc antibody with western-blotting, showed in a band around 93 kDa (gH) (Fig. 5a). Similarly, the glycoproteins in experiment group pulled down by anti-Myc antibody and detected by anti-gL antibody, showed in a band around 27 kDa (gL) (Fig. 5a). No visible band was observed in either of the control groups (Fig. 5a).

Further, the input of either gL or gH was tested by Western-blot. The results showed a band around 93 kDa (gH) and a band around 27 kDa (gL), confirming the expression of gH and gL respectively in COS-7 cells (Fig. 5b). Thus, both western-blotting results of gH and gL confirmed the interaction between them, a phenomenon seen in line with our expectations.

Discussion

The interaction between glycoprotein H and L has been confirmed in HSV, HCMV and EBV, and gH/gL is

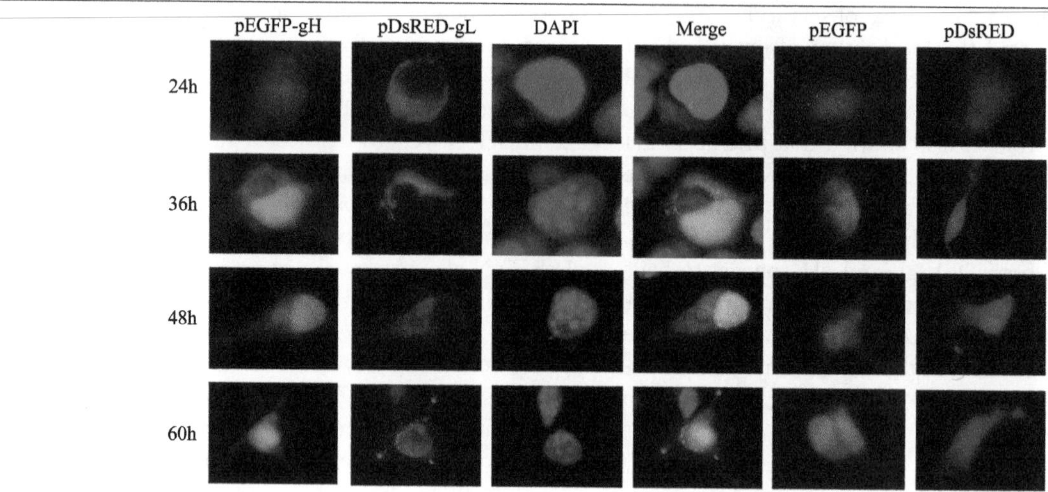

Fig. 3 Co-localization of DEV gH and gL in COS-7 cells. In the first column, green shows the localization of pEGFP-N-gH. In the second column, red shows the localization of pDsRED-N-gL. In the third column, blue shows the nuclear stained with DAPI. The merged imagine of co-localization of gH and gL was showed in the fourth column. The vectors pEGFP-N and pDsRED-N were showed as negative controls in the fifth and sixth columns

conserved in the herpesvirus family [3, 7, 29]. Since DEV belongs to α-herpesvirus subfamily, the interaction of gH and gL was hypothesized in advance [3]. An analysis of the sequencings in GenBank and the prediction results generated using bioinformatics suggest that glycoprotein H is encoded by UL22 gene and has 834 amino acids (around 92.5 kDa) with one signal peptide and two transmembranes, while glycoprotein L, is encoded by UL1 gene, and has 234 amino acids (around 26.2 kDa) with one signal peptide but without a transmembrane region (data not shown).

An analysis of the similarities between DEV, HSV-2 and EBV was done by aligning the DNA sequence of both DEV gL and gH with either HSV-2 or EBV using the online alignment website (https://www.ebi.ac.uk/Tools/psa/emboss_needle/nucleotide.html). The alignment results showed that the percentage of similarity of DEV gH with HSV-2 and EBV is 39.4 and 34.4% respectively, while the percentage of similarity of DEV gL with HSV-2 and EBV is 42.6 and 33% respectively. It was been suggested that either gH or gL of DEV show some similarities with gH or gL of HSV-2 and EBV (data not shown). In addition, previous studies have demonstrated that the interaction regions of gH and gL in both HSV-2 (H1 domain) and EBV (Domain-1) are located in the N terminal of gH [8, 26]. Based on the analysis of both DNA similarity and protein construction, the location of the interaction domain of DEV gH and gL was hypothesized as being somewhere in the N-terminal of gH. However, further studies are needed to localized the interaction sites of the DEV heterodimer [12].

Fluorescence proteins can be used to study the intracellular co-localization between two different proteins [22, 24]. IFA is also a valid technique to analyze the expression and co-localization between two proteins at present [35]. However, compared to the fluorescence method used in this study, IFA is more complicated in that requires antibody responses. The detection of intracellular localization is significant for studying both the co-localization and also the interaction between two proteins, because proteins which may have interaction must attach to each other [30]. Based on the results of co-transfection of pEGFP-N-gH and pDsRED-N-gL into COS-7 cells in this study, a number of co-localization sites were detected with the fluorescence microscope. Therefore, the possibility of an interaction between gH and gL in cytoplasm is plausible based on finding of the partial co-localization within the cell. The results also show that gL essentially localizes in the cytoplasm after transfection for more than 12 h, and gH is expressed in both the cytoplasm and nucleus over the same period. Theoretically, either gH or gL belongs to membrane glycoprotein, which should play an important role in the surface of host cells [19]. In this study, however, a transfection of recombinant plasmids into mammalian cells, instead of the naturally viral infection (in vivo) was conducted. Thus, the co-localization of gH and gL might provide a clue about the interaction, but it is not enough to fully confirm the protein-protein interaction.

Co-IP is an available and useful approach to test the interaction between two different proteins currently [25]. Some research has proved the protein-protein interaction through Co-IP assay [15, 37], although the quality of specific antibodies are significantly essential [25]. Basically, massive cellular protein-protein interactions remained, after non-denaturing cell lysis. Once protein B interacts with protein A, protein B is be drawn to the side of the tube together with protein A

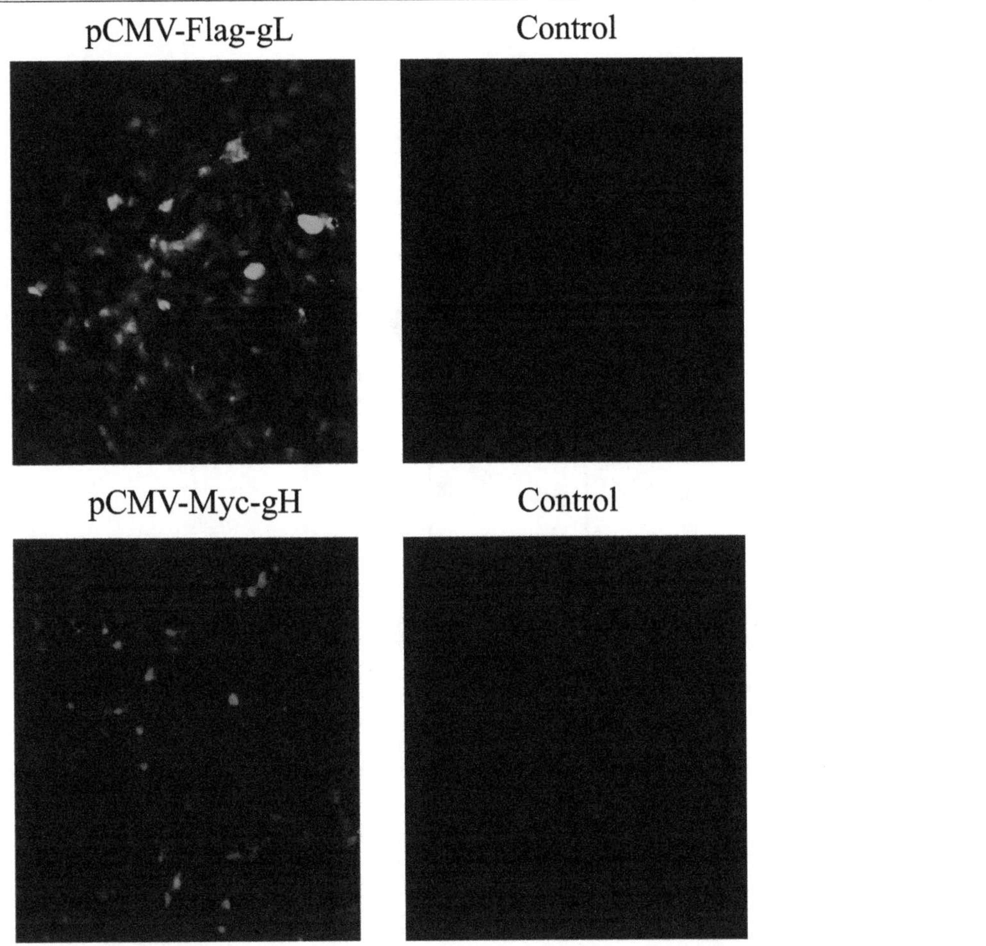

Fig. 4 Identification of expressions of pCMV-Flag-gL and pCMV-Myc-gH in COS-7 cells through IFA. Recombinant plasmids pCMV-Myc-gH and pCMV-Flag-gL were co-transfected into COS-7, and identified the expressions through IFA. **a** Identification of expressions of pCMV-Flag-gL through anti-gL mouse polyclonal antibody; **b** Identification of expressions of pCMV-Myc-gH through anti-myc rabbit polyclonal antibody; Control: Incubated COS-7 cells as negative control

by the Magnetic Racks, after the tethering of protein A with specific antibodies [13]. Thereafter, the existence of protein B with SDS-PAGE is tested, in order to confirm the interaction between protein A and B. In this paper, two experimental groups and two corresponding control groups were set respectively during the Co-IP assay. Contrary to the negative result in the control group, gH was detected when anti-gL antibody was used in Co-IP. Similarly, gL was detected when anti-Myc antibody (Myc-gH fusion protein) was used. Thus, detecting gH and gL respectively through SDS-PAGE, confirmed the interaction between gH and gL, on the basis of results of co-localization. The results from Co-IP are plausible because the bond between gH and gL is formed in a natural environment. It could be used to test not only the interaction between two or more proteins, but also the possible companion of specific receptors [13, 25].

In order to support the specific antibody during Co-IP experiment, the anti-gL mouse polyclonal antibody was prepared as a first step. Because gH protein could not be purified (data not shown), the Myc-gH tagged fusion protein was prepared for the subsequent test. Secondly, as both the gL and gH are membrane proteins of DEV, their interaction should occur on the cellular surface in vivo. However, instead of anti-gH antibody a Myc-tagged rabbit polyclonal antibody was used in the test, thus the Co-IP could not be conducted with both anti-gL and anti-gH specific antibodies. In other words, the interaction between gH and gL in natural state could not be tested. Although co-transfection of plasmids into mammalian cells could test the interaction of gH and gL as well, it neglects the possible scenario that the binding of gH and gL requires the intervention of a third protein [36], or the presence of some specific receptors [6, 16–18]. Based on the Co-IP results, however, the interaction was detected by

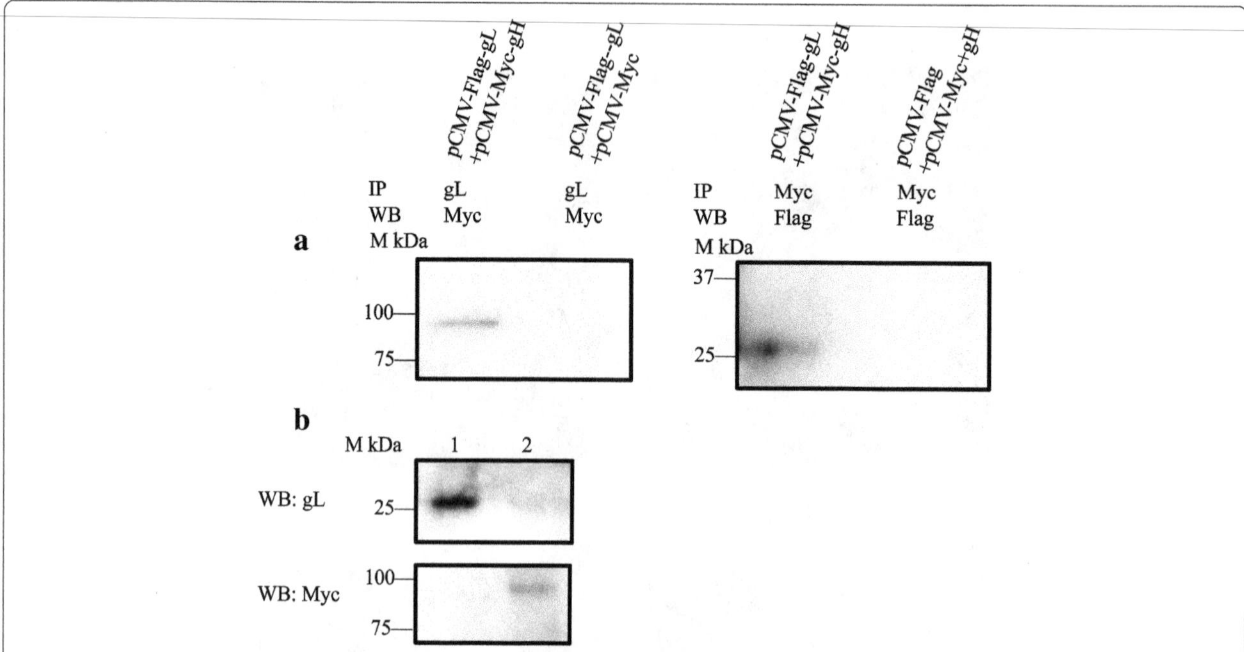

Fig. 5 Co-immunoprecipitation of pCMV-Flag-gL and pCMV-Myc-gH. **a** COS-7 cells were co-transfected with experimental group (pCMV-Flag-gL + pCMV-Myc-gH) and control groups (pCMV-Flag-gL + pCMV-Myc or pCMV-Flag-gL + pCMV-Myc-gH) respectively. The uses of antibodies were as following: IP: anti-gL or anti-Myc polyclonal antibody for immunoprecipitation; WB: anti-Myc or anti-Flag polyclonal antibody for Western-blotting. **b** Recombinant plasmids pCMV-Flag-gL and pCMV-Myc-gH were co-transfected into COS-7 cells. Line 1: cellular extracts that analyzed by Western-blotting through anti-gL mouse polyclonal antibody; Line 2: Extracts that analyzed through anti-Myc rabbit polyclonal antibody

co-transfection with two recombinant plasmids (in vitro), suggesting that gH can interact with gL without other proteins or specific receptors. But it would be better if both anti-gH and anti-gL specific antibody could be used in the experiment.

Moreover, considering the important function of membrane proteins such as gH, gL and gB during the entry of herpesviruses into host cells, the interactions among these proteins with binding their specific receptors are indispensable [11, 18, 21]. For example, in human herpesviruses, gH/gL can bind with gD, gp42, gQ or UL128–131 before the viral entry into host cells [1, 2, 24, 26]. During membrane fusion between HSV and host cells, gH/gL is initiated by receptor-binding gD at the first step, before activating the fusogenic potential of gB [16]. Also, glycoprotein D is found to be one of the essential membrane proteins in DEV, and has a similar function to that known in other herpesviruses [11]. Therefore, the hypothesis that DEV gD may interact with gH/gL during the entry into host cells could be accepted. Since gH and gL are essential proteins for DEV, it could be hypothesized that gH/gL would bind with other glycoproteins to promote DEV entry into host cells. A further study of the structure and function of gH/gL will be helpful to the study of protein-protein interaction in DEV, and also to the study of cell-cell fusion between herpesviruses and host cells.

Conclusions

This research paper describes the co-localization of some of DEV gH and gL in COS-7, as well as the interaction between them through co-immunoprecipitation. The results provide an insight to the study of protein-protein interaction in DEV. In addition, because gH and gL are essential proteins for DEV, further study of the function or construction of gH/gL could be helpful not only to the study of membrane fusion, but also to the study of immunology and prevention of DEV.

Abbreviations

Co-IP: Co-Immunoprecipitation; COS-7: African green monkey kidney cells; DEFs: Duck embryonic fibroblasts; DEV: Duck enteritis virus; DMEM: Dulbecco's modified eagle's medium; EBV: Epstein-Barr Virus; FBS: Fetal bovine serum; HSV: Herpes simplex virus; IFA: Indirect immunofluorescence assay; IPTG: Isopropyl-β-d-thiogalactoside; PBS: Phosphate buffered saline solution; SDS-PAGE: Sodium dodecyl sulfate polyacrylamide gel electrophoresis; TBST: Tris-buffered saline and tween 20

Acknowledgments

We would like to thank our funding sources.

Funding

Design of the study was supported by National Key Research and Development Program of China (2017YFD0500800), preparation of material was supported by National Key R &D Program (2016YFD0500800) and Sichuan Province Research Programs (2017JY0014/2017HH0026), collection, analysis, and interpretation of data were supported by China Agricultural Research System (CARS-43-8) and authors of this paper receive financially supported by Sichuan Province Research Programs (2017JY0014/2017HH0026).

Authors' contributions
DSF and RYJ conceived and designed the experiments. DSF and MC performed the experiments. DSF, MC and RYJ analyzed the data. DSF drafted the manuscript; SYL and RYJ has revised the manuscript. ZQY, QY DKZ, XXZ and YW has contributed reagents, materials and analysis tools, SC and MFL contributed for English proofreading; MSW and ACC were responsible for revising the manuscript critically for expert content. All of the authors read and approved the final manuscript.

Competing interests
The authors declare that they have no competing interests.

Author details
[1]Research Center of Avian Disease, College of Veterinary Medicine, Sichuan Agricultural University, Chengdu 611130, Sichuan, China. [2]Institute of Preventive Veterinary Medicine, Sichuan Agricultural University, Chengdu 611130, Sichuan, China. [3]Key Laboratory of Animal Disease and Human Health of Sichuan Province, Chengdu 611130, Sichuan, China.

References

1. Atanasiu D, Saw WT, Cohen GH, Eisenberg RJ. Cascade of events governing cell-cell fusion induced by herpes simplex virus glycoproteins gD, gH/gL, and gB. J Virol. 2010;84:12292–9.
2. Beyer AR, Bann DV, Rice B, Pultz IS, Kane M, Goff SP, Golovkina TV, Parent LJ. Nucleolar trafficking of the mouse mammary tumor virus gag protein induced by interaction with ribosomal protein L9. J Virol. 2013;87:1069–82.
3. Browne H. The role of glycoprotein H in herpesvirus membrane fusion. Protein Pept Lett. 2009;16:760–5.
4. Campadelli-Fiume G, Menotti L, Avitabile E, Gianni T. Viral and cellular contributions to herpes simplex virus entry into the cell. Curr Opin Virol. 2012;2:28–36.
5. Chang Y, Cesarman E, Pessin MS, Lee F, Culpepper J, Knowles DM, Moore PS. Identification of herpesvirus-like DNA sequences in AIDS-associated Kaposi's sarcoma. Science. 1994;266:1865–9.
6. Cheshenko N, Trepanier JB, González PA, Eugenin EA, Jacobs WR, Herold BC. Herpes simplex virus type 2 glycoprotein H interacts with integrin αvβ3 to facilitate viral entry and calcium signaling in human genital tract epithelial cells. J Virol. 2014;88:10026–38.
7. Chesnokova LS, Nishimura SL, Hutt-Fletcher LM. Fusion of epithelial cells by Epstein–Barr virus proteins is triggered by binding of viral glycoproteins gHgL to integrins αvβ6 or αvβ8. P Natl Acad Sci. 2009;106:20464–9.
8. Chowdary TK, Cairns TM, Atanasiu D, Cohen GH, Eisenberg RJ, Heldwein EE. Crystal structure of the conserved herpesvirus fusion regulator complex gH-gL. Nat Struct Mol Biol. 2010;17:882–8.
9. Ciferri C, Chandramouli S, Donnarumma D, Nikitin PA, Cianfrocco MA, Gerrein R, Feire AL, Barnett SW, Lilja AE, Rappuoli R. Structural and biochemical studies of HCMV gH/gL/gO and Pentamer reveal mutually exclusive cell entry complexes. Proc Natl Acad Sci. 2015;112(6):1762–72.
10. Connolly SA, Jackson JO, Jardetzky TS, Longnecker R. Fusing structure and function: a structural view of the herpesvirus entry machinery. Nat Rev Microbiol. 2011;9:369–81.
11. Eisenberg RJ, Atanasiu D, Cairns TM, Gallagher JR, Krummenacher C, Cohen GH. Herpes virus fusion and entry: a story with many characters. Viruses. 2012;4:800–32.
12. Fan Q, Longnecker R, Connolly SA. A functional interaction between herpes simplex virus 1 glycoproteins gH/gL domains I-II and gD is defined using α-herpesvirus gH and gL chimeras. J Virol. 2015; JVI. 00740–00715
13. Foltman M, Sanchez-Diaz A (2016) Studying protein–protein interactions in budding yeast using co-immunoprecipitation. Yeast cytokinesis. Springer, pp 239-256.
14. Free RB, Hazelwood LA, Sibley DR. Identifying novel protein-protein interactions using co-immunoprecipitation and mass spectroscopy. Cur Protocol Neur. 2009; 5.28. 21–25.28. 14
15. Gao X, Jia R, Wang M, Yang Q, Chen S, Liu M, Yin Z, Cheng A. Duck enteritis virus (DEV) UL54 protein, a novel partner, interacts with DEV UL24 protein. Virol J. 2017;14:166.
16. Gianni T, Cerretani A, DuBois R, Salvioli S, Blystone SS, Rey F, Campadelli-Fiume G. Herpes simplex virus glycoproteins H/L bind to cells independently of αVβ3 integrin and inhibit virus entry, and their constitutive expression restricts infection. J Virol. 2010;84:4013–25.
17. Gianni T, Salvioli S, Chesnokova LS, Hutt-Fletcher LM, Campadelli-Fiume G. αvβ6-and αvβ8-integrins serve as interchangeable receptors for HSV gH/gL to promote endocytosis and activation of membrane fusion. Plos Pathog 9. 12 (2013): e1003806.
18. Gianni T, Massaro R, Campadelli-Fiume G. Dissociation of HSV gL from gH by αvβ6-or αvβ8-integrin promotes gH activation and virus entry. P Natl Acad Sci. 2015;112:E3901–10.
19. Heldwein E, Krummenacher C. Entry of herpesviruses into mammalian cells. Cell Mol Life Sci. 2008;65:1653–68.
20. Hutchinson L, Browne H, Wargent V, Davis-Poynter N, Primorac S, Goldsmith K, Minson A, Johnson D. A novel herpes simplex virus glycoprotein, gL, forms a complex with glycoprotein H (gH) and affects normal folding and surface expression of gH. J Virol. 1992;66:2240–50.
21. Hutt-Fletcher LM. Epstein-Barr virus entry. J Virol. 2007;81:7825–32.
22. Koyama-Honda I, Ritchie K, Fujiwara T, Iino R, Murakoshi H, Kasai RS, Kusumi A. Fluorescence imaging for monitoring the colocalization of two single molecules in living cells. Biophys J. 2005;88:2126–36.
23. Li Y, Bing H, Ma X, Jing W, Feng L, Wu A, Song M, Yang H. Molecular characterization of the genome of duck enteritis virus. Virology. 2009;391: 151–61.
24. Lin M, Jia R, Wang M, Gao X, Zhu D, Chen S, Liu M, Yin Z, Wang Y, Chen X. Molecular characterization of duck enteritis virus CHv strain UL49. 5 protein and its colocalization with glycoprotein M. J Vet Sci. 2014;15:389–98.
25. Masters SC. Co-immunoprecipitation from transfected cells. Methods Mol Biol. 2004;261:337.
26. Matsuura H, Kirschner AN, Longnecker R, Jardetzky TS. Crystal structure of the Epstein-Barr virus (EBV) glycoprotein H/glycoprotein L (gH/gL) complex. Proc Natl Acad Sci. 2010;107:22641–6.
27. McGeoch DJ, Rixon FJ, Davison AJ. Topics in herpesvirus genomics and evolution. Virus Res. 2006;117:90–104.
28. Mettenleiter TC, Klupp BG, Granzow H. Herpesvirus assembly: a tale of two membranes. Curr Opin Microbiol. 2006;9:423–9.
29. Molesworth SJ, Lake CM, Borza CM, Turk SM, Hutt-Fletcher LM. Epstein-Barr virus gH is essential for penetration of B cells but also plays a role in attachment of virus to epithelial cells. J Virol. 2000;74:6324–32.
30. Nair R, Rost B. LOC3D: annotate sub-cellular localization for protein structures. Nucleic Acids Res. 2003;31:3337–40.
31. Poncelet P, Carayon P. Cytofluorometric quantification of cell-surface antigens by indirect immunofluorescence using monoclonal antibodies. J Immunol Methods. 1985;85:65–74.
32. Roizman B, Pellett P. The family Herpesviridae: a brief introduction. Fields virology. 2001;2:2381–97.
33. Roop C, Hutchinson L, Johnson DC. A mutant herpes simplex virus type 1 unable to express glycoprotein L cannot enter cells, and its particles lack glycoprotein H. J Virol. 1993;67:2285–97.
34. Shawky S, Sandhu T, Shivaprasad H. Pathogenicity of a low-virulence duck virus enteritis isolate with apparent immunosuppressive ability. Avian Dis. 2000:590–9.
35. Surapureddi S, Svartz J, Magnusson KE, Hammarström S, Söderström M. Colocalization of leukotriene C synthase and microsomal glutathione S-transferase elucidated by indirect immunofluorescence analysis. FEBS Lett. 2000;480:239–43.
36. Turner A, Bruun B, Minson T, Browne H. Glycoproteins gB, gD, and gHgL of herpes simplex virus type 1 are necessary and sufficient to mediate membrane fusion in a cos cell transfection system. J Virol. 1998;72:873–5.
37. Verhelst J, De VD, Saelens X. Co-immunoprecipitation of the mouse Mx1 protein with the influenza a virus nucleoprotein. J Vis Exp. 2015;2015 e52871-e52871
38. Wu Y, Li Q, Chen X-Z. Detecting protein-protein interactions by far western blotting. Nat Protocol. 2007;2:3278.

First detection of *European bat lyssavirus* type 2 (EBLV-2) in Norway

Torfinn Moldal[1*], Turid Vikøren[1], Florence Cliquet[2], Denise A. Marston[3], Jeroen van der Kooij[4], Knut Madslien[1] and Irene Ørpetveit[1]

Abstract

Background: In Europe, bat rabies is primarily attributed to *European bat lyssavirus* type 1 (EBLV-1) and *European bat lyssavirus* type 2 (EBLV-2) which are both strongly host-specific. Approximately thirty cases of infection with EBLV-2 in Daubenton's bats (*Myotis daubentonii*) and pond bats (*M. dasycneme*) have been reported. Two human cases of rabies caused by EBLV-2 have also been confirmed during the last thirty years, while natural spill-over to other non-flying mammals has never been reported. Rabies has never been diagnosed in mainland Norway previously.

Case presentation: In late September 2015, a subadult male Daubenton's bat was found in a poor condition 800 m above sea level in the southern part of Norway. The bat was brought to the national Bat Care Centre where it eventually displayed signs of neurological disease and died after two days. EBLV-2 was detected in brain tissues by polymerase chain reaction (PCR) followed by sequencing of a part of the nucleoprotein gene, and lyssavirus was isolated in neuroblastoma cells.

Conclusions: The detection of EBLV-2 in a bat in Norway broadens the knowledge on the occurrence of this zoonotic agent. Since Norway is considered free of rabies, adequate information to the general public regarding the possibility of human cases of bat-associated rabies should be given. No extensive surveillance of lyssavirus infections in bats has been conducted in the country, and a passive surveillance network to assess rabies prevalence and bat epidemiology is highly desired.

Keywords: Rabies, Daubenton's bat (Myotis daubentonii), European bat lyssavirus type 2 (EBLV-2), fluorescent antibody test (FAT), polymerase chain reaction (PCR), rabies tissue culture infection test (RTCIT)

Background

Rabies is a fatal zoonotic neurological disease caused by RNA viruses belonging to the genus *Lyssavirus* in the family *Rhabdoviridae. Rabies virus* (RABV) is the prototype of the genus and causes approximately 59,000 deaths in humans yearly [1]. Red fox (*Vulpes vulpes*) is the main reservoir for RABV in Europe, where the disease has been eliminated in many member states of the European Union due to oral vaccination programs [2]. Rabies in non-flying mammals has never been diagnosed in mainland Norway [3]. However, in the Svalbard archipelago, which is under Norwegian jurisdiction, several occasional detections and two rabies outbreaks during the last 35 years have been reported and are related to arctic strains of RABV in Arctic fox (*Vulpes lagopus*), Svalbard reindeer (*Rangifer tarandus platyrhyncus*) and one ringed seal (*Pusa hispida*) [4–6].

The genus Lyssavirus consists of 14 recognized species [7], and all but two have been isolated from bats [8]. The different species display distinct features regarding geographical distribution and host specificity. In Europe, bat rabies is primarily attributed to *European bat lyssavirus* type 1 (EBLV-1) and *European bat lyssavirus* type 2 (EBLV-2) [9]. Most reported bat rabies cases are caused by EBLV-1 which is almost exclusively found in serotine bats (*Eptesicus serotinus*) and Isabelline serotine bats (*E. isabellinus*), while EBLV-2 is detected mainly in Daubenton's bats (*Myotis daubentonii*) and to a lesser extent in pond bats (*M. dasycneme*) [9]. Evolutionary studies suggest that EBLV-2 diverged from other lyssaviruses more than 8000 years ago and that the current

* Correspondence: torfinn.moldal@vetinst.no
[1]Norwegian Veterinary Institute, Postbox 750, Sentrum, 0106 Oslo, Norway
Full list of author information is available at the end of the article

diversity of EBLV-2 has built up during the last 2000 years [10].

Hitherto, EBLV-2 has been detected in bats in the United Kingdom, the Netherlands, Germany, Switzerland, Finland and Denmark [11–16], while antibodies against EBLV-2 have been found in bats in Sweden [17]. Two cases of human rabies associated with EBLV-2 have been reported in Finland [18] and in Scotland [19]. To date, no cases of natural spill-over to other non-flying mammals have been reported, but EBLV-2 has successfully been transferred to sheep and foxes in experimental studies, even though the susceptibility seems to be low in these species [20, 21].

Among the 12 species of bats regularly reported in Norway, the northern bat *(E. nilssonii)*, the soprano pipistrelle (*Pipistrellus pygmaeus*) and the Daubenton's bat are the most common and widespread species [22–24]. The serotine bat has been recorded once and is not considered as a resident species [25]. The range of the Daubenton's bat reaches to 63° N in Norway [24]. The results of ringing experiments in other countries suggest that this species does not migrate over long distances [26], which is likely also the case in Norway. The Daubenton's bat is specialized in hunting insects and spiders, which it mainly catches with its feet from the water surface [27], but it is also found searching for food in forests [24]. Daubenton's bats seldom occur above the timberline and roost mainly in tree cavities, but also use crevices in bridges and cliffs. Roosts in buildings are very rare [24, 28] and so are contacts with the general public [29].

There is no official surveillance program for lyssavirus in bats in Norway. According to the annals of the Norwegian Veterinary Institute (NVI), a total of 27 bats (swabs from 18 bats sampled alive and brain tissues from nine carcasses) were examined for lyssavirus in the period 1998–2015. Here, we report the first detection of EBLV-2 in Norway.

Case presentation

Clinical signs and treatment

On September 29th 2015, a bat was discovered close to a cabin in Valdres in the county of Oppland. The locality is situated just below the timberline at around 800 m above sea level at 60° 59′ N in the southern part of Norway. The landscape is undulating with open spaces and mixed forest dominated by Norway spruce (*Picea abies*) and mountain birch (*Betula pubescens ssp. czerepanovii*), lakes and small meadows. The cabin owner discovered the animal clinging to a stone when she removed a tarpaulin. Three days later, the bat was still there, and the cabin owner called the National Bat Helpline, which advised her to bring the bat to the national Bat Care Centre of the Norwegian Zoological Society (NZI). The centre is approved and partly funded by the Norwegian Environmental Agency (NEA) and the Norwegian Food Safety Authority (NFSA) [29].

The bat arrived at the Bat Care Centre the next day, on October 3rd 2016, and was identified as a male Daubenton's bat based on external characteristics, i.e. short tragus, large feet and the attachment of the wing membrane at the middle of the foot, and as a subadult individual based on the brownish face and the presence of a chin-spot (Fig. 1) [30]. The bat weighed 6.6 g on arrival and had a forearm length of 38.3 mm, hence the body condition index (mass g/forearm mm) was 0.17, which indicated low energy reserves [31]. It was weak, and the tongue was dark red and swollen. It had dark plaque on its teeth, and there was a heavy load of mites of the genus *Spinturnix* on the wing membrane surface. As a safety precaution, the diseased bat was routinely placed in a clean plastic container (35x45x40 cm) separated from other bats in the facility.

The bat drank and ate mealworm (*Tenebrio molitor*) on arrival. It walked around inside the container and on one occasion lunged against the bat carer. After a few hours the bat grew stronger, groomed itself and produced some excrement. The following day, on October 4th 2016, it gradually refused to eat, displayed difficulty when swallowing water and got weaker. The bat was force-fed several times, received fluid subcutaneously and eventually had increased locomotor activity. In the late evening it displayed mild ataxia. The following morning, the bat's condition deteriorated. The bat walked around, but its locomotion was less agile and it was not able to groom itself. The bat became even weaker towards the evening in spite of treatment with

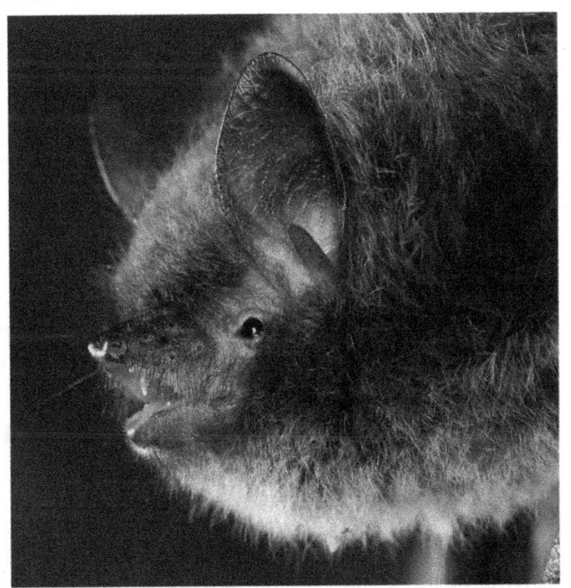

Fig. 1 The bat was identified as a subadult male Daubenton's bat based on external characteristics. Photographer: Jeroen van der Kooij

antibiotics and fluid, and when force-fed with water or fluids, it showed signs of vomiting and spread both wings and legs. It also tried to bite, but without force. The bat died in the evening of October 5th 2016 and was subsequently frozen.

Necropsy, sampling and histology

On October 6th 2015, the bat was submitted to the NVI in Oslo for laboratory examinations on suspicion of rabies. The necropsy revealed that the bat was in a poor condition with no visible body fat. The tongue was dark red and dry. The internal organs were congested with no specific gross findings.

Brain impressions on glass slides were fixed in acetone baths for fluorescent antibody test (FAT), and tissues from different parts of the brain were put on empty tubes and tubes with lysis buffer for rabies tissue culture infection test (RTCIT) and detection of viral RNA by reverse transcription and polymerase chain reaction (RT-PCR), respectively. The remains of the brain and the head including salivary glands, lung, heart, liver, kidney, spleen, gastrointestinal tract and pancreas were fixed in 10% buffered formalin for two weeks, routinely processed and embedded in paraffin before cutting ultrathin sections that were mounted on glass slides and stained with haematoxylin-eosin for histological examination.

Only small pieces of brain tissue were available for histology, and no lesions were detected. In the lungs, multifocal areas of foreign material aspiration with infiltration of inflammatory cells were seen, and a similar small focal inflammation was found in the salivary gland tissue. A few parasitic structures without any associated inflammation were detected in the liver.

FAT and RTCIT

The laboratory tests used for rabies diagnosis are recommended by WHO and OIE. FAT, in which virus antigens in brain tissues are detected, is the gold standard for diagnosing rabies [32]. Fixed brain impressions were stained with FITC Anti-Rabies Monoclonal Globulin (Fujirebio Diagnostics Inc.) according to the manufacturer's instructions. A brain impression from a rabid cow served as a positive control, while a brain impression from a healthy fox served as a negative control. All slides were investigated in a fluorescence microscope (Leitz) under ultraviolet light. Virus antigen was detected in the positive control, but not in the negative control or any of the brain impressions from the bat.

The detection of infectious particles in brain homogenates was performed by RTCIT on neuroblastoma cells with an incubation period of 48 h at 37 °C as previously described [33]. The staining was performed using a rabies anti-nucleocapsid FITC-conjugated antibody (BioRad). The brain tissues from the bat were found positive for the presence of infectious virus by RTCIT.

Nucleic acid extraction, RT-PCR and partial sequencing of the nucleoprotein gene

Total nucleic acids were extracted using the NucliSens® easyMAG™ (bioMerieux Inc.) according to the manufacturer's instructions for the off-board protocol. Nucleic acids were eluted in 55 µl buffer. Primers targeting the nucleoprotein gene were applied in a two-step RT-PCR. RT was performed with Invitrogen SuperScript® III Reverse Transcriptase (Thermo Fisher Scientific) with the primer LYSSA-NMA (5′-ATGTAACACCYCTACAATG-'3) that is modified from primer JW12 in [34].

PCR was performed with Qiagen HotStarTaq® DNA Polymerase (Qiagen) according to the manufacturer's instructions. The forward primer, LYSSA-NMB (5′-ATG TAACACCYCTACAATGGA-'3) was used in two separate PCR reactions with either LYSSA-NGA (5'-TGACTC CAGTTRGCRCACAT-'3) or LYSSA-NGC (5'-GGGTAC TTGTACTCATAYTGRTC-'3) as reverse primers, yielding amplicons of 612 bp and 108 bp respectively. The primers are modified from the primers JW12, JW6 and SB1 in [34] respectively. Amplicons were separated on 1% agarose gel at 90 V for 90 min and visualized by DNA Gel Loading Dye (Thermo Fisher Scientific).

The PCR products yielding bands of the expected size in the agarose gel electrophoresis were further analysed by DNA sequencing. Following enzymatic PCR clean up with illustra™ ExoStar™ (GE Healthcare Life Sciences), DNA sequencing was performed using the BigDye® Terminator v3.1 Cycle Sequencing Kit (Thermo Fisher Scientific) according to the manufacturer's instructions. The products were analysed in an ABI PRISM® 3100 Genetic Analyzer (Thermo Fisher Scientific) according to the manufacturer's instructions and with Sequencher® version 5.3 sequence analysis software (Gene Codes Corporation).

A sequence with 567 nucleotides (GenBank accession number KX644889) was obtained with the primers LYSSA-NMB and LYSSA-NGA. BLAST search against the GenBank database showed 94–96% identity with EBLV-2 (Table 1). The sequence was aligned with sequences for EBLV-2 published in GenBank to evaluate the genetic diversity and the relationship to other bat-associated lyssaviruses, and phylogenetic trees were constructed with MEGA 6 based on an alignment of 391 nucleotides applying the Maximum Likelihood and Neighbor-Joining algorithms with 1000 bootstrap replicates and different substitution models (Fig. 2) [35]. The trees showed similar results, and although the node is not strongly supported, the results indicate that the sequence from the Norwegian virus did not cluster closely with any other published sequences of EBLV-2.

Table 1 The site and year of collection, host species and GenBank accession number for sequences for a part of the nucleoprotein gene of *European bat lyssavirus* 2 used to generate a phylogenetic tree

Place	Country	Year	Host species	GenBank accession number	Reference(s)
Helsinki	Finland	1985	*Homo sapiens*	AY062091	[18, 61]
Wommels	Netherlands	1986	*Myotis dasycneme*	U22847	[12, 62]
Tjerkwerd	Netherlands	1987	*Myotis dasycneme*	U89480	[12, 62]
Andijk	Netherlands	1989	*Myotis dasycneme*	U89481	[12, 62]
Plaffeien	Switzerland	1992	*Myotis daubentonii*	AY212117	[14, 46]
Roden	Netherlands	1993	*Myotis dasycneme*	U89482	[12, 62]
Versoix	Switzerland	1993	*Myotis daubentonii*	U89479	[62]
Sussex	United Kingdom	1996	*Myotis daubentonii*	U89478	[11, 62]
Lancashire	United Kingdom	2002	*Myotis daubentonii*	AY212120	[46]
Angus	United Kingdom	2002	*Homo sapiens*	AY247650	[19]
Geneva	Switzerland	2002	*Myotis daubentonii*	AY863408	[63]
Surrey	United Kingdom	2004	*Myotis daubentonii*	JQ796807	[9, 47]
Lancashire	United Kingdom	2004	*Myotis daubentonii*	JQ796808	[9, 64]
Oxfordshire	United Kingdom	2006	*Myotis daubentonii*	JQ796809	[9, 48]
Magdeburg	Germany	2006	*Myotis daubentonii*	JQ796805	[9, 50]
Schwansee	Germany	2006	*Myotis daubentonii*	KF826115	[65]
Bad Buchau	Germany	2007	*Myotis daubentonii*	GU227648	[13]
Shropshire	United Kingdom	2007	*Myotis daubentonii*	JQ796810	[9, 49]
Surrey	United Kingdom	2008	*Myotis daubentonii*	JQ796811	[9]
Shropshire	United Kingdom	2008	*Myotis daubentonii*	JQ796812	[9, 66]
West Lothian	United Kingdom	2009	*Myotis daubentonii*	JQ796806	[9, 67]
Turku	Finland	2009	*Myotis daubentonii*	GU002399	[15]
Gießen	Germany	2013	*Myotis daubentonii*	KF826149	[65]
Valdres	Norway	2015	*Myotis daubentonii*	KX644889	This study

Whole genome sequencing

Total RNA was depleted of host genomic DNA (gDNA) and ribosomal RNA (rRNA) following methods described previously [36, 37]. Briefly, gDNA was depleted using the on-column DNase digestion protocol in RNeasy plus mini kit (Qiagen) following manufacturer's instructions, eluting in 30 μl molecular grade water. Subsequently, rRNA was depleted, using Terminator 5′-phosphate-dependent exonuclease (Epicentre Biotechnologies). Briefly, 30 μl of gDNA depleted RNA was mixed with 3 μl of Buffer A, 0.5 μl of RNAsin Ribonuclease inhibitor (20–40 U/μl) and incubated at 30 °C for 60 min. The depleted RNA was purified to remove the enzyme using the RNeasy plus mini kit as above, without the DNase digestion, eluting in 30 μl of molecular grade water. Double stranded cDNA (ds-cDNA) was synthesised using random hexamers, and a cDNA synthesis kit (Roche) following manufacturer's instructions. The resulting ds-cDNA was purified using AMPure XP magnetic beads (Beckman Coulter), quantified using Quantifluor (Promega) and approximately 1 ng of the ds-cDNA library was used in a 'tagmentation' reaction mix using a Nextera XT DNA sample preparation kit (Illumina) following manufacturer's instructions – without the bead normalization step. The DNA library was quantified using Quantifluor (Promega) and sequenced as 2 × 150 bp paired-end reads on an Illumina MiSeq platform.

Short reads were mapped to the most genetically related full length EBLV-2 genome available (Germany 2012 – GenBank accession number KY688149). Reads were mapped using the Burrow-Wheeler Aligner (BWA version 0.7.5a–r405) [38] and visualized in Tablet [39]. A modified SAMtools/vcfutils [40] script was used to generate an intermediate consensus sequence in which any indels and SNPs relative to the original reference sequence were appropriately called. The intermediate consensus sequence was used as the reference for four subsequent iterations of mapping and consensus calling, described previously [41].

A total of 38,288 EBLV-2 specific reads were mapped from 5,839,126 total reads (0.66%). The average depth of coverage was 373.6 with a maximum depth of coverage of 1900. Sequencing resulted in complete genomic

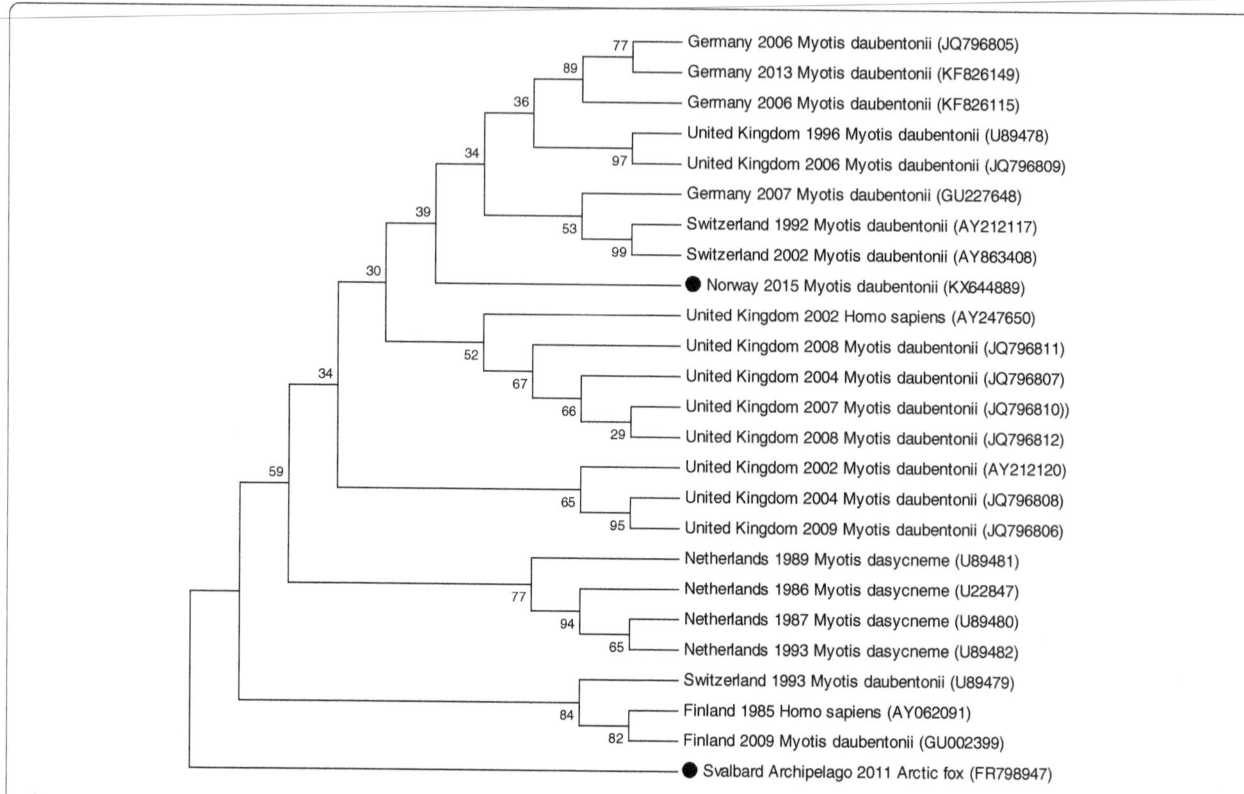

Fig. 2 Phylogenetic relationship between the Norwegian EBLV-2 sequence (●) and other EBLV-2 sequences based on 391 nucleotides from the nucleoprotein gene. The tree was constructed with the Maximum Likelihood algorithm with 1000 bootstrap replicates and the Tamura-Nei substitution model. The number at each branch of the phylogenetic tree represents the likelihood in percentage that the sequences cluster together. RABV from an Arctic fox is used as outgroup (●). The country and year of collection, host species and GenBank accession number for sequences are given

coverage, apart from the first 9 nucleotides. However, the lyssavirus genomic ends are highly conserved between species and palindromic [42, 43]. Therefore the first 9 nucleotides of the 3′ UTR sequence was deduced using the 5′ end sequence and matched 100% with other EBLV-2 genomes available. The total genomic length is 11,928 nucleotides (Table 2), conforming to other EBLV-2 genome sequences (11,924–30), where all regions are identical in length except for the non-coding regions of M-G and G-L. The complete genome sequence was submitted to GenBank (accession number KY688154).

Discussion and conclusions

Here, we report the first detection of EBLV-2 in Norway. The detection of EBLV-2 in a Daubenton's bat is in accordance with previous studies that have revealed that EBLV-2 is mainly found in this species [9]. As the bat was a subadult and ringing experiments suggest that this species does not migrate over long distances [26], it can be assumed that the bat was native to the area where it was found and that EBLV-2 is present in Norway. The latter is supported by the phylogenetic analyses, which indicate that the Norwegian isolate differs from other published isolates, further suggesting that the Norwegian isolate may have evolved separately from a common European ancestor [10]. Very few bats have been examined for rabies in Norway, and we have no knowledge of the prevalence and epidemiology of EBLV. In neighbouring countries,

Table 2 Length of coding (bold) and non-coding regions of EBLV-2 in nucleotides (nts)

Region	Length (nts)
3′ UTR	70
N protein	1356
N-P	101
P protein	894
P-M	88
M protein	609
M-G	210
G protein	1575
G-L	510
L protein	6384
5′ UTR	131
Total	**11,928**

EBLV-2 was detected in diseased Daubenton's bats in Finland in 2009 and 2016 [15, 44], while active surveillance has revealed viral RNA from mouth swabs from Daubenton's bat in Denmark and seropositive Daubenton's bats in both Sweden and Finland [16, 17, 45].

The FAT performed on brain impressions at the NVI was negative, and unfortunately, due to insufficient material, it could not be repeated at the Nancy OIE/WHO/EU Laboratory for Rabies and Wildlife (EURL). However, the EURL confirmed the presence of lyssavirus in the RTCIT. The negative result for FAT does not concur with findings in other Daubenton's bats naturally infected with EBLV-2 [11, 13–15, 44, 46–51]. The annual proficiency tests for rabies diagnosis techniques organized by the EURL have demonstrated the difficulty for laboratories to reliably detect EBLV strains when using the FAT [52], with results depending on the rabies virus antibody conjugate and even the batch used [53]. However, the same batch used in the FAT for the present case has successfully been applied at the NVI for detection of this virus species in two proficiency tests (unpublished results). A polyclonal antibody was used to confirm the presence of lyssavirus in the RTCIT, and a difference between the antibodies regarding the ability to detect the current virus cannot be excluded.

The aspiration pneumonia might have been caused by a dysfunctional swallow reflex or as a result of the force-feeding. The bat had a low state of nutrition, displayed several signs of weakness and, in the end, an inability to fly. This could be caused by the EBLV-2 infection or due to general weakness as a consequence of either shortage of feed at the high altitude or inexperience in hunting since being a subadult. Also, a poor body condition could have made the bat more susceptible for EBLV-2 infection. Weight loss, the inability to fly and death within 14 days followed by detection of viral antigen in brain tissue is reported in Daubenton's bats after intracerebral inoculation with EBLV-2 [54]. In that study the bats were fed ad libitum, so the weight loss could not be related to a decrease in hunting success. During a natural EBLV-2 infection, the course of the disease is probably longer than that found in the referred experiment, and thus the ability to hunt and eat could deteriorate over some time span, resulting in gradual loss of body condition. Poor body condition in bats can also be influenced by heavy parasite loads [55]. Other possible signs of the disease, like the inability to fly, were registered only up to 48 h before death in the experiment [54]. Most of the bats which were inoculated either intramuscularly or subdermally survived the study period of 123 days [54]. These findings highlight that there is a high intra-species barrier in transmission of EBLV-2 and that the incubation period under natural conditions is probably several weeks to months.

Two human cases of rabies caused by EBLV-2 have been reported, in Finland in 1985 and in Scotland in 2002 [18, 19]. Both persons were in close contact with bats over time and had been bitten several times without any history of immunization prior to exposure or post-exposure prophylaxis. Spill-over to other non-flying mammals under natural conditions has not been reported, and the detection of EBLV-2 does not pose an immediate risk to humans as Daubenton's bats are rarely in contact with the public [29]. However, the possibility of human cases of bat-associated rabies acquired in Norway, which is considered free of rabies, cannot be neglected; hence adequate information should be given to the general public as well as to people and professionals who come into contact with bats.

All bat species in Norway are strictly protected under both national laws and international commitments to preserve bats [56, 57]. Culling bats for lyssavirus testing is therefore not a feasible strategy. Passive surveillance by testing bats found sick or dead is the most appropriate way of assessing the incidence of rabies as compared with active surveillance by testing swabs and/or sera from live bats from natural populations [58, 59]. Passive surveillance of bat rabies is therefore strongly recommended by international organizations [58, 60].

Monitoring of EBLV in bats has so far not been prioritized by the Norwegian authorities, probably due to few human cases of bat-associated rabies being reported in Europe. According to the annals of the NVI, only seven carcasses of bats including the EBLV-2 positive case were examined during the last three years. In 2014 and 2015, the NVI in collaboration with the NZS collected and tested nasopharyngeal and cloacal swabs from totally 16 free-ranging Norwegian bats for lyssaviruses with a negative result (unpublished data). The experience gained in this pilot study and from surveillance in other countries is important for future monitoring of bat rabies in Norway.

The authors believe that testing bats found sick or dead should get priority through a national rabies surveillance network in line with the recent recommendations of EUROBATS and EFSA [58, 60]. Additionally, an active surveillance program, over a limited number of years and in defined areas to obtain baseline data on the prevalence of bat rabies in Norway, is also recommended. Finally, we emphasize the importance of good cooperation between bat biologists and laboratory workers to ensure access to representative samples of good quality.

Abbreviations

EBLV-1: *European bat lyssavirus* type 1; EBLV-2: *European bat lyssavirus* type 2; EURL: Nancy OIE/WHO/EU Laboratory for Rabies and Wildlife; FAT: Fluorescent antibody test; gDNA: Genomic DNA; NEA: Norwegian Environmental Agency; NFSA: Norwegian Food Safety Authority;

NVI: Norwegian Veterinary Institute; NZI: Norwegian Zoological Society; PCR: Polymerase chain reaction; RABV: *Rabies virus*; rRNA: Ribosomal RNA; RT: Reverse transcription; RTCIT: Rabies tissue culture infection test

Acknowledgements

The authors want to thank the cabin owner who found the bat and brought it to the Bat Care Centre and Anke Kirkeby for practical care and additional information on the behaviour of the bat. We are grateful to Marianne Heum, Kristin Stangeland Soetaert, Lene Hermansen, Faisal Suhel, Rosa Ferreira Fristad, Britt Gjerset and Hilde Sindre at the NVI, Alexandre Servat and Evelyne Picard-Meyer at the EURL and Richard J. Ellis at Animal and Plant Health Agency.

Funding

The Bat Help Line and the Bat Care Centre receive financial funding from the NEA. The necropsy, histological examination, FAT, RT-PCR and the partial sequencing of the nucleoprotein gene were funded by the NVI, while the RTCIT and the whole genome sequencing were funded by the EURL and the European Union H2020-grant 'European Virus Archive Global (EVAg)' (H2020-grant agreement number 653316) respectively.

Authors' contributions

TM was responsible for the RT-PCR and the partial sequencing of the nucleoprotein gene and coordinated the drafting of the manuscript, TV performed the necropsy, sampling and histological examination, FC was responsible for the RTCIT, DM was responsible for the whole genome sequencing, JvdK had the primary contact with the cabin owner and was responsible for the handling of the bat at the Bat Care Centre, KM coordinated the submission of the bat for necropsy, and IØ supervised the RT-PCR and the partial sequencing of the nucleoprotein gene. All authors took part in writing the manuscript and have approved the final version.

Consent for publication

Not applicable.

Competing interests

The authors declare that they have no competing interests.

Author details

[1]Norwegian Veterinary Institute, Postbox 750, Sentrum, 0106 Oslo, Norway. [2]Nancy OIE/WHO/EU Laboratory for Rabies and Wildlife, French Agency for Food, Environmental and Occupational Health & Safety, CS 40009, 54220 Malzéville, France. [3]Animal and Plant Health Agency, New Haw, Addlestone, Surrey KT15 3NB, UK. [4]Norwegian Zoological Society's Bat Care Centre, Rudsteinveien 67, 1480 Slattum, Norway.

References

1. Hampson K, Coudeville L, Lembo T, Sambo M, Kieffer A, Attlan M, et al. Estimating the global burden of endemic canine rabies. PLoS Negl Trop Dis. 2015; doi:10.1371/journal.pntd.0003709.
2. Müller T, Freuling CM, Wysocki P, Roumiantzeff M, Freney J, Mettenleiter TC, et al. Terrestrial rabies control in the European Union: historical achievements and challenges ahead. Vet J. 2015;203:10–7.
3. Heier BT, Lange H, Hauge K, Hofshagen M. Norway 2014 – Trends and source of zoonoses and zoonotic agents in humans foodstuffs, animals and feedingstuffs. Oslo: Norwegian Veterinary Institute; 2015.
4. Ødegaard ØA, Krogsrud J. Rabies in Svalbard: infection diagnosed in arctic fox, reindeer and seal. Vet Rec. 1981;109:141–2.
5. Ørpetveit I, Ytrehus B, Vikøren T, Handeland K, Mjøs A, Nissen S, et al. Rabies in an Arctic fox on the Svalbard archipelago, Norway, January 2011. Euro Surveill. 2011;16:2–3.
6. Mørk T, Bohlin J, Fuglei E, Åsbakk K, Tryland M. Rabies in the arctic fox population, Svalbard. Norway J Wildl Dis. 2011;47:945–57.
7. International Committee on Taxonomy of Viruses (ICTV). Virus Taxonomy: 2015 Release. https://talk.ictvonline.org/taxonomy/. Accessed 03 Aug 2016.
8. Banyard AC, Evans JS, Luo TR, Fooks AR. Lyssaviruses and bats: emergence and zoonotic threat. Viruses. 2014;6:2974–90.
9. McElhinney LM, Marston DA, Leech S, Freuling CM, van der Poel WHM, Echevarria J, et al. Molecular epidemiology of bat lyssaviruses in Europe. Zoonoses Public Health. 2013;60:35–45.
10. Jakava-Viljanen M, Nokireki T, Sironen T, Vapalahti O, Sihvonen L, Liisa S, et al. Evolutionary trends of European bat lyssavirus type 2 including genetic characterization of Finnish strains of human and bat origin 24 years apart. Arch Virol. 2015;160:1489–98.
11. Whitby JE, Heaton PR, Black EM, Wooldridge M, McElhinney LM, Johnstone P. First isolation of a rabies-related virus from a Daubenton's bat in the United Kingdom. Vet Rec. 2000;147:385–8.
12. Van der Poel WHM, Van der Heide R, Verstraten ERAM, Takumi K, Lina PHC, Kramps JA. European bat lyssaviruses, The Netherlands. Emerg Infect Dis. 2005;11:1854–9.
13. Freuling C, Grossmann E, Conraths FJ, Schameitat A, Kliemt J, Auer E, et al. First isolation of EBLV-2 in Germany. Vet Microbiol. 2008;131:26–34.
14. Megali A, Yannic G, Zahno ML, Brügger D, Bertoni G, Christe P, et al. Surveillance for European bat lyssavirus in Swiss bats. Arch Virol. 2010;155: 1655–62.
15. Jakava-Viljanen M, Lilley T, Kyheroinen EM, Huovilainen A. First encounter of European bat lyssavirus type 2 (EBLV-2) in a bat in Finland. Epidemiol Infect. 2010;138:1581–5.
16. Rasmussen TB, Chriél M, Baagøe HJ, Fjederholt E, Kooi EA, Belsham GJ, et al. Detection of European bat lyssavirus type 2 in Danish Daubenton's bats using a molecular diagnostic strategy. In Proceedings of 8th annual EPIZONE meeting. Copenhagen; 2014.
17. Carlsson U, Lahti E, Elvander M. Surveillance of infectious diseases in animals and humans in Sweden 2012. Uppsala: National Veterinary Institute; 2013.
18. Lumio J, Hillbom M, Roine R, Ketonen L, Haltia M, Valle M, et al. Human rabies of bat origin in Europe. Lancet. 1986;1:378.
19. Fooks AR, McElhinney LM, Pounder DJ, Finnegan CJ, Mansfield K, Johnson N, et al. Case report: isolation of a European bat lyssavirus type 2a from a fatal human case of rabies encephalitis. J Med Virol. 2003;71:281–9.
20. Brookes SM, Klopfleisch R, Müller T, Healy DM, Teifke JP, Lange E, et al. Susceptibility of sheep to European bat lyssavirus type-1 and -2 infection: a clinical pathogenesis study. Vet Microbiol. 2007;125:210–23.
21. Cliquet F, Picard-Meyer E, Barrat J, Brookes SM, Healy DM, Wasniewski M, et al. Experimental infection of foxes with European Bat Lyssaviruses type-1 and 2. BMC Vet Res. 2009;5:19.
22. Syvertsen P, Isaksen K. Rare and potentially new bat species in Norway. Fauna. 2007;60:109–19.
23. Sunding M. The most common bat species in Norway. Fauna. 2007;60:104–8.
24. Isaksen K, Klann M, van der Kooij J, Michaelsen T, Olsen K, Starholm T, et al. Flaggermus i Norge. 2009.
25. Michaelsen T. Movements of bats in western Norway. Fauna. 2011;64:31–43.
26. Hutterer R, Ivanova T, Meyer-Cords C, Rodrigues L. Bat Migrations in Europe: A Review of Banding Data and Literature. 2005.
27. Kalko E, Schnitzler U. The echolocation and hunting behaviour of Daubenton's bat, *Myotis daubentonii*. Behav Ecol Sociobiol. 1989;24:225–38.
28. Michaelsen T. Bat activity along a Norwegian fiord. Fauna. 2011;64:80–3.
29. Olsen K. Kunnskapsstatus for flaggermus i Norge. 1996.
30. van der Kooij J. The Norwegian Zoological Society's bat care centre – five years of practise. Fauna. 2007;60:183–9.
31. Dietz C, Kiefer, A. Die Fledermäuse Europas. Kennen, bestimmen, schützen. Stuttgart: Kosmos Verlag; 2014.
32. Kokurewicz T. Sex and age related habitat selection and mass dynamics of daubenton's bats *Myotis daubentonii* (Kuhl, 1817) hibernating in natural conditions. Acta Chiropterologica. 2004;6:121–44.

33. The World Organisation for Animal Health (OIE). OIE Terrestrial Manual. Chapter 2.1.13 Rabies. 2013.
34. Servat A, Picard-Meyer E, Robardet E, Muzniece Z, Must K, Cliquet F. Evaluation of a Rapid Immunochromatographic Diagnostic Test for the detection of rabies from brain material of European mammals. Biologicals. 2012;40:61–6.
35. Tamura K, Stecher G, Peterson D, Filipski A, Kumar S. MEGA6. Molecular evolutionary genetics analysis version 6.0. Mol Biol Evol. 2013;30:2725–9.
36. Marston DA, McElhinney LM, Ellis RJ, Horton DL, Wise EL, Leech SL, et al. Next generation sequencing of viral RNA genomes. BMC Genomics. 2013;14:444.
37. Brunker K, Marston DA, Horton DL, Cleaveland S, Fooks AR, Kazwala R, et al. Elucidating the phylodynamics of endemic rabies virus in eastern Africa using whole-genome sequencing. Virus Evol. 2015; doi:10.1093/ve/vev011.
38. Li H, Durbin R. Fast and accurate long-read alignment with Burrows-Wheeler transform. Bioinformatics. 2010;26:589–95.
39. Milne I, Stephen G, Bayer M, Cock PJ, Pritchard L, Cardle L, et al. Using Tablet for visual exploration of second-generation sequencing data. Brief Bioinform. 2013;14:193–202.
40. Li H, Handsaker B, Wysoker A, Fennell T, Ruan J, Homer N, et al. The sequence alignment/map format and SAMtools. Bioinformatics. 2009;25:2078–9.
41. Marston DA, Wise EL, Ellis RJ, McElhinney LM, Banyard AC, Johnson N, et al. Complete genomic sequence of rabies virus from an ethiopian wolf. Genome Announc. 2015; doi:10.1128/genomeA.00157-15.
42. Marston DA, McElhinney LM, Johnson N, Müller T, Conzelmann KK, Tordo N, et al. Comparative analysis of the full genome sequence of European bat lyssavirus type 1 and type 2 with other lyssaviruses and evidence for a conserved transcription termination and polyadenylation motif in the G-L 3' non-translated region. J Gen Virol. 2007;88:1302–14.
43. Kuzmin IV, Wu X, Tordo N, Rupprecht CE. Complete genomes of Aravan, Khujand, Irkut and West Caucasian bat viruses, with special attention to the polymerase gene and non-coding regions. Virus Res. 2008;136:81–90.
44. Tiina N. Rabies (EBLV-2), bat - Finland: Daubenton's bat. International Society for Infectious Diseases. Archive Number: 20161018.4568558. 2016.
45. Nokireki T, Huovilainen A, Lilley T, Kyheroinen EM, Ek-Kommonen C, Sihvonen L, et al. Bat rabies surveillance in Finland. BMC Vet Res. 2013;9:174.
46. Johnson N, Selden D, Parsons G, Healy D, Brookes SM, McElhinney LM, et al. Isolation of a European bat lyssavirus type 2 from a Daubenton's bat in the United Kingdom. Vet Rec. 2003;152:383–7.
47. Fooks AR, McElhinney LM, Marston DA, Selden D, Jolliffe TA, Wakeley PR, et al. Identification of a European bat lyssavirus type 2 in a Daubenton's bat found in Staines, Surrey, UK. Vet Rec. 2004;155:434–5.
48. Fooks AR, Marston D, Parsons G, Earl D, Dicker A, Brookes SM. Isolation of EBLV-2 in a Daubenton's bat (*Myotis daubentonii*) found in Oxfordshire. Vet Rec. 2006;159:534–5.
49. Harris SL, Mansfield K, Marston DA, Johnson N, Pajamo K, O'brien N, et al. Isolation of European bat lyssavirus type 2 from a Daubenton's bat (*Myotis daubentonii*) in Shropshire. Vet Rec. 2007;161:384–6.
50. Freuling CM, Kliemt J, Schares S, Heidecke D, Driechciarz R, Schatz J, et al. Detection of European bat lyssavirus 2 (EBLV-2) in a Daubenton's bat (*Myotis daubentonii*) from Magdeburg, Germany. Berl Münch Tierärztl Wschr. 2012;125:255–8.
51. Johnson N, Goddard TM, Goharriz H, Wise E, Jennings D, Selden D, et al. Two EBLV-2 infected Daubenton's bats detected in the north of England. Vet Rec. 2016;179:311–2.
52. Robardet E, Picard-Meyer E, Andrieu S, Servat A, Cliquet F. International interlaboratory trials on rabies diagnosis: an overview of results and variation in reference diagnosis techniques (fluorescent antibody test, rabies tissue culture infection test, mouse inoculation test) and molecular biology techniques. J Virol Methods. 2011;177:15–25.
53. Robardet E, Andrieu S, Rasmussen TB, Dobrostana M, Horton DL, Hostnik P, et al. Comparative assay of fluorescent antibody test results among twelve European National Reference Laboratories using various anti-rabies conjugates. J Virol Methods. 2013;191:88–94.
54. Johnson N, Vos A, Neubert L, Freuling C, Mansfield KL, Kaipf I, Denzinger A, Hicks D, Nunez A, Franka R, et al. Experimental study of European bat lyssavirus type-2 infection in Daubenton's bats (*Myotis daubentonii*). J Gen Virol. 2008;89:2662–72.
55. Lucan RK. Relationships between the parasitic mite *Spinturnix andegavinus* (*Acari: Spinturnicidae*) and its bat host, *Myotis daubentonii* (*Chiroptera: Vespertilionidae*): seasonal, sex- and age-related variation in infestation and possible impact of the parasite on the host condition and roosting behaviour. Folia Parasitol (Praha). 2006;53:147–52.
56. Lov om jakt og fangst av vilt (viltloven). https://lovdata.no/dokument/NL/lov/1981-05-29-38. Accessed 03 Aug 2016.
57. Agreement on the Conservation of Populations of European Bats. http://www.eurobats.org. Accessed 03 Aug 2016.
58. Cliquet F, Freuling C, Smreczak M, Van der Poel WHM, Horton D, Fooks AR, et al. Development of harmonised schemes for monitoring and reporting of rabies in animals in the European Union. 2010.
59. Schatz J, Fooks AR, McElhinney L, Horton D, Echevarria J, Vazquez-Moron S, et al. Bat rabies surveillance in Europe. Zoonoses Public Health. 2013;60:22–34.
60. Battersby J. Guidelines for Surveillance and Monitoring of European Bats. Bonn: UNEP/EUROBATS; 2010.
61. Johnson N, McElhinney LM, Smith J, Lowings P, Fooks AR. Phylogenetic comparison of the genus Lyssavirus using distal coding sequences of the glycoprotein and nucleoprotein genes. Arch Virol. 2002;147:2111–23.
62. Amengual B, Whitby JE, King A, Cobo JS, Bourhy H. Evolution of European bat lyssaviruses. J Gen Virol. 1997;78:2319–28.
63. Davis PL, Holmes EC, Larrous F, Van der Poel WHM, Tjørnehøj K, Alonso WJ, et al. Phylogeography, population dynamics, and molecular evolution of European bat lyssaviruses. J Virol. 2005;79:10487–97.
64. Fooks AR, Selden D, Brookes SM, Johnson N, Marston DA, Jolliffe TA, et al. Identification of a European bat lyssavirus type 2 in a Daubenton's bat found in Lancashire. Vet Rec. 2004;155:606–7.
65. Schatz J, Freuling CM, Auer E, Goharriz H, Harbusch C, Johnson N, et al. Enhanced passive bat rabies surveillance in indigenous bat species from Germany - a retrospective study. PLoS Negl Trop Dis. 2014; doi:10.1371/journal.pntd.0002835.
66. Banyard AC, Johnson N, Voller K, Hicks D, Nunez A, Hartley M, et al. Repeated detection of European bat lyssavirus type 2 in dead bats found at a single roost site in the UK. Arch Virol. 2009;154:1847–50.
67. Horton DL, Voller K, Haxton B, Johnson N, Leech S, Goddard T, et al. European bat lyssavirus type 2 in a Daubenton's bat in Scotland. Vet Rec. 2009;165:383–4.

New insight into dolphin morbillivirus phylogeny and epidemiology in the northeast Atlantic: opportunistic study in cetaceans stranded along the Portuguese and Galician coasts

Maria Carolina Rocha de Medeiros Bento[1*], Catarina Isabel Costa Simões Eira[2,3], José Vitor Vingada[3,4], Ana Luisa Marçalo[2,3], Marisa Cláudia Teixeira Ferreira[3,5], Alfredo Lopez Fernandez[2,6], Luís Manuel Morgado Tavares[1] and Ana Isabel Simões Pereira Duarte[1]

Abstract

Background: Screening Atlantic cetacean populations for Cetacean Morbillivirus (CeMV) is essential to understand the epidemiology of the disease. In Europe, Portugal and Spain have the highest cetacean stranding rates, mostly due to the vast extension of coastline. Morbillivirus infection has been associated with high morbidity and mortality in cetaceans, especially in outbreaks reported in the Mediterranean Sea. However, scarce information is available regarding this disease in cetaceans from the North-East Atlantic populations. The presence of CeMV genomic RNA was investigated by reverse transcription-quantitative PCR in samples from 279 specimens stranded along the Portuguese and Galician coastlines collected between 2004 and 2015.

Results: A total of sixteen animals ($n = 16/279$, 5.7 %) were positive. The highest prevalence of DMV was registered in striped dolphins (*Stenella coeruleoalba*) ($n = 14/69$; 20.3 %), slightly higher in those collected in Galicia ($n = 8/33$; 24.2 %) than in Portugal ($n = 6/36$; 16.7 %).

Conclusions: Phylogenetic analysis revealed that, despite the low genetic distances between samples, the high posterior probability (PP) values obtained strongly support the separation of the Portuguese and Galician sequences in an independent branch, separately from samples from the Mediterranean and the Canary Islands. Furthermore, evidence suggests an endemic rather than an epidemic situation in the striped dolphin populations from Portugal and Galicia, since no outbreaks have been detected and positive samples have been detected annually since 2007, indicating that this virus is actively circulating in these populations and reaching prevalence values as high as 24 % among the Galician samples tested.

Keywords: Cetacean morbillivirus, Dolphin morbillivirus, Striped dolphins, Eastern Atlantic

Abbreviations: BP, *Balaenoptera acutorostrata*; CDV, Canine distemper virus; CEMMA, Coordinadora para o estudo dos mamiferos mariños; CeMV, Cetacean morbillivirus; DD, *Delphinus delphis*; DMV, Dolphin morbillivirus; GM, Globicephala melas; ICNF, Instituto de conservação da natureza e florestas; KB, *Kogia breviceps*; MATBs, Marine animals tissue banks; MMi, *Mesoplodon mirus*; MV, Mealses virus; PDV, Phocine distemper virus; PDV-1, Phocine

(Continued on next page)

* Correspondence: mcarolinabento@fmv.ulisboa.pt
[1]Centre for Interdisciplinary Research in Animal Health, Faculty of Veterinary Medicine, University of Lisbon, 1300-477 Lisbon, Portugal
Full list of author information is available at the end of the article

(Continued from previous page)
distemper virus -1; PMV, Porpoise morbillivirus; PP, Posterior probability; PWMV, Pilot whale morbillivirus; RT-PCR, Reverse transcription PCR; RT-qPCR, Reverse transcription quantitative PCR; SC, *Stenella coeruleoalba*; SPVS, Sociedade portuguesa de vida selvagem; TT, *Tursiops truncatus*

Background

Morbillivirus infection affects mainly the upper respiratory tract, central nervous system and the immune system [1, 2] and has been identified as a cause of death and stranding in marine mammals [3]. In odontecetes, infection has been associated with high mortality rates occurring during disease outbreaks in different parts of the world [4]. The mortality rate in striped dolphins from the Mediterranean Sea in the beginning of the nineties was the highest recorded so far [2, 5, 6]. Further studies are needed to deepen the knowledge about this disease. An integrated approach taking into consideration epidemiological and environmental parameters should provide a better picture of the ecology and evolution of Cetacean Morbillivirus (CeMV) in free-ranging cetaceans [2].

CeMV includes three well characterized viral strains [7]: porpoise morbillivirus (PMV), dolphin morbillivirus (DMV) and pilot whale morbillivirus (PWMV); three novel cetacean morbillivirus strains were recently reported [8–10], adding to the genetic diversity of these viruses.

Morbilliviruses affecting cetaceans have been described in the last decades [2] after the initial detection of viral antigens in these species in the late eighties. The first evidence of morbillivirus infection in cetaceans occurred in 1988 during a PMV outbreak, when the viral antigen was detected in harbour porpoises (*Phocoena phocoena*) stranded in Ireland [11]. In the early nineties, dolphin morbillivirus (DMV) was isolated from striped dolphins from the Mediterranean [12, 13]; in 2000 PWMV was first described in a long-finned pilot whale (*Globicephala melas*) from the US coast [14] and later, in 2011, from a short-finned pilot whale (*Globicephala macrorynchus*) in the Canary Islands [15].

Due to the virus pathogenic impact on cetacean populations, further information about morbillivirus infection in cetaceans worldwide is relevant to understand its epidemiology in these animals. Studying infectious diseases in these species is important, especially considering that additional non-infectious aggressions, mainly due to human activities, render these populations even more susceptible to disease. An annual average of 200 stranded cetaceans were registered between 2010 and 2012, considering the Algarve and the Northern region of the Portuguese continental coast [16] and fisheries bycatch was identified as the most significant cause of death. To this date, no molecular data was published on morbillivirus infection in animals stranded in Portugal or northern Spain. In 2014, dolphin morbillivirus infection was reported in a retrospective study affecting striped dolphins and a common dolphin from the Canary Islands [17], causing non-suppurative meningoencephalitis. Also, a fatal systemic morbillivirus infection was detected in a bottlenose dolphin stranded in 2005 in the Canary Islands [18]. It was suggested that DMV was not endemic in harbour porpoises and common dolphins (*Delphinus delphis*) from the NE Atlantic (British Isles) in the period 1996–1999 [19], as low antibodies titres were detected in animals from Spain and the North Sea.

DMV infection apparently did not persist as an endemic infection in Mediterranean striped dolphins after the 1990–92 epidemic [19]. Both epidemics in the Mediterranean Sea (1990–92 and 2006–07) started near the Gibraltar Strait [20] and it has been suggested that DMV-infected cetaceans may have entered the Strait of Gibraltar and infected striped dolphins, the most common cetacean at the time [7, 21]. Pilot whales had been already proposed as reservoirs in 1995 [22, 23]. Later, in 2006 several long-finned pilot whales were found stranded along the coast of the Alborean Sea,and morbillivirus infection was detected [24]. In this epidemic, deaths were first detected close to the Gibraltar Strait and spread further into the Mediterranean Sea. Recently described sequences found in striped dolphins from the Canary Islands show high identity with sequences from the Mediterranean outbreaks, indicating the possible circulation of viruses between the Atlantic and the Mediterranean [17]. The role of other cetacean species as reservoirs needs to be further assessed.

The objective of the present study was to clarify not only the prevalence of DMV in cetacean populations from the eastern Atlantic, but also to investigate the relationship between the dolphin morbillivirus circulating in the eastern Atlantic and elsewhere in the world, especially in the Mediterranean.

Methods

Sample collection

Stranded cetaceans were collected by the Sociedade Portuguesa de Vida Selvagem (SPVS) in Northern Portugal and the Algarve within the Marine Animal Stranding Network, managed by the Instituto para a Conservação da Natureza e Florestas (ICNF) and in

Galicia by the Coordinadora para o Estudo dos Mamiferos Mariños (CEMMA). Permission was issued by the National Authority (ICNF) to SPVS technicians to collect wildlife samples within the national territory according to laws n.140/99, n.49/2005, n.156-A/2013, and n.316/89. Also, SPVS is a registered CITES scientific research institution (code PT009). CEMMA holds a permit from the Conselleria de Medio Ambiente, Territorio e Infraestruturas de Xunta de Galicia (Spain) to collect and maintain cetacean samples according to law 42/2007 and law 9/2001.

The animals were assigned a decomposition code (1 to 5) according to already established protocols [25]. Animals with a score ranging from to 1 to 3 (fresh to moderate decomposition) were surveyed in the present study. During necropsy, tissue samples were collected from 279 cetaceans: brain, lung, pulmonary lymph node, mesenteric lymph node, spleen, kidney and liver, whenever possible. For animals from Galicia the only available sample was the lung. Samples collected in Portugal were stored in vials with RNAlater® at −20 °C and samples collected in Galicia were frozen at −20 °C. All samples were kept in the marine animals' tissue banks (MATBs) of SPVS and CEMMA. Samples from different species were collected between 2004 and 2015: common dolphins (DD), striped dolphins (SC), bottlenose dolphins (*Tursiops truncatus*; TT), long-finned pilot whales (*Globicephala melas*; GM), Pigmy Sperm Whale (*Kogia breviceps*; KB), True's Beaked Whale (*Mesoplodon mirus*; MMi) and Fin whale (*Balaenoptera physalus*; BP) (Table 1). Samples were identified with a code composed by the species identification (e.g., DD, SC, TT), a number attributed to each stranding, and the year of stranding. From cetaceans stranded in the Portuguese coastline 91 animals from 2011, 56 from 2012, 33 from 2013, 49 from 2014 and 7 from 2015 were tested. From Galicia, a total of 33 lung samples from striped dolphins were tested. Available tissue samples from 10 animals stranded in Portugal from previous years were also included in this study (6 striped dolphins and 4 pilot whales from 2004 to 2009).

Total RNA extraction

Total RNA was extracted from a pool of tissue homogenates using RNeasy mini kit (Qiagen, GmbH, Germany), according to the manufacturer's instructions. The pool included, whenever possible: lung, brain, pulmonary lymph node and mesenteric lymph node. Total RNA quantification and purity was determined using a Nanodrop 2000C spectrophotometer (ThermoScientific, USA) and stored at −80 °C until used.

Detection of dolphin morbillivirus genomic RNA by reverse transcription-quantitative PCR (RT-qPCR)

The detection of viral RNA for the DMV strain of CeMV was performed by RT-qPCR in a StepOnePlus thermocycler (Applied Biosystems), using primers (Stabvida genomics lab, Portugal) and probe (Eurogentec), targeting the N gene of DMV, as previously described [26] (Table 2). A previously detected positive sample for DMV was used as a positive control of the PCR reaction. Negative reaction controls were always included.

One step RT-qPCR assays were performed using 100 ng of the template RNA, in a total reaction volume of 20 μL containing: 10 μL of 1-step qPCR-ROX Mix (2×); 1 μL of RT enhancer; 0,2 μL of Verso Enzyme Mix (Verso 1-Step qRT-PCR ROX kit, ThermoScientific®); 0.4 μM of each primer and 0.25 μM of probe. For positive samples, total RNA was extracted individually for each of the available organs and the infection was evaluated individually in the different organs. The amplified DMV fragment was cloned into a plasmid vector (Pgem Teasy – Promega) and serial tenfold dilutions of the recombinant plasmid DNA were used to construct the standard curve (Fig. 1). The results showed a high correlation ($R^2 = 0.997$) with a calculated efficiency of 81 %. The primers and probe could detect viral RNA copies down to 10^2, and the limit of detection was 224 copies.

Conventional PCR for amplification of DMV genes

Additional sequences were amplified from the positive samples by conventional reverse transcription-PCR

Table 1 Number of stranded cetaceans tested for DMV *per* year

		2004	2005	2006	2007	2008	2009	2010	2011	2012	2013	2014	2015	Total
Portugal	Common dolphin (DD)	–	–	–	–	–	–	–	84	45	29	29	6	193
	Striped dolphin (SC)	1	–	1	4	–	–	–	5	6	3	16	–	36
	Pilot whale (GM)	–	–	–	–	2	2	–	–	1	–	–	–	5
	Bottlenose dolphin (TT)	–	–	–	–	–	–	–	1	2	1	3	–	7
	True's beaked whale (MMi)	–	–	–	–	–	–	–	1	–	–	–	–	1
	Pigmy sperm whale (KB)	–	–	–	–	–	–	–	–	1	–	1	–	2
	Fin whale (BP)	–	–	–	–	–	–	–	–	1	–	–	1	2
Galicia	Striped dolphin (SC)	2	1	1	2	3	6	4	2	3	5	4		33

Table 2 Primers and probe set used in RT-qPCR assays

	5′ Fluorophore	3′ Quencher	Sequences (5′–3′)	Amplicon size	Annealing (°C)	
DMV-N-FP	–	–	TGCCAGTACTCCAGGGAACATCCTTC	173	60	[26]
DMV-N-RP	–	–	TTGGGTCGTCAGTGTTGTCGGACCGTT	173	60	
DMV-N-probe	Cy3	BHQ1	A + CA + CCAAA + AGGGA + CA	–	60	

(RT-PCR) using previously described primers (Table 3) purchased from Stabvida genomics lab (Portugal). Primers were used in different combinations, targeting different genomic regions (Fig. 2).

The obtained amplicons were used to perform a phylogenetic analysis of the DMV sequences, along with sequences retrieved from NCBI for the same genes. L and M genes were not targeted in the conventional RT-PCR since very few sequences were available at the NCBI database.

The amplicons were directly sequenced by Sanger sequencing at Stabvida, Portugal and the specificity of the nucleotide sequences was compared by Blast analysis http://blast.ncbi.nlm.nih.gov/Blast.cgi with CeMV sequences available in the GenBank.

Phylogenetic analysis

The nucleotide sequences of the Portuguese and Galician sequence datasets available in the GenBank (National Center for Biotechnology Information) repository,with the following accession numbers KP835987; KP835991; KP835995; KP835999; KP836003; KP835986; KP835990; KP835994; KP835997; KP836002; KP836006; KP835985; KP835989; KP835993; KP835996; KP836001; KP836005; KP835984; KP835988; KP835992; KP835998; KP836000; KP836004; KP835983; KT878649; KT878650; KT878651; KT878652; KT878653; KT878654; KT878655; KT878656; KT878657; KT878658; KT878659; KT878660; KT878661, were compared with the available CeMV sequences and outgroup taxa (Canine Distemper Virus [CDV], Phocine Distemper Virus [PDV] and Measles Virus [MV]), retrieved from GenBank (Table 4), according to their primary structure similarity using the multiple alignment ClustalW program [27].

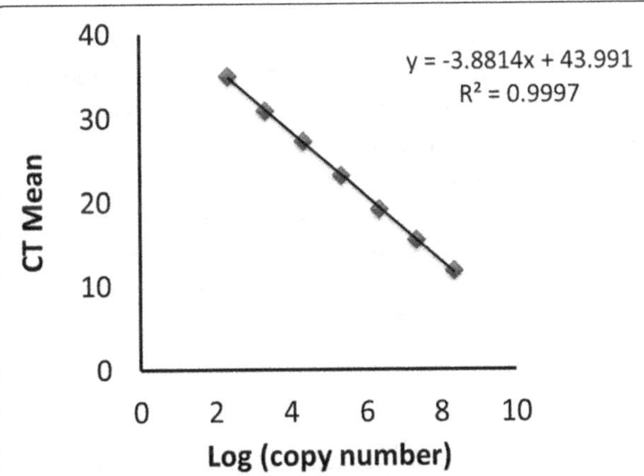

Fig. 1 Standard curve and equation for the determination of the efficiency of the RT-qPCR for the molecular detection of DMV. The N gene fragment obtained in the RT-qPCR reaction was cloned into a plasmid vector (Pgem Teasy – Promega) and serial tenfold dilutions of the recombinant plasmid DNA were amplified by qPCR in duplicate reactions and used to construct the standard curve. *Y axis* represents the mean CT values obtained from the duplicates and *X axis* represents the LOG10 of calculated copy numbers (ranging from 2.24E + 08 to 224 copies). Calculated efficiency of 81 % was determined using the formula: Efficiency = 10(-1/slope) x (-1). Results showed a high correlation (*R2* = 0.997)

Six sets of alignments were considered for the phylogenetic analysis: nucleotide sequence alignments for genes N, P, F and H, composed by sequences of 218, 342, 449 and 316 base pairs, respectively; concatenated sequence of amino acids (540 aa) and nucleotides (1446 bps). Due to heterogeneity of the available DMV sequences it was not feasible to maintain the same set of DMV sequences in the alignment for each gene. In the concatenated alignment only the sequences with all partial genomic regions were included. The multiple sequence alignments were manually corrected with Jalview, Version 2.0.1 [28] removing long internal gaps and unmatched ends to maximize genetic similarities and phylogenetic trees were inferred by Bayesian methods (MrBayes v.3.2.1) [29, 30].

For the Bayesian analysis a Markov chain Monte Carlo (mcmc) simulation technique was carried out to approximate the PP of trees [30]. The evolutionary GTR (nucleotides) and LG (amino acids) models were selected with gamma-distributed rate variation across sites and a proportion of invariable sites (rates = invgamma). The analysis was initiated using a random tree from the dataset with four chains running simultaneously for 20×10^6 generations, sampling every 100 generations. The first 25 % trees were discarded and a majority rule consensus tree was generated from the remaining trees.

The graphical representation and edition of the phylogenetic tree were performed with FigTree v1.3.1. Only support values equal or greater than 0.70 of PP are shown in the trees.

Statistical analysis

Chi-square test of association was performed to assess if the difference in prevalence was statistically significant between different species (DD and SC) and between

Table 3 Primers used in conventional PCR assays

Primer	Target gene	Sequence (5′–3′) (sense)	Tm	Genome position	Reference
CeMV-He1	H	CRTTGATACTYGTGGGTGTG (+)	59	7194–7213	[15]
CeMV-He2	H	TGTTAACTTCTGGGGCATCC (−)	59	7407–7426	
DMVFu-F	F	GGCACCATAATTAGCCAGGA (+)	51	6483–6502	
DMVFu-R	F	GCCCAGATTTGTGCCTACAT (−)	51	6655–6674	
DMV-C	P	ATGTTTATGATCACAGCGGT (+)	51	2132–2151	[35]
DMV-P2	P	ATTGGGTTGCACCACTTGTC (−)	51	2541–2560	
NgeneF	N	CCHAGRATYGCTGAAATGATHTGTGA (+)	48	849–874	[14]
NgeneR	N	AACTTGTTCTGRATWGAGTTYTC (−)	48	1056–1078	

animals from different origins (Portugal and Galicia). For this analysis an online website for statistical computation was used url: http://vassarstats.net/. A confidence interval (CI) of 95 % (for a p value ≤0.05) was considered for all the statistical analysis.

Results

A total of 16 DMV positive cetaceans were identified by RT-qPCR, representing a prevalence of 5.7 % (IC95 %: 3.42;9.32). With respect to the Portuguese coastline, 8 positive animals were detected, including 6 striped dolphins (SC) and 2 common dolphins (DD) [SC/15/2007, SC/257/2011, SC/221/2012, DD/302/2012, SC/11/2013, DD/191/2013, SC/193/2014 and SC/290/2014]. In Galicia, 8 positive striped dolphins were detected [SC/21/2007, SC/24/2008, SC/31/2009, SC/42/2010, SC/49/2011, SC/51/2012, SC/53/2012 and SC/55/2012]. Among all cetacean species, striped dolphins ($n = 69$) revealed a significantly higher DMV prevalence reaching 20.3 % (IC95 %: 11.92; 32.02), whereas common dolphins ($n = 139$) recorded a prevalence of 1.0 % (IC95 %: 0.18; 4.09) (P value *0.00*). Positive striped dolphins were detected every year (from 2007 to 2014) while positive common dolphins were only detected in 2012 and 2013. The DMV prevalence in striped dolphins stranded in Galicia was 24.2 % (IC95 %: 11.74; 42.63) whereas in Portugal the DMV prevalence was 16.7 % (IC95 %: 6.97; 33.47). From the positive animals stranded along the Portuguese coastline, each organ included in the tissue pool was tested individually for viral RNA. Two animals tested positive in all available organs; four were positive for viral RNA only in the brain and one animal tested positive in the lung, and in the pulmonary and mesenteric lymph node (Table 5). Lung was the only available sample to test in samples from Galicia.

For samples SC/15/2007, SC/257/2011, SC/53/2012, SC/55/2012, DD/302/2012, SC/290/2014, SC/11/2013, SC/31/2009, SC/51/2012, SC/21/2007, SC/221/2012 and DD/191/2013 longer genomic regions were amplified by one step RT-conventional PCR with the primers described previously (Table 3). For samples SC/24/2008, SC/42/2010, SC/49/2011 and SC/193/2014 no fragments were amplified by conventional RT-PCR.

For the nucleotide sequences of each genomic region, phylogenetic trees were inferred by Bayesian methods. All trees exhibited a similar sequence topology, supported by robust PP values, regardless of the total number of sequences in each tree.

In the tree of the concatenated nucleotide sequences (Fig. 3) the CeMV sequences were distributed in three

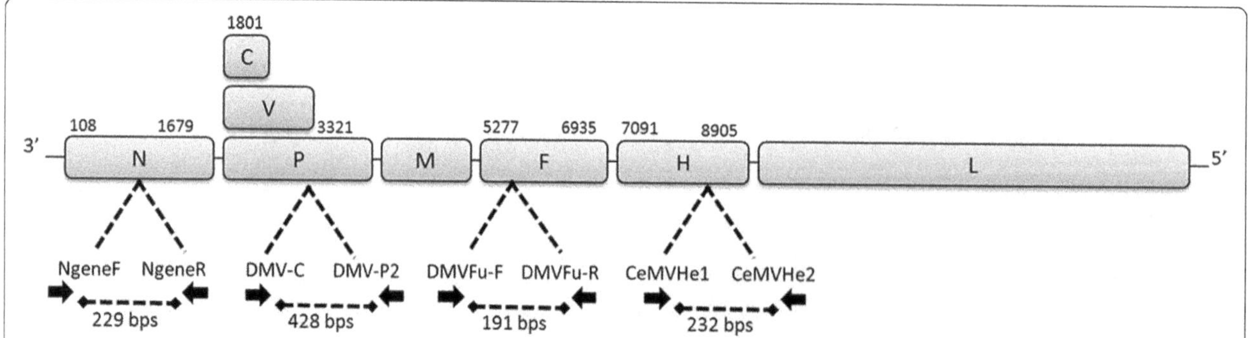

Fig. 2 Schematic representation of the primers used to amplifiy different genomic regions by conventional RT-PCR. Schematic representation of the DMV genome and location of the primers used in the conventional RT-PCR reactions, targeting: the N gene (NgeneF and NgeneR) to amplify a fragment of 229 basepairs (bps); the P gene (DMV-C and DMV-P2) for a fragment of 428 bps; the F gene (DMVFu-F and DMVFu-R) (191 bps) and for a fragment of 232 bps from the H gene (CeMVHe1 and CeMVHe2)

Table 4 Accession number for GenBank sequences used to the phylogenetic analysis and corresponding description

Complete genomes and common sequences	NC_0014981	Measles Virus	AY649446	Canine Distemper Virus
	KC802221	Phocine Distemper Virus	AJ608288	Dolphin Morbillivirus complete genome
	HQ829973	Striped dolphin 2007 SP (Med)	HQ829972	Long-finned pilot whale 2007 SP (Med)
Gene N	X84739	Porpoise 1988 IRL	AF200818	Long-finned pilot whale 1999 USA
	FJ842380	Short-finned pilot whale 1996 SP (Can Isl)		
Gene P	KF695110	Bottlenose dolphin 2005 SP (Can Isl)	JX195718	Longman's beaked whale 2010 USA
	EU039963	Long-finned pilot whale 2007 SP	KF650727	Porpoise 1990 NL
	EF451565	White-beaked dolphin 2007 GM	AF200817	Long-finned pilot whale 1999 USA
	AF333347	Pigmy sperm whale 2001 TW	KJ139451	Striped dolphin 2002 SP (Can Isl)
	KJ139452	Striped dolphin 2007 SP (Can Isl)	JN210891	Striped dolphin 2011 SP (Med)
	KF711855	Guiana dolphin 2010 BR	KJ139454	Striped dolphin 2011 SP (Can Isl)
	KC572861	Striped dolphin 2012 SP (Med)	KJ139453	Striped dolphin 2009 SP (Can Isl)
	KR337460	Fin whale 2013 IT	KR704575	Longman's beaked whale 2013 NC
	KC888945	White-beaked dolphin 2011 NL		
Gene F	AJ224704	Striped dolphin 90's SP	Z30086	DMV 1994
	FJ842382	Short-finned pilot whale 1996 SP		
Gene H	FJ648457	Porpoise MV 1988 IRL	AJ224705	Striped dolphin 90's SP
	Z36978	DMV 1994	FJ842382	Short-finned pilot whale 1996 SP (Can Isl)

SP Spain, *(MED)* Mediterranean, *(Can Isl)* Canary Islands, *IRL* Ireland, *USA* United States of America, *NL* Netherlands, *GM* Germany, *TW* Taiwan, *BR* Brazil, *IT* Italy, *NC* New Caledonia

main branches supported by high PP values. Portuguese and Galician samples from 2011, 2012 and 2013 were included in one branch; sequences from the Mediterranean from 2007, early nineties (AJ608288) and the Portuguese sequence SC/15/2007 in another branch. The only PWMV included in this tree is isolated in a third branch. The tree of the amino acid concatenated sequences presented a similar pattern (Fig. 4), although with a rearrangement within the older sequences branch ([SC/15/2007, GM/2007, Med, SC/2007/Med]; [SC/1990/Med]).

In the nucleotide tree for the F gene (Additional file 1) additional available sequences from the early nineties were included. Samples collected in the Atlantic during the 2011-2013 period clustered in the same branch; samples from the nineties clustered in a separate branch, and samples from the Mediterranean from 2007 clustered in a third branch, together with the sample SC/15/2007, similarly to the distribution of the concatenated trees. The PWMV was included in a unique branch. All branches were supported with a high PP values.

Table 5 Mapping of DMV infection in the available organs in Portuguese samples

	Lung	Brain	Pulmonary LN	Kidney	Spleen	Liver	Mesenteric LN
SC/257/2011		X					
SC/221/2012	X	X	X	X	X	X	X
DD/302/2012	X		X				X
SC/11/2013		X					
DD/191/2013	X	X	X	X	X	X	X
SC/290/2014		X					
SC/193/2014		X					
SC/15/2007*	X						

Tested organs are shown in grey and positive organs are marked with an (X)
* only lung samples were available

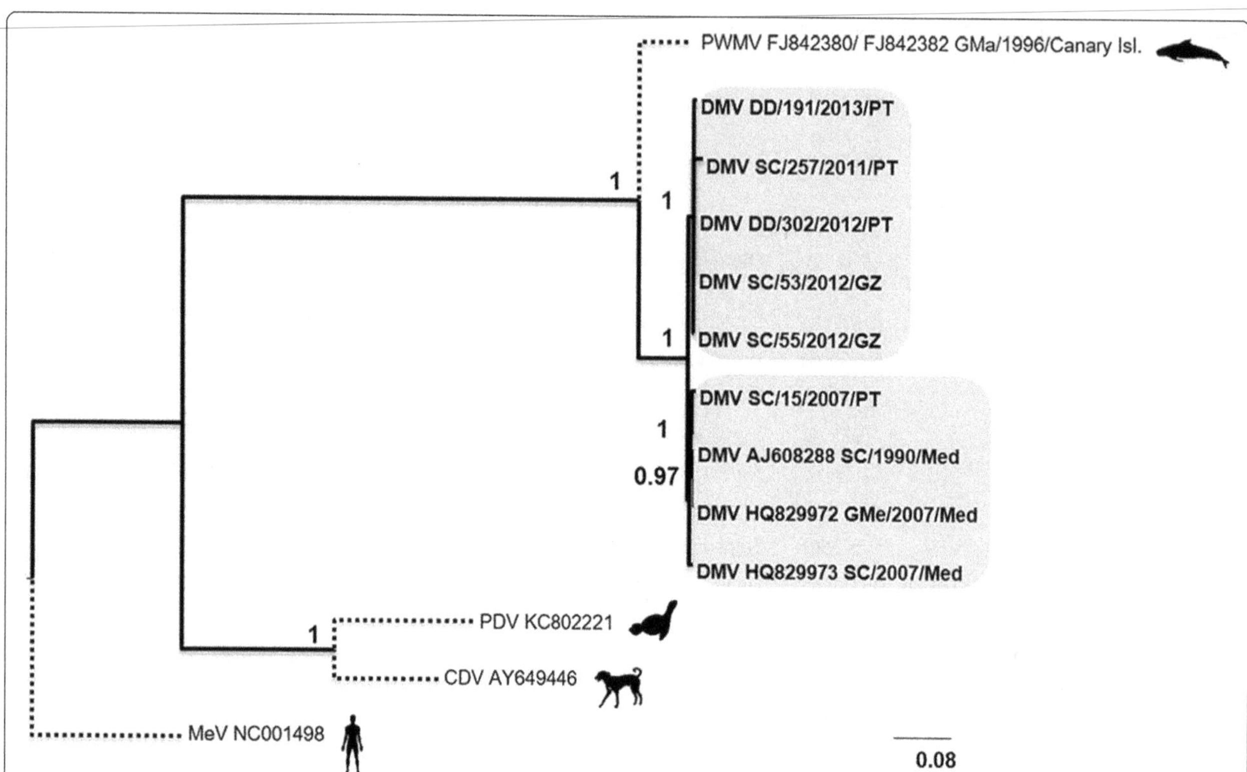

Fig. 3 Phylogenetic tree for the concatenated nucleotide sequences. Phylogenetic tree generated with concatenated nucleotide sequences alignment, inferred by Bayesian methods. Sequences for the outgroup taxa were retrieved from NCBI for Pilot Whale Morbillivirus (PWMV [accession numbers FJ842380 and FJ842382]); Phocine Distemper Virus (PDV [accession number KC802221]); Canine Distemper Virus (CDV [accession number AY649446]) and Measles Virus (MV [accession number NC001498]). Three DMV sequences were also retrieved from NCBI: one isolate from a pilot whale [HQ829972] and one from a striped dolphin [HQ829973], both from 2007; one isolate from 1990, also from a striped dolphin [AJ608288]. Sequences obtained for animals Sc/257/2011 (KP835983; KP835984; KP835985; KP835986), Dd/191/2013 (KP836003; KP836004; KP836005; KP836006), Sc/53/2012 (KP835991; KP835992; KP835993; KP835994), Dd/302/2012 (KP835999; KP836000; KP836001; KP836002) and Sc/55/2012 (KP835987; KP835988; KP835989; KP835990), Sc/15/2007 (KP835995; KP835996; KP835997; KP835998) were also included in this tree

For the H gene nucleotide tree (Additional file 2), a higher number of sequences were included. A set of sequences (9) from Portugal and Galicia ranging from 2009 to 2014 clustered in the same branch, supported by a PP value of 0.98. The SC/15/2007 sequence still clustered with Mediterranean samples from 2007 and samples from the early nineties were grouped in a separate branch. The new sequence for PMV included in this tree, branches out from the DMV samples, similarly to the PWMV sequence (PP value of 0.9).

The nucleotide tree for the P gene (Fig. 5) contained the higher number of sequences (35). One sequence from a guiana dolphin (*Sotalia guianensis*) collected in 2010 in Brazil appeared to be a distinct strain from the already characterized strains of CeMV (PMV, PWMV and DMV. The two PWMV samples clustered in the same branch and the only PMV included in the tree was isolated from all the other sequences. All these strains were supported by high PP values. The DMV sequences included in this tree were all similar, including sequences from distinct geographic origins, such as Germany, Taiwan or the Mediterranean. Two sequences obtained from white-beaked dolphins in Germany and the Netherlands in different years (2007 and 2011 respectively) clustered together with a PP value of 0.99. Sequence AJ608288 from a striped dolphin collected in 1990 in the Mediterranean and sequence AF333347 from a pigmy sperm whale from Taiwan collected in 2001 also clustered together (PP 0.79). Samples from the Canary Islands collected in 2005, 2007 and 2009 clustered with samples from the Mediterranean (2007 and 2011), one sample from New Caledonia and one sample from Portugal (SC/15/2007). The remaining Portuguese and Galician samples clustered in the same clade with two samples with a different origin (KJ139454, Canary Islands and KC572861, Mediterranean).

The N gene nucleotide tree (Additional file 3) showed a dislocation of sequences between branches. One branch included Atlantic samples from 2011 to 2013 grouped with the SC/15/2007 sequence and with

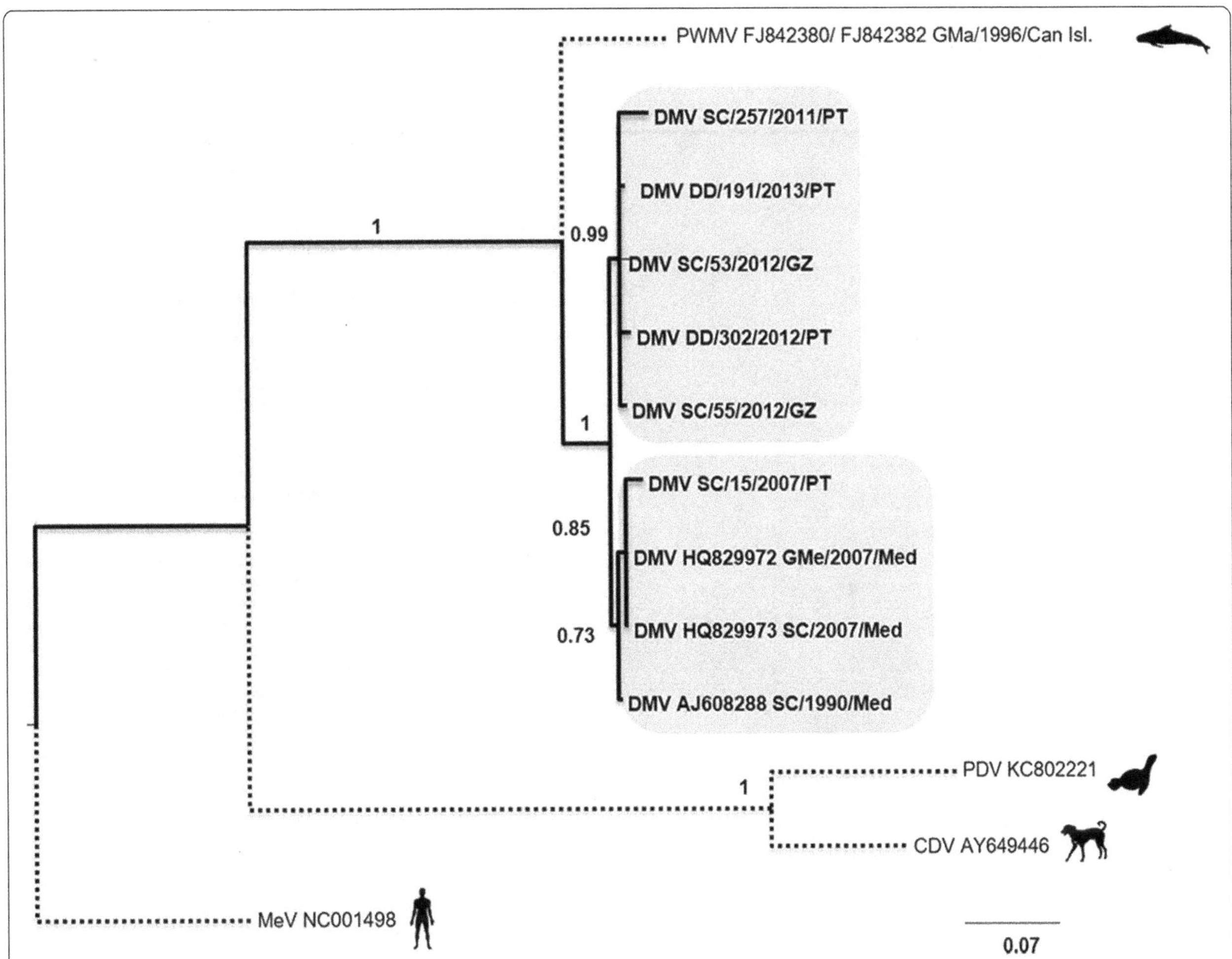

Fig. 4 Phylogenetic tree for the concatenated amino acid sequences. Phylogenetic tree generated with the concatenated amino acid sequences alignment, inferred by Bayesian methods. Sequences for the outgroup taxa were retrieved from NCBI for Pilot Whale Morbillivirus (PWMV [accession numbers FJ842380 and FJ842382]); Phocine Distemper Virus (PDV [accession number KC802221]); Canine Distemper Virus (CDV [accession number AY649446]) and Measles Virus (MV [accession number NC001498]). Three DMV sequences were also retrieved from NCBI: one isolate from a pilot whale [HQ829972] and one from a striped dolphin [HQ829973], both from 2007; one isolate from 1990, also from a striped dolphin [AJ608288]. Sequences obtained for animals Sc/257/2011 (KP835983; KP835984; KP835985; KP835986), Dd/191/2013 (KP836003; KP836004; KP836005; KP836006), Sc/53/2012 (KP835991; KP835992; KP835993; KP835994), Dd/302/2012 (KP835999; KP836000; KP836001; KP836002) and Sc/55/2012 (KP835987; KP835988; KP835989; KP835990), Sc/15/2007 (KP835995; KP835996; KP835997; KP835998) were also included in this tree

sequences from the Mediterranean (1990 and 2007); Atlantic sequences also from 2012 to 2014, were grouped separately. The remaining PWMV and PMV sequences appeared as two different outgroups.

Discussion

In this study we surveyed 279 animals and our results indicate a higher prevalence of DMV among stranded striped dolphins (20.6 %) when compared to stranded common dolphins (1 %) from the Atlantic based populations. Similar results had been previously described in the Mediterranean during the 1990–92 and 2006–08 CeMV breakouts, when striped dolphins presented higher death and stranding rates than other species [2, 31, 32]. Several theories have been hypothesized for this higher mortality rate amongst striped dolphins in the Mediterranean: they were the most numerous species in the Mediterranean and serological studies suggested that, prior to the 2006–08 outbreak, antibody levels were low in this population rendering them more susceptible to the CeMV infection [19]; also, the fact that they are highly gregarious and tend to live in large pods could contribute to the spread of CeMV infection [33]; high polychlorinated biphenyl (PCB) levels were also detected in the affected animals, leading to the hypothesis that an impaired immune system might have facilitated the infection by CeMV; finally, genetic susceptibility as a result of inbreeding in the Mediterranean population [33], which had already been

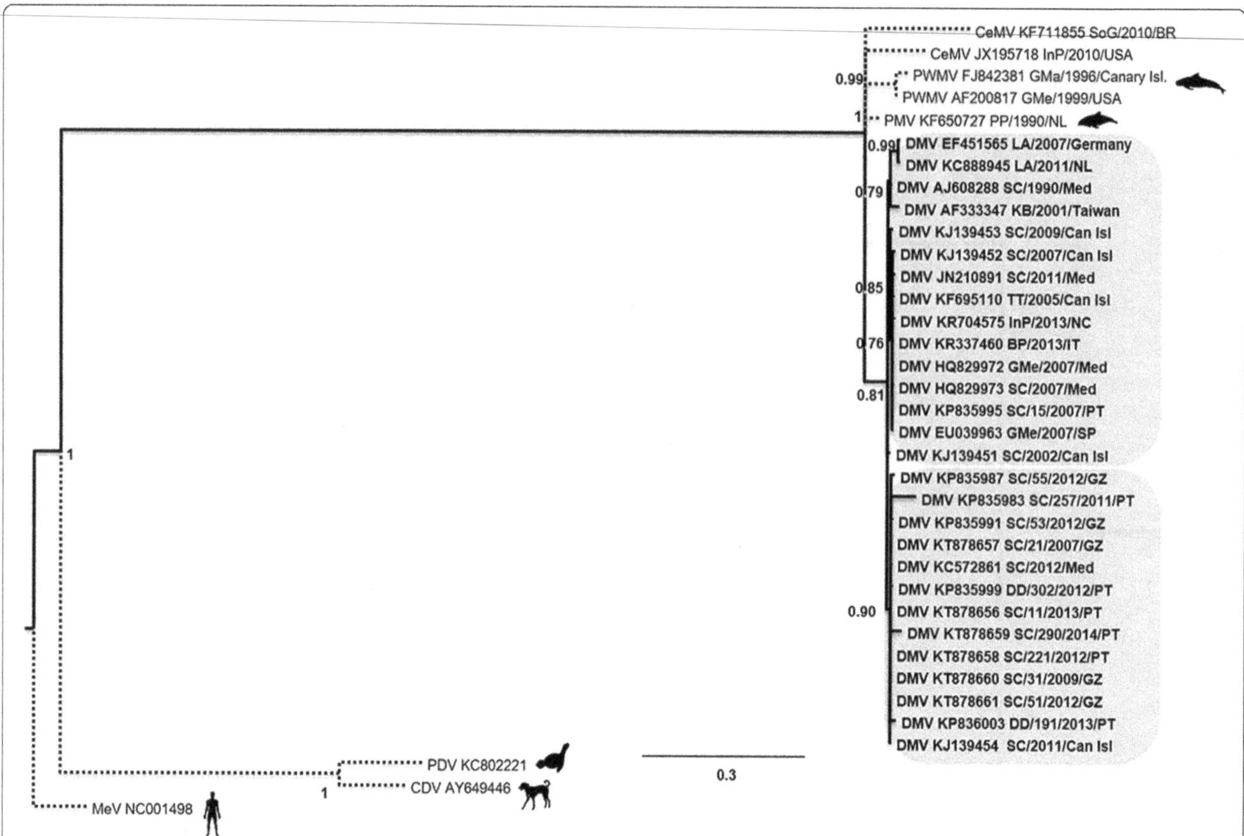

Fig. 5 Phylogenetic tree for the P gene nucleotidic sequences. Phylogenetic tree generated with the aligned sequences for the P gene, inferred by Bayesian methods. Sequences for the outgroup taxa were retrieved from NCBI for Pilot Whale Morbillivirus (PWMV [accession numbers FJ842380 and AF200817]); Porpoise Morbillivirus (PMV [accession number FJ650727]); Phocine Distemper Virus (PDV [accession number KC802221]); Canine Distemper Virus (CDV [accession number AY649446]) and Measles Virus (MV [accession number NC001498]). Two recently described sequences of CeMV (JX195718 and KF711855) were also included, along with: one isolate of DMV from 2002 collected in the Canary Islands (KJ139451), one collected in Taiwan in 2001 (AF333347), five sequences from 2007 collected in Spain and Germany (EF451565, HQ829972, HQ829973, EU039963, KJ139452), one sequence from the 90's (AJ608288), one from 2005 (KF695110) and one from 2011 collected in the Mediterranean (JN210891). Sequences obtained for the P gene of Portuguese and Galician isolates were also included for animals Sc/290/2014 (KT878659), Sc/11/2013 (KT878656), Sc/31/2009 (KT878660), Dd/191/2013 (KP836003), Sc/55/2012 (KP835987), Sc/51/2012 (KT878661), Dd/302/2012 (KP835999), Sc/21/2007 (KT878657), Sc/221/2012 (KT878658), Sc/257/2011 (KP835983), Sc/53/2012 (KP835991), and Sc/15/2007 (KP835995)

reported as relatively isolated from the Atlantic populations [34].

Prevalence among striped dolphins from Galicia was 24.2 % while prevalence in striped dolphins stranded in Portugal was 16.7 %. Although this difference was not statistically significant, it is important to highlight that prevalence among striped dolphin samples from Galicia was probably underestimated since only lung samples were tested. Samples from the Portuguese coastline allowed testing several organs and antigen was only detected in brain samples of four individuals out of the 6 positive striped dolphins. It is therefore possible that the prevalence in striped dolphins from Galicia is being strongly underestimated. Previous studies from the Atlantic based populations were performed in the western part of the Atlantic, along the USA coast, and bottlenose dolphins were the most affected cetaceans in that area. In the Canary Islands a retrospective study was published in 2014 and 6 animals were positive for CeMV (5 striped dolphins and 1 common dolphin) [17]. In this study striped dolphins seem to be the most affected species sampled from the East Atlantic.

In four animals it was not possible to amplify viral genomic fragments by conventional RT-PCR. These samples recorded high CT values in the RT-qPCR, corresponding to a low target copy number (ranging from 105 to 943 copies), which would present a downside using a less sensitive conventional assay. Also, three of the four samples were collected in animals from Galicia originally stored at −20 °C, which may possibly imply RNA degradation hampering the amplification of longer genomic fragments, by conventional RT-PCR.

The genetic distances between samples were low among all sequences included in the phylogenetic trees.

Nonetheless, PP values were high and consistent in all trees particularly in the DNA concatenated tree, adding robustness to the phylogenetic arrangement.

In the phylogenetic trees for the concatenated nucleotides the grouping of viral sequences followed a temporal arrangement, with samples collected since 2007 forming different clades. When a higher number of sequences was added to the trees (P gene tree) a phylogeographic arrangement becomes clear: all samples from Portugal and Galicia cluster together (with isolates ranging from 2007 to 2014), further away from the samples from the Mediterranean. The only exception seems to be the sequence from the animal SC/15/2007,clustering with samples from the Mediterranean, as well as with samples from the Canary Islands. Even samples from animals stranded in the south of Portugal (Algarve), such as SC/11/2013, clustered separately from samples obtained in the Mediterranean. This suggests that these populations may be relatively isolated from each other, which is supported by previous findings by other authors [34]. It is worth noticing that only one sample from a striped dolphin collected in the Canary Islands clusters closer to the Portuguese and Galician samples. All the other samples from the Canary Islands are closer to Mediterranean samples.

Positive samples for DMV antigen were detected annually since 2007 to 2013, showing that the virus is circulating in cetacean populations from the Atlantic off the coast of Portugal and northern Spain and both striped dolphins and common dolphins were found to be positive to viral infection. The infection was mapped in the available organs and positive lung samples were detected without association to higher mortality or stranding rates. Further studies would be necessary to determine if these animals had an acute, sub-acute or chronic infection and if the DMV infection was the cause of death. Animals DD/191/2013, DD/302/2012, SC/21/2007, SC/51/2012, SC/53/2012, SC/11/2013 and SC/257/2011 stranded alive and were in general emaciated and with high parasite loads, suggesting a sub-acute or chronic systemic infection. Histological and immunohistochemical studies should be performed to further characterize the necropsy findings. Four animals (SC) were positive only in brain samples, which might imply the development of chronic localized encephalitis after a systemic infection.

The two common dolphins positive for viral antigen (DD/302/2012 and DD/191/2013) were both alive at the time of stranding and presented high parasite loads and poor body condition. Animals DD/302/2012 and SC/221/2012 are also positive for cetacean gamma herpesvirus (unpublished observations Bento, C.) with viral antigen detected systemically. Unlike Mediterranean populations of striped dolphin [2], morbillivirus infection seems to be endemic in the population of striped dolphins from the Atlantic. This correlates to the serological survey conducted in 2011 in which 21.6 % ($n = 37$) of the analysed cetaceans cross reacted with Canine Distemper Virus antigen in a commercially available ELISA kit (unpublished observations Bento, C.). To date, the harbour porpoise was reported as the most affected species with morbillivirus infection in the north-eastern Atlantic, although infection is probably not endemic considering porpoises' solitary behaviour [2]. Large populations are needed to maintain morbillivirus infections as endemic [19] and although striped dolphin abundance has increased over the last years in the Portuguese Continental coast it is still a rather small population if compared to the common dolphin population (Araújo, H. personal communication). Notwithstanding, evidence suggests an endemic situation rather than an epidemic, since no outbreaks have been detected in the striped dolphin population of the Atlantic. Moreover, positive samples have been detected annually since 2007, indicating that this virus is actively circulating in this population reaching prevalence values as high as 24 % in the Galician samples. In 1999, dolphins stranded along the Atlantic coast of Spain had low antibody titres for CeMV. Considering the results obtained in this study, further serological studies are needed to deepen the knowledge about the epidemiology of this disease in striped dolphins.

Unlike striped dolphins, the prevalence of stranded common dolphins positive for viral antigen is much lower (1 %). The difference in CeMV prevalence between stranded common and striped dolphins needs to be fully assessed and further studies are needed to clarify the virus impact on cetacean populations and why do striped dolphins appear to be more susceptible to DMV infection. New approaches should be considered: viral enrichment and random amplification techniques associated with next generation sequencing could contribute to deepen the knowledge on this virus and its interaction with other pathogens.

Surveys are a unique tool to provide information on viral epidemiology, especially in free-ranging cetaceans.

Conclusion

Our results suggest that DMV infection is endemic in striped dolphin populations of the Eastern Atlantic. Since it was first reported in cetaceans in the early nineties subtle but consistent changes in the reported viral sequences suggest that the Atlantic and the Mediterranean populations are relatively isolated from each other, as suggested by other authors. The prevalence of infection in stranded common dolphins is very low when compared to striped dolphins, and our results are in agreement with previous reports that point to a higher susceptibility of striped dolphins to CeMV. Reasons for differences in susceptibility to this viral infection in different species should be further investigated and serological surveys should also be performed to assess their protection level towards CeMV infection.

Additional files

Additional file 1: Phylogenetic tree for the F gene nucleotidic sequences. Phylogenetic tree generated with the aligned sequences for the F gene, inferred by Bayesian methods. Sequences for the outgroup taxa were retrieved from NCBI for Pilot Whale Morbillivirus (PWMV [accession number FJ842382]); Phocine Distemper Virus (PDV [accession number KC802221]); Canine Distemper Virus (CDV [accession number AY649446]) and Measles Virus (MV [accession number NC001498]). Three DMV sequences from the 90's were retrieved from NCBI and included in the trees (Z30086; AJ224704 and AJ608288) along with two sequences from 2007 (HQ829972 and HQ829973). Sequences obtained for the F gene of Portuguese and Galician isolates were also included for animals Sc/257/2011 (KP835986), Dd/302/2012 (KP836002), Dd/191/2013 (KP836006), Sc/53/2012 (KP835994), Sc/55/2012 (KP835990) and Sc/15/2007 (KP835997). (TIF 5123 kb)

Additional file 2: Phylogenetic tree for the H gene nucleotidic sequences. Phylogenetic tree generated with the aligned sequences for the H gene, inferred by Bayesian methods. Sequences for the outgroup taxa were retrieved from NCBI for Pilot Whale Morbillivirus (PWMV [accession number FJ842382]); Porpoise Morbillivirus (PMV [accession number FJ648457]); Phocine Distemper Virus (PDV [accession number KC802221]); Canine Distemper Virus (CDV [accession number AY649446]) and Measles Virus (MV [accession number NC001498]). Three DMV sequences from the 90's were retrieved from NCBI and included in the trees (Z36778; AJ224705 and AJ608288) along with two sequences from 2007 (HQ829972 and HQ829973). Sequences obtained for the H gene of Portuguese and Galician isolates were also included for animals Sc/31/2009 (KT878652), Sc/290/2014 (KT878651), Sc/221/2012 (KT878650), Sc/55/2012 (KP835989), Sc/11/2013 (KT878649), Dd/302/2012 (KP836001), Dd/191/2013 (KP836005), Sc/257/2011 (KP835985), Sc/53/2012 (KP835993), and Sc/15/2007 (KP835996). (TIF 3936 kb)

Additional file: 3 Phylogenetic tree for the N gene nucleotidic sequences. Phylogenetic tree generated with the aligned sequences for the N gene, inferred by Bayesian methods. Sequences for the outgroup taxa were retrieved from NCBI for Pilot Whale Morbillivirus (PWMV [accession number FJ842380]); Porpoise Morbillivirus (PMV [accession number X84739]); Phocine Distemper Virus (PDV [accession number KC802221]); Canine Distemper Virus (CDV [accession number AY649446]) and Measles Virus (MV [accession number NC001498]). Two sequences from 2007 (HQ829972 and HQ829973) and one from the 90's (AJ608288) were also included in this tree. Sequences obtained for the N gene of Portuguese and Galician isolates were also included for animals Sc/15/2007 (KP835998), Dd/302/2012 (KP836000), Sc/257/2011 (KP835984), Dd/191/2013 (KP836004), Sc/53/2012 (KP835992), Sc/55/2012 (KP835988), Sc/11/2013 (KT878653), Sc/290/2014 (KT878655) and Sc/221/2012 (KT878654). (TIF 4079 kb)

Acknowledgements
The authors thank SPVS and CEMMA for assistance with data and sample collection. The authors further thank to CIISA at FMV-Ulisboa, where the laboratorial work was developed. We also recognize Margarida Duarte (PhD) for the revision of the manuscript from Instituto Nacional de Investigação Agrária e Veterinária, I.P., Laboratório de Virologia and Isabel Marques, from the Bioinformatic Unit of Instituto Gulbenkian de Ciência, Portugal, for the phylogeny analysis.

Funding
Sample collection was partially supported by SafeSea (EEAGrants PT0039), MarPro (Life09 NAT/PT/000038 co-funded by the EU) and CetSenti RECI/AAG-GLO/0470/2012 (FCOMP-01-0124-FEDER-027472), FCT/MCTES (PIDDAC) and FEDER - COMPETE (POFC). C. Bento was supported by PhD grant from Project CetSenti (RECI/AAG-GLO/0470/2012) and CIISA (UID/CVT/00276/2013), C. Eira was supported by the Portuguese Science Foundation (FCT) through CESAM (UID/AMB/50017/2013),A. Marçao, M. Ferreira and A. Lopez were supported by FCT grants (SFRH/BPD/64889/2009, SFRH/BD/30240/2006 and SFRH/BPD/82407/2011, respectively).

Availability of data and materials
The datasets supporting the conclusions of this article are available in the GenBank (National Center for Biotechnology Information) repository in http://ncbi.nlm.nih.gov. Access numbers for the sequences: KP835987; KP835991; KP835995; KP835999; KP836003; KP835986; KP835990; KP835994; KP835997; KP836002; KP836006; KP835985; KP835989; KP835993; KP835996; KP836001; KP836005; KP835984; KP835988; KP835992; KP835998; KP836000; KP836004; KP835983; KT878649; KT878650; KT878651; KT878652; KT878653; KT878654; KT878655; KT878656; KT878657; KT878658; KT878659; KT878660; KT878661. Phylogenetic data was submitted to TreeBase (submission number 19466) and is available at: http://purl.org/phylo/treebase/phylows/study/TB2:S19466.

Authors' contributions
CB, CE, AD were responsible for the conception and the study design and actively participated in the analysis and data interpretation. CE, MF, AL, AM, LT and JV actively participated in the sample collection and data interpretation. CB, CE and AD were also involved in the drafting and revision of the article. All authors have read and approved the final manuscript.

Competing interests
The authors declare that they have no competing interests.

Consent for publication
Not applicable.

Author details
[1]Centre for Interdisciplinary Research in Animal Health, Faculty of Veterinary Medicine, University of Lisbon, 1300-477 Lisbon, Portugal. [2]Department of Biology and CESAM, University of Aveiro, 3810-193 Aveiro, Portugal. [3]Portuguese Wildlife Society, Department of Biology, Minho University, 4710-057 Braga, Portugal. [4]Department of Biology and CESAM, Minho University, 4710-057 Braga, Portugal. [5]Department of Biology and CBMA, Minho University, 4710-057 Braga, Portugal. [6]Coordinadora para o Estudo dos Mamíferos Mariños, 36380 Gondomar, Pontevedra, Spain.

References
1. Beineke A, Siebert U, Wohlsein P, Baumgärtner W. Immunology of whales and dolphins. Vet Immunol Immunopathol. 2010;133:81–94.
2. Van Bressem MF, Duignan P, Banyard A, Barbieri M, Colegrove K, De Guise S, Di Guardo G, Dobson A, Domingo M, Fauquier D, Fernandez A, Goldstein T, Grenfell B, Groch K, Gulland F, Jensen B, Jepson P, Hall A, Kuiken T, Mazzariol S, Morris S, Nielsen O, Raga J, Rowles T, Saliki J, Sierra E, Stephens N, Stone B, Tomo I, Wang J, et al. Cetacean morbillivirus: current knowledge and future directions. Viruses. 2014;6:5145–81.
3. Domingo M, Visa J, Pumarola M, Marco AJ, Ferrer L, Rabanal R, Kennedy S. Pathologic and immunocytochemical studies of morbillivirus infection in striped dolphins (Stenella coeruleoalba). Vet Pathol. 1992;29:1–10.
4. Van Bressem MF, Van Waerebeek K, Raga JA. A review of virus infections of cetaceans and the potential impact of morbilliviruses, poxviruses and papillomaviruses on host population dynamics. Dis Aquat Organ. 1999;38:53–65.
5. Aguilar A, Raga J. The striped dolphin epizootic in the Mediterranean Sea. Ambio. 1993;22:524–8.
6. Forcada J, Aguilar A, Hammond PS, Pastor X, Aguilar R. Distribution and numbers of striped dolphins in the western mediterranean sea after the 1990 epizootic outbreak. Mar Mamm Sci. 1994;10:137–50.
7. Van Bressem MF, Raga JA, Guardo G, Jepson P, Duignan P, Barrett T, César M, Santos DO, Moreno I, Siciliano S, Aguilar A. Emerging and recurring diseases in cetaceans worldwide and the role of environmental stressors. Dis Aquat Organ. 2009;86:143–57.
8. Groch KR, Colosio AC, Marcondes MCC, Zucca D, Díaz-Delgado J, Niemeyer C, Marigo J, Brandão PE, Fernández A, Catão-Dias JL. Novel cetacean morbillivirus in Guiana Dolphin, Brazil. Emerg Infect Dis. 2014;20:511–3.
9. Stephens N, Duignan PJ, Wang J, Bingham J, Finn H, Bejder L, Patterson IAP, Holyoake C. Cetacean morbillivirus in coastal indo-pacific bottlenose dolphins, Western Australia. Emerg Infect Dis. 2014;20:666–70.
10. West KL, Sanchez S, Rotstein D, Robertson KM, Dennison S, Levine G, Davis N, Schofield D, Potter CW, Jensen B. A Longman's beaked whale

(*Indopacetus pacificus*) strands in Maui, Hawaii, with first case of morbillivirus in the central Pacific. Mar Mamm Sci. 2012;767–76.

11. Kennedy S, Smyth JA, Cush PF, McCullough SJ, Allan GM, McQuaid S. Viral distemper now found in porpoises. Nature. 1988;336:21.
12. Van Bressem MF, Visser I, Van de Bildt M, Teppema J, Raga J, Osterhaus A. Morbillivirus infection in Mediterranean striped dolphins (Stenella coeruleoalba). Vet Rec. 1991;129:471–2.
13. Domingo M, Ferrer L, Pumarola M, Marco A, Plana J, Kennedy S, McAliskey M, Rima BK. Morbillivirus in dolphins. Nature. 1990;348:21.
14. Taubenberger JK, Tsai MM, Atkin TJ, Fanning TG, Krafft AE, Moeller RB, Kodsi SE, Mense MG, Lipscomb TP. Molecular genetic evidence of a novel morbillivirus in a long-finned pilot whale (Globicephalus melas). Emerg Infect Dis. 2000;6:42–5.
15. Bellière EN, Esperón F, Fernández A, Arbelo M, Muñoz MJ, Sánchez-Vizcaíno JM. Phylogenetic analysis of a new Cetacean morbillivirus from a short-finned pilot whale stranded in the Canary Islands. Res Vet Sci. 2011;90:324–8.
16. Vingada J, Marçalo A, Ferreira M, Eira C, Henriques A, Miodonsky J, Oliveira N, Marujo D, Almeida A, Barros N, Oliveira I, Monteiro S, Araújo H, Santos J: Capítulo I. Interações Entre as Espécies-Alvo E as Pescas. Anexo Ao Relatório Intercalar Do Projecto LIFE MarPro NAT/PT/00038. 2012.
17. Sierra E, Sanchez S, Saliki JT, Blas-Machado U, Arbelo M, Zucca D, Fernandez A. Retrospective study of etiologic agents associated with nonsuppurative meningoencephalitis in stranded cetaceans in the canary islands. J Clin Microbiol. 2014;52:2390–7.
18. Sierra E, Zucca D, Arbelo M, García-Álvarez N, Andrada M, Déniz S, Fernández A. Fatal systemic morbillivirus infection in bottlenose dolphin, Canary Islands, Spain. Emerg Infect Dis. 2014;20:269–71.
19. Van Bressem MF, Waerebeek KV, Jepson PD, Raga JA, Duignan PJ, Nielsen O, Di Beneditto AP, Siciliano S, Ramos R, Kant W, Peddemors V, Kinoshita R, Ross PS, López-Fernandez A, Evans K, Crespo E, Barrett T. An insight into the epidemiology of dolphin morbillivirus worldwide. Vet Microbiol. 2001;81:287–304.
20. Raga J, Banyard A, Domingo M, Corteyn M, Van Bressem M, Fernández M, Aznar F, Barrett T. Morbillivirus. Emerg Infect Dis. 2008;14:471–3.
21. Di Sciara GN, Venturino MC, Zanardelli M, Bearzi G, Borsani FJ, Cavalloni B. Cetaceans in the central Mediterranean Sea: distribution and sighting frequencies. Bolletino di Zool. 1993;60:131–8.
22. Duignan PJ, House C, Geraci JR, Duffy N, Rima BK, Walsh MT, Early G, St Aubin DJ, Sadove S, Koopman H. Morbillivirus infection in cetaceans of the western Atlantic. Vet Microbiol. 1995;44:241–9.
23. Duignan PJ, House C, Geraci JR, Early G, Copland HG, Walsh MT, Bossart GD, Cray C, Sadove S, Aubin DJST, Moore M. Morbillivirus infection in two species of pilot whale (Globicephala sp.) from the western Atlantic. Mar Mamm Sci. 1995;11:150–62.
24. Fernández A, Esperón F, Herraéz P, de Los Monteros AE, Clavel C, Bernabé A, Sánchez-Vizcaino JM, Verborgh P, DeStephanis R, Toledano F, Bayón A. Morbillivirus and pilot whale deaths, Mediterranean Sea. Emerg Infect Dis. 2008;14:792–4.
25. Geraci JR, Lounsbury VJ. Marine mammals ashore: a field guide for strandings. Galveston: Texas A&M Sea Grant College Program; 1993.
26. Grant RJ, Banyard AC, Barrett T, Saliki JT, Romero CH. Real-time RT-PCR assays for the rapid and differential detection of dolphin and porpoise morbilliviruses. J Virol Methods. 2009;156:117–23.
27. Thompson JD, Higgins DG, Gibson TJ. CLUSTAL W: improving the sensitivity of progressive multiple sequence alignment through sequence weighting, position-specific gap penalties and weight matrix choice. Nucleic Acids Res. 1994;22:4673–80.
28. Waterhouse AM, Procter JB, Martin DM, Clamp M, Barton GJ. Jalview Version 2-a multiple sequence alignment editor and analysis workbench. Bioinformatics. 2009;25:1189–91.
29. Huelsenbeck JP, Ronquist F. MRBAYES: Bayesian inference of phylogenetic trees. Bioinformatics. 2001;17:754–5.
30. Ronquist F, Huelsenbeck JP. MrBayes 3: Bayesian phylogenetic inference under mixed models. Bioinformatics. 2003;19:1572–4.
31. Rubio-Guerri C, Melero M, Esperón F, Bellière EN, Arbelo M, Crespo JL, Sierra E, García-Párraga D, Sánchez-Vizcaíno JM. Unusual striped dolphin mass mortality episode related to cetacean morbillivirus in the Spanish Mediterranean Sea. BMC Vet Res. 2013;9:106.
32. Bellière EN, Esperón F, Sánchez-Vizcaíno JM. Genetic comparison among dolphin morbillivirus in the 1990–1992 and 2006–2008 Mediterranean outbreaks. Infect Genet Evol. 2011;11:1913–20.
33. Valsecchi E, Amos W, Raga JA, Podestà M, Sherwin W. The effects of inbreeding on mortality during a morbillivirus outbreak in the Mediterranean striped dolphin (Stenella coeruleoalba). Anim Conserv. 2004;7:139–46.
34. Bourret VJR, Macé MRJM, Crouau-Roy B. Genetic variation and population structure of western Mediterranean and northern Atlantic Stenella coeruleoalba populations inferred from microsatellite data. J Mar Biol Assoc UK. 2007;87:265.
35. Barrett T, Visser IK, Mamaev L, Goatley L, Van Bressem MF, Osterhaus ADME. Dolphin and porpoise morbillivirus are genetically distinct from phocine distemper virus. Virology. 1993;193:1010–2.

Identification of novel reassortant mammalian orthoreoviruses from bats in Slovenia

Tina Naglič[1*], Danijela Rihtarič[2], Peter Hostnik[2], Nataša Toplak[3], Simon Koren[3], Urška Kuhar[2], Urška Jamnikar-Ciglenečki[4], Denis Kutnjak[5] and Andrej Steyer[1]

Abstract

Background: Recently, mammalian orthoreoviruses (MRVs) were detected for the first time in European bats, and the closely related strain SI-MRV01 was isolated from a child with severe diarrhoea in Slovenia. Genetically similar strains have also been reported from other mammals, which reveals their wide host distribution. The aim of this study was to retrospectively investigate the occurrence and genetic diversity of MRVs in bats in Slovenia, from samples obtained throughout the country in 2008 to 2010, and in 2012 and to investigate the occurrence of the novel SI-MRV01 MRV variant in Slovenian bats.

Results: The detection of MRVs in bat guano was based on broad-range RT-PCR and specific bat MRV real-time RT-PCR. Subsequently, MRV isolates were obtained from cell culture propagation, with detailed molecular characterisation through whole-genome sequencing.

Overall, bat MRVs were detected in 1.9% to 3.8% of bats in 2008, 2009 and 2012. However, in 2010 the prevalence was 33.0%, which defined an outbreak of the single SI-MRV01 strain. Here, we report on the identification of five MRV isolates of different serotypes that are designated as SI-MRV02, SI-MRV03, SI-MRV04, SI-MRV05 and SI-MRV06. There is high genetic variability between these characterised isolates, with evident genome reassortment seen across their genome segments.

Conclusions: In conclusion, we have confirmed the presence of the SI-MRV01 strain in a Slovenian bat population. Moreover, according to genetic characterisation of S1 genome segment, all three MRV serotypes were present in the bat population. In this study, five independent MRV isolates were obtained and detailed whole genome analysis revealed high diversity between them. This study generates new information about the epidemiology and molecular characteristics of emerging bat MRV variants, and provides important molecular data for further studies of their pathogenesis and evolution.

Keywords: Bats, Genome reassortment, Mammalian orthoreovirus, Whole genome sequencing

Background

Mammalian orthoreoviruses (MRVs) are type species of the genus *Orthoreovirus*, subfamily *Spinareovirinae*, family *Reoviridae* and can infect nearly all mammals [1]. Since the first description of MRVs in the 1950s [2], they have been thoroughly studied not only from the genetics and structure perspectives, but also in terms of their epidemiology and pathogenicity [1]. They were initially isolated from the respiratory and gastrointestinal tracts in humans, but were rarely associated with severe medical conditions [2]. In the past few years, there have been several reports of novel MRV variants that can cause severe illness, such as haemorrhagic enteritis, upper respiratory tract infections and encephalitis, in humans and other animals [3, 4].

The MRV genome contains 10 dsRNA segments that are designated as the large (L, three segments), medium (M, three segments) and small (S, four segments) segments,

* Correspondence: tina.naglic@mf.uni-lj.si
[1]Institute of Microbiology and Immunology, Faculty of Medicine, University of Ljubljana, Zaloška cesta 4, SI-1000 Ljubljana, Slovenia
Full list of author information is available at the end of the article

based on their electrophoretic mobilities [5]. Neutralisation and haemagglutinin activities are restricted to the S1 gene segment [6], which encodes the σ1 protein that is located on the outer capsid of the virion. The σ1 protein is responsible for viral attachment to cellular receptors, and it defines the MRV serotype [7]. The other genome segments show no correlations to viral serotype, which suggests that MRVs evolved independently of their serotypes [8]. The segmented nature of the MRV genome poses risks for the formation of novel reassortant viruses with unpredictable biological properties. Indeed, isolation of reassortant MRVs has been described previously [9–13].

In our previous study, a distinct MRV strain (SI-MRV01) was detected in 2012 in a child hospitalised for severe gastroenteritis, with some further symptoms, including red and swollen gums, with oral ulcers [12]. The genome nucleotide sequence of isolate SI-MRV01 shared more than 97% identity with MRVs that were isolated at the same time from bats of the genera *Pipistrellus* spp. and *Myotis* spp. in Italy and Germany [14, 15]. These were the first reports of non-pteropine MRVs in bats. Both of these bat genera are also present in Slovenia. Isolate SI-MRV01 was also the first reported bat MRV variant found in humans. More recently, a MRV with high similarity to SI-MRV01 was detected in an immunocompromised child in Switzerland [16]. Based on the literature data, we consider that isolate SI-MRV01 and other similar isolates represent novel zoonotic MRV variants that are present in European bats. To date, the biological characteristics of this virus have not been investigated in detail.

Bats are the most abundant, assorted and geographically dispersed vertebrates, and they are increasingly known as reservoirs of viruses that can cross species barriers to infect other domestic and wild animals, and humans [17]. Human intervention in nature can lead to interaction with bats, which could be source of infection with emerging pathogens. In Slovenia, there are 30 insectivorous bat species [18]. Recently, detection of virus species in bats has increased rapidly not only due to the development of suitable molecular methods but also due to increasing sampling effort and interest in finding different viruses with zoonotic potential.

The aim of the present study was to retrospectively investigate the occurrence and genetic diversity of MRVs in bats in Slovenia, from samples obtained throughout the country in 2008 to 2010, and in 2012. Our aim was also to investigate the occurrence of the novel SI-MRV01 MRV variant in Slovenian bats, as this might represent the source of infection for humans. An important goal of the present study was also to obtain infectious virus isolates in cell culture for further studies on virus biology. In summary, this study generates new information about the epidemiology and molecular characteristics of emerging bat MRV variants, and provides important molecular data for further studies of their pathogenesis and evolution.

Results

MRV positives

In total, 44 out of 443 individual guano samples (9.9%) were positive for MRV RNA, regardless of the molecular screening method (Table 1). Among the positive samples, there was one pooled sample from a lactating female and a juvenile male (See Additional file 3). The prevalence of MRV RNA in sampling year 2010 stood out in particular, and this was over 10-fold higher than for the other three sampling years, at 33.0% (Table 2). Moreover, all of the 15 samples in 2010 in which the partial MRV L3 genome segment was sequenced shared 99% nucleotide sequence identity, and highest similarity with strain, SI-MRV01, a bat MRV isolated from a child with severe gastroenteritis in Slovenia in 2012 [12].

The most numerous bat species that was positive for MRV RNA in 2010 was the serotine bat, *Eptesicus serotinus*, as 18 positive out of the 94 tested. Overall, six bat species were positive for MRV RNA across all of the sampled years (Table 1). Based on gender, age and lactation status, the more highly positive samples were from lactating females (19 positive of 141 individual samples tested), followed by adult males (10 positive out of 99 individual samples tested). Nine samples were from juvenile females, three from juvenile males, and one from an adult female. Two MRV RNA positive samples were from samples of unknown origin (Table 1).

Statistical analysis showed association between MRV RNA presence in guano samples and sampling years

Table 1 Bat characteristics of the individual guano samples

Gender	MRV RNA positive samples/Bats tested [n/n(%)]				
	Total	Juvenile	Adult	Lactating female	Gravid female
Female	29/268 (10.8)	9/72 (12.5)	1/36 (2.8)	19/141 (13.5)	0/19 (0)
Male	13/129 (10.1)	3/30 (10.0)	10/99 (10.1)	na	na
Unknown	2/46 (4.3)	na	na	na	na
Total	44/443 (9.9)	12/102 (11.8)	11/135 (8.1)	19/141 (13.5)	0/19

na not applicable

Table 2 Characteristics of MRV RNA positive guano samples

Year	Samples tested (*n*)	MRV RNA		Bat species positive for MRV RNA (*n*)
		Positives (real-time RT-PCR/RT-PCR)	Prevalence (%)	
2008	108	2 (0/2)	1.9	*Rhinolophus hipposideros* (1), *Myotis myotis* (1)
2009	186	7 (7/3)	3.8	*Myotis emarginatus* (3), *Eptesicus serotinus* (2), *Myotis daubentonii* (1), unknown (1)
2010	103	34 (34/15)	33.0	*Eptesicus serotinus* (18), *Miniopterus schreibersii* (8), *Myotis myotis* (4), *Myotis daubentonii* (2), *Rhinolophus hipposideros* (1), unknown (1)
2012	54	2 (0/2)	3.7	*Myotis daubentonii* (2)

($p = 0.0023$; OR = 1.44; 95% CI 1.14, 1.82). The association between MRV RNA presence in guano samples and bat gender, age, species and sampling region was not significant.

MRV isolates

From 45 MRV RNA positive guano samples, nine were selected for virus isolation by cell culture. This virus isolation was successful from five guano samples, where the cytopathic effects developed within 3 to 8 days post inoculation. The reoviral morphology from the cell culture supernatants was confirmed under electron microscopy. The MRV isolates were denoted as SI-MRV02, SI-MRV03, SI-MRV04, SI-MRV05 and SI-MRV06 (Table 3), and were then molecularly characterised in depth, to preliminary classify the virus into serotype, based on genetic characterisation of S1 genome segment. All of the isolates characterised in this study were deposited with the European Virus Archive (EVAg; https://www.european-virus-archive.com/), and are available upon request.

Molecular analysis

The whole genomes of all of these five MRV isolates from bats were analysed in detail. For the whole sequences for all 10 of the genome segments, mapping to reference mammalian orthoreovirus was successful for only three of the MRV isolates: SI-MRV02, SI-MRV05 and SI-MRV06. For the other two MRV isolates, SI-MRV03 and SI-MRV04, mapping to several reference MRV isolates of different serotypes and origins did not result in whole-genome assembly. The problematic genome segment was always S1, which was subsequently assembled using the *de-novo* assembly approach, and the generated contigs were compared for similarity against all of the virus sequences deposited in the NCBI GenBank *nt* database, using BLASTn.

The most problematic isolate in terms of whole-genome assembly was isolate SI-MRV03, with only 78% nucleotide and 83% amino-acid identities to the most similar MRV isolate from NCBI GenBank, T1/T28/KM/2013 from the tupaia tree shrew in China, based on an S1 genome segment. The nucleotide and deduced amino-acid identities of the other genome segments varied from 79 to 84% and 91% to 98%, respectively, compared to the isolate with the highest similarity in GenBank (Table 4). Phylogenetic analysis for the S1 genome segment revealed that isolate SI-MRV03 clustered within MRV serotype 1 (Fig. 1). The strain SI-MRV03 was isolated from Daubentoni's bat, *Myotis daubentonii*, in 2012.

SI-MRV02 was isolated from *E. serotinus*, in 2010. The whole-genome analysis showed in 8 MRV genome segments 99% nucleotide and 99% to 100% amino-acid identities to the first of the MRV isolates here, SI-MRV01, which was detected in 2012 in the same region (Osrednjaslovenska) [12]. Other two segments, M1 and S2, showed the highest similarity to the isolate mew716_MRV-3, a MRV with high similarity to SI-MRV01, from a child with a primary immunodeficiency in Switzerland. Based on phylogenetic analysis

Table 3 Characteristics of Slovenian MRV strains successfully isolated from the bats using cell culture

MRV isolate	Year	Guano sample code	Appearance of cytopathic effect		Bat host species	MRV serotype[b]	EVAg Ref. No.[a]
			Days post inoculation	Passage number			
SI-MRV02	2010	SLO1A 4566/10	3	First	*Eptesicus serotinus*	3	007 V-02717
SI-MRV03	2012	SLO1A 2361/12	5	First	*Myotis daubentonii*	1	007 V-02718
SI-MRV04	2009	SLO1A 4449/09	4	First	*Eptesicus serotinus*	1	007 V-02719
SI-MRV05	2008	SLO1A 4273/08	8	First	*Myotis myotis*	2	007 V-02720
SI-MRV06	2009	SLO1A 2200/09	3	Second	*Myotis emarginatus*	1	007 V-02721

[a], All isolates are deposited with the European Virus Archive (EVAg; https://www.european-virus-archive.com/)
[b], The serotype is determined by genetic characterisation of S1 genome segment

Table 4 The highest nucleotide and amino-acid identity of genome ORFs of five Slovenian MRV isolates from bats compared to MRV isolates from GenBank

	SI-MRV02	Identitiy (%)		SI-MRV03	Identitiy (%)		SI-MRV04	Identitiy (%)		SI-MRV05	Identitiy (%)		SI-MRV06	Identitiy (%)	
		Nucl.	Aa		Nucl.	Aa		Nucl.	Aa		Nucl.	Aa		Nucl.	Aa
L1	SI-MRV01	99	99	5515–2/2012	82	98	5515–3/2012	99	99	BatMRV1-IT2011	99	99	BatMRV1-IT2011	99	99
L2	SI-MRV01	99	100	Neth/85	79	94	5515–3/2012	99	99	MORV/47 Ma/06	95	99	BatMRV1-IT2011	99	99
L3	SI-MRV01	99	99	224,660–4/2015	79	95	MRV2Tou05	97	99	MORV/47 Ma/06	98	99	BatMRV1-IT2011	99	99
M1	mew716_MRV-3	99	99	Netherlands 84	81	93	5515–3/2012	99	99	MORV/47 Ma/06	98	98	BatMRV1-IT2011	99	99
M2	SI-MRV01	99	99	WIV7	83	97	5515–3/2012	99	99	MORV/47 Ma/06	95	99	BatMRV1-IT2011	99	99
M3	SI-MRV01	99	100	BatMRV1-IT2011	79	91	5515–3/2012	99	99	MORV/47 Ma/06	99	99	BatMRV1-IT2011	99	99
S1	SI-MRV01	99	99	T1/T28/KM/2013	78	83	HB-A	98	99	MORV/47 Ma/06	93	95	BatMRV1-IT2011	99	99
S2	mew716_MRV-3	99	100	WIV2	81	97	5515–3/2012	99	99	MORV/47 Ma/06	97	99	BatMRV1-IT2011	99	99
S3	SI-MRV01	99	100	342/08	83	97	5515–3/2012	99	99	MORV/47 Ma/06	93	98	BatMRV1-IT2011	99	99
S4	SI-MRV01	99	100	WIV2	84	94	5515–3/2012	99	99	MORV/47 Ma/06	97	98	BatMRV1-IT2011	99	99

GenBank designations of MRV isolates are indicated
Nucl. nucleotide, *Aa* amino-acid

of all MRV genome segments (see Additional file 1), the three mentioned isolates (SI-MRV01, SI-MRV02 and mew716_MRV-3) cluster in the same group in all 10 genome segments. Moreover, the nucleotide and amino-acid identities are 99–100% (Table 4). According to the sequence analysis of the short L3 region of MRV RNA positives, no other MRV isolates rather than these similar to SI-MRV01 were detected in 2010.

SI-MRV04 was also isolated from *E. serotinus*, here in 2009, and this shared 98% nucleotide and 99% amino-acid identities to the MRV isolate HB-A from mink in China, based on the S1 genome segment. According to the phylogenetic analysis of the S1 genome segment, SI-MRV04 clustered within the MRV serotype 1 group (Fig. 1). The other segments, with the exception of the L3 genome segment, shared 99% nucleotide and amino-acid identities to an Italian isolate, 5515–3/2012, from Kuhl's pipistrelle, *Pipistrellus kuhlii*, which was identified as MRV serotype 2. Analysis of the L3 genome segment revealed 97% nucleotide and 99% amino-acid identities to a French isolate from a child with acute necrotising encephalopathy, MRV2Tou05, which clustered within the MRV serotype 2.

SI-MRV05 was isolated from a mouse-eared bat, *M. myotis*, in 2008, and it shared in all of the genome segments, except for L1, with 95% to 99% nucleotide and 98% to 99% amino-acid identities to a Hungarian isolate, MORV/47/Ma/06. This strain was isolated from the common vole, *Microtus arvalis*, and was described as a reassortant virus [13]. The BLASTn analysis of genome segment S1 revealed that the Hungarian isolate was the only score in GenBank and shared 93% nucleotide identity. According to phylogenetic analysis of genome segment S1, the isolate SI-MRV05 clustered together with the Hungarian isolate within the MRV serotype 2 group (Fig. 1). The L1 genome segment shared the highest similarity to the Italian isolate BatMRV1-IT2011 from the lesser horseshoe bat, *Rhinolophus hipposideros*, which was also the highest score in all of the genome segments for the strain SI-MRV06, which was isolated from Geoffroy's bat, *M. emarginatus*, in 2009. Based on the whole genome analysis, isolate BatMRV1-IT2011 was recognized as a reassortant virus [11].

The nucleotide sequences obtained in this study have been deposited in GenBank under the following accession numbers: SI-MRV02, MG457078-MG457087; SI-MRV03, MG457088-MG457097; SI-MRV04, MG457098-MG457107; SI-MRV05, MG457108-MG457117; SI-MRV06, MG457118-MG457126.

Discussion

In this study the prevalence of MRVs in Slovenian bats was investigated with particular focus on the occurrence of the novel MRV bat variant SI-MRV01, which was initially described in a child with diarrhoea [12] and represents a novel bat MRV variant. This might at least partially explain the enzootic properties of the novel MRV, as well as its zoonotic potential. Most of the MRV-positive guano samples were recorded in 2010. Moreover, all of the positives from 2010 shared 99% nucleotide identity for the partial L3 genome segment with the isolate SI-MRV01, and no scores other than isolate SI-MRV01 were obtained. This was relatively surprising for us, in terms of one year that stands out so evidently for MRV positives. This one year of high prevalence can be explained as a probable outbreak of the SI-MRV01-like isolate among the bat populations. According to our data, the outbreak was not limited to one sampling location nor to one bat species, as in 2010, positive samples were detected for eight out of 17 sampling sites across Slovenia. However, this prediction of a single-strain outbreak was based on the partial L3 genome segment analysis, which does not

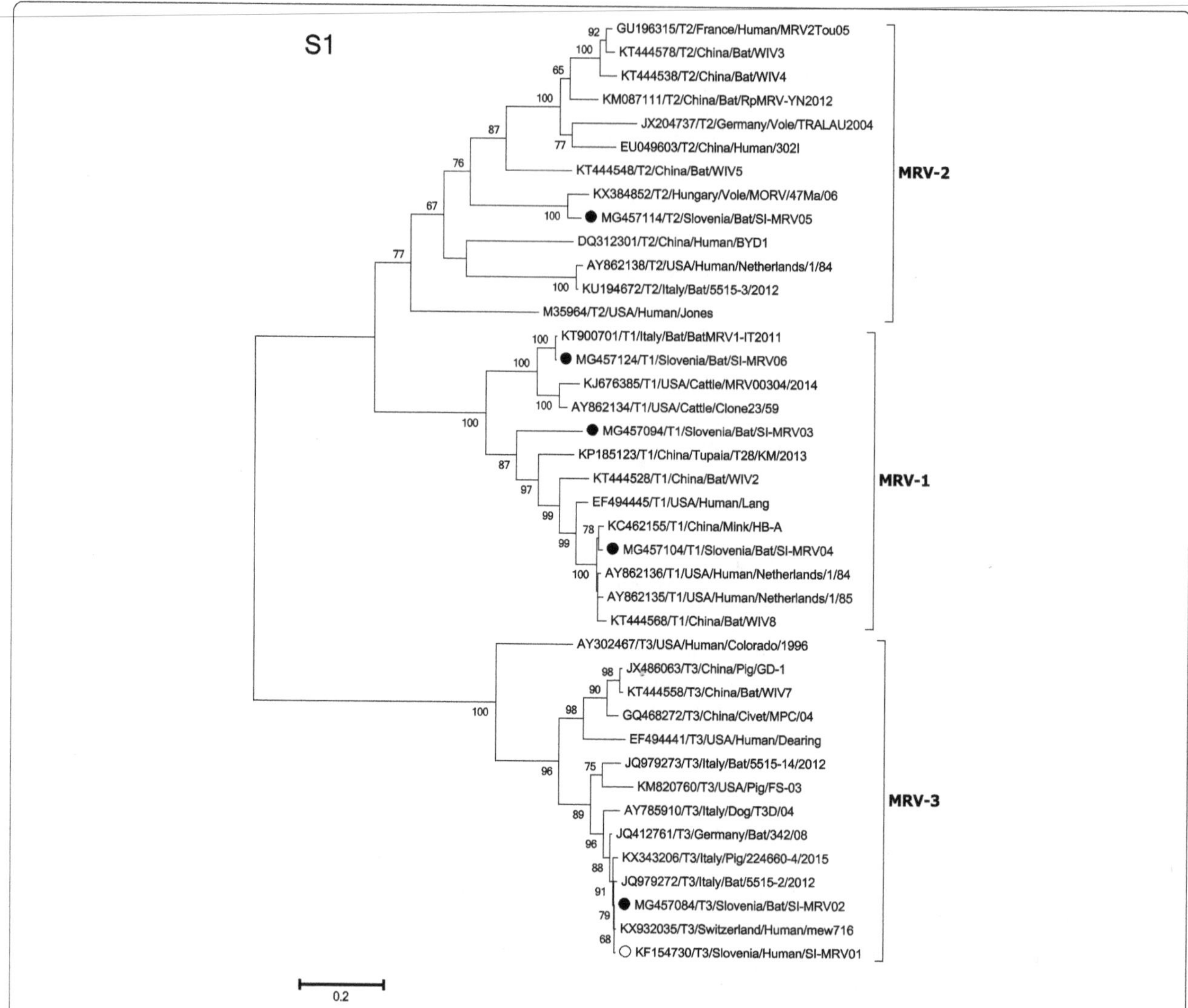

Fig. 1 Phylogenetic tree of the 40 mammalian orthoreovirus 1389 nt ORFs of S1 genome segments. Black dots (•), label sequences of five Slovenian MRV isolates from bats from this study. White dot (○), label sequence of the Slovenian MRV isolate from a child with severe gastroenteritis [12]. The phylogenetic calculations were carried out using maximum likelihood, based on the Tamura-Nei model [24] and applying the best-fit models with 1000 bootstrap replicates. Bootstraps values < 50 are not shown. The scale bar represents the substitutions per site and is proportional to the genetic distance. Evolutionary analyses were conducted in MEGA 6.0 [23]

guarantee identical strains in all of these positive bat species. As shown in previous studies [9–13], and as also confirmed in the present study, MRVs are genetically diverse, with frequent reassortments that can result in different strain variants. Thus, this one-segment comparison between these strains might not reflect the same situation for the genome similarity in other genome segments.

The *E. serotinus*, was evidently the species that was most frequently infected with MRVs, followed by *M. daubentonii*, and *M. myotis*. Based on age, gender and lactation status, the most MRV-prevalent bat group was lactating females, with 13.5% showing positive samples, which represented 7.1% of all of the female bats included. On the contrary, the other female groups (i.e., adults, juveniles) had lower prevalence of MRV RNA, with no positives in the gestating females. The group of lactating female bats was also the largest group, as it represented approximately 50% of all of the tested female bats, and was more than the adult and juvenile male bats combined. The relative high prevalence in these lactating female bats might be due to the grouping of the lactating females in nurseries, where contact transmission of the virus might be more frequent. In the population of male bats, the most prevalent group was adult bats (10.1% MRV RNA prevalence).

In total five independent virus isolates were obtained from different bat species. The whole genome analysis revealed the relatively interesting isolate SI-MRV03, which was very different from the MRV strains in GenBank, and which shared only 73% nucleotide identity for the S1 genome segment, compared to the most similar MRV isolate T1/T28/KM/2013 from the tupaia tree shrew in China. The SI-MRV03 isolate clustered phylogenetically as a distant branch within the MRV serotype 1 clade, and is the third described serotype 1 MRV found in bats. Lelli et al. [11] previously detected a divergent serotype 1 MRV isolate BatMRV1-IT2011 in *R. hipposideros*, in the northern part of the Italy, and Yang et al. [19] described two reassortant MRV serotype 2 isolates WIV-2 and WIV-8 in Chinese *Myotis* sp. and *Hipposideros* sp., respectively. Isolate SI-MRV03 from the present study and the isolate BatMRV1-IT2011 share only 79% whole-genome identity. However, isolate SI-MRV06 from the present study, which also clusters within the MRV serotype 1 group, shares 99% whole-genome identity to the Italian bat MRV serotype 1 isolate BatMRV1-IT2011. The Italian isolate has been identified as reassortant [11]. Based on high amino-acid and nucleotide identity between isolate SI-MRV06 and the Italian isolate, we speculate that those two isolates originated from the common ancestor of reassortant MRV.

In the present study, according to genetic characterisation of S1 genome segment another preliminary MRV serotype 1 isolate was obtained, SI-MRV04, which shared 79% whole-genome identity to isolate SI-MRV03, and 87% whole-genome identity to isolates SI-MRV06 and BatMRV1-IT2011. Based on these findings, we speculate that there is evident high diversity within the MRV serotype 1 strains that circulate in bats.

Here, we also report the identification of the isolate SI-MRV05, which is, according to genetic characterisation of S1 genome segment, the third description of serotype 2 MRV in bats. Wang et al. [10] previously identified a reassortant serotype 2 MRV in the lesser horseshoe bat, *R. pusillu*, in China. Yang et al. [19] described three bat MRV serotype 2 isolates WIV-3, WIV-4 and WIV-5 in *Hipposideros* sp. in China. The SI-MRV05 isolate in the present study was most similar to the Hungarian isolate MORV/47 Ma/06 from the common vole [13], with 94.5% whole-genome identity.

The isolate SI-MRV02 was obtained in 2010, when a probable outbreak of the SI-MRV01-like isolate among the bat populations occurred. The whole genome analysis showed 99% nucleotide and 99–100% amino-acid identities in to the first of the MRV isolates here, SI-MRV01 [12] and to the isolate mew716_MRV-3, a MRV with high similarity to SI-MRV01. Based on clustering in the same phylogenetic group in all 10 MRV genome segments (see Additional file 1) and high nucleotide and amino-acid identities between these three isolates (SI-MRV01, SI-MRV02 and mew716_MRV-3), we speculate that they are sufficiently similar to originate from a common ancestor.

Considering previous reports from European groups [14, 15] and the data from the present study, we can speculate that MRV serotype 3 is the most prevalent serotype in bat species in the European bat population. However, with the detected serotypes 2 in the present study and serotype 1 in this and the study of Lelli et al. [11], the occurrence of MRVs in bats appears not to be serotype specific. However, the serotypes of the isolates in the present study were predicted through phylogenetic clustering, and so they do not reflect antigen reactivities of serotype-specific antibodies. Moreover, to the best of our knowledge, there has not been any precise characterisation study of the antigen epitopes of the S1 protein, which might make reliable serotype characterisation based on nucleotide and/or amino-acid sequence identities possible. Thus, cross-reactivities of serotype-specific antibodies should be performed in future to provide specific serotyping.

The present study of bat guano was performed retrospectively in order to monitor the period when this bat-like SI-MRV01 strain was found in a child with diarrhoea. As shown here, this strain was indeed the most prevalent in 2010, which was 2 years before its detection in this child. We would expect that the circulation of isolates similar to SI-MRV01 would also have been very common in 2012, when this SI-MRV01 infection of the child occurred. However, no positives similar to isolate SI-MRV01 were detected in the present testing of the guano samples from 2012. This might be due to the sampling of the wrong bat target population, or alternatively, the circulation of this virus in 2012 was relatively low and the infection of the child happened by chance. Another factor that might have contributed to detection of a lack of isolates similar to SI-MRV01 in 2012 was the limited number of guano samples that year, which was almost half the number of samples in each of the other 3 years considered here (i.e., 2008, 2009, 2010). However, strain SI-MRV01 might have been circulating already in some other mammals, and it might also have been transmitted to the child through close contact with other domestic animals. Indeed, the testing of domestic animals should be performed to clarify the complete enzootic situation of this and the other bat MRV strains, which will be indicative of the possible risk factors for the introduction of this virus into the human population.

Limitation of our study was uneven distribution of sampling each year, such as variation in sampling location throughout the study years, which could influence statistical analysis. Minor influence on results could present 10% of samples with unknown sampling data (month, location and bat species), which were excluded from statistical analysis. Statistical analysis of bat MRV RNA prevalence distribution between sampling regions and bats species has poor test power due to small group size.

MRV isolates from this study and their molecular characterisation provide the basis for further research into the pathogenesis and molecular epidemiology of these novel bat MRV variants. Indeed, detailed characterisation of these novel bat MRV variants is crucial for the development of diagnostic methods and potential antiviral drugs and vaccines. However, the pathogenic potential of these bat MRV variants still needs to be evaluated, at least in a laboratory mouse model, to define their tissue tropism and dissemination to other organs after local inoculation. Furthermore, early detection of any novel disease agent is essential to the control of such emerging microorganisms.

Conclusions

In conclusion, we have confirmed the presence of the SI-MRV01 strain in a Slovenian bat population. Moreover, according to genetic characterisation of S1 genome segment, all three MRV serotypes were present in the bat population. In this study, five independent MRV isolates were obtained and detailed whole-genome analysis revealed high diversity between them. This study generates new information about the epidemiology and molecular characteristics of emerging bat MRV variants, and provides important molecular data for further studies of their pathogenesis and evolution.

Methods

Bat guano samples

Bat guano was obtained from the Veterinary Faculty, University of Ljubljana. Analysis of bat guano was performed as described by Rihtaric et al. [20]. Briefly, the samples were collected individually from clinically healthy bats in bat roosting sites, from May to October. The bats were sampled mostly in the last gravidity phase of female bats and in the time, when juveniles were old enough to feed by themselves. This periods differ among bat species, for example at the end of June the juveniles of *M. myotis* were quite grown-up, but the females of *R. hipposideros* were only in a phase of delivery. The bats were classified by bat biologists from Centre for Cartography Fauna and Flora, Slovenia according to species, gender, age, gravidity and lactation status, and typed morphological criteria. The bat biologist determined bat species exclusively based on morphologic criteria (without biopsy and genetic classification) [18]. Bats age was estimated based on bats weight, ulnar length and colour of the fur [21]. Altogether, there were 68 sampling sites that were distributed throughout all 8 Slovenian regions (Osrednjeslovenska, Gorenjska, Goriška, Primorska, Dolenjska, Savinjska, Podravska, Pomurska), with 18 bat species identified. The sampling sites for individual sampling could be very close together but within the same region. There might be minor differences in the individual sampling year (for instance, region Gorenjska was included only in 2009) but throughout the study all 8 regions were represented. (see Additional file 2). For 47 samples, the sampling locations and bat characteristics were not known. Altogether, 443 individual and eight pooled guano samples were analysed (Table 1, Additional file 3). The eight pooled samples consisted of two to seven individual bat samples. Hence, guano from 466 bats was included in the analysis. To carry out this sampling and the disturbance of protected bat species, a permit was obtained from the Slovenian Environment Agency (contracts no. 35701–80/2004, 35,601–35/2010–6). The samples were analysed retrospectively from the sampling years of 2008, 2009, 2010 and 2012.

Molecular analysis

For the molecular analysis and virus isolation, the guano samples were resuspended in Eagle's minimum essential medium and centrifuged at 12,600 x *g* for 10 min. Two-hundred microlitres of each supernatant was used for nucleic acids extraction (iPrep Virus DNA/RNA kits; Invitrogen, Thermo Fisher Scientific). Total nucleic acids were eluted in 100 μL elution buffer and were then used for two RT-PCRs: (a) for detection of various MRV strains from different host and origin: broad-spectrum RT-PCR to target the conserved region in L3 genome segment, using the L3–1 and L3–5 primers [4]; (b) for a sensitive and specific detection of the new bat MRV variant: specific real-time RT-PCR to target the region in L1 genome segment of SI-MRV01 isolate variants [15].

For the broad-spectrum RT-PCR, 2 μL RNA was mixed with 3.5 μL nuclease-free water and 0.5 μL 20 μM L3–5 primer; denaturation was at 95 °C for 5 min. Then, 19 μL of this RT-PCR mix was added to denaturated RNA, as: 5.5 μL nuclease-free water; 12.5 μL 2× reaction mix; 0.5 μL 20 μM L3–1 primer; and 0.5 μL RT/Platinum Taq Mix (SuperScript One-Step RT-PCR with Platinum Taq; Invitrogen, Thermo Fisher Scientific). The RT-PCR was carried out with an initial reverse transcription step at 45 °C for 30 min, followed by the PCR activation step at 94 °C for 5 min, 40 cycles of amplification (94 °C for 30 s; 50 °C for 30 s; 72 °C for 1 min), and final extension step for 10 min at 72 °C in a thermal cycler (GeneAmp PCR System 9700; Applied Biosystems, Thermo Fisher Scientific). For detection and sequencing, the PCR products were run on 1% agarose gels that contained 1× SYBR Safe DNA gel stain (Invitrogen, Thermo Fisher Scientific). The products were approximately 512 bp in size, and were purified (Qiaex II Gel Extraction kits; Qiagen), and sequenced through the Sanger sequencing method using BigDye Terminator v3.1 cycle sequencing reaction kits on a genetic analyser (ABI 3500; Applied Biosystem, Thermo Fisher Scientific). The sequencing was performed with the same RT-PCR forward and reverse primers. The sequences were analysed using the CLC Main Workbench 7 (Qiagen), and the contigs were uploaded to the Basic Local Alignment Search Tool (BLASTn) to determine the highest similarities to the sequences in the NCBI GenBank *nt* database.

For the a sensitive and specific detection of the new bat MRV variant, primers and the BatReo probe were used, as described by Kohl et al. [15], with minor modifications of the probe (modified BatReoProbeM 5'-6FAM-CCCAgTCgCggTCAT**T**ACCA**C**TCCg-BBQ-3′, modified positions in bold and underlined). The following real-time RT-PCR reaction was performed: 2 µL RNA was mixed with 1.35 µL nuclease-free water and denaturated at 95 °C for 5 min. Then 6.65 µL RT-PCR mix was added, which comprised 5 µL 2× Reaction Mix, 0.5 µL 10 µM BatReoF primer, 0.5 µL 10 µM BatReoR primer, 0.25 µL 10 µM BatReoProbeM, 0.4 µL Polymerase Mix (AgPath-ID One-Step RT-PCR kits; Ambion, Thermo Fisher Scientific). The real-time RT-PCR was carried out with an initial reverse transcription step at 45 °C for 10 min, followed by the PCR activation step at 95 °C for 10 min, and 45 cycles of amplification (95 °C for 15 s, 60 °C for 45 s). Real-time amplification was run on a StepOne Real-Time PCR system (Applied Biosystems; Thermo Fisher Scientific).

Virus isolation

The isolation of the viruses from selected positive samples was performed using the LLC-MK2 cell line (kidney cells from rhesus monkey, *Macaca mulatta*). The guano samples for virus isolation were selected based on the following criteria: (i) sequences of partial L3 genome segment in BLASTn scored as an MRV strain, or bat MRV-specific real-time RT-PCR resulted in cycle threshold (Ct) value < 34; (ii) if more samples scored as the same MRV strain (i.e., nearly 100% nucleotide sequence similarity), only a few guano samples were selected for the virus isolation. Guano suspensions were vortexed and centrifuged at 9000 x *g* for 10 min to obtain clear supernatants. From these supernatants, 150 µL was transferred to 1.5 mL Eagle's minimum essential medium. The inoculum was passed through 0.2 µm filters and transferred to an 80% confluent cell monolayer. The cells were incubated for 1 h at 37 °C and 5% CO_2, to allow the binding of the viruses to the cells. After this incubation, the inoculum was discarded from the cell layer, and 7 mL Eagle's minimum essential medium with 10% foetal bovine serum was added. The cells were further incubated under conditions specified above, and observed daily for the development of the cytopathic effect, for 14 days. After the onset of the cytopathic effect, the viruses were harvested as follows: after two freeze/ thaw cycles of the infected cell cultures and centrifugation of the cell debris, the total virus was harvested and stored at – 80 °C. In the absence of the cytopathic effect, the cryolysates were sub-cultured twice onto fresh monolayers. Virus isolates from cell cultures were examined under electron microscopy after negative staining with 2% phosphotungstic acid (pH 4.5). Electron micrograph grids were screened at 120 kV in a transmission electron microscope (JEM 1400 Plus; Jeol, Tokyo, Japan). Viral particles were identified based on their morphological characteristics. For safety reasons, handling with bat guanos and propagation of viruses isolated from bat guanos was performed at Biosafety Level 3 at the Institute of Microbiology and Immunology (Faculty of Medicine, University of Ljubljana).

Complete genome sequencing

The whole genomes of virus isolates were determined using next-generation sequencing on the Ion Torrent PGM platform, as described by Steyer et al. [12] and Jamnikar-Ciglenecki et al. [22]. The raw data were analysed using the Geneious 8.1.8 software. The complete genomes were obtained by mapping reads to the reference MRV genomes, obtained from GenBank (MRV Lang 1: Acc. No. M24734, AF378003, AF129820, AF461682, AF490617, AF174382, EF494445, L19774, M14325, M13139; BatMRV1-IT2011: Acc. No. KT900695 - KT900704; BDY1: Acc. No. DQ664184 - DQ664191, DQ318037, DQ312301; MRV 729: Acc. No. JN799419 - JN799428; MORV/47 Ma/06: Acc. No. KX384846 - KX384855; MRV HLJ 2007: Acc. No. HQ642769 - HQ642778; SI-MRV01: KF154724 - KF154733). When mapping to the reference was not achieved, *de-novo* assembly was performed using the Geneious 8.1.8 software with default settings. The contigs generated were then compared for similarity against all of the virus sequences deposited in the NCBI GenBank *nt* database, using BLASTn. Consensus sequences of genome segments were analysed for their ORFs, and the deduced amino-acid sequences were obtained. Phylogenetic and evolutionary analyses were conducted on all 10 MRV genome segments (ORFs) using the MEGA 6.0 software [23]. Alignment was performed with the ClustalW algorithm, followed by construction of the maximum likelihood phylogenetic tree.

Statistical analysis

Descriptive statistics were used to characterize the study population and it's characteristics on sampling year, gender, age, species and location of bats. For the association between virus presence in guano sample and bat's gender, age, species as well as year and region of sample collection, a univariate logistic regression was performed separately for each of the five parameters. The level of significance for statistical tests was set to $\alpha = 0.05$. Statistical analysis was carried out with the R system for statistical computing.

Additional files

Additional file 1: Phylogenetic trees of mammalian orthoreovirus L, M and S genome segments ORFs. Black dots (•), label sequences of five Slovenian MRV isolates from bats from this study. White dot (○), label sequence of the Slovenian MRV isolate from a child with severe gastroenteritis [12]. The phylogenetic calculations were carried out using maximum likelihood, based on the Tamura-Nei model [24] and applying the best-fit models with 1000 bootstrap replicates. Bootstraps values < 50 are not shown. The scale bar represents the substitutions per site and is proportional to the genetic distance. Evolutionary analyses were conducted in MEGA 6.0 [23]. (PDF 167 kb)

Additional file 2: Details of bat guano samples included in this study. (DOCX 44 kb)

Additional file 3: Bat and sample characteristics of the pooled guano samples. (DOCX 13 kb)

Abbreviations

DNA: Deoxyribonucleic acid; MRV: Mammalian orthoreovirus; RNA: Ribonucleic acid; RT-PCR: Reverse transcription polymerase chain reaction

Acknowledgements

The sampling of the bat guano and classification of the bats according to species, gender, age, gravidity and lactation status was performed in cooperation with Primož Presetnik and other bat biologists from the Centre for Cartography of Fauna and Flora, Slovenia. The statistical analysis was conducted by Naja Bohanec, CREA pro d.o.o., Slovenia.

Funding

This study was financially supported by the Slovenian Research Agency (contracts no. P3-0083 and P4-0092).

Availability of data and materials

All data generated or analysed during this study are included in this published article [and its supplementary information files]. The nucleotide sequences obtained in this study have been deposited in GenBank under the following accession numbers: SI-MRV02, MG457078-MG457087; SI-MRV03, MG457088-MG457097; SI-MRV04, MG457098-MG457107; SI-MRV05, MG457108-MG457117; SI-MRV06, MG457118-MG457126. All of the isolates characterised in this study were deposited with the European Virus Archive (EVAg; https://www.european-virus-archive.com/), and are available upon request.

Authors' contributions

TN and AS designed and performed the study. DR and PH were involved in bat guano sampling. NT, SK, UK and UJC performed next-generation sequencing. DK, TN and AS analysed the next-generation sequencing data and assembled the MRV genomes. TN and AS were involved in writing the manuscript. AS supervised the study. All authors critically revised the manuscript. All authors read and approved the final manuscript.

Consent for publication

Not applicable for this study.

Competing interests

The authors declare that they have no competing interests.

Author details

[1]Institute of Microbiology and Immunology, Faculty of Medicine, University of Ljubljana, Zaloška cesta 4, SI-1000 Ljubljana, Slovenia. [2]Institute of Microbiology and Parasitology, Veterinary Faculty, University of Ljubljana, Gerbičeva ulica 60, Ljubljana, Slovenia. [3]Omega d.o.o, Dolinškova ulica 8, Ljubljana, Slovenia. [4]Institute of Food Safety, Feed and Environment, Veterinary Faculty, University of Ljubljana, Gerbičeva ulica 60, Ljubljana, Slovenia. [5]National Institute of Biology, Večna pot, 111 Ljubljana, Slovenia.

References

1. Dermody TS, Parker JSL, Sherry B. In: Knipe DM, Howly PM, editors. Orthoreoviruses. In: *Fields Virology. Volume 2*, 6th edn. Philadelphia: Lippincott Williams & Wilkins; 2013. p. 1304–46.
2. Sabin AB. Reoviruses. A new group of respiratory and enteric viruses formerly classified as ECHO type 10 is described. Science. 1959;130(3386): 1387–9.
3. Decaro N, Campolo M, Desario C, Ricci D, Camero M, Lorusso E, Elia G, Lavazza A, Martella V, Buonavoglia C. Virological and molecular characterization of a mammalian orthoreovirus type 3 strain isolated from a dog in Italy. Vet Microbiol. 2005;109(1–2):19–27.
4. Ouattara LA, Barin F, Barthez MA, Bonnaud B, Roingeard P, Goudeau A, Castelnau P, Vernet G, Paranhos-Baccala G, Komurian-Pradel F. Novel human reovirus isolated from children with acute necrotizing encephalopathy. Emerg Infect Dis. 2011;17(8):1436–44.
5. Nibert ML, Dermody TS, Fields BN. Structure of the reovirus cell-attachment protein: a model for the domain organization of sigma 1. J Virol. 1990;64(6): 2976–89.
6. Weiner HL, Fields BN. Neutralization of reovirus: the gene responsible for the neutralization antigen. J Exp Med. 1977;146(5):1305–10.
7. Lee PWK, Hayes EC, Joklik WK. Protein σ1 is the reovirus cell attachment protein. Virology. 1981;108(1):156–63.
8. Leary TP, Erker JC, Chalmers ML, Cruz AT, Wetzel JD, Desai SM, Mushahwar IK, Dermody TS. Detection of mammalian reovirus RNA by using reverse transcription-PCR: sequence diversity within the lambda3-encoding L1 gene. J Clin Microbiol. 2002;40(4):1368–75.
9. Thimmasandra Narayanappa A, Sooryanarain H, Deventhiran J, Cao D, Ammayappan Venkatachalam B, Kambiranda D, LeRoith T, Heffron CL, Lindstrom N, Hall K, et al. A novel pathogenic mammalian orthoreovirus from diarrheic pigs and swine blood meal in the United States. MBio. 2015; 6(3):e00593–15.
10. Wang L, Fu S, Cao L, Lei W, Cao Y, Song J, Tang Q, Zhang H, Feng Y, Yang W, et al. Isolation and identification of a natural reassortant mammalian orthoreovirus from least horseshoe bat in China. PLoS One. 2015;10(3): e0118598.
11. Lelli D, Moreno A, Steyer A, Naglic T, Chiapponi C, Prosperi A, Faccin F, Sozzi E, Lavazza A. Detection and characterization of a novel Reassortant mammalian Orthoreovirus in bats in Europe. Viruses. 2015;7(11):5844–54.
12. Steyer A, Gutierrez-Aguire I, Kolenc M, Koren S, Kutnjak D, Pokorn M, Poljsak-Prijatelj M, Racki N, Ravnikar M, Sagadin M, et al. High similarity of novel Orthoreovirus detected in a child hospitalized with acute gastroenteritis to mammalian Orthoreoviruses found in bats in Europe. J Clin Microbiol. 2013; 51(11):3818–25.
13. Feher E, Kemenesi G, Oldal M, Kurucz K, Kugler R, Farkas SL, Marton S, Horvath G, Banyai K, Jakab F. Isolation and complete genome characterization of novel reassortant orthoreovirus from common vole (Microtus arvalis). Virus Genes. 2017;53(2):307–11.
14. Lelli D, Moreno A, Lavazza A, Bresaola M, Canelli E, Boniotti MB, Cordioli P. Identification of mammalian orthoreovirus type 3 in Italian bats. Zoonoses Public Health. 2013;60(1):84–92.
15. Kohl C, Lesnik R, Brinkmann A, Ebinger A, Radonic A, Nitsche A, Muhldorfer K, Wibbelt G, Kurth A. Isolation and characterization of three mammalian orthoreoviruses from European bats. PLoS One. 2012;7(8):e43106.
16. Lewandowska DW, Capaul R, Prader S, Zagordi O, Geissberger FD, Kugler M, Knorr M, Berger C, Gungor T, Reichenbach J, et al. Persistent mammalian orthoreovirus, coxsackievirus and adenovirus co-infection in a child with a primary immunodeficiency detected by metagenomic sequencing: a case report. BMC Infect Dis. 2018;18(1):33.

17. Calisher CH, Childs JE, Field HE, Holmes KV, Schountz T. Bats: important reservoir hosts of emerging viruses. Clin Microbiol Rev. 2006;19(3):531–45.
18. Presetnik P, Koselj K, Zagmajster M, Zupančič N, Jazbez K, Žibrat U, Petrinjak A, Hudoklin A: Atlas netopirjev (Chiroptera) Slovenije [Atlas of bats (Chiroptera) of Slovenia]. In: *Atlas faunae et florae Sloveniae 2.* Miklavž na Dravskem polju: Center za kartografijo favne in flore; 2009: 152.
19. Yang XL, Tan B, Wang B, Li W, Wang N, Luo CM, Wang MN, Zhang W, Li B, Peng C, et al. Isolation and identification of bat viruses closely related to human porcine and mink orthoreoviruses. J Gen Virol. 2015;96(12):3525–31.
20. Rihtaric D, Hostnik P, Steyer A, Grom J, Toplak I. Identification of SARS-like coronaviruses in horseshoe bats (Rhinolophus hipposideros) in Slovenia. Arch Virol. 2010;155(4):507–14.
21. Brunet-Rossinni AK, Wilkinson GS: Methods for Age Estimation and the Study of Senescence in Bats. In: Ecological and Behavioral Methods for the Study of Bats. EDN Edited by Kunz TH, Parsons S: Johns Hopkins University Press; 2009.
22. Jamnikar-Ciglenecki U, Toplak I, Kuhar U. Complete genome of chronic bee paralysis virus strain SLO/M92/2010, detected from Apis mellifera carnica. Genome Announc. 2017;5(26):e00602-17.
23. Tamura K, Stecher G, Peterson D, Filipski A, Kumar S. MEGA6: molecular evolutionary genetics analysis version 6.0. Mol Biol Evol. 2013;30(12):2725–9.
24. Tamura K, Nei M. Estimation of the number of nucleotide substitutions in the control region of mitochondrial DNA in humans and chimpanzees. Mol Biol Evol. 1993;10(3):512–26.

Goats as sentinel hosts for the detection of tick-borne encephalitis risk areas in the Canton of Valais, Switzerland

Nadia Rieille[1,4], Christine Klaus[2*], Donata Hoffmann[3], Olivier Péter[1] and Maarten J. Voordouw[4]

Abstract

Background: Tick-borne encephalitis (TBE) is an important tick-borne disease in Europe. Detection of the TBE virus (TBEV) in local populations of *Ixodes ricinus* ticks is the most reliable proof that a given area is at risk for TBE, but this approach is time-consuming and expensive. A cheaper and simpler approach is to use immunology-based methods to screen vertebrate hosts for TBEV-specific antibodies and subsequently test the tick populations at locations with seropositive animals.

Results: The purpose of the present study was to use goats as sentinel animals to identify new risk areas for TBE in the canton of Valais in Switzerland. A total of 4114 individual goat sera were screened for TBEV-specific antibodies using immunological methods. According to our ELISA assay, 175 goat sera reacted strongly with TBEV antigen, resulting in a seroprevalence rate of 4.3%. The serum neutralization test confirmed that 70 of the 173 ELISA-positive sera had neutralizing antibodies against TBEV. Most of the 26 seropositive goat flocks were detected in the known risk areas in the canton of Valais, with some spread into the connecting valley of Saas and to the east of the town of Brig. One seropositive site was 60 km to the west of the known TBEV-endemic area. At two of the three locations where goats were seropositive, the local tick populations also tested positive for TBEV.

Conclusion: The combined approach of screening vertebrate hosts for TBEV-specific antibodies followed by testing the local tick population for TBEV allowed us to detect two new TBEV foci in the canton of Valais. The present study showed that goats are useful sentinel animals for the detection of new TBEV risk areas.

Keywords: ELISA, Flavivirus, Goats, *Ixodes ricinus*, Sentinel host, Seroprevalence, Switzerland, Tick-borne encephalitis virus, Vector-borne disease

Background

Tick-borne encephalitis (TBE) is the most important viral tick-borne zoonosis in Europe and causes between 5352 (in 2008) and 12,733 (in 1996) human cases per year in Europe and parts of Asia, especially in the Siberian part of Russia [1]. In Switzerland, where TBE has been treated as a notifiable disease since 1984 [2], about 100–130 cases are reported each year (the maximum was 244 cases in 2006). The reasons for these annual fluctuations in the incidence of TBE are not well understood. Across Europe, geographic variation in the prevalence of TBE in humans is largely dependent on climate factors that influence the questing activity of ticks [3]. In addition, anthropomorphic changes in agriculture and outdoor and leisure activities influence the risk that humans will contract TBE [4]. Theoretical models that examine how climate change will influence tick ecology predict that the TBE virus (TBEV) will spread to the north and to higher altitudes over the next decades [5]. Field studies have confirmed that TBEV has spread northwards in Norway [6] and to higher altitudes in the Czech Republic and Austria [7–10]. The spread of TBEV to higher altitudes is also an important concern in Switzerland.

TBEV is a member of the Flavivirus genus that includes the yellow fever virus and the dengue virus [11]. In Central Europe, the main vector for TBEV is the hard tick *Ixodes ricinus*, which has three blood-feeding stages: larva, nymph, and adult. The larvae and nymphs maintain TBEV in nature because they feed on the same group of TBEV-competent reservoir hosts, mainly wild rodents [5]. Larval ticks acquire the virus after feeding

* Correspondence: christine.klaus@fli.de
[2]Friedrich-Loeffler-Institut, Institute of Bacterial Infections and Zoonoses, Naumburger Str. 96a, D-07743 Jena, Germany
Full list of author information is available at the end of the article

on an infected rodent, but this mode of transmission is relatively inefficient because the duration of infectivity to ticks is short (2-3 days) [12]. Other studies have shown that TBEV can be found in rodent tissues at 10 to 50 days post-infection [13]. Compared to other tick-borne pathogens, the prevalence of TBEV in *I. ricinus* populations is generally very low (< 1.0%) [14, 15]. Theoretical models have shown that the short duration of infectivity is the main reason why TBEV has such a low prevalence in nature [16–18]. Larval ticks can also acquire TBEV by co-feeding transmission where they feed in close proximity to an infected nymph on the same reservoir host [12, 18, 19]. Co-feeding transmission is a fragile mode of transmission because it depends on the synchronized questing activity of larval and nymphal ticks, which in turn, depend on a particular set of climatic conditions [5]. This particular set of climatic conditions is one reason why TBEV has a patchy geographic distribution across Europe [5]. Even in areas where TBEV is endemic, the presence of the virus in ticks and reservoir hosts is often highly focal [20, 21]. Pawlovskij pointed out that a natural TBEV focus depends on a number of botanical, zoological, climatical and geo-ecological conditions [22].

In veterinary medicine, clinical cases of TBE are rare, but have been reported in horses [23] and dogs [24, 25]. Other species like goats, sheep and cattle develop antibody titres without exhibiting clinical signs. These species are of high relevance for the so-called alimentary TBE. During viraemia, TBEV is excreted into the milk and can be ingested via consumption of raw milk or raw milk products such as cheese. While TBE in humans is mostly caused by tick bites, cases of alimentary TBE have been reported in recent years from Slovakia [26], Estonia [27], the Czech Republic [28], Austria [9], and Hungary [29].

Many studies have surveyed populations of wild *I. ricinus* ticks for the prevalence of TBEV [30–32]. The advantage of this approach is the direct detection of the virus in the tick vector. The disadvantage is that testing ticks for TBEV is time-consuming and expensive. Due to the low prevalence of TBEV in ticks (0.1–5.0% [14, 15]), and the high variability of TBEV prevalence in space and time, there is much interest in developing alternative methods to assess the human risk of TBE [33]. One such method is the detection of TBEV-specific antibodies in sentinel vertebrate hosts. Various wild and domestic vertebrates such as rodents [13, 34, 35], roe deer [36], goats [37, 38], sheep [39], dogs [40, 41], and even brown bears [42] have been used as sentinels in endemic areas. A disadvantage of this method is that detection of antibodies does not provide any information on the time and place of infection. An advantage of this method is that vertebrate hosts can feed many ticks and therefore "amplify" the TBEV signal in a given area. Another advantage is that screening vertebrate serum samples for TBEV-specific antibodies is fast and cheap. Goats, sheep and horses may be especially well-suited as sentinel hosts because they graze in meadows over long periods each year and are therefore potentially exposed to many TBEV-infected ticks. In addition, small ruminants and horses are kept in locations that are well known to the owner, which provides information on the location of TBEV infection. These locations can then be sampled for ticks to test for the existence of an endemic TBEV focus [43, 44].

Since 2013, the Swiss Federal Office for Public Health (FOPH) has edited two different maps with respect to TBEV. One map shows the risk of TBE infection and is based on the frequency of human TBE cases. The other map indicates the areas where the FOPH recommends prophylactic vaccination against TBEV. This map describes a risk area by using all relevant information, including human cases of TBE and TBEV-positive ticks (Bull OFSP 18/2013). A recent area of interest in Switzerland with respect to TBEV is the canton of Valais, which is located in the south of Switzerland. In this canton, 17 human cases have been described within the past 10 years, 15 of them since 2010. In 2009, two foci of TBEV were identified in Valais by a large nation-wide survey that screened populations of *I. ricinus* ticks [30]. Over the following years, the presence of TBEV in these areas was confirmed. A large tick survey that sampled more than 19,000 ticks across the canton of Valais found four new foci close to the two original ones [45, 46]. The aim of the present study was to use goats as sentinels to confirm existing TBEV risk areas and to detect new ones. The value of goats as sentinel animals was confirmed by analyzing ticks collected from areas identified by sero-positive goats. Our study demonstrates that testing antibodies in goats is an effective method for detecting new foci of TBEV.

Methods

Serum collection from goats

Goat sera were collected as part of a national survey supervised by the Swiss Veterinary Service on caprine arthritis encephalitis (CAE), a viral disease that occurs exclusively in goats. A total of 4114 individual goat sera were collected between October 2011 and March 2012. Only goats older than 6 months were sampled. The Cantonal Veterinary Service of the Canton of Valais kindly provided us with these goat serum samples, which were used in the present study.

ELISA procedure

We adapted the Serion ELISA classic TBE virus IgG (quantitative) test for humans (Serion GmbH, Germany) for veterinary use. Here solid phase compounds of the ELISA (coated plates) were used, while solutions were prepared in the laboratory (see below). The positive control was provided by the Institute of Bacterial Infections and Zoonoses, Friedrich-Loeffler-Institut, Jena, Germany. This goat was vaccinated four times at weeks 1, 2, 4, and 12 with FSME-Immun for adults (Baxter, Deutschland GmbH, Germany), as described by Klaus et al. [47]. The reagents for the ELISA protocol were prepared as follows: the dilution buffer contained 0.5% of BSA diluted in TBS solution, and the washing buffer consisted of TBS solution (pH 7.5) containing 0.5% Tween 20. For the secondary antibody, a rabbit-derived anti-goat immunoglobulin G conjugated to recombinant horseradish peroxidase (Invitrogen by Thermo Fischer Scientific, Rockford, IL 61105 USA) was used. This secondary antibody was used at a dilution of 1:7500 and the corresponding peroxidase substrate (Invitrogen) was added following the manufacturer's instructions. The stopping solution was 0.5 M sulfuric acid.

All ELISA tests were performed using a DSX workstation (Dynex technologies, Worthing, UK). The following protocol was used: 200 µl of dilution buffer and 10 µl of serum corresponding to a dilution of 1:21 were added to each well and incubated at room temperature (20–25 °C) for 75 min. Wells were washed four times with 300 µl of washing buffer before adding 100 µl of the secondary antibody at room temperature for 25 min. Wells were washed four times with 300 µl of washing buffer before adding 100 µl of substrate solution. After incubating the plate for 5 min in the dark at room temperature, the reaction was stopped by the addition of 100 µl of H_2SO_4. The optical density (OD) was measured at 405 nm.

Determination of threshold values

We established an optical density (OD) value that we considered as the threshold between TBEV-seropositive and seronegative goats. The data used to calculate the threshold were based on 200 serum samples from goats living in a part of the canton where the virus is presumed to be absent. The threshold was calculated as three standard deviations from the mean OD value (mean OD = 0.032, standard deviation = 0.34, threshold = 1.01). Goats with OD values >1 and <1 were considered seropositive and seronegative for TBEV, respectively.

Serum neutralization test

All goat serum samples that tested positive in the ELISA (OD value >1) were retested using the serum neutralization test (SNT). The SNT is considered as the gold standard for establishing whether vertebrate hosts have been exposed to TBEV or not. In addition, we randomly selected 100 serum samples and tested them using SNT to demonstrate that the ELISA was an efficient prescreen for detecting TBEV-positive serum samples. The SNT described by Klaus et al. [37] was used: the avirulent TBEV strain Langat was used with 100 $TCID_{50}$/well. The virus titre used was confirmed by back-titrations. Serum samples were inactivated by freezing and thawing as a first step and by incubating them at 56 °C for 30 min as a second step. They were titrated in duplicate starting at a dilution of 1:5 in MEM Earle's medium. After an incubation period of 24 h at 37 °C, a BHK-21 cell suspension was added and incubated for four additional days. Virus replication was detected using immunofluorescence analysis and a TBEV-specific rabbit-antiserum. Titres were expressed as the reciprocal of the dilution that caused 50% neutralization (ND50).

Absorption test

An absorption test (AT) was done for all serum samples that tested positive for the ELISA (OD value >1). To determine the optimal serum dilution for the AT, the OD values of 9 positive samples (OD > 1) and 6 borderline negative samples (0.8 < OD < 1) were retested over a range of dilutions: 1:100 to 1:25,600. The protocol for the AT included the creation of an absorbed mixture and a control mixture for each tested serum sample. The absorbed mixture was prepared by incubating 50 µl of 1:125 pre-diluted sera in the presence of 50 µl of 1:2 pre-diluted TBEV antigen (25 µl of TBE vaccine (FSME-Immun CC, Baxter) diluted in 25 µl of dilution buffer). The control mixture was prepared by incubating 50 µl of 1:125 pre-diluted sera in the presence of 50 µl of dilution buffer. The absorbed and control mixtures were incubated at room temperature for 3 h before performing the classical ELISA test described above. For each serum sample, we obtained an OD value for the control mixture (ODc) and the absorbed mixture (ODa) and calculated the extinction rate (Er) as follows: Er = 100*(ODc - ODa)/ODc. Serum samples with an Er value >55% and <45% were considered as seropositive and seronegative for TBEV antibodies, respectively. Serum samples exhibiting 45% < Er <55% were considered doubtful.

Data collection

The names and residences of the goat owners were obtained from the Veterinary Office of the Canton of Valais. The Swiss Federation of Goat Rearing (SFGR) kindly provided additional information from member goat owners with respect to origin, sex, breed and age of the goats. All goat owners of SNT-positive animals and eight owners of SNT-negative flocks were contacted. Owners who agreed to participate in our study were asked where they pastured their goats. Owners who were not members of the SFGR were asked to provide information about their goats. The distances between the owners' residences and the pasture sites were measured using the tools of Google Maps Labs.

Tick collection

The detection of TBEV-seropositive flocks (see results) allowed us to select three sites in the canton of Valais that are potentially new TBEV foci. The three sites were located (1) near the town of Brig, (2) near the town of Gampel on the north side of the Rhone River, and (3) in an isolated area to the west of the municipality of Finhaut (Fig. 1). To confirm whether these three sites were true TBEV foci, *I. ricinus* ticks were sampled from the different pasture sites of these TBEV-seropositive flocks by dragging a white cotton flag over the ground in forested areas [45]. The ticks were frozen at −80 °C until use. For each sample site, ticks were identified and separated according to species, sex, and developmental stage. Pools of 10 adult ticks or 50 nymphs were combined for lysis and DNA/RNA extraction. Quantitative realtime RT-PCR was performed according to the method described by Gäumann et al. [30].

Analysis of data

The serum neutralization test was considered as the gold standard for estimating the sensitivity and specificity of the ELISA and absorption test. Only serum samples confirmed by the serum neutralization test were used to create the maps of TBEV-positive goats.

Results

Goats

ELISA test

The 4114 goats examined in this study represent 73.7% (4114/5583) of all goats older than 6 months living in the canton of Valais. These goats belonged to 277 owners whose places of residence were distributed over 105 localities covering all 13 districts of the canton of Valais. Almost half of the goat serum samples (2048) were collected from two of these 13 districts (Brig and Visp) (Table 1). Of the 4114 goat serum samples, 175 (4.25%) samples were positive according to the ELISA test. The 175 sero-positive goats came from 88 different flocks. Of these 88 flocks, 55 had only one positive goat.

Absorption test and serum neutralization test

All ELISA-positive samples were tested with the absorption test (n = 175) and the serum neutralization test

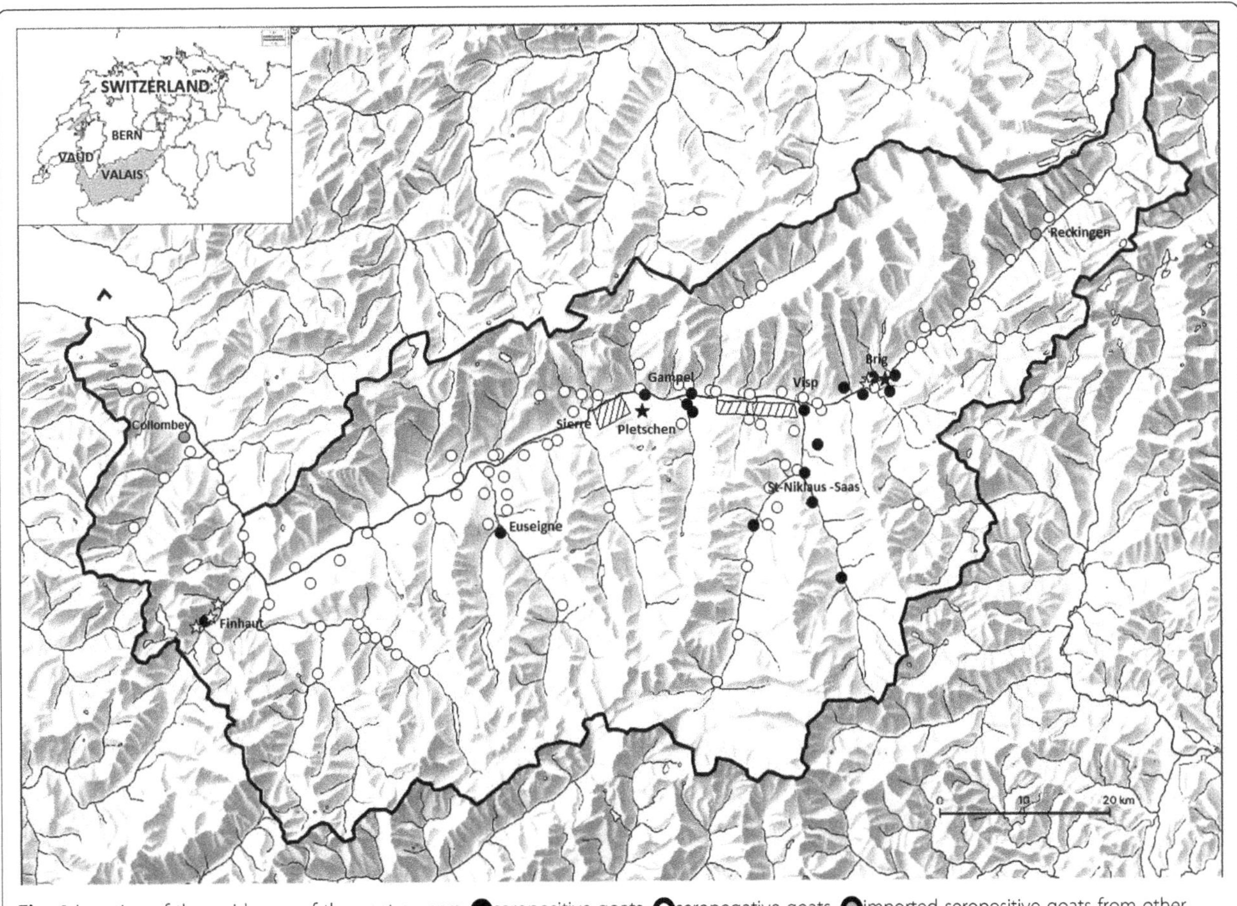

Fig. 1 Location of the residences of the goat owners. ● seropositive goats, ○ seronegative goats, ◐ imported seropositive goats from other Swiss cantons, ★ TBE positive tick pool, ☆ TBE negative tick pool, ▨ known endemic areas for TBE virus

Table 1 Location of the goat flocks that were tested for IgG antibodies against the tick-borne encephalitis virus[a]

District	Location	Latitude	Longitude	n.flock	n.sera
Brig	Brig	46°18′60″ N	7°59′23″ E	1	8
	Eggerberg	46°18′25″ N	7°52′51″ E	1	14
	Glis	46°18′33″ N	7°58′21″ E	4	38
	Mund	46°18′59″ N	7°56′34″ E	12	250
	Naters	46°19′33″ N	7°59′17″ E	18	368
	Ried-Brig	46°18′44″ N	8°00′52″ E	2	13
	Simplon Dorf	46°11′45″ N	8°03′22″ E	2	6
	Termen	46°19′40″ N	8°01′20″ E	3	10
Total				43 (15.52%)	707 (17.19%)
Conthey	Nendaz	46°11′13″ N	7°18′10″ E	2	6
Total				2 (0.73%)	6 (0.15%)
Entremont	Bruson	46°03′57″ N	7°13′06″ E	1	7
	Fionnay	46°01′58″ N	7°18′27″ E	1	16
	Le Châble	46°04′50″ N	7°12′32″ E	3	18
	Lourtier	46°02′57″ N	7°15′57″ E	2	30
	Orsières	46°01′50″ N	7°08′47″ E	2	22
	Sarreyer	46°03′46″ N	7°15′03″ E	1	2
	Sembrancher	46°04′42″ N	7°09′02″ E	2	38
	Versegères	46°03′57″ N	7°14′02″ E	4	23
Total				16 (5.78%)	156 (3.79%)
Goms	Binn	46°21′51″ N	8°11′02″ E	1	24
	Blitzingen	46°26′36″ N	8°12′08″ E	1	25
	Ernen	46°23′55″ N	8°08′45″ E	3	33
	Fieschertal	46°25′16″ N	8°08′34″ E	2	77
	Lax	46°23′20″ N	8°07′12″ E	1	13
	Münster	46°29′10″ N	8°15′44″ E	1	17
	Obergesteln	46°30′49″ N	8°19′25″ E	1	4
	Reckingen	46°28′10″ N	8°14′31″ E	2	103
Total				12 (4.33%)	296 (7.19%)
Hérens	Euseigne	46°10′19″ N	7°25′22″ E	2	57
	Hérémence	46°10′53″ N	7°24′16″ E	1	24
	La sage	46°05′55″ N	7°30′54″ E	3	20
	Mase	46°11′42″ N	7°25′60″ E	1	30
	Nax	46°13′42″ N	7°25′42″ E	2	14
	Vernamiège	46°12′38″ N	7°25′57″ E	2	69
	Vex	46°12′41″ N	7°23′53″ E	1	5
Total				12 (4.33%)	219 (5.32%)
Leuk	Albinen	46°20′31″ N	7°37′59″ E	1	66
	Bratsch	46°19′15″ N	7°42′26″ E	1	2
	Ergisch	46°17′35″ N	7°42′49″ E	1	6
	Erschmatt	46°19′17″ N	7°41′32″ E	2	27
	Gampel	46°18′56″ N	7°44′24″ E	1	22

Table 1 Location of the goat flocks that were tested for IgG antibodies against the tick-borne encephalitis virus[a] *(Continued)*

	Leuk Stadt	46°19′03″ N	7°38′06″ E	3	25
	Leukerbad	46°22′47″ N	7°37′40″ E	2	11
	Niedergampel	46°18′45″ N	7°42′43″ E	3	41
	Oberems	46°16′54″ N	7°41′44″ E	2	13
	Salgesch	46°18′42″ N	7°34′14″ E	2	21
	Susten	46°18′39″ N	7°38′30″ E	8	66
	Turtmann	46°18′05″ N	7°42′16″ E	1	2
Total				27 (9.75%)	302 (7.34%)
Martigny	Charrat	46°07′19″ N	7°08′10″ E	3	8
	Fully	46°08′15″ N	7°06′51″ E	5	78
	Martigny	46°06′02″ N	7°04′26″ E	5	50
	Riddes	46°10′23″ N	7°13′21″ E	1	2
	Saxon	46°08′43″ N	7°10′49″ E	1	2
	Trient	46°03′22″ N	6°59′41″ E	2	16
Total				17 (6.14%)	156 (3.79%)
Monthey	Champéry	46°10′43″ N	6°52′12″ E	4	45
	Collombey	46°16′16″ N	6°56′44″ E	1	37
	Monthey	46°15′20″ N	6°57′17″ E	1	1
	Torgon	46°19′14″ N	6°52′36″ E	1	26
	Troistorrents	46°13′42″ N	6°54′58″ E	1	24
	Vionnaz	46°18′41″ N	6°53′58″ E	1	2
	Vouvry	46°20′11″ N	6°53′28″ E	1	3
Total				10 (3.61%)	138 (3.35%)
Oestlich Raron	Betten	46°22′35″ N	8°04′09″ E	3	15
	Bister	46°21′37″ N	8°03′53″ E	2	66
	Grengiols	46°22′20″ N	8°05′35″ E	3	29
	Mörel	46°21′23″ N	8°02′50″ E	4	93
Total				12 (4.33%)	203 (4.93%)
Sierre	Chalais	46°15′57″ N	7°30′32″ E	2	61
	Crans-Montana	46°18′41″ N	7°29′01″ E	1	2
	Grône	46°15′03″ N	7°27′31″ E	1	12
	Miège	46°18′43″ N	7°32′50″ E	1	2
	Mollens	46°18′56″ N	7°31′14″ E	2	5
	Réchy	46°15′41″ N	7°29′43″ E	2	14
	Sierre	46°17′39″ N	7°32′00″ E	4	29
	St-Jean	46°11′49″ N	7°35′09″ E	3	75
	St-Léonard	46°15′05″ N	7°25′11″ E	2	7
Total				18 (6.50%)	207 (5.03%)
Sion	Bramois	46°14′00″ N	7°24′20″ E	1	4
	Salins	46°12′38″ N	7°21′25″ E	1	8
	Sion	46°13′40″ N	7°21′33″ E	1	7
	St-Germain	46°15′01″ N	7°20′59″ E	2	25
	Uvrier	46°15′02″ N	7°24′45″ E	1	2

Table 1 Location of the goat flocks that were tested for IgG antibodies against the tick-borne encephalitis virus[a] *(Continued)*

Total				6 (2.17%)	46 (1.12%)
St-Maurice	Collonges	46°10′14″ N	7°02′06″ E	1	3
	Dorénaz	46°08′50″ N	7°02′39″ E	5	33
	Finhaut	46°05′00″ N	6°58′37″ E	1	26
	Massongex	46°14′33″ N	6°59′22″ E	2	13
	Salvan	46°07′14″ N	7°01′15″ E	2	18
	St-Maurice	46°12′60″ N	7°00′07″ E	2	7
Total				13 (4.69%)	100 (2.43%)
Visp	Eisten	46°12′02″ N	7°53′36″ E	2	12
	Embd	46°12′54″ N	7°49′42″ E	2	33
	Eyholz	46°17′38″ N	7°54′32″ E	1	5
	Gasenried	46°10′43″ N	7°49′29″ E	1	18
	Grächen	46°11′43″ N	7°50′18″ E	8	76
	Herbriggen	46°08′07″ N	7°47′33″ E	3	76
	Lalden	46°18′01″ N	7°54′13″ E	3	39
	Saas-Grund	46°07′22″ N	7°56′11″ E	3	16
	St-Niklaus	46°10′40″ N	7°48′11″ E	23	542
	Stalden	46°13′59″ N	7°52′14″ E	1	1
	Staldenried	46°13′49″ N	7°52′59″ E	3	43
	Täsch	46°04′01″ N	7°46′42″ E	1	26
	Törbel	46°14′16″ N	7°51′06″ E	6	96
	Visp	46°17′39″ N	7°52′56″ E	5	83
	Visperterminen	46°15′32″ N	7°54′09″ E	9	205
	Zeneggen	46°16′22″ N	7°51′58″ E	1	8
	Zermatt	46°01′11″ N	7°44′46″ E	4	40
Total				76 (27.44%)	1319 (32.06%)
Westlich Raron	Ausserberg	46°18′47″ N	7°50′57″ E	3	88
	Blatten(Lötschen)	46°25′13″ N	7°49′10″ E	1	34
	Bürchen	46°16′50″ N	7°48′56″ E	1	1
	Raron	46°18′41″ N	7°47′59″ E	2	22
	Steg	46°18′49″ N	7°44′56″ E	1	7
	Unterbäch	46°17′08″ N	7°47′50″ E	2	20
	Wiler (Lötschen)	46°24′15″ N	7°47′04″ E	3	87
Total				13 (4.69%)	259 (6.30%)
TOTAL	105			277 (100.00%)	4114 (100.00%)

[a]For each location, the following information is provided: name of the district, name of the location, latitude, longitude, number of flocks (n.flock), and the number of goat sera (n.sera) analyzed in the study

(n = 173; for two samples, one AT-positive and one AT-negative, there was not enough serum); results for individual goats are in Table S1 of Additional file 1. For the absorption test there were 50 positive, 6 equivocal and 117 negative serum samples. Thus, of the 175 serum samples, 28.6% (50/175) or 32.0% (56/175) tested positive on the AT depending on whether the 6 equivocal samples were treated as negative or positive. For the serum neutralization test, there were 70 positive and 103 negative serum samples. Thus, of the 173 serum samples, 40.4% (70/173) tested positive on the SNT. The 70 SNT-positive samples came from 26 different flocks of goats. To calculate the specificity and sensitivity of the absorption test, we used the serum neutralization test as the gold standard. When the 6 equivocal AT serum samples were classified as negative, the specificity of the absorption test was 100.0% and the sensitivity was 71.4%. When the 6 equivocal AT serum samples were classified as positive, the specificity was 98.1% and the sensitivity was 77.1% (Table 2). Of the 100 randomly selected goat sera, 97 were negative for both ELISA and SNT, one sample was negative for ELISA (just below the cut-off) but positive for SNT, and two samples were positive for both ELISA and SNT.

Origin of goats

For 3653 animals belonging to 249 owners, the current locality of the owner and the locality from which the goats were purchased were known. For 2543 (69.6%) goats, the current locality and the locality of origin were the same and for the remaining 1110 goats (33.3%), these two localities were different. 3201 (87.6%) goats were born in the canton of Valais and the remaining 452 goats originated from 23 other Swiss cantons. Of the 452 goats born outside the canton of Valais, the majority (54.2%) came from the neighboring cantons of Bern, Fribourg and Vaud.

The geographical distribution map of TBEV-seropositive goats was created based on the results of serum neutralization test as this is the gold standard. Seropositive goats originating from other Swiss cantons were marked specifically, like the goat flock of Reckingen (Fig. 1, Table 3). Most of the seropositive goats were located in the known TBE risk area between the towns of Sierre and Visp. Other seropositive goats were found to the east of this area around the town of Brig and to the south of this area in the adjoining valley of St-Niklaus-Saas (Fig. 1). One flock of 26 goats, of which seven were positive for TBEV antibodies (according to the SNT), was discovered in the municipality of Finhaut, which is located 60 km west of the known TBEV-endemic area (Fig. 1).

Sex, age and breed of the goats

Individual information about the goats including origin, sex, age and breed were obtained for 1372 animals. Of these 1372 animals, 92.0% (1262/1372) were females and 8.0% (110/1372) were males. The age of the goats ranged from 0.5 to 11 years with a mean of 2.8 years (95% confidence limits (CL): 2.7–2.9 years). The most common goat breeds were col noir 63.1% (866/1372), chamoisée 21.5% (295/1373), Gessenay 5.4% (74/1372), and Grisonne 3.3% (45/1372). The other breeds (n = 92) included Boer, Verzasca, Toggenburg, Naine, Paon, Appenzell, Botée, and hybrids.

Of the 173 serum samples tested by SNT, the age of the goat was known for 115 individuals. The mean age of the seropositive goats (n = 52, mean = 3.4 years; 95% CL = 2.8–4.1 years) was almost twice as high as that of the seronegative goats (n = 63, mean = 1.6 years; 95% CL = 1.2–2.1 years) and this difference was statistically significant (χ^2 = 29.85, df = 10, p = 0.0009). Of the 70 SNT-positive serum samples, the sex and breed of the goat were known for 52 individuals. Among the 52 SNT-positive goats, 51 were female and one was male. Of the 52 SNT-positive goats, 80.8% (42/52) were col noir, 9.6% (5/52) were Gessenay, 3.9% (2/52) were Appenzell, one was Chamoisée (1.9%) and 3.9% (2/52) were hybrids.

Sites of pasture were known for 24 flocks, the average distance between the pasture site and the owner's place of residence was 3 km (95% CL = 1.9–4.1 km) and the mean number of pasture sites per owner was 2 (95% CL = 1.6–2.4 sites per owner).

As TBEV foci can be very small [9], the exact location of where goats were pastured was very important in cases where we sampled the local tick population for TBEV. For creating the maps, the nearby residences of the owners could be used because of the low average distance of 3 km.

Table 2 Comparison between the serum neutralization test (SNT) and the absorption test (AT) for 173 goat serum samples that tested positive for TBEV-specific antibodies on a preliminary ELISA

	SNT positive	SNT negative	Total
AT positive	50	0	50
AT equivocal	4	2	6
AT negative	16	101	117
Total	70	103	173

The association of the TBEV-positive or TBEV-negative status was highly significant between the SNT and the AT (p < 0.001). The SNT is considered as the gold standard for deciding whether a goat was exposed to tick-borne encephalitis virus

Ticks and TBEV

The three sites in the canton of Valais that had TBEV-seropositive flocks were located (1) near the town of Brig, (2) near the town of Gampel on the north side of the Rhone River, and (3) in an isolated area to the west of the municipality of Finhaut (Fig. 1). A total of 2045 *I. ricinus* ticks (adults and nymphs) were tested for TBEV using quantitative realtime RT-PCR [45]. TBEV-positive pools of ticks were detected at the site near Brig and the site near Gampel. With respect to the Gampel site, TBEV-positive pools of ticks were sampled from a pasture site that was located near Pletschen, which is 12 km from the town of Gampel and is located on the south side of the Rhone River. None of the tick pools from the two pasture sites near Finhaut tested positive for TBEV, despite a large sample size (1263 ticks).

Discussion

The most important result was that we were able to detect two new TBEV foci out of three potential sites

Table 3 Seroprevalence of TBEV-specific IgG antibodies in seropositive goat flocks collected between October 2011 and March 2012 in the canton of Valais

District	Municipality of flock	Total # of sera	SNT-positive sera	SNT Sero-prev (%)	Canton of origin	Municipality of origin	TBEV-endemic area[a]
Brig	Glis	13	9	69.23	Valais	Glis	No
					Lucerne	Büron	Yes
	Mund	62	2	3.23	Bern	Zweisimmen	No
					Valais	Ausserberg	No
	Naters	30	6	20.00	Valais	Naters	No
					Valais	Glis	No
		25	1	4.00	Valais	Naters	No
	Ried-Brig	8	1	12.50	St-Gall	Salez	No
	Termen	6	1	16.67	Valais	Naters	No
Total		144	20	20.94			
Goms	Reckingen	99	5	5.05		Oberwil im Simmental	Yes
						Oensingen	Yes
						Bettlach	No
Total		99	5	5.05			
Hérens	Euseigne	51	1	1.96		Euseigne	No
Total		51	1	1.96			
Leuk	Ergisch	6	1	16.67		Ergisch	No
	Niedergampel	10	1	10.00		Niedergampel	No
	Susten	5	1	20.00		Niedergampel	No
	Susten	16	6	37.50		St-Niklaus	No
						Susten	No
						Turtmann	No
	Susten	12	10	83.33		Gonten	No
						Stalden	No
						Susten	No
	Turtmann	2	1	50.00		Susten	No
Total		51	20	36.25			
Monthey	Collombey	37	1	2.70		Bex	No
Total		37	1	2.70			
St-Maurice	Finhaut	26	7	26.92		Finhaut	No
Total		26	7	26.92			
Visp	Eisten	9	1	11.11		Eisten	No
	Saas-Grund	9	3	33.33		Blumenstein	Yes
						Saas-Grund	No
	St-Niklaus	3	1	33.33		St-Niklaus	No
		22	0	0.00			
		20	0	0.00			
	Staldenried	24	3	12.50		Stalden	No
	Visp	52	3	5.77		Glis	No
						Visp	No
	Visp	6	3	50.00		Unknown	

Table 3 Seroprevalence of TBEV-specific IgG antibodies in seropositive goat flocks collected between October 2011 and March 2012 in the canton of Valais *(Continued)*

	Visperterminen	36	1	2.78		Visperterminen	No
		38	1	2.63		Visperterminen	No
Total		219	16	15.15			
TOTAL		627	70	11.16			

[a]Zones where TBE vaccination is recommended by the Federal Office of Public Health
For each seropositive goat flock, the following information is given: the district and the municipality where the flock is located, the total number of sera tested, the number of SNT-positive sera, the SNT seroprevalence (%), the canton and municipality from which the goats were originally obtained, and whether the site of origin was a TBEV-endemic area

based on the initial detection of seropositive goats and the subsequent confirmation of TBEV-positive ticks [45, 46]. The identification of TBEV foci in *I. ricinus* tick populations is difficult for a number of reasons. The geographic distribution of TBEV foci is highly patchy, and the foci are often very small with an area of about 100 m^2 [20, 21]. At such foci, the percentage of infected ticks is generally very low (<1%). Identification of new TBEV foci in the tick population therefore depends on strong a priori evidence that such a focus exists. Goats are interesting sentinel animals to use for sero-epidemiological surveys of TBEV. In Switzerland, goats are kept in the same two or three enclosed pasture sites year after year. If a TBEV focus is present, the goats will encounter infected ticks over the years and develop an antibody response. In summary, the fact that goats are kept in small, well-defined pasture sites in Switzerland makes them ideal sentinel hosts for screening for TBEV.

From an epidemiological point of view, the method used in the present study should be effective at identifying seropositive flocks without necessarily identifying every seropositive goat in that flock. In our study, we were confronted with some unexplained results. For example, one confirmed positive goat was detected in a flock of 51 goats in the village of Euseigne. Checking the origin of the seropositive goat confirmed that it was born and raised in Euseigne. This observation suggests that there may be a TBEV focus in the pasture sites of Euseigne even though this village is not located in the known endemic area of TBEV. An alternative explanation is that this serum sample produced a false positive on both the ELISA and the SNT, but such a result would be unlikely. As shown in a district in Thuringia, Germany, a single positive goat can provide evidence of a TBE risk area [38]. A sero-survey of the flock in Euseigne should be initiated in the future. Our study also showed the importance of knowing the origin of the goats. In a flock of goats in Reckingen, five sera tested positive, but all of these goats originated from the cantons of Bern and Solothurn, where TBEV is endemic. Thus, the conservative conclusion is that these goats were exposed to TBEV in these other cantons before being transported to the canton of Valais.

Statistical analyses are difficult due to the patchy pattern of TBEV foci and because in the present study, 50% of the goats originated from the known endemic area between Sierre and Visp. Data on sex and breed show that TBEV-positive goats are typical of the 'average' goat in the canton of Valais. Most of the goats in the canton of Valais are female (92.0% = 1262/1372) and most belong to the col noir breed (63.1% = 866/1372). Similarly, most of the TBEV-positive goats in the present study were female (51/52) and belonged to the col noir breed (80.8% = 42/52). This breed of goats is specific to the German-speaking region, which includes the known TBEV foci, and we experienced some difficulties in obtaining complete data from some of these goat flocks. It is unlikely that a TBEV infection is associated with a specific breed or sex. In contrast, breed is important in dogs where individuals with long and pale hair were more often infected by TBEV [48]. Seroprevalence increased with age as for many other infectious diseases. The mean age of the 52 positive goats was 3.4 years compared to the mean age of 2.8 years for the entire sample of goats. These findings were in agreement with other studies on sheep [47], where older animals with more than one season on the pasture were more likely to have TBEV antibodies than younger animals. The obvious explanation is that older animals are more likely to have been exposed to TBEV-infected ticks than younger animals.

The design of our ELISA allowed for quick screening of >4000 individual goat serum samples. A subsample of 175 of the most reactive goat serum samples was studied further using the absorption test (AT) and the serum neutralization test (SNT). Of the subsample of goat serum samples that were highly reactive on the ELISA, 26.9–33.1% tested positive on the AT (depending on how the equivocal samples were treated) and 40.4% tested positive on the SNT. Comparison between the AT and the SNT revealed that the AT was highly specific (98.1–100.0%), but with moderate sensitivity (71.4–77.1%; Table 2). The testing of 100 randomly selected serum samples by SNT demonstrated that the results of our ELISA were good. ELISA is the best and easiest assay to screen a large number of sera. Positive sera need to be checked using the gold standard SNT to confirm TBEV-seropositive status. A possible explanation for serum samples that test positive on the ELISA but negative on the SNT is infection with other flaviviruses that cross-react with the antigens of the ELISA assay. Possible flaviviruses that could result in cross-reactivity include the louping ill virus and the "ruminant" TBEV-like viruses that have been identified recently [49, 50].

The tick-borne encephalitis virus is classified into three subtypes that are found in different geographical areas: European (TBEV-Eu), Siberian (TBEV-Sib), and Far Eastern (TBEV-FEa) [11, 51, 52]. The European subtype is the least virulent subtype and is responsible for all human cases of TBE in Western Europe. In Switzerland, TBEV isolates circulating in ticks collected at 39 foci were closely related and all of them belonged to the European subtype [53]. In our previous study, we had shown that all tick-derived TBEV isolates from the canton of Valais belonged to the European subtype [45]. In vivo and in vitro studies of Swiss TBEV isolates suggest a high number of avirulent isolates, which is in agreement with a high proportion of subclinical or mild TBE infections in the Swiss public [53].

Conclusions

Our present study confirmed the utility of goats as sentinel animals in sero-epidemiological surveys of TBEV. Sero-surveys help to identify candidate sites where ticks should be collected to detect new TBEV foci. ELISA is an effective and simple tool to screen large numbers of sentinel hosts for antibodies against TBEV. Confirmation of ELISA-positive results is necessary and the SNT is the recommended gold standard. Knowledge of the origin of seropositive goats is essential to further define the potential presence of a new TBEV focus. Our sero-survey of goats in the canton of Valais revealed several hot spots that should be investigated in the future.

Abbreviations

AT: absorption test; BHK: baby hamster kidney; BSA: bovine serum albumin; CAE: caprine arthritis encephalitis; ELISA: enzyme - linked immunosorbent assay; Er: extinction rate; FOPH: Swiss Federal Office for Public Health (english name for OFSP); I: *Ixodes*; ICHV: Institut Central des Hôpitaux de Valais; Ig: immunoglobulin; ND: neutralizing dose; OD: optical density; OFSP: Office fédéral de la santé publique (french name for FOPH); RT-PCR: reverse transcriptase polymerase chain reaction; SFGR: Swiss Federation of Goat Rearing; SNT: serum neutralization test; TBE: tick-borne encephalitis; TBEV: tick-borne encephalitis virus; TBS: Tris-buffered saline solution; TCID: tissue culture infectious dose

Acknowledgements

We want to thank the Institut Galli Valerio (Lausanne, Switzerland) and the Veterinary Service of the Canton of Valais for their active collaboration in this project and for providing us with the goat serum samples for the present study. We would also like to thank the Swiss Federation of Goat Rearing (SFGR) and all goat owners who provided information about their animals and allowed us to use it for this study. The authors are very grateful to Doreen Schulz and Mareen Lange for excellent technical assistance.

Funding

The Cantonal Service of Public Health and the Institut Central des Hôpitaux de Valais (ICHV) provided financial support for this study (funds for research and development). There is no grant number for this funding dedicated to the TBE survey in the canton of Valais.

Authors' contributions

NR collected sera and data from goats, carried out the ELISA, analysed the results, and wrote parts of the manuscript. CK helped design the study, provided sera from immunized animals, and wrote parts of the manuscript. DH carried out the SNT, and wrote parts of the manuscript. OP designed the study, analyzed and interpreted the data, and wrote parts of the manuscript. MJV helped design the study and edited the manuscript. All authors helped to revise the manuscript and approved the final version.

Ethics approval and consent to participate

Ethics statement and animal experimentation permits:
All goat sera were collected during regular animal screening by professional veterinarians for caprine arthritis encephalitis virus (CAEV). This survey was mandated by the Swiss Federal Veterinary Office and was conducted in accordance with Swiss legislation on animal handling. The Veterinary Service of the Canton of Valais kindly allowed us to use the goat serum samples from this CAEV survey in the present study. No animal experimentation was done by the authors of this study. The Swiss Federation of Goat Rearing (SFGR) and private goat owners provided information about the animals and allowed us to use it for this study.

Consent for publication

Not applicable.

Competing interests

The authors declare that they have no competing interests.

Author details

[1]Central Institute of Valais Hospitals, Infectious diseases, Av Grand Champsec 86, -1950 Sion, CH, Switzerland. [2]Friedrich-Loeffler-Institut, Institute of Bacterial Infections and Zoonoses, Naumburger Str. 96a, D-07743 Jena, Germany. [3]Friedrich-Loeffler-Institut, Institute of Diagnostic Virology, Südufer 10, D-17493 Greifswald-Insel Riems, Germany. [4]Institute of Biology, Laboratory of Ecology and Evolution of parasites, University of Neuchâtel, Rue Emile-Argand 11, 2000 Neuchâtel, Neuchâtel, Switzerland.

References

1. Süss J. Tick-borne encephalitis 2010: Epidemiology, risk areas, and virus strains in Europe and Asia – an overview. Ticks Tickborne Dis. 2011, 2:2–15.
2. Altpeter E, Zimmermann H, Oberreich J, Péter O, Dvořák C. Tick related diseases in Switzerland, 2008 to 2011. Swiss Med Wkly. 2013; doi:10.4414/smw.2013.13725.

3. Randolph SE. Tick-borne encephalitis incidence in central and Eastern Europe: consequences of political transition. Microbes Infect. 2008;10:209–16.
4. Randolph SE. To what extent has climate change contributed to the recent epidemiology of tick-borne diseases? Vet Parasitol. 2010;167:92–4.
5. Randolph SE, Rogers DJ. Fragile transmission cycles of tick-borne encephalitis virus may be disrupted by predicted climate change. Proc Biol Sci. 2000;267:1741–4.
6. Skarpaas T, Golovljova I, Vene S, Ljøstad U, Sjursen H, Plyusnin A, Lundkvist Å. Tickborne encephalitis virus. Norway and Denmark Emerg Infect Dis. 2006;12:1136–8.
7. Materna J, Daniel M, Metelka L, Harčarik J. The vertical distribution, density and development of the tick *Ixodes ricinus* in mountain areas influenced by climate change (the Krkonoše Mts. Czech Republic). Int. J Med Microbiol. 2008;298(S1):25–37.
8. Danielová V, Kliegrová S, Daniel M, Beneš C. Influence of climate warming on tickborne encephalitis expansion to higher altitudes over the last decade (1997-2006) in the highland region (Czech Republic). Cent Eur J Public Health. 2008;16:4–11.
9. Holzmann H, Aberle SW, Stiasny K, Werner P, Mischak A, Zainer B, Netzer M, Koppi S, Bechter E, Heinz FX. Tick-borne encephalitis from eating goat cheese in a mountain region of Austria. Emerg Infect Dis. 2009;15:1671–3.
10. Daniel M, Danielová V, Kříž B, Růžek D, Fialová A, Malý M, Materna J, Pejčoch M, Erhart J. The occurrence of *Ixodes ricinus* ticks and important tick-borne pathogens in areas with high tick-borne encephalitis prevalence in different altitudinal levels of the Czech Republic part I. *Ixodes ricinus* ticks and tick-borne encephalitis virus. Epidemiol Mikrobiol Imunol. 2016;65:118–28.
11. Pierson TC, Diamond MS. Flavivirus. In: Knipe EM, Howley PM, Cohen JI, Griffin DE, Lamb RA, Martin MA, Racaniello VR, Roizman B, editors. Fields Virology. Sixth ed. Philadelphia: Wolters Kluwer Health/Lippincott Williams & Wilkins; 2013. p. 747–94.
12. Randolph SE. Transmission of tick-borne pathogens between co-feeding ticks: Milan Labuda's enduring paradigm. Ticks Tickborne Dis. 2011;2:179–82.
13. Achazi K, Růžek D, Donoso-Mantke D, Schlegel M, Ali HS, Wenk M, Schmidt-Chanasit J, Ohlmeyer L, Rühe F, Vor T, Kiffner T, Kallies R, Ulrich RG, Niedrig M. Rodents as sentinels for the prevalence of tick-borne encephalitis virus. Vector Borne Zoonotic Dis. 2011;11:641–7.
14. Randolph SE. The shifting landscape of tick-borne zoonoses: tick-borne encephalitis and Lyme borreliosis in Europe. Philos Trans R Soc Lond Ser B Biol Sci. 2001;356:1045–56.
15. Kunz C. Tick-borne encephalitis in Europe. Acta Leiden. 1992;60:1–14.
16. Hartemink NA, Randolph SE, Davis SA, Heesterbeek JAP. The basic reproduction number for complex disease systems: defining R-0 for tick-borne infections. Am Nat. 2008;171:743–54.
17. Harrison A, Bennett N. The importance of the aggregation of ticks on small mammal hosts for the establishment and persistence of tick-borne pathogens: an investigation using the R-0 model. Parasitology. 2012;139:1605–13.
18. Randolph SE, Gern L, Nuttall PA. Co-feeding ticks: epidemiological significance for tick-borne pathogen transmission. Parasitol Today. 1996;12:472–9.
19. Labuda M, Jones LD, Williams T, Danielova V, Nuttall PA. Efficient transmission of tick-borne encephalitis virus between cofeeding ticks. J Med Entomol. 1993;30:295–9.
20. Korenberg E. Recent epidemiology of tick-borne encephalitis an effect of climate change? Adv Virus Res. 2009;74:123–44.
21. Kupča AM, Essbauer S, Zoeller G, de Mendonça PG, Brey R, Rinder M, Pfister K, Spiegel M, Doerrbecker B, Pfeffer M, Dobler G. Isolation and molecular characterization of a tick-borne encephalitis virus strain from a new tick-borne encephalitis focus with severe cases in Bavaria. Germany Ticks Tick Borne Dis. 2010;1:44–51.
22. Korenberg E. Some contemporary aspects of natural focality and epidemiology of tick-borne encephalitis. Folia Parasitol. 1976;23:159–62.
23. Waldvogel K, Matile H, Wegmann C, Wyler R, Kunz C. Zeckenenzephalitis beim Pferd. Tick-borne encephalitis in horses. Schweiz Arch Tierheilkd. 1981;123:227–33. (in German)
24. Tipold AR, Fatzer R, Holzmann H. Zentraleuropäische Zeckenenzephalitis beim Hund. Central-European tick-borne encephalitis in dogs. Kleintierpraxis. 1993;38:619–28. in German, with English abstract
25. Leschnik MW, Kirtz GC, Thalhammer JG. Tick-borne encephalitis (TBE) in dogs. Int J Med Microbiol. 2002;291(Suppl 33):66–9.
26. Labuda M, Elecková E, Licková M, Sabó A. Tick-borne encephalitis virus foci in Slovakia. Int J Med Microbiol. 2002;291(Suppl 33):43–7.
27. Kerbo N, Donchenko I, Kutsar K, Vasilenko V. Tickborne encephalitis outbreak in Estonia linked to raw goat milk, may-June 2005. Euro Surveill. 2005;10:2–4.
28. Kříž B, Beneš C, Daniel M. Alimentary transmission of tick-borne encephalitis in the Czech Republic (1997-2008). Epidemiol Mikrobiol Imunol. 2009;58:98–103.
29. Balogh Z, Ferenczi E, Szeles K, Stefanoff P, Gut W, Szomor K, Takacs M, Berencsi G. Tick-borne encephalitis outbreak in Hungary due to consumption of raw goat milk. J Virol Methods. 2010;163:481–5.
30. Gäumann R, Mühlemann K, Strasser M, Beuret CM. High-throughput procedure for tick surveys of tick-borne encephalitis virus and its application in a national surveillance study in Switzerland. Appl Environ Microbiol. 2010;76:4241–9.
31. Reye AL, Hübschen JM, Sausy A, Muller CP. Prevalence and seasonality of tick-borne pathogens in questing *Ixodes ricinus* ticks from Luxembourg. Appl Environ Microbiol. 2010;76:2923–31.
32. Lommano E, Burri C, Maeder G, Guerne M, Bastic V, Patalas E, Gern L. Prevalence and genotyping of tick-borne encephalitis virus in questing *Ixodes ricinus* ticks in a new endemic area in western Switzerland. J Med Entomol. 2012;49:156–64.
33. Stefanoff P, Pfeffer M, Hellenbrand W, Rogalska J, Rühe F, Makówka A, Michalik J, Wodecka B, Rymaszewska A, Kiewra D, Baumann-Popczyk A, Dobler G. Virus detection in questing ticks is not a sensitive indicator for risk assessment of tick-borne encephalitis in humans. Zoonoses Public Health. 2013;60:215–26. doi:10.1111/j.1863-2378.2012.01517.x.
34. Knap N, Korva M, Dolinšek V, Sekirnik M, Trillar T, Avšič-Županc T. Patterns of tick-borne encephalitis virus infection in rodents in Slovenia. Vector Borne Zoonotic Dis. 2012;3:236–42.
35. Burri C, Korva M, Bastic V, Knap N, Avšič-Županc T, Gern L. Serological evidence of tick-borne encephalitis virus infection in rodents captured at four sites in Switzerland. J Med Entomol. 2012;49:436–9.
36. Gerth HJ, Grimshandl D, Stage B, Döller G, Kunz C. Roe deer as sentinels for endemicity of tick-borne encephalitis virus. Epidemiol Infect. 1995;115:355–65.
37. Klaus C, Hoffmann B, Moog U, Schau U, Beer M, Süss J. Can goats be used as sentinels for tick-borne encephalitis (TBE) in non-endemic areas? Experimental studies and epizootiological observations. Berl Muench Tieraerztl Wochenschr. 2010;123:441–5.
38. Klaus C, Beer M, Saier R, Schau U, Moog U, Hoffmann B, Diller R, Süss J. Goats and sheep as sentinels for tick-borne encephalitis (TBE) virus - epidemiological studies in areas endemic and non-endemic for TBE virus in Germany. Ticks Tickborne Dis. 2012;3:27–37.
39. Hubálek Z, Mitterpák J, Prokopic J, Juricová Z, Kilík J. A serological survey for Bhanja and tick-borne encephalitis viruses in sheep of eastern Slovakia. Folia Parasitol. 1985;32:279–83.
40. Csángó PA, Blakstad E, Kirtz GC, Pedersen JE, Czettel B. Tick-borne encephalitis in southern Norway. Emerg Infect Dis. 2004;10:533–4.
41. Pfeffer M, Dobler G. Tick-borne encephalitis virus in dogs - is this an issue? Parasit Vectors. 2011;1:59. doi:10.1186/1756-3305-4-59.
42. Paillard L, Jones KL, Evans AL, Berret J, Jacquet M, Lienhard R, Bouzelboudjen M, Arnemo JM, Swenson JE, Voordouw MJ. Serological signature of tick-borne pathogens in Scandinavian brown bears over two decades. Parasit Vectors. 2015;8:398. doi:10.1186/s13071-015-0967-2.
43. Klaus C, Hörügel U, Hoffmann B, Beer M. Tick-borne encephalitis virus (TBEV) infection in horses: clinical and laboratory findings and epidemiological investigations. Vet Microbiol. 2013;163:368–72.
44. Klaus C, Ziegler U, Kalthoff D, Hoffmann B, Beer M. Tick-borne encephalitis virus (TBEV) - findings on cross reactivity and longevity of TBEV antibodies in animal sera. BMC Vet Res. 2014;10:78. doi:10.1186/1746-6148-10-78.
45. Rieille N, Bressanelli S, Freire CC, Arcioni S, Gern L, Péter O, Voordouw MJ. Prevalence and phylogenetic analysis of tick-borne encephalitis virus (TBEV) in field-collected ticks (*Ixodes ricinus*) in southern Switzerland. Parasit Vectors. 2014;7:443. doi:10.1186/1756-3305-7-443.
46. Rieille N, Bally F, Péter O. Tick-borne encephalitis: first autochtonous case and epidemiological surveillance in canton Valais, Switzerland. Rev Med Suisse. 2012;8:1916–20. in French, with English abstract
47. Klaus C, Beer M, Saier R, Schubert H, Bischoff S, Süss J. Evaluation of serological tests for detecting tick-borne encephalitis virus (TBEV) antibodies in animals. Berl Muench Tieraerztl Wochenschr. 2011;124:443–9.
48. Janitza-Futterer D. [Serological investigations of the endemic situation of TBEV infection in a South-Baden horse and dog population.] Diss., Tieraerztl.

Fak, Ludwig-Maximilians-Universität, München, 2003. (in German, with English abstract).

49. Grard G, Moureau G, Charrel RN, Lemasson JJ, Gonzalez JP, Gallian P, Gritsun TS, Holmes EC, Gould EA, de Lamballerie X. Genetic characterization of tick-borne flaviviruses: new insights into evolution, pathogenetic determinants and taxonomy. Virology. 2007;361:80–92.
50. Mansfield KL, Morales AB, Johnson N, Ayllon N, Hofle U, Alberdi P, Fernandez de Mera IG, Marin JF, Gortazar C, de la Fuente J, Fooks AR. Identification and characterization of a novel tick-borne flavivirus sub-type in goats (*Capra hircus*) in Spain. J Gen Virol. 2015;96:1676–81.
51. Ecker M, Allison SL, Meixner T, Heinz FX. Sequence analysis and genetic classification of tick-borne encephalitis viruses from Europe and Asia. J Gen Virol. 1999;80:179–85.
52. Mansfield KL, Johnson N, Phipps LP, Stephenson JR, Fooks AR, Solomon T. Tick-borne encephalitis virus - a review of an emerging zoonosis. J Gen.Virol 2009, 90:1781–1794.
53. Gäumann R, Růžek D, Mühlemann K, Strasser M, Beuret CM. Phylogenetic and virulence analysis of tick-borne encephalitis virus field isolates from Switzerland. J Med Virol. 2011;83:853–63.

14

Molecular evolution of type 2 porcine reproductive and respiratory syndrome viruses circulating in Vietnam from 2007 to 2015

Hai Quynh Do[1], Dinh Thau Trinh[1], Thi Lan Nguyen[1], Thi Thu Hang Vu[2], Duc Duong Than[2], Thi Van Lo[2], Minjoo Yeom[3], Daesub Song[3], SeEun Choe[4], Dong-Jun An[4] and Van Phan Le[1*]

Abstract

Background: Porcine respiratory and reproductive syndrome (PRRS) virus is one of the most economically significant pathogens in the Vietnamese swine industry. ORF5, which participates in many functional processes, including virion assembly, entry of the virus into the host cell, and viral adaptation to the host immune response, has been widely used in molecular evolution and phylogeny studies. Knowing of molecular evolution of PRRSV fields strains might contribute to PRRS control in Vietnam.

Results: The results showed that phylogenetic analysis indicated that all strains belonged to sub-lineages 8.7 and 5.1. The nucleotide and amino acid identities between strains were 84.5–100% and 82–100%, respectively. Furthermore, the results revealed differences in nucleotide and amino acid identities between the 2 sub-lineage groups. N-glycosylation prediction identified 7 potential N-glycosylation sites and 11 glycotypes. Analyses of the GP5 sequences, revealed 7 sites under positive selective pressure and 25 under negative selective pressure.

Conclusions: Phylogenetic analysis based on ORF5 sequence indicated the diversity of PRRSV in Vietnam. Furthermore, the variance of N-glycosylation sites and position under selective pressure were demonstrated. This study expands existing knowledge on the genetic diversity and evolution of PRRSV in Vietnam and assists the effective strategies for PRRS vaccine development in Vietnam.

Keywords: PRRSV, Vietnam, ORF5, Phylogeny

Background

Porcine reproductive and respiratory syndrome (PRRS) is a major infectious disease affecting pork industries worldwide. Its outbreaks were first reported in the USA and EU in the late 1980s and early 1990s, respectively [4, 5, 42]. The main clinical signs of the disease are respiratory problems in pigs of all ages and reproductive failure in pregnant sows. In Vietnam, PRRS outbreaks have continuously occurred since 2007 [8, 23, 28]. PRRS viruses, the causative agents of the disease, can be divided into two distinct genotypes, type I (EU type) and type II (American type), which present with identical disease symptoms, despite their genetic differences [24]. PRRSV is a mono-partite, linear, positive-sense single stranded RNA virus belonging to the *Arterviridae* family [5]. Its genome of approximately15 kb in size is organized into 10 open reading frames (ORFs) [24, 38]. Two large ORF1a and ORF1b genes encode non-structural proteins that play important roles in viral replication and virulence [13, 18]. The other ORFs encode for structural proteins that are necessary for production of infectious virions [44]. ORF5, which participates in many functional processes, including virion assembly [44], entry of the virus into the host cell [7], and viral adaptation to the host immune response [41], has been widely used in molecular evolution and phylogeny studies [30, 34, 35].

* Correspondence: letranphan@vnua.edu.vn
[1]Faculty of Veterinary Medicine, Vietnam National University of Agriculture, Hanoi, Vietnam
Full list of author information is available at the end of the article

Evolutionary studies indicate that PRRSV diverged long before the first detected outbreaks of the disease. Evolutional analyses based on ORF5, as well as serological evidence, indicated that PRRSV type 2 first appeared around the 1980s [3, 35, 48]. In contrast, type 1 PRRSV originated approximately 100 years ago [30]. Further analysis of the whole PRRSV genome shows that the two types of PRRSV diverged from a common ancestor about 800 years ago [46]. Furthermore, genetic analyses indicate that the evolutionary trends, antigenic characteristics, and genetic diversity of PRRSV in different regions have distinct patterns [6, 11, 17, 32, 36, 40].

Thus far, type 2 PRRSV has been divided into 10 sub-lineages, including 9 old sub-lineages [34] occurring worldwide, and a new sub-lineage, which recently appeared in Thailand [40]. In Vietnam, several studies show that the circulating PRRSV strains belong to a highly pathogenic (HP) variant that recently emerged in China and South East Asian countries [12, 28]. However, few studies have focused on the evolutionary trends and characterization of PRRSV presenting in Vietnam. Thus, the aim of this study was to investigate the genetic diversity, selective pressure, and glycosylation patterns in GP5 of PRRSV strains that appeared in Vietnam during 2007–2015.

Methods

Sample collection

For this study, we used 40 PRRS-positive sera or tissue samples, as confirmed by RT-PCR; the samples were collected from pigs in provinces in North Vietnam during 2011–2015. Total PRRSV RNA was extracted using TRIzol Reagent (Invitrogen, USA) according to the manufacturer's instruction. Reverse transcription was performed using SuperScript™ III First-Strand Synthesis SuperMix (Thermo Fisher, USA). ORF5 sequences were amplified by RT-PCR using previously described primers [12]. PCR products were directly sequenced (Macrogen, Seoul, Korea). The raw sequences were assembled and aligned using BioEditv7.2.5 [14] against the corresponding ORF5 sequences from GenBank to construct the complete ORF5 sequence. Additional 104 Vietnamese ORF5 reference sequences from field isolates collected from GenBank were also used in this study (Additional file 1: Table S1).

Phylogenetic analysis and classification

In order to identify the lineage classifications for all the PRRSV strains circulating in Vietnam, an ORF5-based phylogeny was reconstructed using a restricted parameter substitution model [35] with IQ-TREE software [27]. The total data set in this study contained 144 Vietnamese ORF5 gene sequences and 612 worldwide ORF5 reference sequences for lineages 1 to 9 [35]. Bootstrap values were obtained using the ultrafast bootstrap approximation method with 1000 replicates [25] (both programs are available at http://iqtree.cibiv.univie.ac.at/).

Bayesian phylogenetic inference of ORF5 from Vietnamese strains

The coalescent-based Bayesian Markov Chain Monte Carlo (MCMC) method was used to investigate the phylogenetic relationship among Vietnamese PRRSV strains based on ORF5 sequences. The SRD06 codon-based model was used as a nucleotide substitution model [29, 31] and combined with (i) 5 molecular clock models (Strict clock, uncorrelated lognormal relaxed clock, uncorrelated exponential relaxed clock, random clock, and fixed local clock) and (ii) 7 demographic coalescent tree models (constant size, exponential growth, logistic growth, expansion growth, Bayesian skyline, extended Bayesian skyline plot, and Bayesian skygrid). In each analysis, the MCMC chain (50 million generations, sampling every 5000 stages) was performed using BEAST v1.8.2 software [9]. Five independent runs were done to verify the distribution in the MCMC run. The corresponding output log files were combined by LogCombiner before subsequently analyzing via Tracer v1.6 to select the best-fit data models for molecular clock and coalescent tree priors using Akaike's information criterion (AICM) analysis with 1000 replicates [2]. A Bayesian phylogenetic tree was selected from combined trees files from the above chosen best-fit models using TreeAnnotator in BEAST package.

Glycosylation site prediction

Glycosylation sites in the Vietnamese PRRSV strains were predicted using the NetNGlyc server web utility (http://www.cbs.dtu.dk/services/NetNGlyc/). Adefault threshold of 0.5 was used to identify potential N-glycosylation sites, followed by additional thresholds of 0.75 and 0.9 to identify the potential N-glycosylation sites with higher confidence levels.

Selective pressure

GP5 sites undergoing positive selection were inferred using 5 algorithms: SLAC, FEL, IFEL, FUBAR, and MEME (available on the Datamonkey web server: www.datamonkey.org). Sites undergoing negative selection were predicted using 4 algorithms: SLAC, FEL, IFEL, and FUBAR. To identify other sites undergoing potential selective pressure, sites were analyzed for either diversifying or purifying selection at *P*-value ≤ 0.1 using SLAC, FEL, IFEL, and MEME methods, or for posterior probability $\geq 90\%$ using the FUBAR method.

Results

To investigate the evolution of Vietnamese PRRSV strains, we analyzed the time scale phylogenetic tree, the genetic diversity among strains, the time of most common ancestor of PRRSV strains in Vietnam as well as the change of N-glycosylation pattern during this time.

Phylogenetic analysis of the ORF5 sequence

Based on the constructed phylogenetic tree, the major PRRSV strains ($n = 138$) isolated in Vietnam could be classified into sub-lineage 8.7, which is closely related to the highly pathogenic PRRSV strains recently isolated in China, including JXA1 and SX2009 [47]. The remaining strains ($n = 6$) were classified into sub-lineage 5.1, which contains VR-2332-related strains [35] (Additional file 2: Figure S1). Further analysis based on Bayesian inference showed that HP-PRRSVs in Vietnam can be divided into two main sub-groups (Fig. 1). Interestingly, most of the PRRSV strains collected in North Vietnam during 2013–2015 belonged to sub-group ii (Fig. 1). Under the best-fit model selected, the substitution rate in the ORF5 gene of the Vietnamese PRRSV strains was about4.459 × 10^{-3} (95% highest posterior density (HPD) intervals: 3.0981 × 10^{-3}–5.8523 × 10^{-3}). In addition, the geometric mean time to the most recent common ancestor (T_{MRCA}) of the HP-PRRSV isolated in Vietnam was approximately 13 years ago and the T_{MRCA} of sub-lineage 5.1 PRRSV strains was more than 16 years ago (95% HPD was 9.2708–18.8409 and 6.9872–32.5917 for the HP-PRRSV group and sub-lineage 5.1 group, respectively).

Genetic diversity of the Vietnamese PRRSV strains during 2007–2015

Genetic comparison of the ORF5 gene of the Vietnamese PRRSV strains collected from 2007 to 2015 showed that 144Vietnamese PRRSV strains in this study shared 81–100% nucleotide identity (Table 1). Furthermore, the similarity among the ORF5 sequences presented in the same year was about 84.5–100%. Especially in 2010, 2012, 2013, and 2014, when the appearance of PRRSV sub-lineage 5.1 strains was recorded, differences among nucleotide sequences was up to 15.5% while in the remaining years, it was just about 2%. Further analysis showed that the similarity among Vietnamese HP-PRRSV strains was of 91.6–100% while the difference among sub-lineage 5.1 strains was up to 9.5%.

The deduced amino acid sequence encoded by the ORF5 gene of 144 Vietnamese PRRSV strains shared 82–100% identity. For each sub-lineage group, the amino acid identity was 90–100% and 86.5–100% for HP-PRRSV and sub-lineage 5.1, respectively (Table 1).

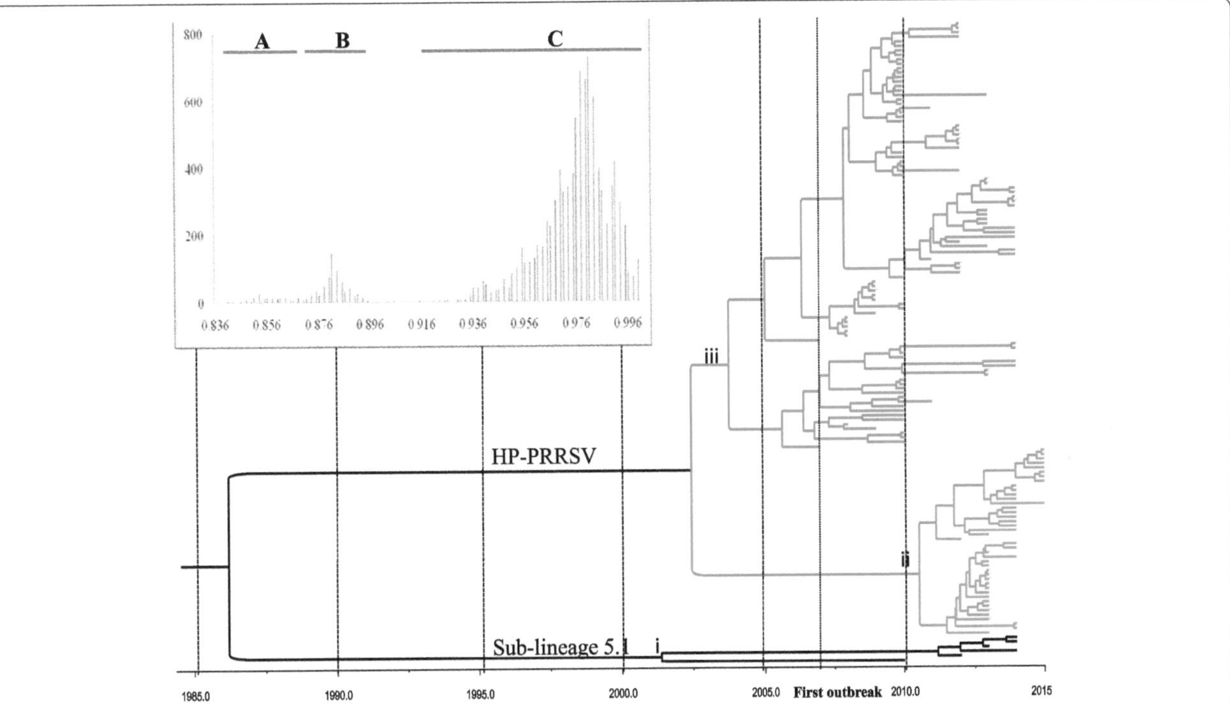

Fig. 1 Phylogenetic tree based on nucleotide sequence of the ORF5 gene of 144 PRRSV strains isolated in Vietnam during 2007–2015. The phylogenetic tree, generated via the MCMC method using BEAST v1.8.2 software, identified three different groups. The inserted histogram illustrates pairwise sequence comparisons of Vietnamese PRRSV type 2 strains. Three distinct nucleotide identity distribution peaks are shown. The time-scale (in years) represented in the tree is indicated by the scale bar

Table 1 Nucleotide and deduced amino acid identities among 144 Vietnamese PRRSV strains

Year		2015	2014	2013	2012	2011	2010	2009	2008	2007
2015	nt	99.3–100	88.1–99.6	88.1–99.5	89.1–99.1	98.1–99.1	86–99.1	98.1–99.1	98.1–99.1	98.5–99.1
	aa	98–100	85.5–95.5	85.5–98.5	88–98.5	96.5–99	84–99	97–99	97–99	97.5–98.5
2014	nt		84.8–100	85.1–100	86–99.5	87.8–99.6	84.1–99.1	87.8–99.1	87.8–99.1	88–99.1
	aa		82–100	82–100	84–99.5	85.5–100	81–99.5	86–99.5	86–99.5	85.5–99
2013	nt			86.5–100	87–99.5	87.8–99.3	83.6–99	87.8–99	87.8–99	88–99
	aa			83.5–100	84.5–99.5	86–99	81.5–99.5	86–99	86–99	85.5–98.5
2012	nt				87.8–100	88.8–99.8	84.6–100	88.8–99.5	88.8–99.5	89–98.6
	aa				87–100	88.5–99.5	81–100	88.5–99.5	88.5–99.5	88–99
2011	nt					98.1–100	85.6–100	98.1–99.5	98.1–99.5	98.1–99.3
	aa					98–100	83–100	98–100	98–100	97.5–99.5
2010	nt						84.5–100	85.3–100	85.3–100	85.5–99.3
	aa						81.5–100	83–100	83–100	83.5–99.5
2009	nt							98.1–100	98.1–100	98.1–99.3
	aa							98.5–100	98.5–100	98–99.5
2008	nt								98.1–100	98.1–99.3
	aa								98.5–100	98–99.5
2007	nt									98.5–100
	aa									98.5–100

Glycosylation site variants

A total of 7 potential N-glycosylation sites (amino acids 30, 32, 33, 34, 35, 44 and 51) were found for the Vietnamese PRRSV strains isolated in Vietnam during the 2007–2015 period. The identified positions and the total numbers of N-glycosylation sites were diverse. Notably, PRRSV strains isolated during the outbreaks in 2014 had the greatest variation in N-glycosylation patterns, followed by those from the outbreaks in 2013, which had 8 and 7 glycotypes (Table 2). Glycosylation site variations were located between amino acids 32 and 35, while N44 and 51 seemed to be conserved in most of the Vietnamese strains, presenting in 97.2% and 100% of strains, respectively. Furthermore, N41 was predicted as a glycosylation site with higher potential (≥0.75). An N-glycosylation pattern of N30, N35, N44, and N51 seems to be the main glycotype in Vietnamese PRRSV strains, accounting for nearly 61%. Interestingly, we observed differences in the frequencies of N-glycosylation positions between sub-lineage 5.1 strains and sub-lineage 8.7 strains. To be specific, N30, N32, N33, N34, and N35 were identified as potential N-glycosylation sites in sub-lineage 8.7, accounting for 92.09, 2.88, 12.95, 18.71 and 82.73% of strains, respectively, whereas only N30, N33, and N34 were predicted as in sub-lineage 5.1 accounting for 33.33, 66.67, and 33.33% of strains, respectively (data not shown). Furthermore, only two Vietnamese PRRSV sub-lineage 5.1 strains had similar N-glycosylation patterns as the vaccine strain VR2332, while the other strains lacked the potential N-glycosylation site at N30. On the other hand, 88 Vietnamese HP-PRRSV strains had the same N-glycosylation pattern as the JXA1 vaccine strains.

Selective pressure in GP5

To identify positions under selective pressure, SLAC, FEL, IFEL, MEME, and FUBAR methods were implemented separately. Since each method utilizes a different algorithm for predicting sites under positive or negative selection, for our study, we considered sites to be undergoing diversifying selection if so predicted by all 5 of the methods, and to be undergoing purifying selection if predicted by 4 of the methods. Consequently, we identified 7 positions as potentially undergoing positive selection (codons 25, 33, 34, 35, 58, 59, and 104). Most of the positive selection sites were located in ecto-domain 1 ($n = 5$), while only 1 site undergoing diversifying selection was found in each ecto-domain 2 and signal domain (Fig. 2).

A different pattern was observed for the negative selection sites. In our study, 25 sites were predicted to be undergoing negative selection (codons 12, 45, 46, 74, 80, 81, 83, 87, 93, 100, 109, 113, 126, 139, 149, 156, 157, 160, 163, 165, 172, 173, 174, 177, and 193). Purifying selection sites were mainly located in the endo-domain ($n = 12$). Furthermore, negative selection sites were detected in trans-membrane 1, trans-membrane 2, and ecto-domain 2 (positions 4, 3, and 3, respectively) (Fig. 2).

Table 2 Glycosylation pattern of PRRSV strains in Vietnam during 2007–2015

Year/Ref Strain	N-glycosylation site							Number of sequence	% of total
	30	32	33	34	35	44	51		
VR-2332	x		x			xxx	x		
JXA1	x				x	xxx	x		
2007	x				x	xx	x	1	0.69 %
	x					xx	x	1	0.69 %
2008	x				x	xx	x	6	4.17 %
2009	x				x	xx	x	7	4.86 %
2010	x				x	xx	x	31	21.53 %
	x			x	x	xx	x	1	0.69 %
	x				x		x	1	0.69 %
				x		xx	x	4	2.78 %
	x		x			xx	x	3	2.08 %
2011	x		x			xx	x	1	0.69 %
	x				x	xx	x	2	1.39 %
2012	x				x	xx	x	11	7.64 %
	x		x			xx	x	2	1.39 %
		x	x			xx	x	3	2.08 %
	x				x		x	1	0.69 %
2013	x			x	x	xx	x	14	9.72 %
	x		x	x		xx	x	2	1.39 %
	x		x			xx	x	1	0.69 %
	x					xx	x	2	1.39 %
	x				x	xx	x	6	4.17 %
			x			xx	x	1	0.69 %
					x	xx	x	3	2.08 %
2014	x				x		x	1	0.69 %
	x		x				x	1	0.69 %
	x			x	x	xx	x	6	4.17 %
	x				x	xx	x	15	10.42 %
	x		x			xx	x	5	3.47 %
		x	x			xx	x	1	0.69 %
			x			xx	x	2	1.39 %
				x		xx	x	1	0.69 %
2015	x				x	xx	x	9	6.25 %

x: indicating the potential N-glycosylation site at cut off value; xx and xxx: indicating the potential N-glycosylation site at additional value (>0.75 and >0.9, respectively)

Discussion

Since first identified in the late 1980s, PRRSV has become the most significant porcine reproductive pathogen. The ORF5 gene is the most diverse gene not only in PRRSV, but also in other arteriviruses, and has been an important target for investigations on the genetic characterization and evolution of PRRSV worldwide [34].

In this study, most of the PRRSV strains isolated in Vietnam during 2007–2015 belonged to sub-lineage 8.1, except for 6 strains that belonged to sub-lineage 5.1 (Additional file 2: Figure S1). This result is consistent with that of previous reports, indicating that most of the PRRSV strains isolated in Vietnam are close related to JXA1 [12, 28]. In fact, attenuated vaccine strains belonging to sub-lineage 5.1, such as VR2332 and BSL-PS, have been approved for use in Vietnam. Furthermore, although HP-PRRSV strains are the main agents of PPRS in countries around Vietnam, such as China and others in South East Asia, strains from other type 2 PRRSV lineages such as lineages 1, 3, and 5 have also circulated [22, 40]. Therefore, these data suggest that the appearance of sub-lineage 5.1 in Vietnam may be due to vaccine descendants or commercial activities. In addition, our study indicates that the HP-PRRSV strains circulating during 2013–2015 were distantly related to the other HP-PRRSV strains. This may have resulted from the introduction of new PRRSV strains into Vietnam. However, the limited number of Vietnamese PRRSV ORF5 sequences used in this study may not exactly reflect the genetic diversity of PRRSVs in Vietnam.

Although, the PRRSV strains circulating in Vietnam clustered within sub-lineage 5.1 and 8.7, their percentages of intra-sub-lineage genetic diversity were 90.5–100% and 91.6–100%, respectively. The intra-sub-lineage genetic diversity in our study was higher than in a previous study [35]. This result might be due to the high substitution rate detected in the ORF5 gene. In addition, the substitution rate in the ORF5 gene of the Vietnamese strains, which was 4459 × 10^{-3}, was slightly faster than the substitution rates observed in common type II PRRSV strains [31] This supports our hypothesis. According to our analysis, the T_{MRCA} of sub-lineage 5.1 strains was approximately 17 years ago, which is supported by serological evidence for anti-PRRSV antibodies in Vietnam during this time. The T_{MRCA} of sub-lineage 8.7 strains was estimated to have occurred in 2002, which is similar to the T_{MRCA} of the HP-PRRSV from China [36].

It is reported that the N-glycosylation positions in GP5 affect the adaptation of PRRSV to the host's immune response and infectivity [1]. In our study, 11 potential N-glycotypes were observed for the GP5 protein of the Vietnamese PRRSV strains isolated between 2007 and 2015 (Table 2). Furthermore, our investigation revealed diversity in the putative N-glycosylation site amino acid positions (7 different positions) and quantity (3 to 5 sites). However, the N-glycotype diversity identified in the Vietnamese PRRSV strains was less than that recorded for the PRRSV strains from Eastern Canada isolated between 1998 and 2009 [6]. Another report shows that the PRRSV strains isolated in China from

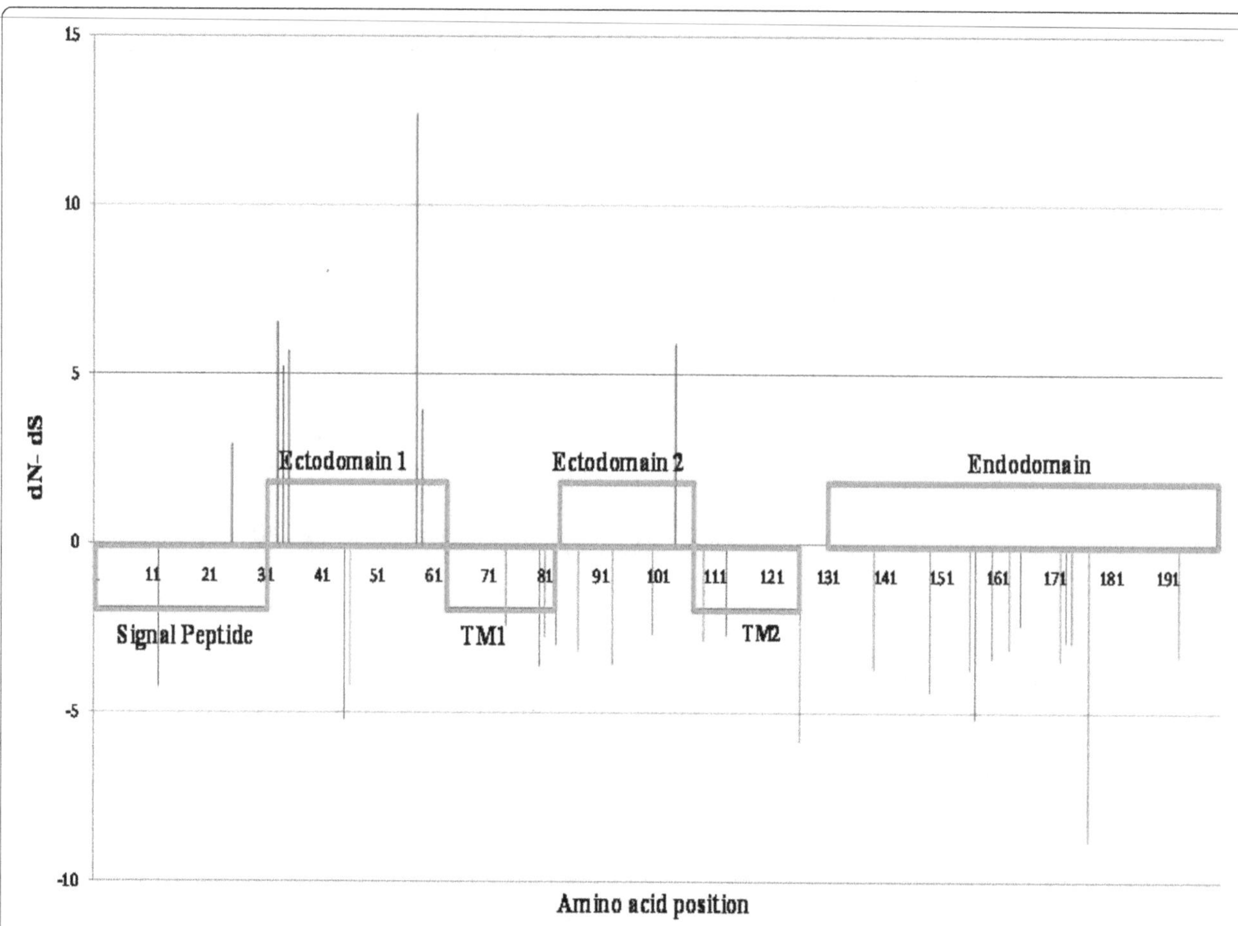

Fig. 2 Amino acids under selective pressure. *Upper rectangles* indicate the ecto-domain 1, ecto-domain 2, and endo-domain, whereas *lower rectangles* indicate the signal peptide, trans-membrane 1, and trans-membrane 2 regions. The *red* lines and *blue* lines indicate sites under significant positive and negative selection, respectively. dN-dS represents the normalized dN–dS value according to the FEL method

2006 to 2009 have the same N-glycosylation sites as the Vietnamese strains [26]. In all positions, the N44 site was predicted at a high confidence level (≥0.75) and seemed to be conserved. These results are consistent with the important role of this residue in infectivity [1]. The N51 glycosylation site has also been demonstrated to affect the growth kinetics of PRRSV [1], and is highly conserved in type 2 PRRSV from many countries [6, 17, 19].

On the other hand, N-glycosylation sites seem to vary at positions 32–35 (Table 2). These sites are located within the hyper-variable region that has been previously described [39]. It is believed that the variations in N-glycosylation in this region may influence viral neutralization [16]. However, not all potential glycosylation sites in this region are glycosylated. Li et al. [20] suggested that only 2 or 3 glycosylation sites in this region are utilized and that their exact positions are still unknown.

Another notable result of this study was the distribution of diversity and purified selection positions throughout the GP5 of Vietnamese PRRSV strains. Nguyen et al. [31] can not conclude the role of selected position in typical PRRSV and HP-PRRSV. In this study, our analysis of sites under selective pressure indicated that most of the sites undergoing positive selection (amino acids 33, 34, 35, 58, and 59) are located in ecto-domain 1, which contains the linear epitope, and an additional positive selection site is also predicted in ecto-domain 2 (amino acid 104). Our results are generally consistent with those of previous studies, with the exception of the positive selection site predicted in the signal peptide [6, 15, 26]. A previous study demonstrated that site-directed mutagenesis of the amino acid residues at 102 and 104 can enhance PRRSV evasion of neutral antibodies in vitro [10]. This supports the results from our current study. In addition, the potential N-glycosylation sites at the N33, N34, and N35 have been previously identified as undergoing positive selection in the HP-PRRSV strains recently isolated in China [26, 45]. In a previous investigation, in vitro neutralization experiments showed that mutation of the N34 Asp (wt) to Asn slightly decreases the neutralizing activity of Asp-34 sera [33].

Our study showed that the main locations under negative selection were in the endo-domain, following by trans-membrane 1, trans-membrane 2, and ecto-domain 2. This agrees with the findings of Xu et al. [45]. Negative selective pressure within the trans-membrane domains may relate to the integrity or functionality of the virion whereas the distribution of sites under purifying selection in the endo-domain could relate to the budding process of PRRSV. A similar function has been observed in alphaviruses, where E2 and the nucleocapsid protein specifically interact with each other [37]. In addition, mutations within endo-domain of glycoprotein E2 affected the biological characteristic of Sindbis virus [21, 43].

Conclusions

This study first describes the molecular evolution of ORF5 of PRRSV occurred in Vietnam since the first outbreaks. Phylogenetic analysis based on ORF5 sequence indicated the diversity of PRRSV in Vietnam. Furthermore, the variance of N-glycosylation sites and position under selective pressure were demonstrated. This study expands existing knowledge on the genetic diversity and evolution of PRRSV in Vietnam and assists the effective strategies for PRRS vaccine development in Vietnam.

Abbreviations

HP: Highly pathogenic; HPD: Highest posterior density; MCMC: Markov Chain Monte Carlo; ORF: Open reading frame; PRRS: Porcine reproductive and respiratory syndrome; PRRSV: Porcine reproductive and respiratory syndrome virus; T_{MRCA}: Time to the most recent common ancestor

Acknowledgements

None.

Funding

This work was supported by the Vietnam National Project under the Project Code No: SPQG.05b.02 and by a grant (Project Code No. 313014-03-1-HD030) from the Korea Institute of Planning & Evaluation for Technology in Food, Agriculture, Forestry & Fisheries, 2013.

Authors' contributions

HQD, SC, MY, TVL conducted experiment, analyzed the data and wrote the paper. TTHV and DDT assisted sample preparation and experiment. DS and TLN shared ideas and discussed the research data. VPL, DTT, DJA contributed to supervision, had the idea for the project, and directed the project. All authors read and approved the final manuscript.

Competing interests

The authors declare that they have no competing interests.

Consent for publication

Not applicable.

Author details

[1]Faculty of Veterinary Medicine, Vietnam National University of Agriculture, Hanoi, Vietnam. [2]Research and Development Laboratory, Avac Vietnam Company Limited (AVAC), Hung Yen, Vietnam. [3]College of Pharmacy, Korea University, Sejong, Republic of Korea. [4]Animal and Plant Quarantine Agency, Gyeonggi-do, Gimcheon, Gyeongsangbukdo, Republic of Korea.

References

1. Ansari IH, Kwon B, Osorio FA, Pattnaik AK. Influence of N-linked glycosylation of porcine reproductive and respiratory syndrome virus GP5 on virus infectivity, antigenicity, and ability to induce neutralizing antibodies. J Virol. 2006;80(8):3994–4004. doi:10.1128/JVI.80.8.3994-4004.2006.
2. Baele G, Lemey P, Bedford T, Rambaut A, Suchard MA, Alekseyenko AV. Improving the accuracy of demographic and molecular clock model comparison while accommodating phylogenetic uncertainty. Mol Biol Evol. 2012;29(9):2157–67.
3. Carman S, Sanford S, Dea S. Assessment of seropositivity to porcine reproductive and respiratory syndrome (PRRS) virus in swine herds in Ontario–1978 to 1982. Can Vet J. 1995;36(12):776.
4. Collins JE, Benfield DA, Christianson WT, Harris L, Hennings JC, Shaw DP, Goyal SM, McCullough S, Morrison RB, Joo HS, et al. Isolation of swine infertility and respiratory syndrome virus (isolate ATCC VR-2332) in North America and experimental reproduction of the disease in gnotobiotic pigs. J Vet Diagn Invest. 1992;4(2):117–26.
5. Conzelmann KK, Visser N, Van Woensel P, Thiel HJ. Molecular characterization of porcine reproductive and respiratory syndrome virus, a member of the arterivirus group. Virology. 1993;193(1):329–39. doi:10.1006/viro.1993.1129.
6. Delisle B, Gagnon CA, Lambert ME, D'Allaire S. Porcine reproductive and respiratory syndrome virus diversity of Eastern Canada swine herds in a large sequence dataset reveals two hypervariable regions under positive selection. Infect Genet Evol. 2012;12(5):1111–9. doi:10.1016/j.meegid.2012.03.015.
7. Delputte P, Costers S, Nauwynck H. Analysis of porcine reproductive and respiratory syndrome virus attachment and internalization: distinctive roles for heparan sulphate and sialoadhesin. J Gen Virol. 2005;86(5):1441–5.
8. Do DT, Park C, Choi K, Jeong J, Nguyen TT, Le DT, Vo KM, Chae C. Nucleotide sequence analysis of Vietnamese highly pathogenic porcine reproductive and respiratory syndrome virus from 2013 to 2014 based on the NSP2 and ORF5 coding regions. Arch Virol. 2016;161(3):669–75. doi:10.1007/s00705-015-2699-1.
9. Drummond AJ, Rambaut A. BEAST: Bayesian evolutionary analysis by sampling trees. BMC Evol Biol. 2007;7(1):214. doi:10.1186/1471-2148-7-214.
10. Fan B, Liu X, Bai J, Zhang T, Zhang Q, Jiang P. The amino acid residues at 102 and 104 in GP5 of porcine reproductive and respiratory syndrome virus regulate viral neutralization susceptibility to the porcine serum neutralizing antibody. Virus Res. 2015;204:21–30.
11. Fang Y, Schneider P, Zhang W, Faaberg K, Nelson E, Rowland R. Diversity and evolution of a newly emerged North American Type 1 porcine arterivirus: analysis of isolates collected between 1999 and 2004. Arch Virol. 2007;152(5):1009–17.
12. Feng Y, Zhao T, Nguyen T, Inui K, Ma Y, Nguyen TH, Nguyen VC, Liu D, Bui QA, To LT, Wang C, Tian K, Gao GF. Porcine respiratory and reproductive syndrome virus variants, Vietnam and China, 2007. Emerg Infect Dis. 2008; 14(11):1774–6. doi:10.3201/eid1411.071676.
13. Grebennikova T, Clouser D, Vorwald A, Musienko M, Mengeling W, Lager K, Wesley R, Biketov S, Zaberezhny A, Aliper T. Genomic characterization of virulent, attenuated, and revertant passages of a North American porcine reproductive and respiratory syndrome virus strain. Virology. 2004;321(2):383–90.
14. Hall T. A. BioEdit: a user-friendly biological sequence alignment editor and analysis program for Windows 95/98/NT. Paper presented at the Nucleic acids symposium series. 1999.
15. Hanada K, Suzuki Y, Nakane T, Hirose O, Gojobori T. The origin and evolution of porcine reproductive and respiratory syndrome viruses. Mol Biol Evol. 2005;22(4):1024–31. doi:10.1093/molbev/msi089.

16. Jiang W, Jiang P, Wang X, Li Y, Wang X, Du Y. Influence of porcine reproductive and respiratory syndrome virus GP5 glycoprotein N-linked glycans on immune responses in mice. Virus Genes. 2007;35(3):663–71.
17. Kvisgaard LK, Hjulsager CK, Brar MS, Leung FC, Larsen LE. Genetic dissection of complete genomes of Type 2 PRRS viruses isolated in Denmark over a period of 15 years. Vet Microbiol. 2013;167(3):334–44.
18. Kwon B, Ansari IH, Pattnaik AK, Osorio FA. Identification of virulence determinants of porcine reproductive and respiratory syndrome virus through construction of chimeric clones. Virology. 2008;380(2):371–8.
19. Li B, Fang L, Guo X, Gao J, Song T, Bi J, He K, Chen H, Xiao S. Epidemiology and evolutionary characteristics of the porcine reproductive and respiratory syndrome virus in China between 2006 and 2010. J Clin Microbiol. 2011;49(9):3175–83.
20. Li J, Tao S, Orlando R, Murtaugh MP. N-glycosylation profiling of porcine reproductive and respiratory syndrome virus envelope glycoprotein 5. Virology. 2015;478:86–98. doi:10.1016/j.virol.2015.02.013.
21. Liu LN, Lee H, Hernandez R, Brown DT. Mutations in the endo domain of Sindbis virus glycoprotein E2 block phosphorylation, reorientation of the endo domain, and nucleocapsid binding. Virology. 1996;222(1):236–46.
22. Lu WH, Tun HM, Sun BL, Mo J, Zhou QF, Deng YX, Xie QM, Bi YZ, Leung FC-C, Ma JY. Re-emerging of porcine respiratory and reproductive syndrome virus (lineage 3) and increased pathogenicity after genomic recombination with vaccine variant. Vet Microbiol. 2015;175(2):332–40.
23. Metwally S, Mohamed F, Faaberg K, Burrage T, Prarat M, Moran K, Bracht A, Mayr G, Berninger M, Koster L, To TL, Nguyen VL, Reising M, Landgraf J, Cox L, Lubroth J, Carrillo C. Pathogenicity and molecular characterization of emerging porcine reproductive and respiratory syndrome virus in Vietnam in 2007. Transbound Emerg Dis. 2010;57(5):315–29. doi:10.1111/j.1865-1682.2010.01152.x.
24. Meulenberg JJ. PRRSV, the virus. Vet Res. 2000;31(1):11–21. doi:10.1051/vetres:2000103.
25. Minh BQ, Nguyen MA, von Haeseler A. Ultrafast approximation for phylogenetic bootstrap. Mol Biol Evol. 2013;30(5):1188–95. doi:10.1093/molbev/mst024.
26. Mu C, Lu X, Duan E, Chen J, Li W, Zhang F, Martin DP, Yang M, Xia P, Cui B. Molecular evolution of porcine reproductive and respiratory syndrome virus isolates from central China. Res Vet Sci. 2013;95(3):908–12. doi:10.1016/j.rvsc.2013.07.029.
27. Nguyen LT, Schmidt HA, von Haeseler A, Minh BQ. IQ-TREE: a fast and effective stochastic algorithm for estimating maximum-likelihood phylogenies. Mol Biol Evol. 2015;32(1):268–74. doi:10.1093/molbev/msu300.
28. Nguyen TDT, Thi Thu N, Son NG, Ha LTT, Hung VK, Nguyen NT, Khoa DVA. Genetic analysis of ORF5 porcine reproductive and respiratory syndrome virus isolated in Vietnam. Microbiol Immunol. 2013;57(7):518–26.
29. Nguyen VG, Kim HK, Moon HJ, Park SJ, Chung HC, Choi MK, Kim AR, Park BK. ORF5-based evolutionary and epidemiological dynamics of the type 1 porcine reproductive and respiratory syndrome virus circulating in Korea. Infect Genet Evol. 2014;21:320–8.
30. Nguyen VG, Kim HK, Moon HJ, Park SJ, Chung HC, Choi MK, Park BK. A Bayesian phylogeographical analysis of type 1 porcine reproductive and respiratory syndrome virus (PRRSV). Transbound Emerg Dis. 2014;61(6):537–45. doi:10.1111/tbed.12058.
31. Nguyen VG, Kim HK, Moon HJ, Park SJ, Chung HC, Choi MK, Park BK. Evolutionary dynamics of a highly pathogenic type 2 porcine reproductive and respiratory syndrome virus: analyses of envelope protein-coding genes. Transbound Emerg Dis. 2015;62(4):411–20. doi:10.1111/tbed.12154.
32. Nilubol D, Tripipat T, Hoonsuwan T, Tipsombatboon P, Piriyapongsa J. Dynamics and evolution of porcine reproductive and respiratory syndrome virus (PRRSV) ORF5 following modified live PRRSV vaccination in a PRRSV-infected herd. Arch Virol. 2014;159(1):17–27.
33. Rowland RR, Steffen M, Ackerman T, Benfield DA. The evolution of porcine reproductive and respiratory syndrome virus: quasispecies and emergence of a virus subpopulation during infection of pigs with VR-2332. Virology. 1999;259(2):262–6.
34. Shi M, Lam TT, Hon CC, Hui RK, Faaberg KS, Wennblom T, Murtaugh MP, Stadejek T, Leung FC. Molecular epidemiology of PRRSV: a phylogenetic perspective. Virus Res. 2010;154(1–2):7–17. doi:10.1016/j.virusres.2010.08.014.
35. Shi M, Lam TT, Hon CC, Murtaugh MP, Davies PR, Hui RK, Li J, Wong LT, Yip CW, Jiang JW, Leung FC. Phylogeny-based evolutionary, demographical, and geographical dissection of North American type 2 porcine reproductive and respiratory syndrome viruses. J Virol. 2010;84(17):8700–11. doi:10.1128/JVI.02551-09.
36. Song J, Shen D, Cui J, Zhao B. Accelerated evolution of PRRSV during recent outbreaks in China. Virus Genes. 2010;41(2):241–5. doi:10.1007/s11262-010-0507-2.
37. Strauss JH, Strauss EG. The alphaviruses: gene expression, replication, and evolution. Microbiol Rev. 1994;58(3):491–562.
38. Sun L, Li Y, Liu R, Wang X, Gao F, Lin T, Huang T, Yao H, Tong G, Fan H. Porcine reproductive and respiratory syndrome virus ORF5a protein is essential for virus viability. Virus Res. 2013;171(1):178–85.
39. Thaa B, Sinhadri BC, Tielesch C, Krause E, Veit M. Signal peptide cleavage from GP5 of PRRSV: a minor fraction of molecules retains the decoy epitope, a presumed molecular cause for viral persistence. PLoS One. 2013;8(6):e65548.
40. Tun HM, Shi M, Wong C, Ayudhya S, Amonsin A, Thanawonguwech R, Leung F. Genetic diversity and multiple introductions of porcine reproductive and respiratory syndrome viruses in Thailand. Virol J. 2011;8:164.
41. Vu HL, Kwon B, Yoon K-J, Laegreid WW, Pattnaik AK, Osorio FA. Immune evasion of porcine reproductive and respiratory syndrome virus through glycan shielding involves both glycoprotein 5 as well as glycoprotein 3. J Virol. 2011;85(11):5555-64.
42. Wensvoort G, Terpstra C, Pol JM, ter Laak EA, Bloemraad M, de Kluyver EP, Kragten C, van Buiten L, den Besten A, Wagenaar F, et al. Mystery swine disease in The Netherlands: the isolation of Lelystad virus. Vet Q. 1991; 13(3):121–30. doi:10.1080/01652176.1991.9694296.
43. West J, Hernandez R, Ferreira D, Brown DT. Mutations in the endodomain of Sindbis virus glycoprotein E2 define sequences critical for virus assembly. J Virol. 2006;80(9):4458–68.
44. Wissink EH, Kroese MV, van Wijk HA, Rijsewijk FA, Meulenberg JJ, Rottier PJ. Envelope protein requirements for the assembly of infectious virions of porcine reproductive and respiratory syndrome virus. J Virol. 2005;79(19):12495–506. doi:10.1128/JVI.79.19.12495-12506.2005.
45. Xu Z, Chang X, Xiao S, Chen H, Zhou R. Evidence for the adaptive evolution of ORF 5 gene of Porcine reproductive and respiratory syndrome virus isolated in China. Acta Virol. 2010;54(4):281–5.
46. Yoon SH, Kim H, Kim J, Lee H-K, Park B, Kim H. Complete genome sequences of porcine reproductive and respiratory syndrome viruses: perspectives on their temporal and spatial dynamics. Mol Biol Rep. 2013;40(12):6843–53.
47. Yu X, Chen N, Wang L, Wu J, Zhou Z, Ni J, Li X, Zhai X, Shi J, Tian K. New genomic characteristics of highly pathogenic porcine reproductive and respiratory syndrome viruses do not lead to significant changes in pathogenicity. Vet Microbiol. 2012;158(3):291–9.
48. Zimmerman J, Yoon K-J, Wills R, Swenson S. General overview of PRRSV: a perspective from the United States. Vet Microbiol. 1997;55(1):187–96.

Bat rabies surveillance in France: first report of unusual mortality among serotine bats

Evelyne Picard-Meyer[1*], Alexandre Servat[1], Marine Wasniewski[1], Matthieu Gaillard[2], Christophe Borel[3] and Florence Cliquet[1]

Abstract

Background: Rabies is a fatal viral encephalitic disease that is caused by lyssaviruses which can affect all mammals, including human and bats. In Europe, bat rabies cases are attributed to five different lyssavirus species, the majority of rabid bats being attributed to European bat 1 lyssavirus (EBLV-1), circulating mainly in serotine bats (*Eptesicus serotinus*). In France, rabies in bats is under surveillance since 1989, with 77 positive cases reported between 1989 and 2016.

Case presentation: In the frame of the bat rabies surveillance, an unusual mortality of serotine bats was reported in 2009 in a village in North-East France. Six juvenile bats from an *E. serotinus* maternity colony counting ~200 individuals were found to be infected with EBLV-1. The active surveillance of the colony by capture sessions of bats from July to September 2009 showed a high detection rate of neutralising EBLV-1 antibodies (≈ 50%) in the colony. Moreover, one out of 111 animals tested was found to shed viable virus in saliva, while lyssavirus RNA was detected by RT-PCR for five individuals.

Conclusion: This study demonstrated that the lyssavirus infection in the serotine maternity colony was followed by a high rate of bat rabies immunity after circulation of the virus in the colony. The ratio of seropositive bats is probably indicative of an efficient virus transmission coupled to a rapid circulation of EBLV-1 in the colony.

Keywords: Rabies, EBLV-1, Bat, Mortality

Background

Rabies is an ancient viral zoonotic disease caused by negative-strand RNA viruses of the genus lyssavirus, family *Rhabdoviridae*, known to affect the central nervous system of mammals including humans. There are currently 14 recognised species in the lyssavirus genus [1] the prototype virus species being rabies lyssavirus (RABV), which is found in a plethora of terrestrial mammals worldwide and in bats in the Americas. With the exception of two species -Mokola lyssavirus and Ikoma lyssavirus- all lyssaviruses have been isolated from bats [2].

In Europe, bat rabies is caused by five different lyssavirus species: European bat 1 lyssavirus (EBLV-1), European bat 2 lyssavirus (EBLV-2), Bokeloh bat lyssavirus (BBLV), West Caucasian bat lyssavirus (WCBV) and Lleida bat lyssavirus (LLEBV), a putative species detected in Spain in *Miniopterus schreibersii* [3]. While EBLV-1 has been mainly associated with serotine bats and reported across much of Europe [4], EBLV-2 has been isolated from Myotis bats (*Myotis dasycneme* and *Myotis daubentonii*). BBLV, recently reported in France and Germany, is mainly associated with *Myotis nattereri* [5, 6], while WCBV has only been isolated once in *Miniopterus schreibersii* [7]. In France, the bat rabies surveillance scheme -involving in particular veterinary services and the national bat conservation network (SFEPM) [8]- reported 45 serotine bats infected with EBLV-1 between 1989 and 2009. In 2009, an unusual case of mortality of bats was reported in the Moselle department located in North-East France, in the frame of the bat rabies surveillance. Neither lyssavirus infection was found in bats originating from the Moselle department between 1989 and 2009, nor unusual mortality except classic mortality of juveniles. We report here an unusual case of mortality in a colony of bats in France, coupled to the presence of EBLV-1 in the colony, as well as an high proportion of seropositive individuals in that colony.

* Correspondence: evelyne.picard-meyer@anses.fr
[1]ANSES Nancy Laboratory for Rabies and Wildlife, European Union Reference Laboratory for Rabies, WHO Collaborating Centre for Research and Management in Zoonoses Control, OIE Reference Laboratory for Rabies, European Union Reference Institute for Rabies Serology, Technopôle agricole et vétérinaire de Pixérécourt, CS 40009, 54220 Malzéville, France
Full list of author information is available at the end of the article

Case presentation

An unusual mortality of juvenile serotine bats was reported in June 2009 in the centre of Ancy sur Moselle in the Moselle department in North-East France (lat. 49.054158, long. 6.058299) (Fig. 1). Approximately 30 to 40 juveniles were found dead in a large luxurious house in June of that year. The 30–40 juveniles were found in a maternity colony of serotine bats consisting out of ~200 pregnant female bats. The maternity colony, identified in attic of the private house since many years, was regularly observed by the owner. No case of unusual mortality was previously recorded in that colony. The 30–40 carcasses were not submitted for rabies diagnosis by the owner who alerted different services, including her veterinarian, the Mayor's office and bat biologists. Three weeks after the report of the unusual mortality among serotines, nine bats were found dead in the attic by the owner. The nine cadavers were collected in the attic of the private house and sent for analysis to ANSES's Nancy Laboratory for Rabies and Wildlife. The laboratory techniques used for routine diagnosis were the ones recommended by WHO and OIE [9, 10]: fluorescent antibody test (FAT) to detect antigens, rabies tissue culture infection test (RTCIT) to detect the infectious virus and in cases of positive rabies diagnosis an additional RT-PCR to detect viral RNA.

The Fluorescent Antibody Test was carried out on brain tissue specimens as described previously [10]. The RTCIT was used to confirm the presence of live virus, as previously described [9]. Viral RNA was extracted from 200 μL of supernatant from a 10% (*w*/*v*) brain suspension, then subjected to the partial Nucleoprotein (N) gene amplification, as previously described [11]. PCR products were bi-directionally sequenced by Beckman Coulter Genomics (Takeley, Essex, United Kingdom) with the same specific primers used in the nested PCR [11]. Sequence alignment of the six EBLV-1 sequences of Ancy sur Moselle and the EBLV-1 sequence of Mars la Tour was generated using the BioEdit software v. 7.2.5. based on the 600 bp region of the nucleoprotein gene.

Table 1 summarises the results of rabies infection in dead bats from the maternity colony and the results of passive surveillance undertaken in the two departments, Moselle and Meurthe et Moselle.

Of the nine cadavers collected on 29 June 2009, four were not analysed due to the advanced decomposition of the carcasses. Of the five analysable specimens, four were tested positive for EBLV-1 by referenced rabies diagnostic methods.

Strengthening of the passive bat rabies surveillance

Passive bat rabies surveillance was reinforced through a prefectural order in the departments of Moselle and Meurthe & Moselle following the report of rabies infection in the colony in Ancy sur Moselle [12]. Two additional positive cases were recorded in the maternity colony in two juvenile bats found dead respectively 7

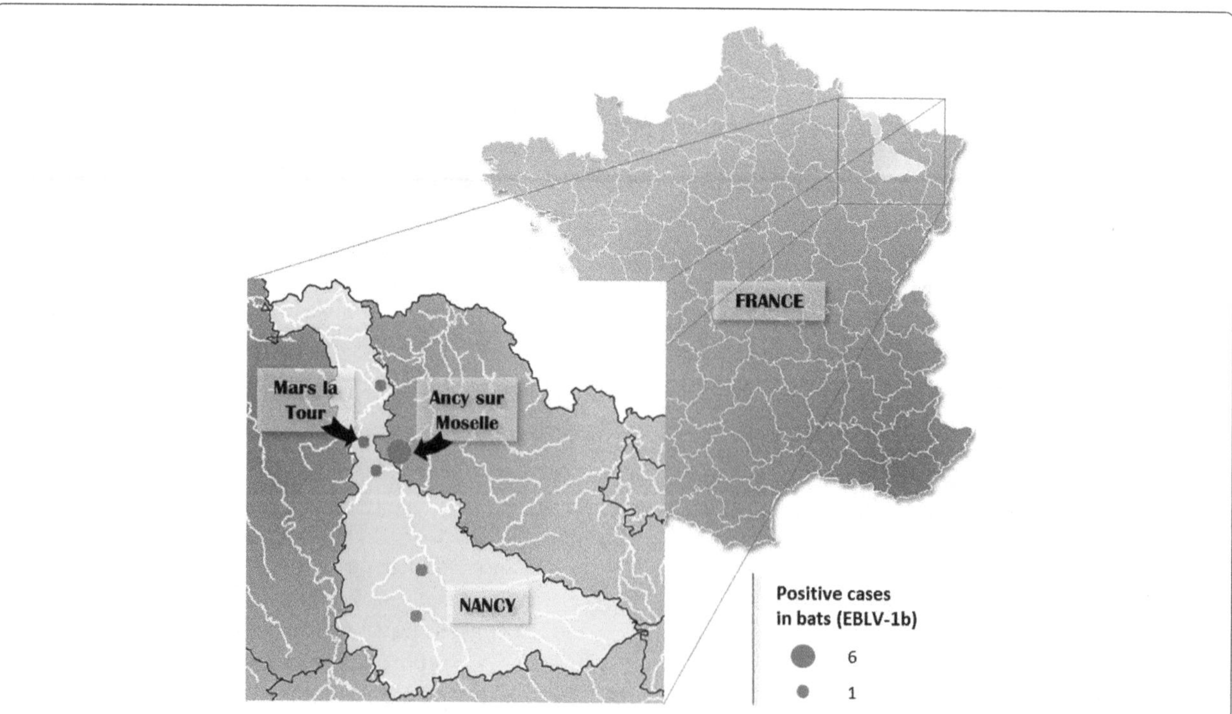

Fig. 1 Map of France showing the location of the serotine bat colony in North-East France with the geographical distribution of cases of bat rabies reported in Moselle and Meurthe & Moselle from 1989 to 2016

Table 1 Results of passive bat rabies surveillance undertaken in Moselle and Meurthe & Moselle

Studied zone	Area (km^2)	Total found dead	Bats not analysed	FAT	RTCIT	hnRT-PCR	Typing of virus
Moselle	6216	28	12	6[a] (10)	6 (10)	7 (9)	EBLV-1
Meurthe & Moselle	5246	12	2	1[b] (9)	1 (9)	1 (9)	EBLV-1

Passive bat rabies surveillance was strengthened in Moselle and Meurthe & Moselle following the report of rabies infection in dead bats found in an *Eptesicus serotinus* maternity colony in Ancy sur Moselle
Analysis covered the period from 29 June to 15 October 2009
Of 28 bats found dead in Moselle, 19 were found dead in the maternity colony of Ancy sur Moselle. Ten were not analysed, six tested positive by FAT and three tested negative by FAT
Values in brackets correspond to the number of negative samples; the number of positive samples is shown in bold and italics
[a] Samples from Ancy sur Moselle (lat. 49.054158, long. 6.058299)
[b] Samples from Mars La Tour (lat. 49.098675, long. 5.887584)

and 11 days after the first four positive cases. Between July and October, a further eight cadavers were collected from the maternity colony. Two were tested negative and six were not analysed, due to the high level of decomposition of the carcasses. The colony's mortality peak occurred between June and July, a further four serotine bats being found dead between August and October.

The strengthening of the passive surveillance scheme in the departments of Moselle and of Meurthe & Moselle led to a further nine and twelve dead bats being reported in the respective departments (Table 1). Sixteen days after the first four cases in Ancy sur Moselle, a third juvenile serotine was reported infected with EBLV-1 in Mars la Tour (lat. 49.098675, long. 5.887584), 14 km away. No maternity serotine colonies were found in the vicinity of Mars la Tour. A nucleotide comparison of the partial nucleoprotein gene sequence (570-nt) amplified from the seven EBLV-1 isolates showed a 100% nucleotide identity, suggesting that the seven isolates (MG334623) could be related to the same viral strain. BLAST analysis showed 100% of nucleotide identity between the consensus sequence of the 7 positive isolates and the sequence AY245833, isolated in 2000 in the North-East of France.

Active bat rabies surveillance

To complement passive surveillance, six sessions of capture were held at nightfall between July and September 2009 after nocturnal counting of bats to evaluate the size of the maternity colony (Table 2). Each captured bat was marked with a lipped bat band positioned on the forearm and after sampling, all bats were immediately released at the capture's site. Oropharyngeal swabs were taken from 111 individual bats to detect any viable rabies virus and lyssavirus RNA. Blood samples were taken from 94 bats to detect neutralising EBLV-1 antibodies [13]. Captures, handling and sampling were undertaken following authorisation from the French Ministry of the Environment [14].

Oral swabs stored in 0.3 mL of culture medium were analysed using RTCIT on murine neuroblastoma cells [9] and by conventional RT-PCR [11]. To detect EBLV-1-specific neutralising antibodies in blood samples, a modified FAVN test was performed [13] with an EBLV-1 virus strain (ANSES, No. 121411) isolated in France in 2000. Samples were tested in quadruplicate, in threefold dilutions on BHK-21 cells with a starting dilution of 1/27. Controls included uninfected BHK-21 cells, OIE positive dog serum, negative dog serum and back-titration of the specific EBLV-1 virus. Levels of virus neutralizing antibodies were expressed in log D50. The threshold of antibody detection was calculated by using the Spearman–Karber formula and set at 1.67 log D50.

Table 2 summarizes the detection of infectious particles, viral RNA and antibodies in serotine bats. For each independent session, we computed the proportion of seropositive bats. Interval confidence of proportions (95% CI) were computed using the Bernoulli exact method. Results showed that one animal – a seropositive juvenile female captured only once— out of the 111 tested was found to shed viable virus in saliva, and five individuals were found positive for lyssavirus RNA. Viral RNA was detected during the first capture for two individual bats and during the second capture for three other individual bats (Table 2). Of the five animals found positive for the detection of RNA, three (2 females—one adult and one juvenile—and one juvenile male) were only captured once. Two animals were recaptured and sampled twice. One adult female was found positive during the first capture and surprisingly was found negative for the detection of RNA in the second capture. By opposite, the second animal —a female adult— was shown negative by RT-PCR in the first capture, and shown positive in the second one. The two animals were found to be seronegative for the detection of EBLV-1 antibodies. Not all animals from the colony developed a detectable neutralising antibody response, 49% of bats seroconverting between July and September with a detection threshold set at 1.67 logD50 [15]. Serology results showed an apparent decrease of the proportion of seropositive bats from July to September 2009, but the differences were not significantly different.

Discussion and conclusions

We described here an unusual mortality among bats with six juveniles found dead in a short interval of time

Table 2 Detection of infectious particles (RTCIT), viral RNA (hnRT-PCR) and antibodies (mFAVNt) in serotine bats

Dates of capture	Nb. counted bats[a]	Genus and age of bats		Detection in bats of:				
				Viable virus	Viral RNA	EBLV-1 antibodies		
				RTCIT	hnRT-PCR	mFAVNt	% with AB	95% CI
7 July	135	A	F	0/36	1/36	12/29	45.5% (15/33)	28.5–63.4
			M	–	–	–		
		I	F	1/2	1/2	2/2		
			M	0/2	0/2	1/2		
10 July	81	A	F	0/33	2/33	18/30	59.5% (22/37)	42.1–75.2
			M	0/1	0/1	0/1		
		I	F	0/2	0/2	1/2		
			M	0/4	1/4	3/4		
23 July	30	A	F	0/8	0/8	4/7	55.5% (5/9)	21.2–86.3
			M	–		–		
		I	F	0/1	0/1	1/1		
			M	0/1	0/1	0/1		
4 August	30	A	F	0/10	0/10	1/6	14.3% (1/7)	0.4–57.9
			M	–	–	–		
		I	F	–	–	–		
			M	–	0/1	0/1		
20 August	21	A	F	0/4	0/4	0/3	0% (0/3)	0.0–70.8
			M	–	–	–		
		I	F	0/1	0/1	–		
			M	–	–	–		
10 September	10	A	F	0/5	0/5	3/5	60% (3/5)	14.7–94.7
			M	–	–	–		
		I	F	–	–	–		
			M	–	–	–		
Total		A		0/97	3/97	38/81	49% (44/94)	36.4–57.4
		I		1/14	2/14	8/13		

Six capture/release sessions were held following the report of rabies infection in the maternity colony between July to September 2009, to complement passive bat rabies surveillance

[a]Nocturnal countings of bats were performed 1 day before each capture to evaluate the colony's size

Values correspond to the number of positive samples out of all the tested samples

Abbreviations: A Adult, *AB* antibody, *F* female, *I* Immature, *M* male

coupled to a high ratio of seropositive bats in a bat colony counting ~ 200 individuals. While the detection of EBLV-1 antibodies is frequently reported in field bat studies, the mortality of numerous bats from one site is unusually reported. The detection of viral RNA and viable virus in oropharyngeal swabs that we reported in the study is in accordance with previous studies [16]. Repeated detection of viral RNA in saliva swabs from recaptured bats was not observed in this study. These results are in accordance with the study of Vasquez-Moron et al. [17], that showed similar results in a serotine bat colony targeted after the detection of a bat rabies case. The majority of positive RT-PCR results are commonly associated with serotine bats. We showed here a high detection rate of neutralising EBLV-1 antibodies in the colony, with 49% of seroconverting bats. The detection of EBLV-1 virus-neutralizing antibodies is frequent in bats [16, 18], with a seroprevalence varying according to the site location, species and time. The high rate of seropositive bats in this study is probably indicative of an efficient virus transmission coupled to a rapid circulation of EBLV-1 in the colony.

Bats and their roosts are protected by French and European legislation. Bats play a key role in regulating insect populations and are one of the best natural indicators of the environment's health. The serotine bat, considered as the reservoir of EBLV-1, is a common and widely distributed species in France generally and the Lorraine region particularly (Fig. 2) [19]. While some

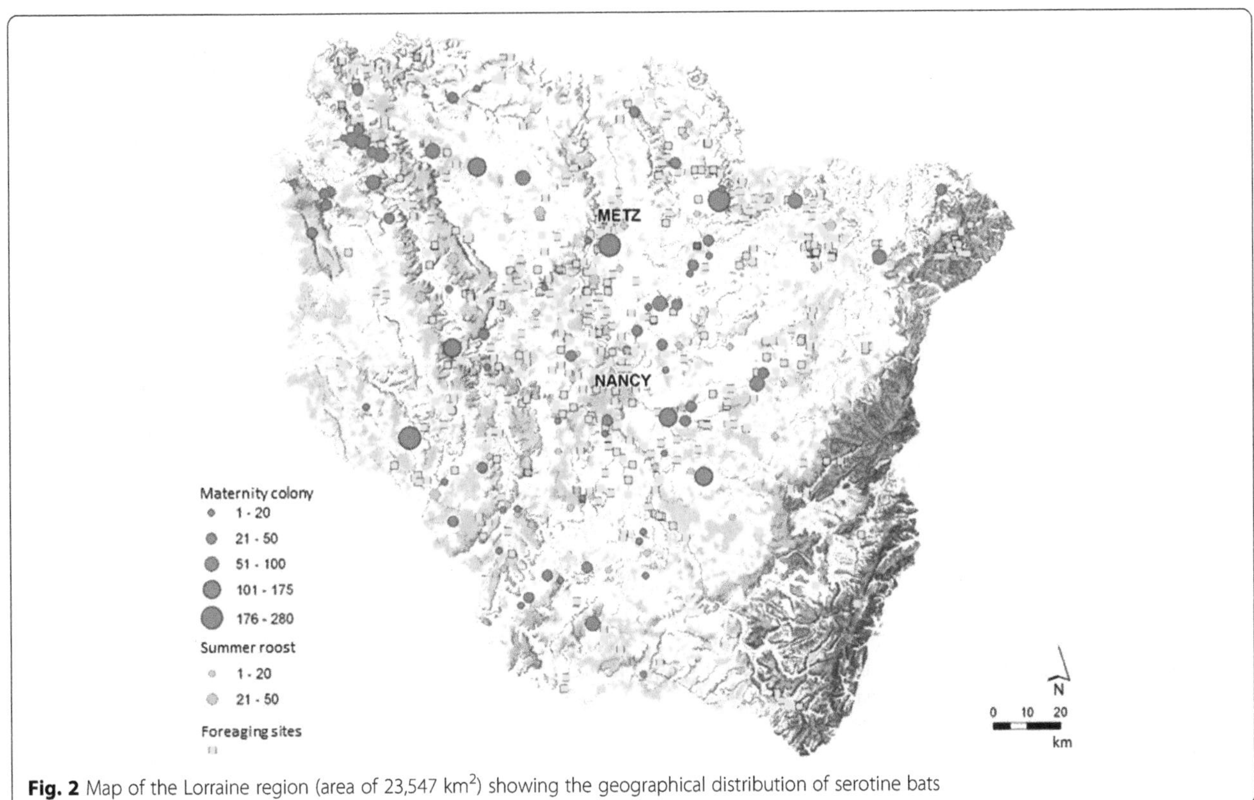

Fig. 2 Map of the Lorraine region (area of 23,547 km^2) showing the geographical distribution of serotine bats

aspects of serotine bat ecology have been studied, including habitat use and roosting behaviour, little is known about the bats' winter ecology, dispersal and seasonal connectivity with other species. It should be noted that three *Myotis emarginatus* bats were observed in August in the roost site. One of them was captured and subjected to lyssavirus analysis, but tested negative by FAT and PCR.

While bat rabies is a public health concern in Europe, the epidemiology and the pathogenicity of EBLV-1 in bats are still unknown as well as the dynamic of the infection and the virus influence on the mortality rate in bat colonies. To better understand the mechanisms by which EBLV(s) are maintained within bat populations, it is necessary to improve the passive surveillance of rabies in bats and further investigate serotine bat ecology through close collaboration between bat biologists and scientists. When possible, a colony of bats found to be naturally infected with EBLV-1 should be preserved and studied to shed light on the ecology of bat diseases. Neither human nor animal contact with the infected bat colony was identified. The owner of the house in Ancy sur Moselle received a post-exposure prophylaxis, although she did not report exposure. The house owner's dog, already identified by microchip and vaccinated against rabies, received a booster vaccination in accordance with French regulations.

Abbreviations

BBLV: Bokeloh bat lyssavirus; EBLV-1: European bat 1 lyssavirus; EBLV-2: European bat 2 lyssavirus; FAT: Fluorescent antibody test; RNA: Ribonucleic acid; RTCIT: Rabies tissue culture infection test; RT-PCR: Reverse-transcription polymerase chain reaction; SFEPM: National bat conservation network; WCBV: West Caucasian bat lyssavirus; WHO: World Health Organisation; OIE: World Organisation for Animal Health

Acknowledgements

We gratefully acknowledge the expert technical support of the diagnostics team (Estelle Litaize and Valère Brogat), the serology team (Anouck Labadie, Laetitia Tribout and Jonathan Rieder), the molecular biology team (Mélanie Biarnais and Jean-Luc Schereffer), the field team (Dr. Franck Boué, Dr. Marie Moinet, Jean-Michel Demerson and Christophe Caillot), Dr. Emmanuelle Robardet and Sylvie Tourdiat for the maps, and Dr. Jacques Barrat for helpful contributions. We would also like to thank all the voluntary members of the CPEPESC Lorraine association, Dr. Anne Dupire (DDPP57) and Dr. Dominique Bemer.
This research was funded by ANSES, the French Agency for Food, Environmental and Occupational Health & Safety.
We would like to thank the SFEPM group, and especially the bat network workers, for their collaboration in the passive and active surveillance of bat rabies.

Funding

This research was funded by ANSES, the French Agency for Food, Environmental and Occupational Health & Safety.

Authors' contributions

All authors were involved in the study. Wrote the manuscript: EPM, FLC; performed laboratory investigations: EPM, AS, MW; performed the bat colony monitoring and collected data: EPM, CB, MG; revised the manuscript: EPM, AS, MW, MG, CB, FLC. All authors read and approved the manuscript.

Consent for publication

The owner of the property where the bats were located gave written consent to publish this case report.

Competing interests

The authors declare that they have no competing interests.

Author details

[1]ANSES Nancy Laboratory for Rabies and Wildlife, European Union Reference Laboratory for Rabies, WHO Collaborating Centre for Research and Management in Zoonoses Control, OIE Reference Laboratory for Rabies, European Union Reference Institute for Rabies Serology, Technopôle agricole et vétérinaire de Pixérécourt, CS 40009, 54220 Malzéville, France. [2]Néomys association, Centre Ariane, 240 rue de Cumène, 54230 Neuves-Maisons, France. [3]CPEPESC-Lorraine, Centre Ariane, 240 rue de Cumène, 54230 Neuves-Maisons, France.

References

1. ICTV virus taxonomy: 2016 release (http://www.ictvonline.org/virustaxonomy.asp). Accessed 02 November 2017.
2. Kuzmin I, Rupprecht C. Bat Lyssaviruses. In: Wang L, Cowled C, editors. Bats and viruses: a new frontier of emerging infectious diseases. New Jersey, US: John Wiley & Sons; 2015. p. 47–97.
3. Arechiga Ceballos N, Vazquez Moron S, Berciano JM, Nicolas O, Aznar Lopez C, Juste J, Rodriguez Nevado C, Aguilar Setien A, Echevarria JE: Novel lyssavirus in bat, Spain Emerg Infect Dis 2013; doi: 10.3201/eid1905.121071.
4. Schatz J, Fooks AR, McElhinney L, Horton D, Echevarria J, Vazquez-Moron S, Kooi EA, Rasmussen TB, Muller T, Freuling CM. Bat rabies surveillance in Europe. Zoonoses Public Health. 2013; doi: 10.1111/zph.12002.
5. Freuling CM, Abendroth B, Beer M, Fischer M, Hanke D, Hoffmann B, Hoper D, Just F, Mettenleiter TC, Schatz J, Muller T: Molecular diagnostics for the detection of Bokeloh bat lyssavirus in a bat from Bavaria, Germany Virus Res 2013; doi: 10.1016/j.virusres.2013.07.021.
6. Picard-Meyer E, Servat A, Robardet E, Moinet M, Borel C, Cliquet F. Isolation of Bokeloh bat lyssavirus in Myotis Nattereri in France. Arch Virol. 2013; doi: 10.1007/s00705-013-1747-y.
7. Botvinkin AD, Poleschuk EM, Kuzmin IV, Borisova TI, Gazaryan SV, Yager P, et al. Novel lyssaviruses isolated from bats in Russia. Emerg Infect Dis 2003; 9(12):1623-1625. Epub 2004/01/15. doi: 10.3201/eid0912.030374. PubMed PMID: 14720408; PubMed Central PMCID: PMC3034350.
8. Picard-Meyer E, Robardet E, Arthur L, Larcher G, Harbusch C, Servat A, Cliquet F. Bat rabies in France: a 24-year retrospective epidemiological study. PLoS One. 2014; doi: 10.1371/journal.pone.0098622.
9. Servat A, Picard-Meyer E, Robardet E, Muzniece Z, Must K, Cliquet F. Evaluation of a rapid Immunochromatographic diagnostic test for the detection of rabies from brain material of European mammals. Biologicals. 2012; doi: 10.1016/j.biologicals.2011.12.011.
10. Meslin F, Kaplan M, Koprowski H. Laboratory techniques in rabies. 4th ed. Geneva: World Health Organization. 1996.
11. Picard-Meyer E, Bruyere V, Barrat J, Tissot E, Barrat MJ, Cliquet F: Development of a hemi-nested RT-PCR method for the specific determination of European Bat Lyssavirus 1. Comparison with other rabies diagnostic methods. Vaccine 2004; doi: 10.1016/j.vaccine.2003.11.015.
12. Treffel J-F. Arrêté de mise sous surveillance d'une population de chauves-souris (sérotines communes) d'où est susceptible d'être issue une chauve-souris porteuse d'un Lyssavirus. Arrêté préfectoral, Préfecture de la région Lorraine 2009; 1-3.
13. Cliquet F, Aubert M, Sagne L. Development of a fluorescent antibody virus neutralisation test (FAVN test) for the quantitation of rabies-neutralising antibody. J Immunol Methods. 1998;212:79–87.
14. Wintergest J. Dérogation ministérielle de capture, prélèvement et transport de chiroptères dans le cadre de la mission d'épidémiosurveillance et de recherche sur la rage des chiroptères 2012; 1–4.
15. Picard-Meyer E, Dubourg-Savage MJ, Arthur L, Barataud M, Becu D, Bracco S, Borel C, Larcher G, Meme-Lafond B, Moinet M, et al. Active surveillance of bat rabies in France: a 5-year study (2004-2009). Vet Microbiol. 2011; doi: 10.1016/j.vetmic.2011.03.034.
16. Schatz J, Ohlendorf B, Busse P, Pelz G, Dolch D, Teubner J, Encarnacao JA, Muhle RU, Fischer M, Hoffmann B, et al. Twenty years of active bat rabies surveillance in Germany: a detailed analysis and future perspectives. Epidemiol Infect. 2014;142:1155–66.
17. Vazquez-Moron S, Juste J, Ibanez C, Ruiz-Villamor E, Avellon A, Vera M, Echevarria JE. Endemic circulation of European bat lyssavirus type 1 in serotine bats, Spain. Emerg Infect Dis. 2008;14:1263–6.
18. Amengual B, Bourhy H, Lopez-Roig M, Serra-Cobo J. Temporal dynamics of European bat Lyssavirus type 1 and survival of Myotis Myotis bats in natural colonies. PLoS One. 2007;2:e566.
19. Plan National d'Actions en faveur des chiroptères : 2009–2013: Bilan technique et financier des 5 ans du PNA 2009–2013, le diagnostic des 34 espèces, Sérotine commune, *Eptesicus serotinus*. http://www.plan-actions-chiropteres.fr/IMG/diagnostic-especes-chiropteres-2eme-PNA-FINAL.pdf (2017). Accessed 30 August 2017.

A deep sequencing reveals significant diversity among dominant variants and evolutionary dynamics of avian leukosis viruses in two infectious ecosystems

Fanfeng Meng[1†], Xuan Dong[1†], Tao Hu[2], Shuang Chang[1], Jianhua Fan[3], Peng Zhao[1*] and Zhizhong Cui[1*]

Abstract

Background: As a typical retrovirus, the evolution of Avian leukosis virus subgroup J (ALV-J) in different infectious ecosystems is not characterized, what we know is there are a cloud of diverse variants, namely quasispecies with considerable genetic diversity. This study is to explore the selection of infectious ecosystems on dominant variants and their evolutionary dynamics of ALV-J between DF1 cells and specific-pathogen-free (SPF) chickens. High-throughput sequencing platforms provide an approach for detecting quasispecies diversity more fully.

Results: An average of about 20,000 valid reads were obtained from two variable regions of *gp85* gene and *LTR-U3* region from each sample in different infectious ecosystems. The top 10 dominant variants among ALV-J from chicken plasmas, DF1 cells and liver tumor were completely different from each other. Also there was a difference of shannon entropy and global selection pressure values (ω) in different infectious ecosystems. In the plasmas of two chickens, a large portion of quasispecies contained a 3-peptides "LSD" repeat insertion that was only less than 0.01% in DF1 cell culture supernatants. In parallel studies, the *LTR-U3* region of ALV-J from the chicken plasmas demonstrated more variants with mutations in their transcription regulatory elements than those from DF1 cells.

Conclusions: Our data taken together suggest that the molecular epidemiology based on isolated ALV-J in cell culture may not represent the true evolution of virus in chicken flocks in the field. The biological significance of the "LSD" insert and mutations in *LTR-U3* needs to be further studied.

Keywords: Subgroup J avian leukosis virus, Infectious ecosystem, 3-peptides LSD repeat insert (LSD^+), Deep sequencing

Background

Avian leukosis virus (ALV) is an oncogenic retrovirus that induced lymphoid tumors in chickens and its genomic structure and molecular characteristics are well defined. It plays a critical role in the discoveries of reverse transcriptase, v-oncogenes and proto-oncogenes [1]. According to the host range, viral envelope interference and cross-neutralization patterns, avian leukosis viruses (ALVs) are classified into six subgroup (A to J) in chickens. ALV-J was first detected in meat-type chickens in the late 1980's [2], and then spread globally [3–8]. So far, ALV-J is more pathogenic and mutate easily than other subgroups [9]. Although the eradication programs on ALV-J have been conducted in meat-type chickens since its discovery, it had spread into egg-type stock and the Chinese local breeds, which caused significant economic losses in China during the last 10 years [10–15].

Proteins gp85 and gp37 are encoded by the envelope gene of ALV, while gp85 protein constitute globular structures on the surface of the virus, which is closely associated with the process of viral binding and determine the specificity of subgroups. To understand molecular epidemiology of ALV-J among different types of chickens with various genetic backgrounds in many parts of the world, more than 200 ALV-J isolates have been subsequently sequenced and compared with gp85 region of

* Correspondence: zhaopeng@sdau.edu.cn; zzcui@sdau.edu.cn
†Equal contributors
[1]College of Veterinary Medicine, Shandong Agricultural University, Taian 271018, China
Full list of author information is available at the end of the article

envelope gene since late 1980s [4, 10, 16–23]. The early study suggested that gp85 sequences of the ALV-J strains isolated from different geographical areas and farms in different years showed highly variable and their similarity varied in the range of 80 – 100%. In terms of the *gp85* identity, the later isolates seemed to deviate gradually from the earliest isolate HPRS-103 [7]. However, many new isolates were obtained from different provinces of China after 1999. We found that no further deviate from HPRS-103 and all of their *gp85* sequences still varied in the same range [4]. There was also no evidence to show further sequence deviation from HPRS-103 even for the ALV-J strains isolated in the recent 10 years from layers or Chinese local breeds of chickens [10, 11, 13, 14, 24]. In addition, among 10 ALV-J isolates from ten individual layers with myelocytomas from the same flock demonstrated that they varied in the range of 80.3–97.1% in *gp85* region [25]. It seemed to suggest that there was no close relationship between ALV-J *gp85* homology levels and its pathogenicity or adaptation to different chicken breeds with different genetic backgrounds, although there were some epidemic phenomena indicated that ALV-J evolved to higher pathogenicity in different breeds of chickens.

In the past 30 years, almost all the molecular epidemiological data have been obtained by sequencing DNA fragments amplified and cloned from ALV-J infected CEF or DF1 cells. Such process would set up a bias for selection of certain quasispecies from the large population of viral particles in the given pathologic materials, for instance tumor tissues. By such selection, some significant variants associated with pathogenicity or adaptation to different genetic breeds may be survived by selective pressures. Wellehan reported that the dominant variants of San Miguel Sea Lion Virus populations altered significantly after its replication ecosystem switched from infected sea lions to cell cultures for 5 passages, the rare variants in sea lions became the dominant ones in cell cultures [26].

In this study, we analyzed and quantitatively compared dominant variants between ALV-J population replicated in infected chickens and cell cultures with the aid of deep sequencing-based method. The purpose of this study is to advance in understanding if cell culture ecosystem would cause selection pressures different from that of chickens with ALV-J infection, and whether the selection pressures would influence the evolution of ALV dominant variants. With these studies, we hope to identify specific epitopes or domains on *gp85*, or other genes, such as in the region of *LTR-U3*, which may associate with the differential selective pressures.

Methods

Sample preparation

The avian cell line DF-1 were obtained from the American Type Culture Collection (Manassas, VA, USA). These cells were grown in Dulbecco's modified Eagle's medium (DMEM; gibco, USA) supplemented with 10% fetal bovine serum and 100 mg/ml of penicillin and streptomycin. A liver with myloid tumors was collected aseptically from clinically hy-line variety brown with spontaneous infections and identified as ALV-J via virus isolation (Genbank: KR049171, KR049172). Tumor homogenate prepared in plasma-free DMEM was lysed, the supernatant was purified by high-speed centrifugation and 0.22-um-pore-size cellulose—acetate filtration. The resultant purified tumor suspension was designated as original liver suspension (Ori) which was used as the viral strain in laboratory experiments, and the concentration was about 1500 $TCID_{50}$/100 ul.

In vitro group, 3000 $TCID_{50}$ Ori were inoculated into DF-1 cells in logarithmic phase and maintained for 5 days as one passage. Then the infected DF1 cells were cultured via serial passages and cell free cultured supernatants were harvested at the 1st and 5th passage (P1 and P5). In vivo group, a total of 10 one-day-old specific pathogen free (SPF) chickens from the SPAFAS Co. (Jinan, China; a joint venture with Charles River Laboratory, Wilmington, MA, USA) were inoculated intraperitoneally with 3000 $TCID_{50}$ Ori. The blood plasma collections were performed for virus isolation, while antibodies of ALV-J were detected at 2, 4 and 6 weeks post inoculation, respectively. Following inoculation, plasma was obtained from whole blood and stored at −80 °C. Two plasma samples free of antibody at 2 weeks of sampling and the cell culture from the 1st and 5th passages were chosen for high throughput sequencing (C1 and C2). The animal infection protocol was reviewed and approved by the Shandong Province Animal Ethics Committee.

RNA extraction, RT-PCR and sequencing

Total viral RNA was extracted from samples of two ecosystems and original liver inoculum (Ori) using MagMAX Viral RNA isolation Kit (Life Technologies, USA) following the manufacturer's instructions. Each sample was amplified using a forward primer with a six-digit error-correcting barcode as described earlier [27]. In addition, a 2-bp GT linker was added between the barcode and the 5′end of the forward primer to avoid a potential match between the barcode and the target sequences. Therefore, the forward primer was barcode-GT-primer, in which the barcode indicates the six barcode sequences that are specific to different samples, then three pairs of primers were designed according to the reference sequence HPRS-103 (Genbank: Z46390), namely gp85-A, gp85-B and LTR-U3 (Additional file 1: Table S1 and Figure S1). ALV-specific RT-PCR targeting the hypervariable region of the gp85 and LTR-U3 genes were then performed on the viral RNA using the two-Step RT-PCR Kit (TAKARA, China) at 42 °C for 45 min, 5 min denaturation at 95 °C, followed by 35 cycles of denaturation at 95 °C for 5 min, 95 °C for 30 s; annealing at 53 °C for 30 s, and extension at 72 °C for 30 s, with a final extension step at 72 °C for 10 min. The PCR products from both rounds were run of this reaction on a 1% agarose gels and scored. Bands of interest in the gels were cut out and the DNA was extracted from the gel using Qiagen Quick Gel Extraction Kit. The Products were quantified with a NanoDropND-1000 spectrophotometer (Thermo Fisher Scientific, Waltham, MA). A mixture of the amplicons was then used for sequencing on Illumina MiSeq platform according to the manufacturer's instructions at the Beijing Genomics Institute (Shenzhen, China). A base-calling pipeline (Sequencing Control Software, SCS; Illumina) was used to process the raw fluorescent images and the call sequences, and data quality assessment were performed on the MiSeq instrument.

Data analysis

Raw nucleotide sequences were filtered, aligned, trimmed and translated using pre-specified criteria applied uniformly. On average, there are 95% of the data above a quality value of Q30, which demonstrates a good quality of the demultiplexed reads. Then switch nucleic acid sequences of A, B fragment into amino acid sequences and threw away sequences with no biological significance for the following analysis (reads appeared more than 2 times were retained). The dominant *gp85* and *LTR-U3* variants of the samples were compared with each other under different infectious ecosystems using the Clustal W algorithm in MegAlign program of the DNASTAR package. Transcriptional regulatory elements in the U3 region were analyzed by the online service system of NSITE (Recognition of Regulatory motifs) of Soft Berry (http://www.softberry.com/berry.phtml). The statistical analysis was done by Duncan's multiple range test.

In order to investigate quasispecies diversity under different ecosystems, we calculated the Shannon entrophy using clean reads of each sample.

Formula followed:

$$H_{shannon} = -\sum_{i=1}^{S_{obs}} \frac{n_i}{N} \ln\frac{n_i}{N}$$

$$\mathrm{var}(H_{shannon}) = \frac{\sum_{i=1}^{S_{obs}} \frac{n_i}{N}\left(\ln\frac{n_i}{N}\right)^2 - H^2{}_{shannon}}{N} + \frac{S_{obs}-1}{2N^2}$$

S_{obs} = The amount of haplotype observed by sequencing
n_i = Number of sequences for haplotype i
N = The total valuable sequence number obtained by sequencing

To minimize potential sampling bias and reduce the computation load, we performed a bootstrapping strategy for the clean reads of each sample. For each re-sampling with replacement, phylogenetic analysis was performed using RaxML [28] with 200 bootstrap replicates, under the GAMMACAT substitution model. All other parameters were set to their default values. Global selection pressure values (ω) were estimated using HyPhy method [29].

Results

MiSeq high throughput sequencing data

After several filtering steps, about 94 to 97% of the nucleic acid sequences (*LTR-U3*) or 86–93% of the amino acid sequences (*gp85*-A and *gp85*-B) from any sample of the raw reads were retained for subsequent analyses. The raw reads and the filtered reads obtained using MiSeq High-throughput Sequencing of the extracted RNA generated a median of more than 20,000 reads per sample (Additional file 1: Table S2).

Comparison of the ratios of haplotypes in different viral infectious ecosystems

The ratios of sequence haplotypes to total valid reads for *gp85*-A, *gp85*-B and *LTR-U3* fragments from both plasmas of two chickens (C1 and C2) and DF1 cell culture supernatants of two different passages (P1 and P5) were decreased significantly as compared to that of the original liver inoculum (Ori) (Additional file 1: Table S3). The results suggested there might be some selective pressures on quasispecies of both *gp85*-A (hr1 and vr2 regions), *gp85*-B (hr2 and vr3 regions) and *LTR-U3* fragments when ALV-J from the Ori replicated in chickens or in DF1 cell cultures. Some variants in Ori were decreased dramatically to undetectable levels when replication ecosystem changed.

Evolutionary dynamics of *gp85*-B (hr2 and vr3) under different infectious ecosystems

The dominant variants of gp85-B altered dramatically after replication under the two different ecosystems. The percentage of the most dominant variant of *gp85*-B in Ori decreased to a very low level and even became undetectable in infected chickens or cell culture supernatants, while some other sub-dominant variants were increased and decreased at a high and low percentages (Fig. 1a). Actually, the top 5 dominant variants in chicken plasmas or cell cultures were rare ones in Ori (Fig. 1b). It suggested that there were some strong selective pressures having influence on the evolution of dominant variants of *gp85*-B from infectious ecosystems.

The most dominant variant (BO0001) accounting for 32.85% in the Ori did not appear among the first 10 dominant quasipecies in either cell culture supernatants or chicken plasmas. We also find that there are two identical variants within the first 10 dominant variants of *gp85*-B from both chicken samples (C1 and C2). Sequences of the first dominant variant (BC1001) from C1

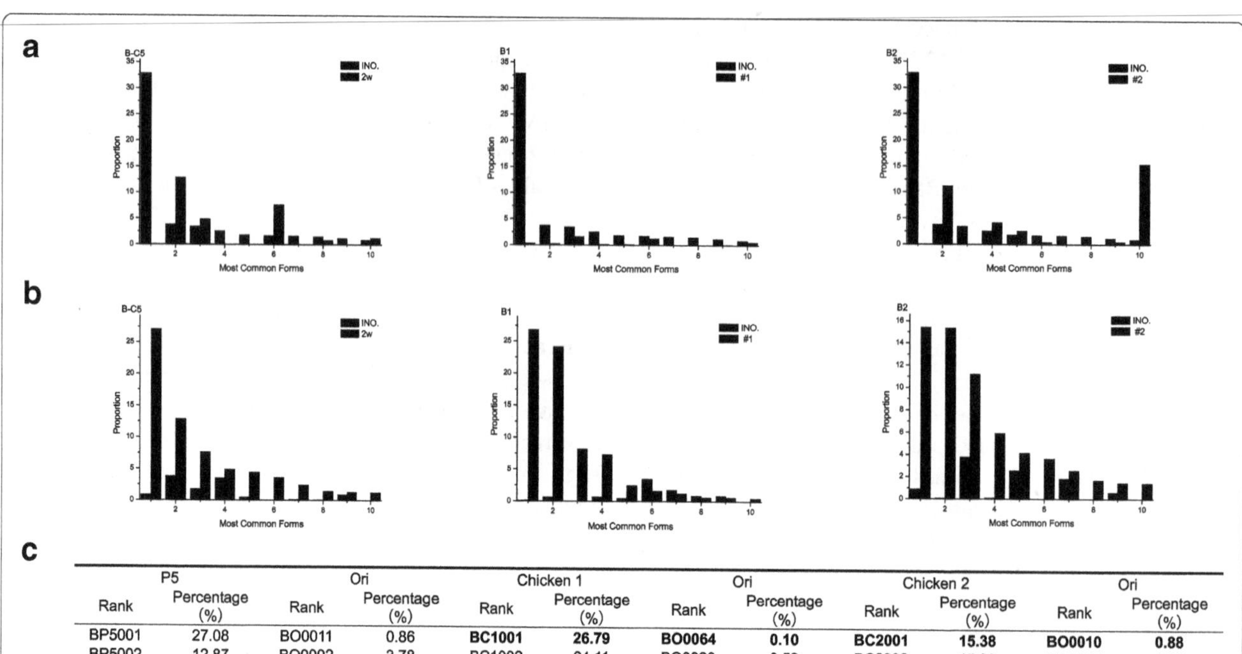

c

P5		Ori		Chicken 1		Ori		Chicken 2		Ori	
Rank	Percentage (%)	Rank	Percentage (%)	Rank	Percentage (%)	Rank	Percentage (%)	Rank	Percentage (%)	Rank	Percentage (%)
BP5001	27.08	BO0011	0.86	**BC1001**	**26.79**	**BO0064**	**0.10**	**BC2001**	**15.38**	**BO0010**	**0.88**
BP5002	12.87	BO0002	3.78	BC1002	24.11	BO0020	0.59	BC2002	15.33	BO0077	0.08
BP5003	7.64	BO0006	1.74	BC1003	8.16	NA	NA	BC2003	11.2	BO0002	3.78
BP5004	4.81	BO0003	3.47	BC1004	7.35	BO0016	0.66	**BC2004**	**5.93**	**BO0064**	**0.10**
BP5005	4.37	BO0022	0.44	BC1005	2.44	BO0022	0.44	BC2005	4.16	BO0004	2.55
BP5006	3.57	BO0185	0.04	BC1006	1.64	BO0003	3.47	BC2006	3.64	BO0978	0.01
BP5007	2.40	BO0053	0.11	BC1007	1.25	BO0006	1.74	BC2007	2.56	BO0005	1.84
BP5008	1.39	BO0118	0.06	BC1008	0.60	BO0011	0.86	BC2008	1.68	BO0732	0.01
BP5009	1.27	BO0010	0.88	**BC1009**	**0.57**	**BO0010**	**0.88**	BC2009	1.46	BO0019	0.60
BP5010	1.25	NA	NA	BC1010	0.48	NA	NA	BC2010	1.42	BO0906	0.01
NA	**NA**	**BO0001**	**32.85**	**BC1012**	**0.32**	**BO0001**	**32.85**	**NA**	**NA**	**BO0001**	**32.85**

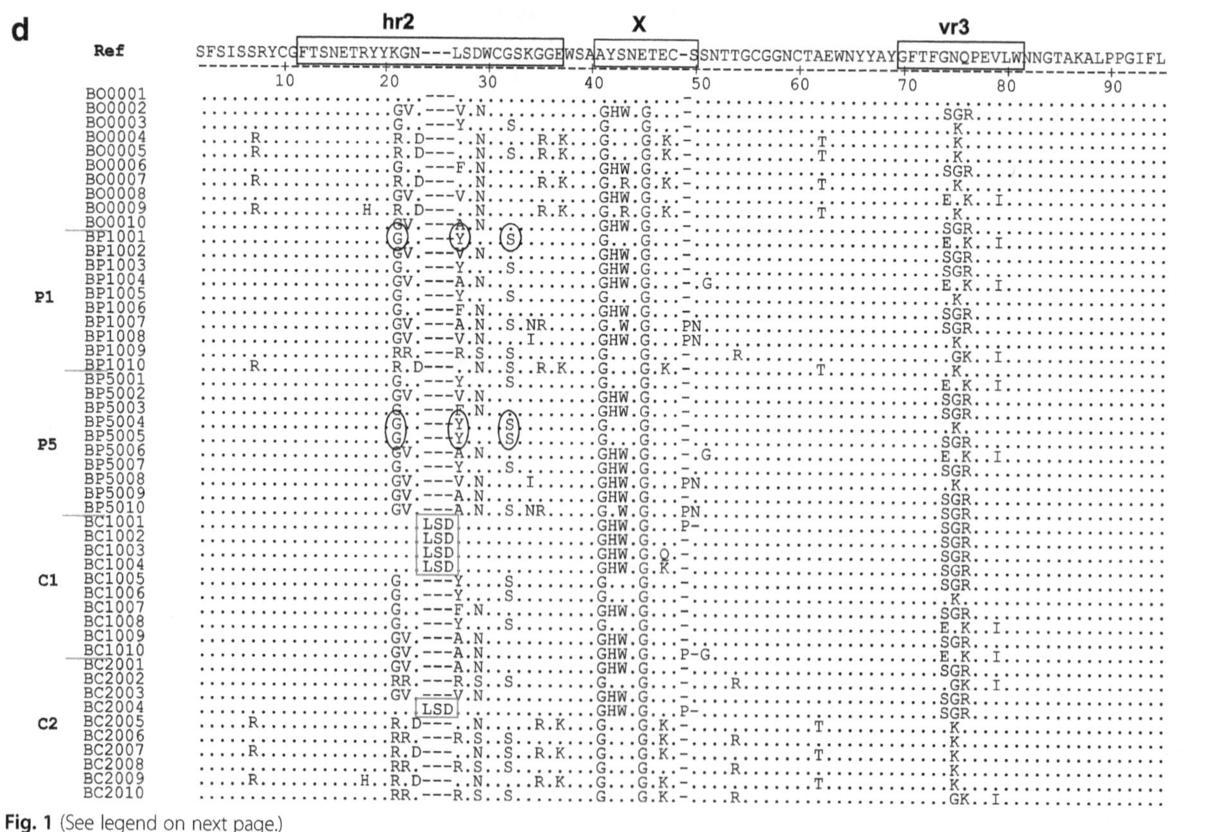

Fig. 1 (See legend on next page.)

(See figure on previous page.)
Fig. 1 Evolutionary dynamics of *gp85*-B dominant quasispecies of ALV-J under different infectious ecosystems. **a** Percentages of the first 10 dominant quasispecies of gp85-B in the ori and their equals in cell culture supernatants of 5th passage (B-C5). Ori = black bars; chickens #1 (B1) and #2 (B2) plasmas = red bars. X-axis represents the first 10 dominant quasispecies of gp85-B in the original inoculum ordered based their ranks. Y-axis represents their percentages from total valid reads in each samples collected under different replication ecosystem. **b** Percentages of the first 10 dominant quasispecies of gp85-B in cell culture supernatants of 5th passage (B-C5) and their equals in the Ori. chickens #1 (B1) and #2 (B2) plasmas = red bars; Ori = black bars. **c** Haplotypes' rank and their percentages of the first 10 dominant quasispecies of gp85-B in cell culture supernatants of 5th passage, plasma samples, and in the Ori. P5 = the cell culture supernatant samples of 5th passages; chicken 1 = plasma of #1 infected chicken; chicken 2 = plasma of #2 infected chicken, Ori = original liver inoculum. **d** The gp85-B amino acid alignment of the first 10 dominant quasispecies for viral samples collected in different replication ecosystyms. The dots indicate identical residues; the letter indicate amino acid substitutions; the dashes indicate gaps produced in the alignment; the blue square indicates two known variable regions (vr3 and hr2) and a new possible variable region X. The most dominant quasispecies (BO0001,32.85%) in the original inoculum (Ori) is used as the reference sequence on the top line corresponding to the amino acids sites #163-264 of *gp85* in ALV-J prototype HPRS-103 [Genbank:Z46390]

and the fourth dominant variant (BC2004) from C2 are 100% identical to the 64th variant (BO0064) in the Ori. Similarly, the most dominant variant (BC2001) from C2 and the 9th dominant variant (BC1009) from C1 are 100% identical to the 10th dominant variant (BO0010) in the Ori (Fig. 1c). Although the other 8 dominant variants are different between the two chicken samples, but they have high homologies and only limited number of different sites than those from cell culture supernatants (Fig. 1d).

Characterization of a specific domain of *gp85*-B associated with selective evolution of dominant variants under different viral replication conditions

The differences of amino acids in *gp85*-B dominant variants under different viral replication conditions are mainly in two known variable regions hr2 (aa#11–36) and vr3 (aa#69–80), or a new possible variable region X (aa#40–48). The most prominent difference is that the major dominant variants from chicken plasmas have a 3-peptides LSD repeat insert (LSD$^+$) when compared to samples from the Ori or cell culture supernatants. Although the evolutionary dynamics of LSD$^+$ in two chickens are not the same. In the plasma of chicken #1 (C1), all the top 4 dominant variants (BC1001, BCC1002, BC1003, BC1004) accounting for 66.32% from total valid reads get LSD$^+$ at 2w post infection which is significantly increased compared to these equals in Ori consisting of only 1.35%. In another chicken (C2), only one as the 4th dominant variant (BC1004) with LSD$^+$ appears in the top 10 dominant variants, which accounting for 5.93% from the total valid reads, that was still a significant increase as the identical sequence haplotype (BO0064) in Ori was only 0.10%. In contrast, there is no LSD$^+$ in the top 10 dominant variants in cell culture, and only 4 haplotypes with LSD$^+$ consisted of only 0.02% in its total 29,986 valid reads are found among 3846 variants haplotypes, which was dramatically declined from 1.35% when compared to that only 163 haplotypes from 4047 variants haplotypes in Ori.

All variants with LSD$^+$ in different ecosystems were compared and analyzed. In C1, the top 10 LSD$^+$ dominant variants accounting for 66.95% of the total valid reads, compared to only 1.45% in the Ori. While in C2, the first 10 dominant variants with LSD$^+$ consisted of 6.62% of the total valid reads, but only 0.70% in the Ori. It indicates that variants with LSD$^+$ were dramatically increased by positive selection after replication in chickens. Specifically, there are three completely identical variants with LSD$^+$ in two chicken plasma samples (BC1001 vs BC2004, BC1002 vs BC2022, and BC1022 vs BC2040). However, evolution of variants with LSD$^+$ are to the opposite direction after replication in DF1 cell cultures, that is, percentages of LSD$^+$ positive variants decline rapidly and even disappear. Some LSD$^+$ positive variants detected in infected chicken plasma are not detectable in the Ori even all reads are analyzed (Additional file 1: Table S4). There are some amino acid alterations in x and vr3 regions, but such variations trend to be convergent (Additional file 1: Figure S2). Also two pairs of LSD$^+$ positive and LSD$^+$ negative variants are compared for their antigenic index by computational analysis. The results indicated that LSD$^+$ significantly increased the antigenic index in the new domain around the LSD insert (Fig. 2).

Mutational analysis of *gp85-B* under different infectious ecosystems

By analyzing the data from deep sequencing, mutational frequency of each site in the LSD$^+$ domain is also compared independently among *gp85*-B quasispecies of ALV-J replicated in different ecosystems (Fig. 3). Each amino acid of L, S and D at the insertion sites (aa#23–#25) appears at the frequency of 2.78–2.79% in Ori, but it dramatically decreases to 0–0.01% after ALV-J is passaged in DF1 cells. In contrast, their frequencies were increased to 85.31–85.34% and 8.58–8.60% respectively in C1 and C2 which were very close to the frequency of 85.31 and 8.58% of the entire "LSD", suggesting that the positive effect in infected chickens and negative selection in cell cultures were mainly associated with the intact 3-peptides insertion of LSD.

Besides, specific mutations of another 3 important sites in *gp85*-B were recognized to be associated with positive

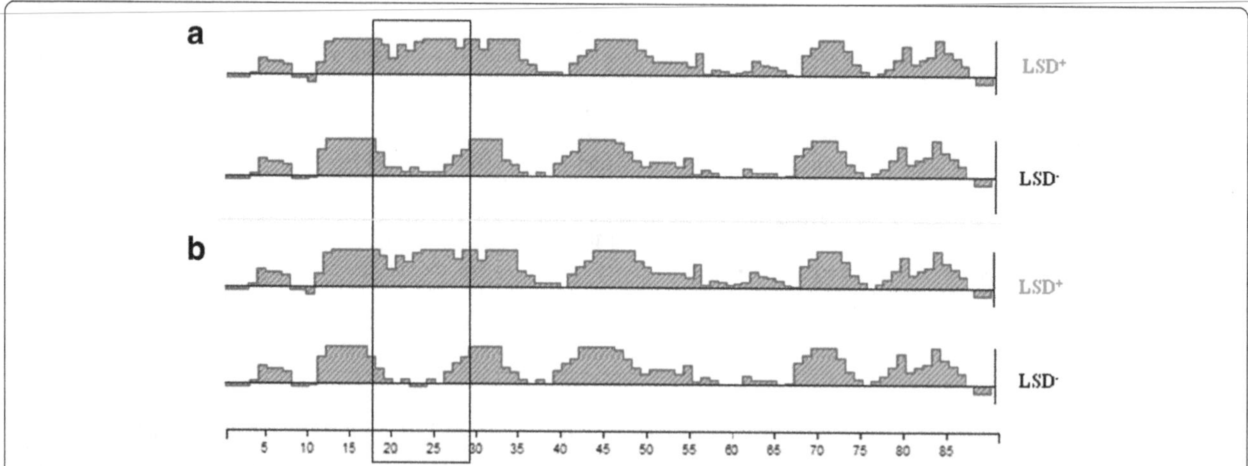

Fig. 2 Comparison of the antigen index between the *gp85*-B quasispecies with or without LSD+. **a** The comparison of antigen index between 102001 (LSD+) and B102007 (LSD−). **b** The comparison of antigen index between B202004 (LSD+) and B202003 (LSD−). Antigenic profiles calculated with Jameson and Wolf (Jameson and Wolf, 1988) algorithm from the linear amino acid sequences, the different area of antigen index with or without LSD insertion mutation was marked with a black frame

selective pressures in DF1 cells. The proportions of variants consisted of the 3 none-successive amino acids G-Y-S at aa positions #20, #26 and #30 were significantly increased, to 90.34, 49.32 and 60.12% in P1 and 99.36, 53.71 and 55.84% in P5 from 30.18, 11.02 and 17.08% in Ori, respectively (Fig. 3). In contrast, the proportions were 12.21, 7.33 and 8.3% in C1 and 39.51, 0.18 and 39.39% in C2. Further analysis of sequence data in Fig. 1 indicated that the 1st, 4th, 5th and 7th ones among the top 10 dominant variants in 5th passage cell culture contained the none-successive G-Y-S and consisted 38.96% of total reads, while two of the first 10 dominant variants in the original liver inoculum contained the G-Y-S which accounting for only 5.21% of total valid reads.

Position / Reference	20 K	26 L	31 G	23 –	24 –	25 –
Ori	K54.80	L67.43	G82.77	_97.21	_97.21	_97.21
	G30.38	Y11.22	S17.08	L2.78	S2.28	D2.79
	R14.50	V10.88	D0.15	P0.01	P0.01	
P1	G90.34	Y49.32	S60.12	_99.99	_99.99	_99.99
	R9.14	A16.36	G39.81			
	_0.45	V14.59	N0.05			
P5	G99.36	Y53.71	S55.84	_99.99	_99.99	_99.99
	R0.44	V20.72	G44.07	L0.01	S0.01	D0.01
	_0.11	F14.13	N0.03			
C1	K86.24	L86.99	G91.62	L85.32	S85.34	D85.31
	G12.21	Y7.33	S8.30	_14.52	_14.52	_14.52
	R1.44	A2.26	D0.06	P0.11	P0.10	G0.11
C2	R51.74	R34.16	G60.55	_91.39	_91.39	_91.38
	G39.53	L26.16	S39.39	L8.58	S8.60	D8.58
	K8.64	A21.48	N0.03	P0.01		G0.01

Fig. 3 Comparison of amino acid changes in six sites on *gp85*-B under different infectious ecosystems. The first 3 frequent amino acids at each site and their percentages were listed for samples collected from the original inoculum. P1 = the 1st passage cell culture; P5 = the 5th passage cell culture; #1 = chicken 1; #2 = chicken 2. The capital letters indicate specific amino acid. The numerical numbers indicate the sites of the gp85-B by use of the most dominant quasispecies in the original inoculum as the reference

Comparisons of the Shannon entropy and global selection pressure values (ω) under different infectious ecosystems

To roughly quantify the pressures that quasispesis underwent under different infectious ecosystems, we calculated the Shannon entropy and the global selection pressure values (ω). Our results showed the Shannon entropy in Ori was the highest, but when inoculated into chickens or DF1 cells different degree of decline were observed (Table 1). On the global selection pressure values (ω), those in cells were relatively stable at about 0.61, bigger than that from the Ori, but there were two different situations in the chickens ecosystem. Specifically, the ω values of the quasispecies in C1 were higher than those in P1 and P5, but the Shannon entropy is lower. Moreover, the ω values of the quasispecies in C2 were lower than those in P1 and P5, however the Shannon entropy is higher.

Table 1 The Shannon entropy and global selection pressure values (ω) under different infectious ecosystems

Ecosystems	The Shannon entropy	Global selection pressure values (ω)
Ori	4.90	0.59 ± 0.03^{A}
C1	3.65	0.63 ± 0.04^{B}
C2	4.52	0.54 ± 0.04^{C}
P1	4.40	0.61 ± 0.04^{D}
P5	4.24	0.61 ± 0.04^{D}

Each value are calculated by 200 bootstrap re-samples of the distribution of variants. Each column in the upper label with different letters mean significant difference on Duncan's multiple range test ($P < 0.01$)

Comparisons of dominant variants in *LTR-U3* region under different infectious ecosystems

The high throughput sequencing of *LTR-U3* region showed that the most dominant variants had not changed in both chicken plasmas and DF1 cell culture supernatants when compared to the Ori. However, the other sub-dominant variants of *U3* region evolved into different directions under different infectious ecosystems, in vivo and cell cultures. There were 7 of the top 10 variants were exactly the same in two chicken plasma samples which consisted of 70.08 and 72.62% of the total valid reads in the two chickens respectively (Fig. 4a). But all the left 9 major dominant variants of *U3* in cell cultures were completely different from those of chicken plasma.

Sequence alignment analysis demonstrated that *U3* region was more conservative in ALV-J replicated in cell cultures than in infected chickens, as the other 9 major sub-dominant variants haplotypes from DF1 cell culture supernatants had fewer mutations than that from 2 chicken plasma samples when compared to the most dominant variant common in different infectious ecosystems. More importantly, the most base alterations were located within some motifs as transcription regulatory elements (Fig. 4b) in samples from two chickens and the

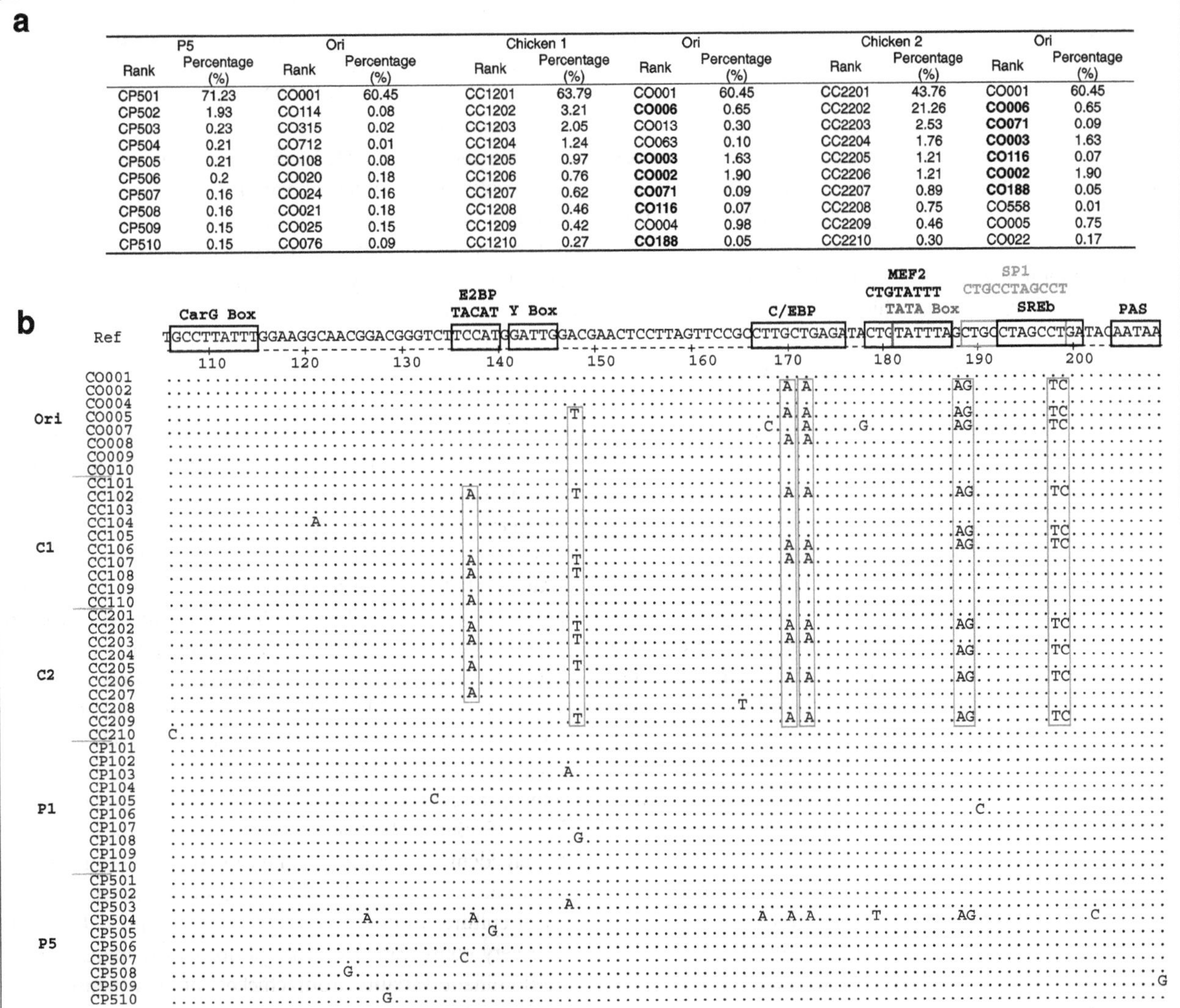

a

P5		Ori		Chicken 1		Ori		Chicken 2		Ori	
Rank	Percentage (%)	Rank	Percentage (%)	Rank	Percentage (%)	Rank	Percentage (%)	Rank	Percentage (%)	Rank	Percentage (%)
CP501	71.23	CO001	60.45	CC1201	63.79	CO001	60.45	CC2201	43.76	CO001	60.45
CP502	1.93	CO114	0.08	CC1202	3.21	**CO006**	0.65	CC2202	21.26	**CO006**	0.65
CP503	0.23	CO315	0.02	CC1203	2.05	CO013	0.30	CC2203	2.53	**CO071**	0.09
CP504	0.21	CO712	0.01	CC1204	1.24	CO063	0.10	CC2204	1.76	**CO003**	1.63
CP505	0.21	CO108	0.08	CC1205	0.97	**CO003**	1.63	CC2205	1.21	**CO116**	0.07
CP506	0.2	CO020	0.18	CC1206	0.76	**CO002**	1.90	CC2206	1.21	**CO002**	1.90
CP507	0.16	CO024	0.16	CC1207	0.62	**CO071**	0.09	CC2207	0.89	**CO188**	0.05
CP508	0.16	CO021	0.18	CC1208	0.46	**CO116**	0.07	CC2208	0.75	CO558	0.01
CP509	0.15	CO025	0.15	CC1209	0.42	CO004	0.98	CC2209	0.46	CO005	0.75
CP510	0.15	CO076	0.09	CC1210	0.27	**CO188**	0.05	CC2210	0.30	CO022	0.17

Fig. 4 Evolutionary dynamics of the first 10 dominant quansispecies of *U3* fragment in different infectious ecosystems. **a** The first 10 dominant quasispecies haplotype (in the order according ranks) and their percentages of total valid reads in DF1 cell cultures (combined two passages, the left part). CC1001 = the quasispecies ranked the first in segment C from chicken 1; O0 = the original liver inoculum; C2 = chicken 2; P1 = passage1; P5 = passage 5 in cell culture. The last 3 numbers represents their ranks in the quasispecies population in each sample. **b** Base sequence alignment of the first 10 dominant quasispecies of U3 detected in each different infectious ecosystems. The reference sequence is the most dominant quasispecies in the original inoculum (CO0001, 60.45%), corresponding to bases #122–225 of LTR of ALV-J prototype strain HPRS-103. The dots indicate identical residues, while the letters indicate amino acid substitutions. The motifs as transcription regulatory elements were labeled on the top

Ori. There were eight common mutation sites and similar replacement in the ecosystems of the two chickens, and then C/EBP, SP1, MEF2 and SREb regulatory elements losing their integrity but made one more motif of E2BP due to the replacement of C to A at the site #137. Obviously, it is due to differences of infectious ecosystems.

Discussion

The object of this study was to understand if there is any influence from different infectious ecosystems on ALV-J dominant variants evolution. Both chickens and DF1 cell cultures were inoculated with the same Ori. The RT-PCR products of each sample were directly sequenced and analyzed by the deep sequencing, it produced extreme large sequence data covering genome variants even at very low frequencies. With the technology progresses on diversity in quasispecies and its evolutionary dynamics were obtained under different immunoselective pressures, antiviral drugs, and various viruses, such as HIV [30–33], and hepatitis B, C, and E viruses [34, 35] and some animal viruses [36].

Deep Sequencing generated a median of more than 20,000 reads per sample, which were large enough to compare and understand the quasispecies diversity from these samples. The ratios of haplotypes/valid reads of 3 fragments from chickens and cell culture samples were decreased compared to the Ori, suggesting that both infectious ecosystems demonstrated a negative selective effect on some quasispecies in the Ori. Among the 3 fragments sequenced, gp85-A fragment was less influenced by selective pressures, but gp85-B fragment was significantly influenced. Some dominant variants in the Ori was dramatically decreased in the inoculated chicken plasma and cell culture supernatant samples, but some very rare variants became the dominant ones. The results provided the direct experimental evidence that the infectious ecosystems would dramatically influenced the evolution of viral quasispecies. It is clear that the Ori was liver suspension and its ALV-J quasispecies mainly replicated in liver-associated cells, however, viruses in chicken plasmas or cell culture supernatant came from all kinds of sensitive cells in the body or replicated only in DF1 cells after the Ori was infected chickens or cell cultures. Bioinformatic analysis results showed the Shannon entropy in Ori was the highest, but when inoculated into chickens or DF1 cells different degree of decline were observed, which indicated there were some pressures in the ecosystems. On the global selection pressure values (ω), there was a big individual difference in chickens, but relatively stable in DF1 cells. There was significant difference ($P < 0.01$) between the two groups from chickens and DF1 cells, which also suggested different selective pressure in the two groups.

The envelope protein gp85 is related to recognition and adhesion to sensitive cells, and also is the major antigen for viral neutralization [9]. The diversity in *gp85* sequence, especially epitopes at certain sites may influence the tropism of virus quasispecies to different types of cells in chicken body. For example, it has been reported that ALV-J prototype HPRS-103 has a low tropism for bursal follicles cells but does replicate well in cultured blood monocytes [9]. The most interesting result in this study is the discovery of the 3-peptides "LSD" repeat insertion (LSD$^+$) in novel dominant variants of gp85-B fragments emerging in chicken plasmas samples, which increased the antigen index in the sub-region. However, there was no LSD$^+$ positive variants among the top 10 dominant variants in two DF1 supernatant samples. Referencing to the principle of site-by-site positive selection analysis using the two rate fixed-effects likelihood (FEL) method [37], after artificial calculation we found that the LSD$^+$ were under positive selection in chickens, while negative selection in DF1 cells. It might help to explain how evolution of different variants with specific epitope could be influenced by some selective pressures from ecosystem or its infectious ecosystems such as different organs, tissues or cell types. We speculate that variation in gp85 sequence similarity may not necessarily reflect its relationship to evolution in terms of higher pathogenicity to different genetic breeds of chickens, but some specific epitopes or domains on gp85 would influence.

LTR-U3 region of ALV has only about 250 bp but contains several biological active motifs and enhancers influencing transcription and virus replication [38, 39], also it is a fragment easy to mutate on the ALV genome. However, analysis of deep sequencing data of *U3* region in different samples demonstrated that the viral population in chicken plasma samples came from ALV-J replicated in different types of cells, organs and tissues of the chicken and experienced quite different ecosystem selection pressures. Chicken plasma samples and DF1 supernatant sample had the same most dominant variants of *U3* as the Ori, which indicated that different infectious ecosystems did not have as high selective pressures on the evolution of *U3* quasispecies as that of *gp85*-B. But in the two chicken plasmas there were 6 absolutely identical variants for the top 10 sub-advantage variants while no reapeat with the top 10 sub-dominant variants in cell culture. Although the first 10 dominant *U3* variants were very conservative in cell culture supernatants, several sub-dominant variants from chicken plasma samples and Ori had mutations in its regulation elements C\EBP, SP1, MEF2 and SREb. Concerning the biological significance, its needs to be further investigated and studied.

Conclusions

In conclusion, this study is the first to explore the replication of ALV-J in different ecosystem using deep sequencing technique. We found that significant differences in dominant variants and their evolution dynamics of gp85 from ALV-J in infected chickens or cell cultures. Especially, a tri-peptides "LSD" insert associated with positive selective pressures in infected chickens and negative selective pressures in DF1 cell cultures in *gp85* were identified. It suggests that the replication ecosystem has a significant influence on the evolution of viruses. The molecular epidemiology studies based on the isolated ALV-J in cell culture may not represent the true evolution of these viruses in infected chicken flocks in the field.

Additional file

Additional file 1: Table S1. Three pairs of primers for Miseq High-throughput Sequencing. **Table S2** The numbers of raw reads and clean reads in each sample. **Table S3** Variations of ratios of sequence haplotype numbers to total valid reads in different replication ecosystems. **Table S4** Dynamics of the first 10 LSD+ positive quasispecies of gp85-B in different replication ecosystems. **Figure S1** The structure of env and LTR in ALV-J. **Figure S2** Amino acid alignment of top 10 LSD+ positive quasispecies of gp85-B in different replication ecosystems. (DOC 311 kb)

Abbreviations

LSD^+: A 3-peptides LSD repeat insert; ALV-J: Avian leukosis virus subgroup J; P1 and P4: The 1st and 5th passage; C1 and C2: Chicken 1 and chicken 2; Ori: Original liver suspension; SPF: Specific pathogen free

Acknowledgements

We thank Lucy F. Lee for helpful discussions. We also thank BGI-Shenzhen for MiSeq sequencing.

Funding

This work was supported by grants from the National Natural Science Foundations of China (grant numbers: 31472216, 31402226) and the Natural Science Fundation of Jiangsu Province (grant No.BK20151317).

Authors' contributions

All authors approved the manuscript. MFF, DX, HT, CS, FJH, ZP and CZZ contributed to study design and data interpretation. MFF was the principal investigator. MFF and DX wrote the manuscript and produced all figures. All authors read, corrected, and approved the final manuscript prior to submission.

Competing interests

The authors declare that they have no competing interests.

Consent for publication

Not applicable.

Author details

[1]College of Veterinary Medicine, Shandong Agricultural University, Taian 271018, China. [2]Institute of Pathogen Biology, Taishan Medical College, Taian, Shandong, China. [3]Poultry Institute, Chinese Academy of Agricultural Sciences, Yangzhou, Jiangsu, China.

References

1. Weiss RA, Vogt PK. 100 years of Rous sarcoma virus. J Exp Med. 2011;208:2351–5.
2. Payne LN, Brown SR, Bumstead N, Howes K, Frazier JA, et al. A novel subgroup of exogenous avian leukosis virus in chickens. J Gen Virol. 1991;72:801–7.
3. Bagust TJ, Fenton SP, Reddy MR. Detection of subgroup J avian leukosis virus infection in Australian meat-type chickens. Aust Vet J. 2004;82:701–6.
4. Cui Z, Du Y, Zhang Z, Silva RF. Comparison of Chinese field strains of avian leukosis subgroup J viruses with prototype strain HPRS-103 and United States strains. Avian Dis. 2003;47:1321–30.
5. Fadly AM, Smith EJ. Isolation and some characteristics of a subgroup J-like avian leukosis virus associated with myeloid leukosis in meat-type chickens in the United States. Avian Dis. 1999;43:391–400.
6. Nakamura K, Ogiso M, Tsukamoto K, Hamazaki N, Hihara H, et al. Lesions of bone and bone marrow in myeloid leukosis occurring naturally in adult broiler breeders. Avian Dis. 2000;44:215–21.
7. Silva RF, Fadly AM, Hunt HD. Hypervariability in the envelope genes of subgroup J avian leukosis viruses obtained from different farms in the United States. Virology. 2000;272:106–11.
8. Zavala G, Cheng S, Jackwood MW. Molecular epidemiology of avian leukosis virus subgroup J and evolutionary history of its 3' untranslated region. Avian Dis. 2007;51:942–53.
9. Payne LN, Nair V. The long view: 40 years of avian leukosis research. Avian Pathol. 2012;41:11–9.
10. Dong X, Zhao P, Li W, Chang S, Li J, et al. Diagnosis and sequence analysis of avian leukosis virus subgroup J isolated from Chinese Partridge Shank chickens. Poult Sci. 2015;94(4):668–72.
11. Gao Y, Guan X, Liu Y, Li X, Yun B, et al. An avian leukosis virus subgroup J isolate with a Rous sarcoma virus-like 5'-LTR shows enhanced replication capability. J Gen Virol. 2015;96:150–8.
12. Gao YL, Qin LT, Pan W, Wang YQ, Le Qi X, et al. Avian leukosis virus subgroup J in layer chickens. China Emerg Infect Dis. 2010;16:1637–8.
13. Li Y, Liu X, Liu H, Xu C, Liao Y, et al. Isolation, identification, and phylogenetic analysis of two avian leukosis virus subgroup J strains associated with hemangioma and myeloid leukosis. Vet Microbiol. 2013;166:356–64.
14. Sun S, Cui Z. Epidemiological and pathological studies of subgroup J avian leukosis virus infections in Chinese local "yellow" chickens. Avian Pathol. 2007;36:221–6.
15. Xu B, Dong W, Yu C, He Z, Lv Y, et al. Occurrence of avian leukosis virus subgroup J in commercial layer flocks in China. Avian Pathol. 2004;33:13–7.
16. Du Y, Cui ZZ, Qin AJ, Silva RF, Lee LF. Isolation of subgroup J avianleukosis viruses and their partial sequence comparison. Chinese Journal of Virology. 2000;16:341–6.
17. Gao Y, Yun B, Qin L, Pan W, Qu Y, et al. Molecular epidemiology of avian leukosis virus subgroup J in layer flocks in China. J Clin Microbiol. 2012;50:953–60.
18. Jiang L, Zeng X, Hua Y, Gao Q, Fan Z, et al. Genetic diversity and phylogenetic analysis of glycoprotein gp85 of avian leukosis virus subgroup J wild-bird isolates from Northeast China. Arch Virol. 2014;159:1821–6.
19. Lai H, Zhang H, Ning Z, Chen R, Zhang W, et al. Isolation and characterization of emerging subgroup J avian leukosis virus associated with hemangioma in egg-type chickens. Vet Microbiol. 2011;151:275–83.
20. Li H, Xue C, Ji J, Chang S, Shang H, et al. Complete genome sequence of a J subgroup avian leukosis virus isolated from local commercial broilers. J Virol. 2012;86:11937–8.
21. Mao Y, Li W, Dong X, Liu J, Zhao P. Different quasispecies with great mutations hide in the same subgroup J field strain of avian leukosis virus. Sci China Life Sci. 2013;56:414–20.
22. Pan W, Gao Y, Qin L, Ni W, Liu Z, et al. Genetic diversity and phylogenetic analysis of glycoprotein GP85 of ALV-J isolates from Mainland China between 1999 and 2010: coexistence of two extremely different subgroups in layers. Vet Microbiol. 2012;156:205–12.
23. Wang H, Cui Z. The identification and sequence analysis of ALV-J isolated from layers. Chin J Virol. 2008;24:369–75.
24. Wang Z, Cui Z. Evolution of gp85 gene of subgroup J avian leukosis virus under the selective pressure of antibodies. Sci China C Life Sci. 2006;49:227–34.
25. Bian XM, Li DQ, Zhao P, Cui ZZ. Continuous observation of subgroup J avian leukosis for three groups of commercial layer chicken. Sci Agric Sin. 2013;46:409–16.
26. Wellehan Jr JF, Yu F, Venn-Watson SK, Jensen ED, Smith CR, et al. Characterization of San Miguel sea lion virus populations using pyrosequencing-based methods. Infect Genet Evol. 2010;10:254–60.

27. Hamady M, Walker JJ, Harris JK, Gold NJ, Knight R. Error-correcting barcoded primers for pyrosequencing hundreds of samples in multiplex. Nat Methods. 2008;5:235–7.
28. Stamatakis A, Ludwig T, Meier H. RAxML-III: a fast program for maximum likelihood-based inference of large phylogenetic trees. Bioinformatics. 2005;21:456–63.
29. Pond SL, Frost SD, Muse SV. HyPhy: hypothesis testing using phylogenies. Bioinformatics. 2005;21:676–9.
30. Bansode V, McCormack GP, Crampin AC, Ngwira B, Shrestha RK, et al. Characterizing the emergence and persistence of drug resistant mutations in HIV-1 subtype C infections using 454 ultra deep pyrosequencing. BMC Infect Dis. 2013;13:52.
31. Henn MR, Boutwell CL, Charlebois P, Lennon NJ, Power KA, et al. Whole genome deep sequencing of HIV-1 reveals the impact of early minor variants upon immune recognition during acute infection. PLoS Pathog. 2012;8:e1002529.
32. Recordon-Pinson P, Raymond S, Bellecave P, Marcelin AG, Soulie C, et al. HIV-1 dynamics and coreceptor usage in Maraviroc-treated patients with ongoing replication. Antimicrob Agents Chemother. 2013;57:930–5.
33. Rozera G, Abbate I, Ciccozzi M, Lo Presti A, Bruselles A, et al. Ultra-deep sequencing reveals hidden HIV-1 minority lineages and shifts of viral population between the main cellular reservoirs of the infection after therapy interruption. J Med Virol. 2012;84:839–44.
34. Rodriguez-Frias F, Tabernero D, Quer J, Esteban JI, Ortega I, et al. Ultra-deep pyrosequencing detects conserved genomic sites and quantifies linkage of drug-resistant amino acid changes in the hepatitis B virus genome. PLoS One. 2012;7:e37874.
35. Miura M, Maekawa S, Takano S, Komatsu N, Tatsumi A, et al. Deep-sequencing analysis of the association between the quasispecies nature of the hepatitis C virus core region and disease progression. J Virol. 2013;87:12541–51.
36. Borucki MK, Chen-Harris H, Lao V, Vanier G, Wadford DA, et al. Ultra-deep sequencing of intra-host rabies virus populations during cross-species transmission. PLoS Negl Trop Dis. 2013;7:e2555.
37. Kosakovsky Pond SL, Frost SD. Not so different after all: a comparison of methods for detecting amino acid sites under selection. Mol Biol Evol. 2005;22:1208–22.
38. Ryden TA, Beemon K. Avian retroviral long terminal repeats bind CCAAT/enhancer-binding protein. Mol Cell Biol. 1989;9:1155–64.
39. Zachow KR, Conklin KF. CArG, CCAAT, and CCAAT-like protein binding sites in avian retrovirus long terminal repeat enhancers. J Virol. 1992;66:1959–70.

17

Molecular and serological survey of lyssaviruses in Croatian bat populations

Ivana Šimić[1*], Ivana Lojkić[1], Nina Krešić[1], Florence Cliquet[2], Evelyne Picard-Meyer[2], Marine Wasniewski[2], Anđela Ćukušić[3], Vida Zrnčić[3] and Tomislav Bedeković[1]

Abstract

Background: Rabies is the only known zoonotic disease of bat origin in Europe. The disease is caused by species belonging to the genus *Lyssavirus*. Five Lyssavirus species, i.e., European bat lyssavirus (EBLV)-1, EBLV-2, Bokeloh bat lyssavirus, Lleida bat lyssavirus, and West Caucasian bat virus, have been identified in European bats. More recently, a proposed sixth species, Kotalahti bat lyssavirus, was detected. Thus, in this study, active surveillance was initiated in order to obtain insights into the prevalence of lyssaviruses in Croatian bat populations and to improve our understanding of the public health threat of infected bats.

Results: In total, 455 bats were caught throughout Continental and Mediterranean Croatia. Antibodies were found in 20 of 350 bats (5.71%, 95% confidence interval 3.73–8.66). The majority of seropositive bats were found in Trbušnjak cave (Continental Croatia, Eastern part), and most seropositive bats belonged to *Myotis myotis* (13/20). All oropharyngeal swabs were negative for the presence of *Lyssavirus*.

Conclusions: The presence of lyssaviruses in bat populations was confirmed for the first time in Croatia and Southeastern Europe. The results of this study suggest the need for further comprehensive analyses of lyssaviruses in bats in this part of Europe.

Keywords: Bat, *Lyssavirus*, European bat lyssavirus-1, Croatia, Antibodies

Background

Rabies is a fatal viral zoonotic disease infecting all warm-blooded mammals, including bats, and is caused by viruses belonging to the genus *Lyssavirus*. The World Health Organization (WHO) reported that 59, 000 human deaths occur annually around the world due to dog-transmitted rabies. In contrast, rabies transmitted from bats causes a small proportion of human cases globally [1]. Currently, 16 *Lyssavirus* species are recognized by the International Committee on the Taxonomy of Viruses [2], all of which have been reported in bats except for two species, Mokola lyssavirus and Ikoma lyssavirus [3, 4]. Recently, two related viruses, i.e., Taiwan bat lyssavirus (TWBLV) and Kotalahti bat lyssavirus (KBLV), were isolated from bats [2, 5, 6].

During the last century, analysis of lyssaviruses in bats has shown that bats play an important role as a reservoir for these viruses. In the Americas (New World), only variants of classical Rabies virus (RABV) are associated with bats, whereas across Africa, Asia, Europe (Old World), and Australia no detection of RABV has been reported in any bat species. However, other lyssaviruses have been detected. The long-term association of lyssaviruses with bats suggests that lyssaviruses are the most important and only confirmed zoonotic pathogen of bat origin in Europe [3, 7].

Lyssaviruses are divided into three phylogroups, among which only phylogroup I viruses are all neutralized by existing rabies vaccines [3]. Rabies in European bat populations is caused by five species and two phylogroups: European bat lyssavirus (EBLV) -1 (phylogroup I), EBLV-2 (phylogroup I), Bokeloh bat lyssavirus (BBLV; phylogroup I), West Caucasian bat lyssavirus (WCBV; phylogroup III), and Lleida bat lyssavirus (LLEBV; confirmed phylogroup III) [7]. Recently, a putative species of KBLV (tentatively phylogroup I) was detected in Finland in *Myotis brandtii* [6].

* Correspondence: simic@veinst.hr
[1]Croatian Veterinary Institute, Savska cesta 143, 10000 Zagreb, Croatia
Full list of author information is available at the end of the article

EBLV-1 and EBLV-2 are the two main lyssavirus species detected in bats in Europe. EBLV-1 is detected in the majority of bat rabies cases and is primarily found in serotine bats (*Eptesicus serotinus*), whereas less than 40 EBLV-2 cases have been recorded in Daubenton's bats (*Myotis daubentonii*) and pond bats (*Myotis dasycneme*) [3, 8]. Few cases of transmissions of EBLV-1 to other terrestrial animals (sheep, stone marten, and domestic cats) and humans have been recorded, confirming that the risk of spillover infection remains low but not negligible [3]. Therefore, additional studies are clearly needed to investigate the distribution and genetic characteristics of lyssaviruses across Europe.

In Europe, bat rabies surveillance is highly heterogeneous in terms of the existing networks of bat biologists, active and passive surveillance, number of bat species submitted for rabies diagnosis and individuals sampled [9]. The passive surveillance is based on the testing of sick bats (bats showing clinical signs or abnormal behaviors linked to rabies) or bats found dead. Active surveillance is based on the monitoring of free-living indigenous bat populations for *Lyssavirus* infections [10]. Some European bat species have never been tested for rabies; thus, their role in the epidemiology of lyssaviruses remains uncertain [9].

The network between bat biologists and rabies scientists in Croatia has been poor and inconsistent, and the number of bats included in passive surveillance was negligible, with only 124 bats submitted for rabies diagnosis from 2010 to 2017. There are 34 insectivorous bat species in Croatia, of which five migrate longer distances [11, 12]. The geographical distribution of each bat species in Croatia is still not clearly defined, despite the efforts of bat biologists. Accordingly, data on bat rabies in Croatia is scarce and not up to date. Initial research on bats and their zoonotic diseases was performed in 1968 for military purposes. The objective was to determine the risk of exposure to zoonotic pathogens in caves, since such underground sites had important roles as hiding places, hospitals, and weapon stores owing to their inaccessibility, constant temperature, and access to water. In these studies, 470 cave-dwelling bats belonging to 11 species (*Myotis myotis, Myotis oxygnatus, Rhinolophu. blasii, R. ferrumequinum, R. hipposideros* minimus and hipposideros, *R. euryale, Myotis emarginatus, Miniopterus schreibersii, Pipistrellus kuhlii,* and *R. mehelyi*) were sampled in 15 caves across Croatia. All collected samples were found negative for rabies by laboratory analysis in Prague by using immunofluorescence on inoculated mouse brain [13]. In 1986, the Croatian Veterinary Institute started a study on bat rabies in Croatia and tested around 30 *Eptesicus serotinus* bats, all of which were found negative for rabies by fluorescence antibody test (FAT). These investigations were stopped because of the Croatian War of Independence. Additionally, between 2008 and 2012, 203 dead bats from six genera (*Miniopterus, Myotis, Nyctalus, Rhinolophus, Pipistrellus, Plecotus, Eptesicus,* and *Hypsugo*) collected on various locations around Croatia during field research for inventory purposes were sampled. All samples were found negative by FAT [14]. In this study, we performed active surveillance to investigate the prevalence of EBLVs in bats across Croatia by detecting EBLV-1 antibodies in blood samples using a modified fluorescent antibody virus neutralization (mFAVN) test. The presence of the *Lyssavirus* genome in oropharyngeal swabs of the tested animals was assessed by reverse transcription polymerase chain reaction (RT-PCR). The main objective of this study was to obtain data on the prevalence of lyssaviruses in apparently healthy bats in Croatia in order to improve our understanding of virus distribution and the public health risk associated with bats in Southeastern Europe (SEE).

Results

In total, 455 bats were caught between 2016 and 2017 (Table 1). Of these bats, 440 bats from seven species (*E. serotinus, Myotis blythii, Myotis emarginatus, Myotis myotis, Myotis nattereri, Miniopterus schreibersii,* and *R. ferrumequinum*) were captured. Fourteen bats were unable to be confidently categorized between *Myotis myotis* and *Myotis blythii* and were therefore designated as *Myotis myotis/blythii*. For one individual, neither species nor sex was determined because the animal escaped. All animals caught in the spring in both years were adults, and only 10 animals caught in autumn of 2017 were subadult. Females ($n = 241$) outnumber males ($n = 213$; Tables 1 and 2.). Most of the trapped bats were *Miniopterus schreibersii* ($n = 255$), followed by *R. ferrumequinum* ($n = 90$) and *Myotis myotis* ($n = 56$). Only one *E. serotinus* and one *Myotis nattereri* were caught (Table 2).

Overall, 195 samples were from four Continental locations, and 260 samples were from seven Mediterranean locations. Most of the samples were collected in location 1 ($n = 111$) and location 5 ($n = 92$; Figs. 1 and 2).

Detection of EBLV-1 antibodies

In this study, 363 of 392 sampled bats were subjected to analysis (Table 2). Readable results were obtained for 350 animals. All samples were tested individually.

Table 1 Number of bats caught through active surveillance

Sex	Spring 2016	Spring 2017	Autumn 2017
Male	56	27	130
Female	120	74	47
Not determined	–	–	1
Total	176	101	178

Table 2 Number of bats tested for virus- neutralizing antibodies per species and percentage of positive bats with confidence intervals (CIs) during active surveillance in 2016 and 2017

Species	Number of sampled bats	M	F	Analyzed blood samples	M	F	Obtained results		
							Overall (%pos) [CI]	M (%pos) [CI]	F (%pos) [CI]
ES	1	1	/	1	1	/	1 (0.00)	1 (0.00)	/
MS	222	109	113	210	100	110	200 (2.50) [1.07–5.72]	96 (2.08) [0.57–7.28]	104 (2.88) [0.99–8.14]
MB	17	15	2	16	14	2	16 (6.25) [1.11–28.33]	14 (7.14) [1.27–31.47]	2 (0.00)
ME	1	/	1	/	/	/	/	/	/
MM	56	2	54	53	2	51	52 (25.00) [15.23–38.21]	2 (50.00) [9.45–90.55]	50 (24.00) [14.30–37.41]
MM/B	7	2	5	3	1	2	2 (50.00) [9.45–90.55]	/	2 (50.00) [9.45–90.55]
MN	1	/	1	1	/	1	1 (0.00)	/	1 (0.00)
RF	86	45	41	78	40	38	77 (0.00)	39 (0.00)	38 (0.00)
Not determined	1	/	/	1	/	/	1 (0.00)	/	/
Total	392	174	217	363	158	204	350 (5.71) [3.73–8.66]	152 (2.63) [1.03–6.57]	197 (8.12) [5.06–12.78]

ES Eptesicus serotinus, MS Miniopterus schreibersii, MB Myotis blythii, ME Myotis emarginatus, MM Myotis myotis, MM/B Myotis myotis/blythii, MN Myotis nattereri, RF Rhinolophus ferrumequinum, M male, *F* female, *pos* positive

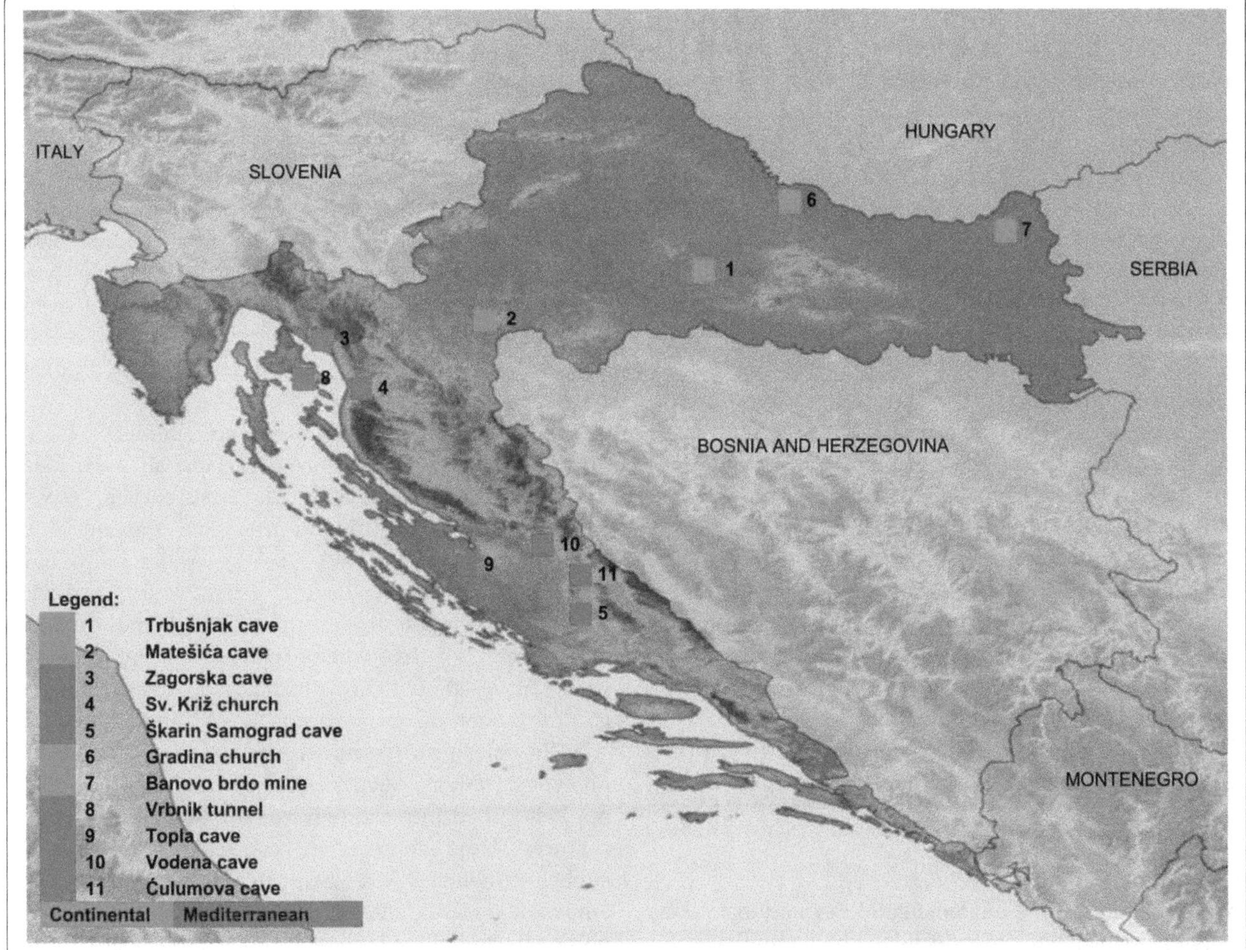

Fig. 1 Locations of bat sampling in Continental (green) and Mediterranean (blue) Croatia.
Source: https://hr.wikipedia.org/wiki/Datoteka:Croatia_map_blank.png

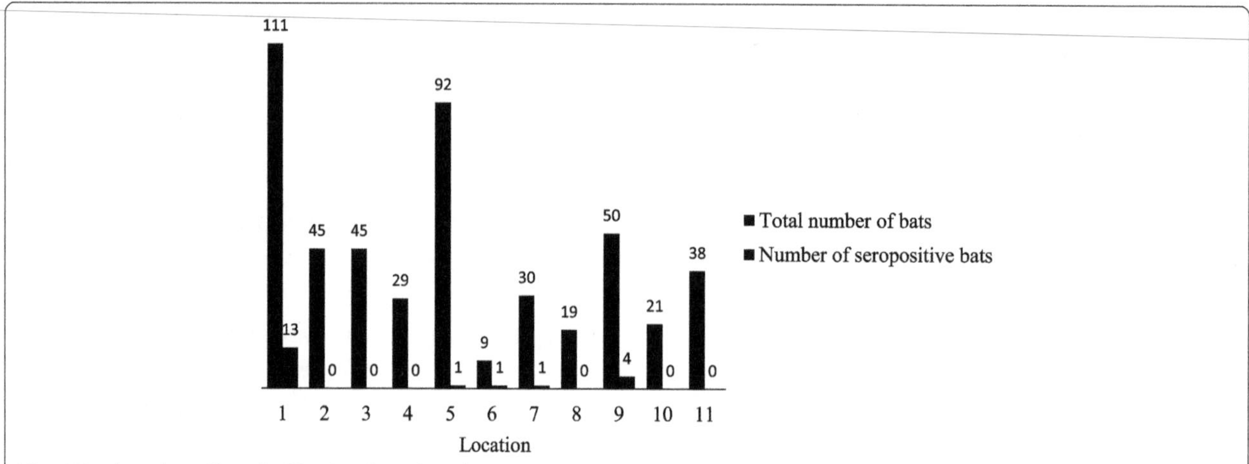

Fig. 2 Total number of bats (red bars) and number of seropositive bats (purple bars) per location (1–11). Location designations are the same as on map (Fig. 1)

In total, 20 serum samples (range: 1.67–2.62 log D_{50}, all ≥1:27) showed detectable levels of neutralizing antibodies against EBLV-1 from 16 females (*Myotis myotis/blythii, Myotis myotis, Miniopterus schreibersii*) and four males (*Myotis blythii, Myotis myotis, Miniopterus schreibersii*; Table 2). Seroprevalence among females was significantly higher than that among males ($p < 0.001$). Among seropositive bats, 65% belonged to *Myotis myotis* (13/20; Table 2), although the majority of bats caught in this survey were *Miniopterus schreibersii*. Seroprevalence was significantly higher in *Myotis myotis* than in *Miniopterus schreibersii* ($p < 0.001$).

Seropositive bats were found in five locations with the majority (15/20) found in Continental Croatia. At location 1 (Fig. 1.), where most bats were sampled (111/392), number of seropositive bats was the highest (13/392). At locations 9, four bats were seropositive, whereas at locations 5, 6 and 7 (Figs. 1 and 2) only one seropositive bat was found per site. In the remaining six locations, all bats were negative on the day of capture (locations 2–4, 8, 10, and 11; Figs. 1 and 2).

Detection of lyssaviral RNA

All 453 oropharyngeal swabs were negative for the presence of lyssaviral RNA, suggesting that none of the bats were excreting virus in saliva at the time of sampling. Two samples were inappropriate for processing.

Beta-actin was detected in all the swabs analyzed (n = 453), indicating that host material was present on the swabs.

Discussion

In Europe, due to the implementation of national rabies programs, which primarily focus on oral rabies vaccination (ORV) of wildlife, the numbers of rabies cases has dramatically decreased in non-flying mammals [15, 16]. However, distinct rabies epidemiological cycles occur in certain European bat species, and the public health impact of bat rabies in Europe should not be underestimated [17, 18]. Bats are the reservoirs for the majority of lyssavirus species, and available rabies vaccines do not confer efficient protection against all of these species. Additionally, minor bite wounds from small insectivorous bats could result in cryptic rabies, which is often reported in North America, although both EBLV-1 and EBLV-2 are less pathogenic than RABV [3].

In Croatia ORV was implemented in 2011, and the last case of rabies was reported in a fox in February 2014. Accordingly, we have focused our research on lyssaviruses circulating on autochthonous bats. An active surveillance program was undertaken to assess the potential public health risk and to elucidate lyssavirus epidemiology in Croatia and SEE.

Approximately 735 dead bats were submitted for rabies diagnosis from 1986 to 2017, and all were found negative by FAT and/or RT-PCR, consistent with reports from several European countries and dependent on number of samples tested [19–21]. Additionally, most tested bats were found dead by bat biologists and were in different stages of decomposition. Some carcasses were frozen for years before testing, or brain tissues were kept in ethanol [14], which could decrease possibility of detection of viral antigens by FAT and/or RT-PCR.

In this study, no lyssavirus RNA was detected in oropharyngeal swabs, suggesting that lyssavirus RNA was not being shed into the saliva of the bats sampled upon capture. These findings are consistent with previous studies conducted elsewhere in Europe [19, 22–24]. Similarly, intermittent excretion of virus in saliva was observed during experimental inoculations [25–27] and may explain the absence of lyssaviral RNA in oropharyngeal swabs.

In Germany, most RT-PCR positive results are associated with *E. serotinus* [20], whereas in Switzerland, most are associated with *Myotis daubentonii* [17]; these species are natural reservoirs for EBLV-1 and EBLV-2, respectively. The under-representation of *E. serotinus* and *Myotis daubentonii* in this study could also explain the absence of EBLV RNA. Although *E. serotinus* is a widespread species in Croatia [28], at the time of sampling, this species had abandoned roost at one location, and we did not have access to sample individuals from another location. In this study, we focused on anthropophilic and cave-dwelling bat species with known roosts; thus *Myotis daubentonii*, as a typical forest species, was not included [29]. Notably, Freuling et al. [25] emphasized that focusing on virus detection in live bats alone has limited effectiveness and should be accompanied by serological surveys.

In this study, for the first time, we confirmed the presence of anti-EBLV-1 antibodies among bats in Croatia and in SEE. Neutralizing antibodies were found in four bat species (*Miniopterus schreibersii, Myotis blythii, Myotis myotis, Myotis myotis/blythii*), with a seroprevalence of 5.71%. Although, for various reasons (challenge virus, test used, cut-off value), it is difficult to compare the results of serological testing between studies, similar seroprevalence rates were observed in Sweden [19], France [24], and Scotland [22].

In contrast, in neighboring countries (Serbia [30] and Slovenia [31, 32]) where active surveillance was conducted, neither virus neutralizing antibodies (VNA) nor virus was detected, although the number of investigated animals was similar to that in our study. In contrast, virus was detected in northern Hungary on a few occasions ($n = 7$) [21].

VNAs have been found in many bat species in several European countries; however, because of cross-reactivity, seropositivity cannot be linked to a specific lyssavirus [19]. In our study, few samples were positive for EBLV-1 but not tested for EBLV-2, RABV, or representatives of phylogroup III (such as LLEBV) due to the low volume of blood collected per bat. Although serological cross-reactivity between members of one phylogroup exists, higher sensitivity of the neutralization test is obtained when using host- specific EBLV-1 as the challenge virus [33]. Therefore, it is possible that we may have missed detection in some bat species because only one test virus was used particularly because *Miniopterus schreibersii* and *Myotis nattereri* have also been associated with WCBV and BBLV, respectively [3, 7]. Furthermore, LLEBV was detected in *Miniopterus schreibersii* found in Spain [34, 35] and recently in France (Picard-Meyer E, Beven V, Hirchaud E, Guillaume C, Larcher G, Robardet E, Servat A, Blanchard Y, Cliquet F: First Isolation of Lleida Bat Lyssavirus from a Schreiber's bat in France, submitted).

Miniopterus schreibersii forms the largest winter and maternity colonies of all Croatian bat species [36] and was the most common bat sampled in this study. This species dwells in four of the five locations where seropositive bats have been found, and as seasonal migrators (> 350 km), they could be one of the dispersion vectors of the disease in Croatia and neighbouring countries [37, 38]. Record of *Miniopterus schreibersii* banded in Slovenia at location 1 confirms that this is possible [39].

However, only five seropositive *Miniopterus schreibersii* were found, and the bat species with the highest prevalence of VNA positivity was *Myotis myotis*. These findings are inconsistent with previous studies [24, 40], in which most records of VNA were found in *E. serotinus*. As described earlier, under-representation of *E. serotinus* in this study could explain these discrepancies. Detection of EBLV-1 VNA in 25% (13/52) of the analyzed *Myotis myotis* samples suggests that bats of this species were infected with EBLV-1 and may be also involved in dispersion of EBLV-1 in countries in SEE, such as Spain [37].

In our study, females were more frequently captured and were more prevalent among seropositive bats. This result could be a consequence of bat ecology and the time of sampling because the majority of sampling was performed in the spring, when maternity colonies consisting of pregnant females are formed. Additionally, pregnancy in bats during spring may change their immune responses with respect to lyssaviruses, which may have affected our capacity to determine detectable antibodies [22].

At location 1, a large number of positive samples ($n = 13$) was observed, likely because the most bats were sampled from this location ($n = 111$) over two consecutive years. This finding confirmed the importance of sampling more at every location and the need for prolonged monitoring of roosts. In this location in 2016, more positive bats were observed ($n = 10$) than in 2017 ($n = 3$). However, lack of previous data and unmarked bats prevented us from making conclusions related to the lyssavirus epidemiological cycle in that colony and emphasized the importance of bat ringing.

Conclusions

In this study, we confirmed the presence of EBLV-1 antibodies in Croatia, suggesting the circulation of EBLV-1 in autochthonous bats, particularly in the continental part of the country. Although *E. serotinus* bats are thought to play a key role in the epidemiology of bat rabies in Europe [3], no conclusion have been made regarding their roles in bat rabies in Croatia.

Whether the lower seroprevalence of lyssaviruses will persist over time remains to be confirmed. Additionaly, testing of other resident bat species in Croatia should be performed, particularly for reservoir species, for species previously not sampled, and by using other lyssavirus species with mFAVN to assess the potential public health risks. All bat biologists should be aware of the risks and be vaccinated to prevent rabies transmission from bats. Education of the general public is strongly suggested, and any contact with bats should be considered a possible exposure.

Methods

Sample collection

In this study, we evaluated seven of 34 bat species present in the country, i.e., greater mouse-eared bat (*Myotis myotis*), lesser mouse-eared bat (*Myotis blythii*), Geoffroy's bat (*Myotis emarginatus*), Schreiber's bent-winged bat (*Miniopterus schreibersii*), greater horseshoe bat (*R. ferrumequinum*), serotine bat (*E. serotinus*), and Natterer's bat (*Myotis nattereri*). Because of the morphological similarity between *Myotis myotis* and *Myotis blythii*, for 14 individuals, species could not be precisely determined. These individuals were designated *Myotis myotis/blythii* (Table 2).

Bats (Table 1) were captured in spring of 2016 and spring and autumn of 2017 at 11 locations in Continental ($n = 4$) and Mediterranean ($n = 7$) Croatia (Figs. 1 and 2). From the selected locations, two were churches (locations 4 and 6), one was a tunnel (location 8), one was a closed mine (location 7), and seven were caves (locations 1–3, 5, 9–11). The locations were selected because they are important underground sites for bats in Croatia (churches excluded) [41], consistent with bat colony behaviors (anthropophilic or cave-dwelling bat species). Bat experts conducted captures using mist nets (Ecotone Mist Nets) at the entrances of caves during night (locations 2, 5, 7, and 10) or using hand nets inside colonies during the day (the other seven locations). In three locations (locations 1, 2, and 5), sampling was conducted repeatedly over two consecutive years. Since bats were not marked, recapture could not be assessed at these three locations.

During sampling, bats were placed individually in cotton bags and were identified by bat biologists according to morphological criteria [42]. Age, body mass, forearm length, sex, and reproductive status were recorded. Capturing, handling, and sampling of bats were approved by the State Institute for Nature Protection (UP/I-612-07/16–48/163).

Blood samples acquired from the uropatagial vein using a 26-G needle (BD Microlance, Becton, Dickinson &Co. Ltd., Drogheda, Ireland) were collected on small pieces of filter papers (Mini Trans-Blot; Bio-Rad, Hercules, CA, USA). A maximum of approximately 23 µl of blood was applied to each piece of filter paper, with the number of pieces varying between one and four per animal based on the size of the animal. Filter papers were dried in the laboratory and stored at − 20 °C until analysis.

Two oropharyngeal swabs were taken from each bat with a dry sterile swab (Copan Italia SpA, Brescia, Italy). One swab was preserved in 500 µL nucleic acid stabilization reagent (DNA/RNA Shield; Zymo Research, Irvine, CA, USA) for RT-PCR, and the second was preserved in 500 µL transport medium (Dulbecco's modified Eagle's medium [DMEM] supplemented with 10% fetal bovine serum [FBS] and 1% antibiotic / antimycotic) for further virus isolation in cases of positive RT-PCR results. The swabs remained in these solutions until processing, at which time the solution was aspirated and used in the assay. In the laboratory, swabs in DNA/RNA Shield were kept at room temperature, whereas swabs in the transport medium were stored at − 20 °C until testing.

After sampling, bats were offered glucose solution orally, and all were successfully released at the location of their capture.

Furthermore, at each location, we searched for potential bat cadavers, but none were found. Brains or other tissues from bats were not collected during this study.

Sample analysis

Detection of anti-EBLV antibodies

Collected blood samples were tested for neutralizing anti-EBLV-1 antibodies with mFAVN tests. Samples soaked on filter papers were first diluted with growth medium, with 65 µL per piece of paper. The mFAVN test was performed according to a previously described protocol [43], except that the virus/ serum mix was distributed on 24- h old BHK-21 monolayers (1×10^5 cells/mL) in 96-well plates (Thermo Fisher Scientific, Roskilde, Denmark). The challenge virus EBLV-1 was diluted at around 100 $TCID_{50}$ per well. The complete growth medium used in the mFAVN test was DMEM (Sigma-Aldrich, St. Louis, MO, USA), supplemented with heat- inactivated FBS (10%; Gibco, US origin, Paisley, UK) and antibiotic / antimycotic (1%; Sigma-Aldrich). Microplates were incubated at 35 °C with 95% relative humidity and 5% CO_2 for 48 h.

Owing to limited sample volume, samples were analyzed in duplicate to determine the presence of anti-EBLV-1 antibodies and serially diluted using a three-fold series. Because positive serum from an EBLV-1- infected bat was not available, a rabies immunoglobulin standard preparation (WHO International Laboratory for Biological Standards, Copenhagen, Denmark) was used as the positive control. FBS was used as negative control. Fluorescein isothiocyanate-conjugated anti-rabies virus monoclonal globulin (Fujirebio Diagnostics, Malvern, PA, USA), diluted according to the manufacturer's instructions, was used as a conjugate. A reciprocal titer of 27 (1.67 log D_{50}) was used as a positive cut-off [22, 24, 37].

Detection of lyssaviral RNA

All collected saliva samples were analyzed for the presence of beta-actin RNA and lyssavirus RNA by real -time and conventional RT-PCR, respectively.

Briefly, oropharyngeal swabs from 453 bats preserved in DNA/RNA Shield were vortexed and centrifuged at 3000 x g for 10 min. RNA was extracted from 230 μL supernatant samples using an iPrep PureLink Virus Kit (Invitrogen, Carlsbad, CA, USA) on an iPrep Purification Instrument according to the manufacturer's instructions. RNA extracts were stored at – 20 °C until used.

To detect *Lyssavirus* RNA, hemi-nested RT-PCR was performed using a SuperScript III One-Step RT-PCR System with Platinum Taq DNA Polymerase (Invitrogen) according to a previously described protocol [44].

All amplifications were performed in a 2720 Thermal Cycler (Applied Biosystems, Foster City, CA, USA). PCR products were visualized under ultraviolet light after gel electrophoresis on 1.5% agarose. Positive (CVS) and negative (phosphate-buffered saline) controls were added for RNA extraction, RT-PCR, and hemi-nested PCR.

To prevent any false negative results due to the absence of oropharyngeal host material or degradation of RNA, a real – time TaqMan RT-PCR (qRT-PCR) was conducted on all samples using specific primers [45] targeting mammalian beta-actin. The qRT-PCR reaction was conducted using a Multiplex Real-Time One-Step RT-PCR Kit according to the manufacturer's instructions (Qiagen, Hilden, Germany) and RotorGene Q (Qiagen).

Statistical analysis

Comparison of sex and species distribution between seropositive bats were performed using non-parametric Wilcoxon Rank-Sum Tests. The 95% confidence intervals of seroprevalence data were calculated using STATA 13.1 (Stata Press, College Station, TX, USA). Results were considered significant when p values were less than 0.001.

Acknowledgments
We thank Mirjana Frljužec for the technical assistance and Dr. Igor Pavlinić and Dr. Maja Đaković for assistance in bat capturing. We also thank Dr. Željko Mihaljević for helping with statistical analysis.

Funding
This work has been fully supported-supported in part by Croatian Science Foundation under the project No.8513 (BatsRabTrack).

The findings and conclusions in this report are those of the authors and do not necessarily represent the views of the funding agency.

Authors' contributions
IŠ participated in sample collection, performed all the experiments, collected and analyzed data, and drafted the manuscript. IL conceived the study and participated in sample collection and manuscript revision. VZ, AĆ and NK performed the bat colony monitoring and sample collection and participated in writing the manuscript. MW participated to the design of the study and analysis of serological data, she has also actively revised the draft manuscript particularly serology section. EPM participated to the design of the study and interpretation of data related to molecular biology work. She has actively revised the draft manuscript for the molecular biology aspects. FC participated to the design of the overall study and in the interpretation of the data. She made important revising during the drafting process. TB conceived the study and participated in its design and coordination. All authors read and approved the final manuscript.

Consent for publication
Not applicable.

Competing interests
The authors declare that they have no competing interests.

Author details
[1]Croatian Veterinary Institute, Savska cesta 143, 10000 Zagreb, Croatia. [2]ANSES - Nancy Laboratory for rabies and wildlife, Batiment H CS 40009, 54220 Malzeville, France. [3]Croatian Biospeleological Society, Demetrova 1, 10000 Zagreb, Croatia.

References

1. World Health Organization. WHO Expert Consultation on Rabies. 2018.
2. International Committee on the Taxonomy of Viruses. The ICTV Online (10th) Report on Virus Taxonomy. 2017. Available: https://talk.ictvonline.org/taxonomy/
3. Banyard AC, Hayman DT, Freuling CM, Muller T, Fooks AR, Johnson N. Bat rabies. In: Jackson AC, editor. Rabies: scientific basis of the disease and its management. Oxford: Academic Press; 2013. p. 215–68.
4. Banyard AC, Hayman D, Johnson N, McElhinney L, Fooks AR. Bats and lyssaviruses. Adv Virus Res. 2011;79:239–89.
5. Tu Y, Chang J, Tsai K, Cheng M. Lyssavirus in Japanese Pipistrelle, Taiwan. Emerg Infect Dis. 2018;24:2016–9.
6. Kokkonen NTU, Gadd TKT. Tentative novel lyssavirus in a bat in Finland. Transbound Emerg Dis. 2018;65:593–6.
7. Kuzmin IV, Rupprecht CE. Bat lyssaviruses. In: Wang L-F, Cowled C, editors. Bats and viruses: a new frontier of emerging infectious disease; 2015. p. 47–98.
8. McElhinney LM, Marston D, Wise E, Freuling CM, Bourhy H, Zanoni R, et al. Molecular epidemiology and evolution of European bat lyssavirus 2. Int J Mol Sci. 2018;19:E156.
9. Schatz J, Fooks AR, McElhinney L, Horton D, Echevarria J, et al. Bat rabies surveillance in Europe. Zoonoses Public Health. 2013;60:22–34.
10. Cliquet F, Freuling C, Smreczak M, Van der Poel WHM, Horton DL, Fooks AR, et al. Development of harmonised schemes for monitoring and reporting of rabies in animals in the European Union. EFSA. 2010;34:7–8.
11. Ministry of Environmental and Nature Protection of the Republic of Croatia, Nature Protection Directorate and State Institute for Nature Protection. Sixth National Report on the Implementation of the Agreement Croatia June 2010 - June 2014. 2014.
12. Tvrtković N. The findings of Mehely's horseshoe bat (*Chiroptera*) in the last century in Croatia were mistakes in identification. Natura Croacia. 2016;25:165–72.
13. Heneberg Đ, Bakić J, Heneberg N, Nikolić B, Agoli B, Hronovsky V, Dusbabek F, et al. Ekološko – medicinska ispitivanja pećina dalmatinskog krša. Zb Vojnomed Akad. 1968;43–6.
14. Pavlinić I, Čač Ž, Lojkić I, Đaković M, Bedeković T, Lojkić M. Šišmiši biološki rezervoari i potencijalni prijenosnici lyssavirusa. Vet Stanica. 2009;40:297–304.
15. European Commission. DG health and food safety overview report - rabies eradication in the EU. 2017.
16. Müller TF, Schröder R, Wysocki P, Mettenleiter TC, Freuling CM. Spatio-temporal use of oral rabies vaccines in fox rabies elimination programmes in Europe. PLoS Negl Trop Dis. 2015;9:1–16.
17. Megali A, Yannic G, Zahno ML, Brügger D, Bertoni G, Christe P, et al. Surveillance for European bat lyssavirus in Swiss bats. Arch Virol. 2010;155:1655–62.

18. Jackson AC. Human rabies: a 2016 update. Curr Infect Dis Rep. 2016;18:38.
19. Hammarin A-L, Treiberg Berndtsson L, Falk K, Professor A, Nedinge M, Olsson G, et al. Lyssavirus-reactive antibodies in Swedish bats. Infect Ecol Epidemiol. 2016;6:31262.
20. Schatz J, Freuling CM, Auer E, Goharriz H, Harbusch C, Johnson N, et al. Enhanced passive bat rabies surveillance in indigenous bat species from Germany - a retrospective study. PLoS Negl Trop Dis. 2014;8:e2835.
21. Rabies - Bulletin - Europe [Internet]. Available: https://www.who-rabies-bulletin.org/site-page/queries. Accessed 19 Feb 2018.
22. Brookes SM, Aegerter JN, Smith GC, Healy DM, Jolliffe TA, Swift SM, et al. European bat lyssavirus in Scottish bats. Emerg Infect Dis. 2005;11:572–8.
23. Nokireki T, Huovilainen A, Sihvonen L, Jakava-Viljanen M. Bat rabies surveillance in Finland. Rabies Bull Eur. 2011;35:8–10.
24. Picard-Meyer E, Dubourg-Savage MJ, Arthur L, Barataud M, Bécu D, Bracco S, et al. Active surveillance of bat rabies in France: a 5-year study (2004-2009). Vet Microbiol. 2011;151:390–5.
25. Freuling C, Vos A, Johnson N, Kaipf I, Denzinger A, Neubert L, et al. Experimental infection of serotine bats (*Eptesicus serotinus*) with European bat lyssavirus type 1a. J Gen Virol. 2009;90:2493–502.
26. Johnson N, Vos A, Neubert L, Freuling C, Mansfield KL, Kaipf I, et al. Experimental study of European bat lyssavirus type-2 infection in Daubenton's bats (*Myotis daubentonii*). J Gen Virol. 2008;5:2662–72.
27. Franka R, Johnson N, Mu T, Vos A, Neubert L, Freuling C, et al. Susceptibility of north American big brown bats (*Eptesicus fuscus*) to infection with European bat lyssavirus type 1. J Gen Virol. 2008;1:1998–2010.
28. Pavlinić I, Đaković M, Tvrtković N. The first records of maternity colonies of the serotine bat, *Eptesicus serotinus* in Croatia. Nat Croat. 2009;20:455–8.
29. Tvrtković N. Bats of Croatia: short research history and identification key. Zagreb/Rijeka: Croatian Natural Historx Museum; 2017. p. 70–1.
30. Vranješ N, Paunović M, Milićević V, Stankov S, Karapandža B, Ungurović U, et al. Passive and active surveillance of lyssaviruses in bats in Serbia. 2010. https://www.researchgate.net/publication/280012289_Passive_and_active_surveillance_of_lyssaviruses_in_bats_in_Serbia.
31. Hostnik P, Rihtarič D, Presetnik P, Podgorelec M, Pavlinič S, Toplak I. Ugotavljanje lisavirusov pri netopirjih v Sloveniji. Determination of bat lyssavirus in Slovenia. Zdrav Vestn. 2009;79:265–71.
32. Presetnik P, Podgorelec M, Hostnik P, Rihtarič D, Toplak I, Wernig J. Active surveillance for lyssaviruses in bats did not reveal the presence of EBLV in Slovenia.2010. https://www.researchgate.net/publication/276272508_Sunny_news_from_the_sunny_side_of_the_Alps_Active_surveillance_for_lyssaviruses_in_bats_did_not_reveal_the_presence_of_EBLV_in_Slovenia
33. Moore SM, Ricke TA, Davis RD, Briggs DJ. The influence of homologous vs. heterologous challenge virus strains on the serological test results of rabies virus neutralizing assays. Biologicals. 2005;33:269–76.
34. Ceballos NA, Morón SV, Berciano JM, Nicolás O, López CA, Juste J, et al. Novel lyssavirus in bat, Spain. Emerg Infect Dis. 2013;19:793–5.
35. Marston D, Ellis R, Wise E, Arechiga-Ceballos N, Freuling CM, Banyard AC, et al. Complete genome sequence of Lleida bat lyssavirus. Genome Announc. 2017;5:e01427–16.
36. Pavlinić I, Đaković M, Tvrtković N. The atlas of Croatian bats (*Chiroptera*) part I. Nat Croat. 2010;19:295–337.
37. Serra-Cobo J, Amengual B, Abellán C, Bourhy H. European bat lyssavirus infection in Spanish bat populations. Emerg Infect Dis. 2002;8:413–20.
38. Hutson AM, Aulagnier S, Benda P, Karataş A, Palmeirim J, Paunović M. *Miniopterus schreibersii*. The IUCN Red List of Threatened Species. 2008: e.T13561A4160556.
39. Presetnik P. Contribution to the knowledge of current migration of Miniopterus schreibersii (Kuhl, 1817) in NW of Panonian basin. http://www.ckff.si/javno/projekti/2009_Mis_Migration_NWPanonia_PPresetnik_poster.pdf.
40. Schatz J, Ohlendorf B, Busse P, Pelz G, Dolch D, Teubner J, et al. Twenty years of active bat rabies surveillance in Germany: a detailed analysis and future perspectives. Epidemiol Infect. 2014;142:1155–66.
41. Eurobats. Conservation of Key Underground sites: the database [Internet]. Available: http://www.eurobats.org/sites/default/files/documents/Underground_sites/Croatia.pdf. Accessed 11 Dec 2017.
42. Dietz C, Kiefer A. Bats of Britain and Europe. 1st ed. London: Bloomsbury Natural History; 2016.
43. Bedeković T, Lemo N, Lojkić I, Mihaljević Ž, Jungić A, Cvetnić Ž, et al. Modification of the fluorescent antibody virus neutralisation test-elimination of the cytotoxic effect for the detection of rabies virus neutralising antibodies. J Virol Methods. 2013;189:204–8.
44. Heaton PR, Johnstone P, Elhinney LMMC, Cowley ROY, Sullivan EO, Whitby JE. Heminested PCR assay for detection of six genotypes of rabies and rabies-related viruses. J Clin Microbiol. 1997;35:2762–6.
45. Toussaint JF, Sailleau C, Breard E, Zientara S, De Clercq K. Bluetongue virus detection by two real-time RT-qPCRs targeting two different genomic segments. J Virol Methods. 2007;140:115–23.

The prevalence and genomic characteristics of hepatitis E virus in murine rodents and house shrews from several regions in China

Wenqiao He, Yuqi Wen, Yiquan Xiong, Minyi Zhang, Mingji Cheng and Qing Chen*

Abstract

Background: Urban rodents and house shrews are closely correlated in terms of location with humans and can transmit many pathogens to them. Hepatitis E has been confirmed to be a zoonotic disease. However, the zoonotic potential of rat HEV is still unclear. The aim of this study was to determine the prevalence and genomic characteristics of hepatitis E virus (HEV) in rodents and house shrews.

Results: We collected a total of 788 animals from four provinces in China. From the 614 collected murine rodents, 20.19% of the liver tissue samples and 45.76% of the fecal samples were positive for HEV. From the 174 house shrews (*Suncus murinus*), 5.17% fecal samples and 0.57% liver tissue samples were positive for HEV. All of the HEV sequences obtained in this study belonged to *Orthohepevirus* C1. However, we observed a lower percentage of identity in the ORF3 region upon comparing the amino acid sequences between *Rattus norvegicus* and *Rattus losea*. HEV derived from house shrews shared a high percentage of identity with rat HEV. Notably, the first near full-length of the HEV genome from *Rattus losea* is described in our study, and we also report the first near full-length rat HEV genomes in *Rattus norvegicus* from China.

Conclusion: HEV is prevalent among the three common species of murine rodents (*Rattus. norvegicus*, *Rattus. tanezumi*, and *Rattus. losea*) in China. HEV sequences detected from house shrews were similar to rat HEV sequences. The high identity of HEV from murine rodents and house shrews suggested that HEV can spread among different animal species.

Keywords: Genomic characteristic, Hepatitis E virus, House shrew, Murine rodent, Prevalence

Background

Hepatitis E virus (HEV) is the causative agent of hepatitis E in humans worldwide. According to 2018 data from the World Health Organization (WHO), there are 20 million HEV infections each year, leading to about 3.3 million symptomatic cases, with approximately one-third of the world's population having been exposed to HEV [1]. HEV can cause hepatitis outbreaks and sporadic hepatitis, which is usually a self-limiting disease [2]. However, in immunosuppressed patients, HEV infection can cause rapidly progressive cirrhosis [3]. In the general population, the mortality rates of HEV infection range from 0.2–1%; however, the mortality rate is higher during pregnancy, especially in developing countries [4]. Therefore, HEV infection is a major global public health concern.

HEV is a positive-sense, single-stranded, non-enveloped RNA virus with a size of 27–34 nm. The genome of HEV contains three overlapping open reading frames (ORF), specifically ORF1 that encodes nonstructural proteins; ORF2 that encodes viral capsid protein, which is responsible for self-assembly of virus particles [5]; and ORF3 that encodes proteins involved in virus morphogenesis and release [6]. ORF4 has been identified in rat and ferret HEV strains; however, its function is still unknown [7–9].

HEV is classified in the genus *Orthohepevirus* in the family *Hepeviridae* [10]. *Orthohepevirus* is divided into four species, as follows: *Orthohepevirus* A, B, C, and D. All of the HEV strains isolated from humans belong to the species *Orthohepevirus* A with four genotypes, genotypes 1 and 2 can only infect humans, while genotypes 3

* Correspondence: 18002270308@163.com
Department of Epidemiology, School of Public Health, Guangdong Provincial Key Laboratory of Tropical Disease Research, Southern Medical University, 1838 Guangzhou North Road, Guangzhou 510515, China

and 4 can infect animals and are considered zoonotic pathogens [11]. *Orthohepevirus* B includes avian HEV strains. *Orthohepevirus* C is divided into two species, one of which is *Orthohepevirus* C1, derived from rats, and the other of which is *Orthohepevirus* C2, derived from ferrets. Bat HEV is classified in *Orthohepevirus* D [2].

The origin of HEV is still unknown. HEV was first considered to be restricted to presentation in humans [12]. The most common transmission route of HEV is the fecal–oral route. However, in the recent years, HEV had been detected in a variety of animal species, such as pigs, wild boars, cattle, cats, rabbits, rodents, and dogs [13]. The consumption of meat or meat products and direct or indirect contact with pigs can be important modes of zoonotic HEV transmission [2, 11]. Hepatitis E has been confirmed to be a zoonotic disease [14]. In Japan, HEV strains isolated from *Rattus norvegicus* (*R. norvegicus*) that had been trapped near a pig farm where HEV was prevalent were genetically identical to the HEV strains derived from pigs [15], indicating that rats may be a reservoir for the HEV that causes infection in pigs.

Rat HEV was first detected in *R. norvegicus* from Germany [8]. Many studies since have reported the detection of HEV in rats. However, it is still inconclusive whether rat HEV can cross transmit to other species. In recent years, HEV has been detected in house shrews (*Suncus murinus*) trapped in China with a high similarity to the rat HEV strains, indicating that rat HEV might be transmitted between rats and shrews [16].

Murine rodents and house shrews are reservoirs of some human pathogens, such as mammarenavirus and hantavirus, owing to their close contact with humans [16]. However, the zoonotic potential of rat HEV is still unclear. Nonetheless, some studies have pointed out the possibility of HEV transmission between humans and rats. For example, recent investigations have shown that HEV genotype 3 has been detected in wild rats; rat HEV can successfully replicate in three human hepatoma cell lines (PLC/PRF/5, HuH-7, and HepG2 cells); and anti-rat HEV antibodies were detected in forest workers [15, 17–19].

It is important to understand the possibility of interspecies transmission of HEV between humans, rodents, and shrews. In China, the first HEV in rats was reported in 2013 [20]. So far, the presence of HEV in rats and shrews has only been reported in the Guangdong and Yunnan Provinces of China [16, 21]. Separately, the detection of full-length genomes of HEV in the Chevrier's field mouse (*Apodemus chevrieri*) and Père David's vole (*Eothenomys melanogaster*) have been reported in China [21]. However, the genomic characteristics of HEV in commensal rodents in China remain unclear.

In the present study, we investigated the prevalence of HEV in murine rodents and house shrews in five regions of four provinces in China. Besides, we analyzed the characteristics of the nucleic acid sequences of HEV obtained from the murine rodents and house shrews.

Results

Prevalence of HEV in murine rodents and shrews

From 2014 to 2017, a total of 788 animals were trapped. Liver and fecal specimens were collected from 456 *R. norvegicus*, 64 *Rattus tanezumi* (*R. tanezumi*), 93 *Rattus losea* (*R. losea*), 1 *Bandicota indica* (*B. indica*), and 174 *Suncus murinus* (*S. murinus*) (Fig. 1 and Table 1).

According to nested broad-spectrum RT-PCR and nested PCR, 20.19% (124/614) of the liver tissue samples and 45.76% (281/614) of the fecal samples from murine rodents were positive for HEV. Of the 281 fecal-HEV-positive murine rodents, 84 (29.89%) also tested positive for the virus in their liver tissue. Nine (5.17%) fecal samples from house shrews were positive for HEV. Among these nine fecal-HEV-positive house shrews, one of them also was positive for HEV in the liver (Table 1).

Among the five regions of sampling, the prevalence of HEV in liver tissue samples ranged from 7.71–26.88%, while that in fecal samples ranged from 24.20–66.67%. Detection rates of HEV in liver samples from different species of animals ranged from 0 to 27.96%. With regard to fecal samples, the positive rates of HEV in different species of animals ranged from 0 to 51.54%. The positive rate of HEV in liver tissue samples from *R. losea* was significantly higher than in other species ($\chi^2 = 4.399$, $P < 0.05$). For fecal samples, the positive rate for HEV was significantly higher in *R. norvegicus* than in other species ($\chi^2 = 86.309$, $P < 0.001$) (Table 1).

Phylogenetic analysis

Partial ORF1 sequences and ORF1–ORF2 sequences were detected from liver tissue samples and fecal samples.

The identity among the representative partial ORF1 sequences obtained in our study ranged from 77.2–100% (Additional file 1: Table S1). Among these sequences, two sequences detected in *R. losea* from Xiamen City were totally identical (XM16 and XM27). Additionally, one sequence detected in house shrew from Xiamen City was identical to another sequence detected in *R. norvegicus* from Yunnan Province (XM54 and MLP51). The identity between the representative sequences obtained in this study and that reported in a previous study were also compared (GQ504010) [8] (Table 2). HEV detected from *R. losea* had a lower percentage of identity than that detected from other species. In this region, the percentage of identity at the amino acid level is greater than the percentage of identity at the nucleotide level.

Phylogenetic trees constructed by using two different methods were similar. Here, we only showed the

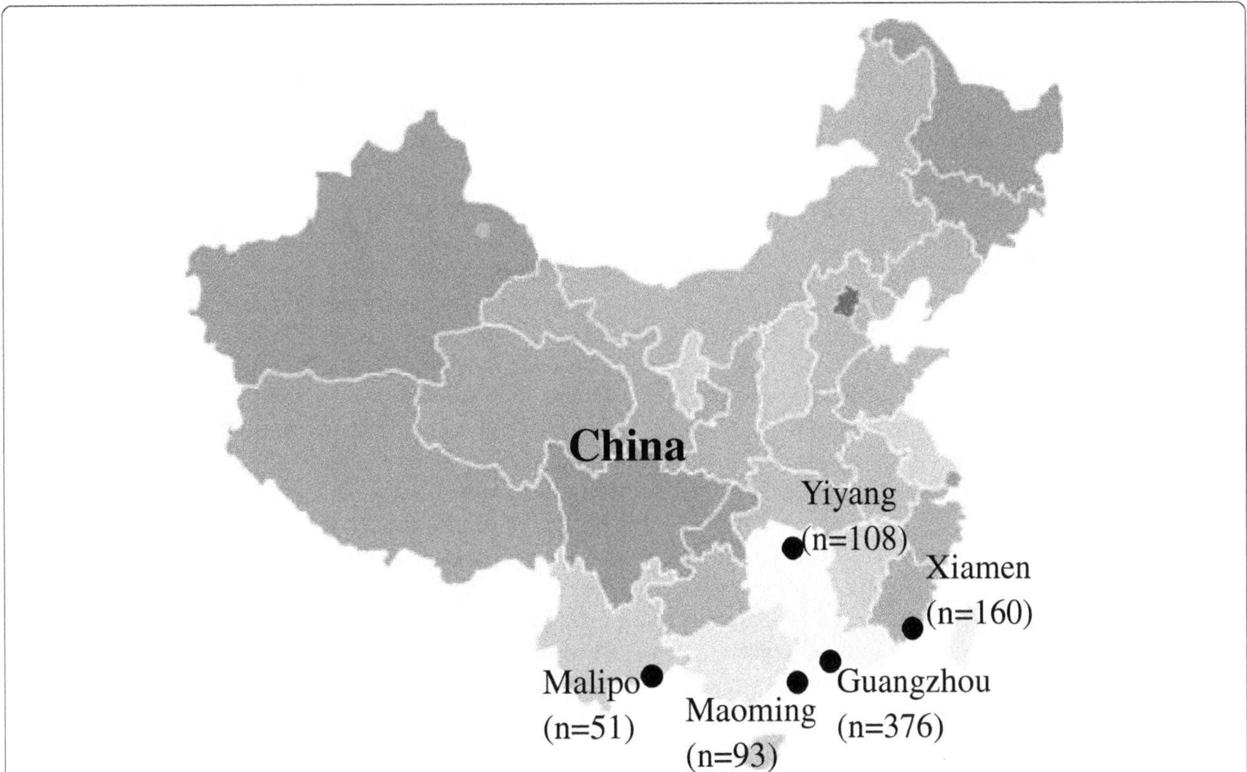

Fig. 1 Location of the trapping sites of the murine rodents and house shrews in China. Number of animals trapped are indicated in brackets. Map source: http://image.so.com/v?src=360pic_normal&z=1&i=0&cmg=9923e970972c38717e105d62081f0181&q=%E4%B8%AD%E5%9B%BD%E5%9C%B0%E5%9B%BE&correct=%E4%B8%AD%E5%9B%BD%E5%9C%B0%E5%9B%BE&cmsid=4372c66d63a3bc1a0c3a77413cee395a&cmran=0&cmras=0&cn=0&gn=0&kn=11#multiple=0&gsrc=1&dataindex=44&id=8c7daffc8205d4592f8e4a6e8c0e25e2&currsn=0&jdx=44&fsn=71

Table 1 Prevalence of HEV in liver and fecal samples (%, n)

Sample	Family	Species	Guangzhou	Yiyang	Xiamen	Malipo	Maoming	Total
Liver	*Muridae*	*Rattus norvegicus*	14.57 (29/199)	30.34 (27/89)	12.50 (4/32)	14.00 (7/50)	29.07 (25/86)	20.18 (92/456)
		Rattus tanezumi	0 (0/13)	10.53 (2/19)	16.67 (4/24)	0 (0/1)	0 (0/7)	9.48 (6/64)
		Rattus losea	-	-	27.96 (26/93)	-	-	27.96 (26/93)
		Bandicota indica	-	-	0 (0/1)	-	-	0 (0/1)
		Subtotal	13.68 (29/212)	26.85 (29/108)	22.67 (34/150)	13.72 (7/51)	26.88 (25/93)	20.19 (124/614)
	Soricidae	*Suncus murinus*	0 (0/164)	-	10.00 (1/10)	-	-	0.57 (1/174)
	Total		7.71 (29/376)	26.85 (29/108)	21.88 (35/160)	13.72 (7/51)	26.88 (25/93)	15.86 (125/788)
Fecal	*Muridae*	*Rattus norvegicus*	42.21 (84/199)	65.17 (58/89)	28.12 (9/32)	66.00 (33/50)	59.30 (51/86)	51.54 (235/456)
		Rattus tanezumi	15.38 (2/13)	68.42 (13/19)	12.50 (3/24)	100 (1/1)	28.57 (2/7)	32.81 (21/64)
		Rattus losea	-	-	26.88 (25/93)	-	-	26.88 (25/93)
		Bandicota indica	-	-	0 (0/1)	-	-	0 (0/1)
		Subtotal	40.57 (86/212)	65.74 (71/108)	24.66 (37/150)	66.67 (34/51)	56.99 (53/93)	45.76 (281/614)
	Soricidae	*Suncus murinus*	3.05 (5/164)	-	40.00 (4/10)	-	-	5.17 (9/174)
	Total		24.20 (91/376)	65.74 (71/108)	25.62 (41/160)	66.67 (34/51)	56.99 (53/93)	36.80 (290/788)

Table 2 Nucleotide (nt) and amino acid sequence identity for the region spanning nt positions 4139–4393 or amino acid positions 1379–1462, based on strain GQ504010

Species	Nt identity (%)	Amino acid identity (%)
Rattus norvegicus (13[a])	81.5–86.6	88.2–96.4
Rattus tanezumi (2[a])	84.3–86.2	95.2
Rattus losea (3[a])	78.4–79.6	94.1
Suncus murinus (1[a])	83.9	95.2

[a]number of the sequences
GQ504010, GenBank accession number of a HEV strain isolated from *R. norvegicus* from Germany

phylogenetic trees constructed with MrBayes (version 3.2) (Figs. 2, 3, 4, 5, 6, 7 and 8). The phylogenetic trees obtained based on the neighbor-joining method are shown in the supplementary information file (Additional file 1: Figures S1-S7).

Phylogenetic trees were generated for the region spanning the nucleotides (nt) 4139 to 4393 (numbering based on a HEV sequence from *R. norvegicus*, GenBank accession number: JN167538), with the selected representative fragment sequences from our study and nucleotide sequences of HEV from GenBank (Fig. 2). From the result of the phylogenetic analysis, all of the novel sequences were found to belong to the species *Orthohepevirus* C1. The HEV sequence detected in house shrew clustered with rat HEV sequences from murine rodents, indicating that shrew HEV was similar to rat HEV. Excluding three sequences detected in *R. losea* from Xiamen City, all selected sequences were clustered with the HEV strains detected in Germany.

Upon analyzing the representative nested PCR products of rat HEV, the similarity ranged from 77.1–99.6% (Additional file 1: Table S2). As compared with the sequence reported in a previous study (GQ504010) [8], the HEV sequences detected in *R. losea* have a lower degree of identity than the sequences detected in other species (Table 3). The amino acid sequences of this region also have a higher percentage of identity than the nucleotide sequences, similar to our findings in the analysis of the partial ORF1 sequences. Phylogenetic tree was generated for the region spanning the nt 4157 to 4925 (numbering based on HEV sequence from *R. norvegicus*, GenBank accession number: JN167538), with the representative fragment sequences from this study and HEV sequences from GenBank. Phylogenetic analysis showed that all of the sequences belonged to the species *Orthohepevirus* C1. Two sequences from *R. norvegicus* and one sequence from *R. losea* clustered together with a rat HEV sequence from Asia, while the other HEV strains clustered with rat sequences from Europe (Fig. 3).

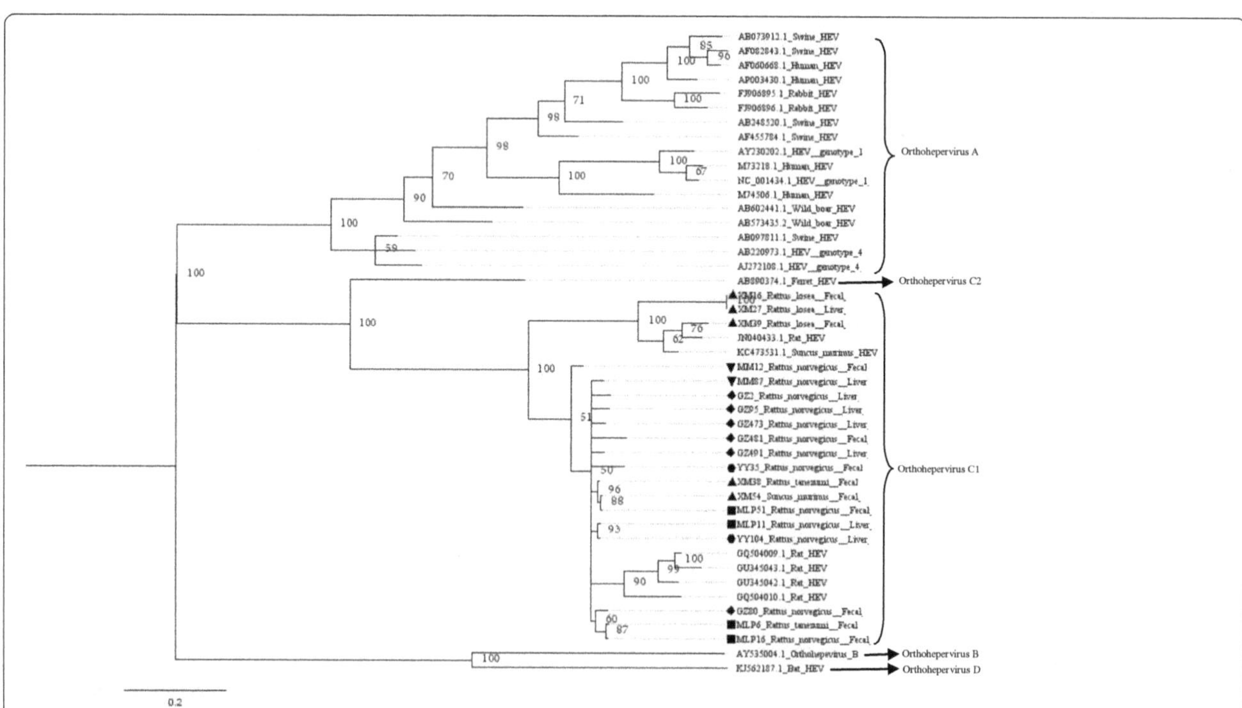

▲ XM, Xiamen City in Fujian Province; ■ MLP, Malipo County in Yunnan Province; ◆ Guangzhou City in Guangdong Province; ● YY, Yiyang City in Hunan Province; ▼ MM, Maoming City in Guangdong Province.

Fig. 2 Phylogenetic tree constructed based on partial nucleotide sequences of partial ORF1 regions (255-nt) of 19 HEV strains (MrBayes, GTR + G + I nucleotide substitution model). Twenty five representative HEV isolates derived from rats, swine, humans, rabbits, wild boars, bat, and ferret are included for comparison. Percentages of the posterior probability (PP) values are indicated

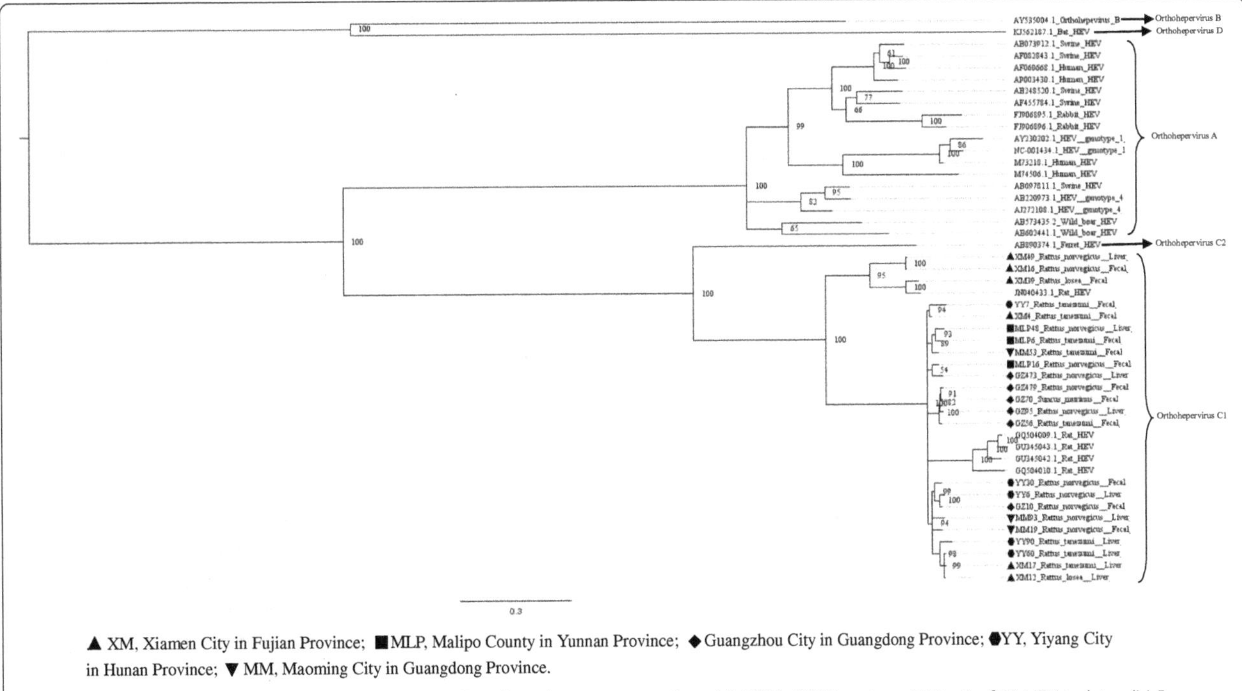

Fig. 3 Phylogenetic tree constructed based on partial nucleotide sequences of partial ORF1–ORF2 regions (769-nt) of 23 HEV isolates (MrBayes, GTR + G + I nucleotide substitution model). Five rat HEV isolates and 20 HEV isolates derived from swine, humans, rabbits, wild boars, bat, and ferret are included for comparison. Percentages of the PP values are indicated

Characterization of the HEV genome

In murine rodents, three near full-length HEV genomes (GenBank accession numbers: MH729810–MH729812) detected in *R. norvegicus* (GZ95 and GZ481) and *R. losea* (XM11) were obtained. The length of these sequences ranged between the nt 6692 and 6790. These sequences were compared with the rat HEV sequence reported in a previous study (GU345042) [22]. The XM11 nucleotide and amino acid sequences had a lower percentage of identity than other HEV sequences. Additionally, the identity of the amino acid sequences of three near full-length HEV genomes in the ORF3 region were lower than other regions (Table 4 and Fig. 9). Phylogenetic trees were constructed based on these near full-length sequences and some representative HEV sequences from humans and other species of animals (Figs. 4, 5, 6, 7 and 8). The results showed that the near full-length HEV genomes obtained in this study clustered with rat HEV strains (Fig. 4). Two sequences from *R. norvegicus* in Guangzhou city clustered together in one branch, while the other HEV sequence from *R. losea* was located in another branch of the phylogenetic trees (Figs. 4, 5, 6, 7 and 8).

Five partial HEV nucleic acid sequences from house shrews, whose length ranged from 778 bp to 825 bp, were obtained (GenBank accession numbers: MH729813–MH729817). The selected representative house shrew sequence was clustered with rat HEV sequences (Figs. 2 and 3).

Discussion

In this study, we investigated HEV in four species of murine rodents and house shrews from four provinces in China via nested broad-spectrum RT-PCR assay and nested PCR method for rat HEV [8, 23]. During the conduct of this study, care was taken to limit contamination and a strict operating standard was followed during each experimental step, with a negative control established for each experiment. Our study evaluated the geographical distribution of HEV in murine rodents and house shrews in China.

HEV-positivity was detected in all of the cities investigated in this study, indicating that HEV is widely distributed among the three common species of murine rodents (*R. norvegicus*, *R. tanezumi*, and *R. losea*) in China. Specifically, 20.19% (124/614) of the liver tissue samples and 45.76% (281/614) of the fecal samples from the murine rodents were positive for HEV. In comparison, HEV-positivity in *R. norvegicus* was found to be only 2.87% in a previous study conducted in China [16]. The reason for the discrepancy in the detection rates of HEV in the present study versus the previous study is unclear. One explanation may be variations in the PCR methods used, which may affect the results obtained. The primers of two PCR methods we selected were designed based on the well-conserved regions of all types of HEV and the

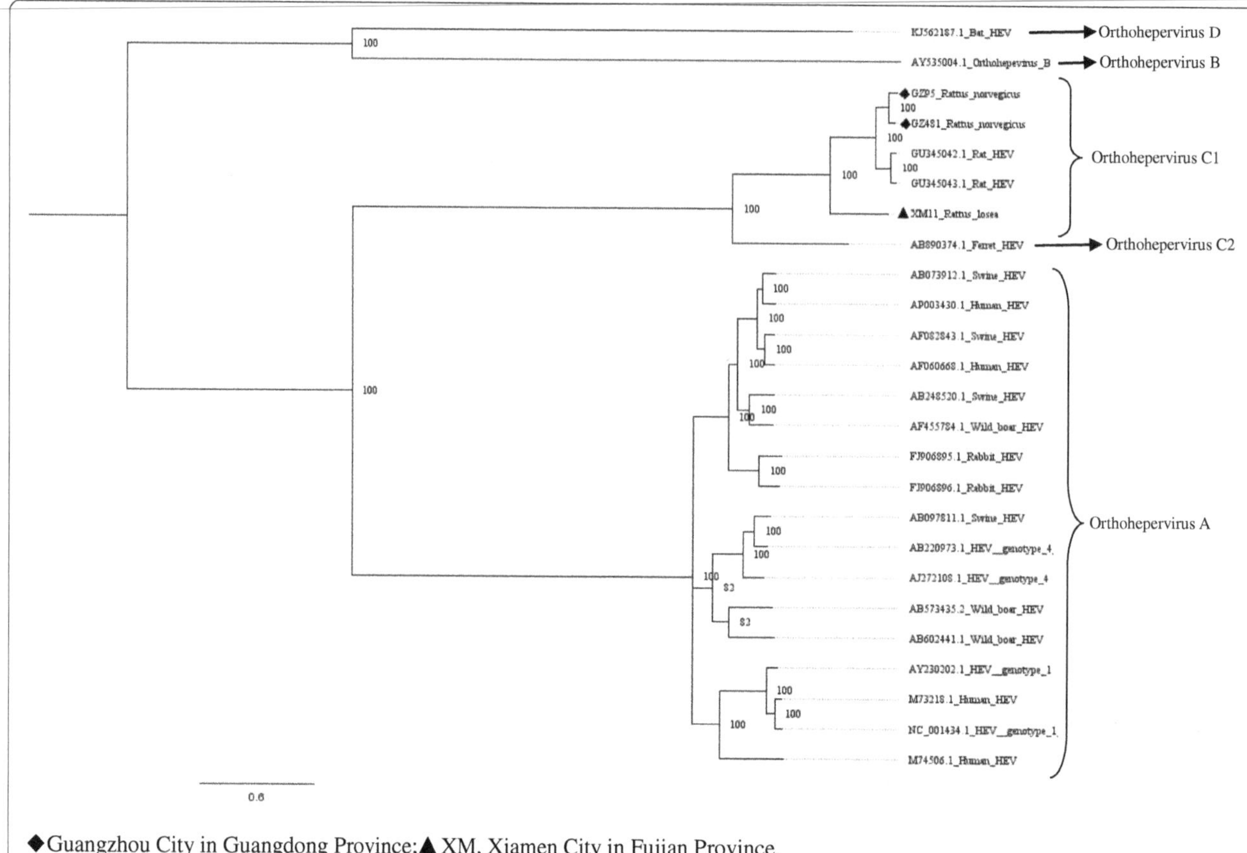

◆Guangzhou City in Guangdong Province;▲ XM, Xiamen City in Fujian Province.

Fig. 4 Phylogenetic tree constructed based on near full-length genomes of HEV (MrBayes, GTR + G + I nucleotide substitution model). Three rat HEV isolates and 22 HEV isolates derived from swine, humans, rabbits, wild boars, bat, and ferret are included for comparison. Percentages of the PP values are indicated

rat HEV genome, which may have contributed to the high detection rates of HEV in this study.

Positive liver samples were detected in all of the murine rodents, except *B. indica*. The positive liver samples indicated that HEV was capable of infecting these animals and subsequently replicating in them. The high prevalence and effective replication of HEV furthermore suggested that these animals are hosts of HEV.

In our study, the detection rate of HEV in house shrews was lower than in other species. This result was similar to that of the previous study conducted in China, in which only one of the 196 house shrews was positive for HEV [16]. Another study conducted in Nepal also reported the low prevalence of anti-HEV Ig-G in house shrews [24]. These indicate that house shrews might not be a natural host of HEV.

HEV has been detected in several species of murine rodents, such as *R. norvegicus*, *R. flavipectus*, and *R. losea*. In our study, the detection rate for HEV in *R. losea* in liver samples was higher than in other species of animals; this result was consistent with the findings of a previous study, indicating that *R. losea* is an important natural host for HEV [25]. *R. norvegicus* had the highest prevalence of HEV in fecal samples (Table 1). This result might be explained by the rats' habitats. *R. norvegicus* mainly inhabits sewers and garbage dumps. Poor sanitation is conducive to the spread of HEV, which might explain the high HEV detection rate in the fecal samples from *R. norvegicus*. Further study is needed to explain why the detection rate in the liver samples from *R. losea* was higher than that in *R. norvegicus*, while the positive rate in fecal samples from *R. losea* was lower than that in *R. norvegicus*.

To our knowledge, we report the first near full-length of HEV genome from *R. losea* in our study. We also present the first near full-length rat HEV genomes in *R. norvegicus* from China. According to the phylogenetic analysis, all of the HEV sequences detected in the trapped murine rodents belonged to rat HEV. This result was in agreement with those of most previous studies [16, 26, 27], indicating that murine rodents are exclusively susceptible to rat HEV. However, some previous studies reported that HEV genotype 3 was detected in *R. norvegicus* [18]. Follow-up studies are necessary to confirm whether other genotypes of HEV can infect murine rodents.

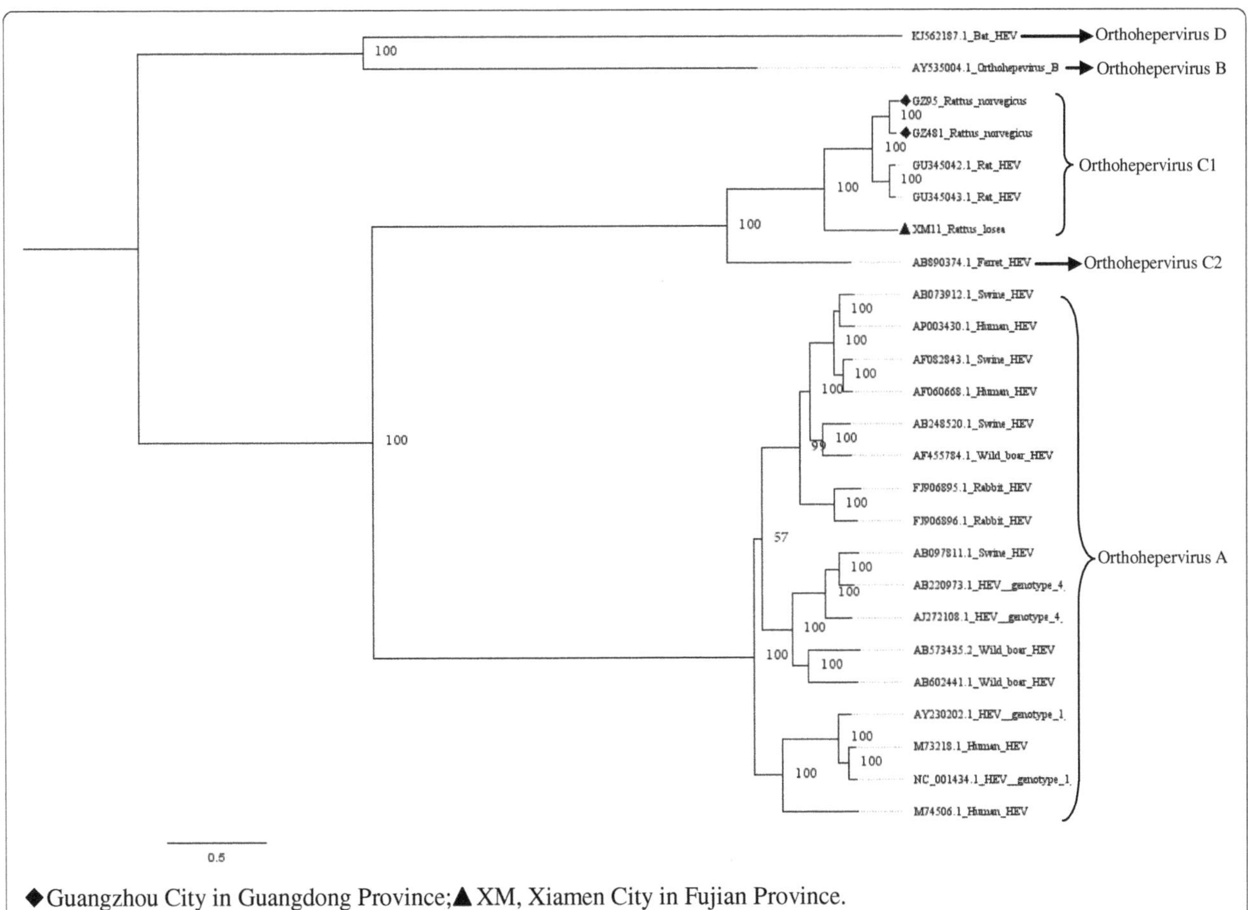

Fig. 5 Phylogenetic tree constructed based on partial ORF1 of the near full-length genomes of HEV (MrBayes, GTR + G + I nucleotide substitution model). Three rat HEV isolates and 22 HEV isolates derived from swine, humans, rabbits, wild boars, bat, and ferret are included for comparison. Percentages of the PP values are indicated

Near full-length genome analysis showed that rat HEV sequences obtained in our study clustered with other rat HEV strains that were reported previously (Fig. 4). When analyzing the identity of the different regions between amino acid sequences, the partial ORF2 region seems to be a relatively conserved region (Fig. 9). The ORF2 region is known to be associated with capsid assembly and has many immune epitopes; thus, the ORF2 protein can induce strong immune responses [28]. In recent years, anti-rat HEV antibodies have been detected in forest workers and febrile patients [17, 29]. This suggests that rat HEV might infect humans and lead to liver diseases. However, further research is necessary to investigate the transmission of HEV between humans and rats.

Partial ORF1 and the ORF1–ORF2 sequences obtained in animals from different regions were clustered together, and we found that two sequences detected in *R. losea* from Xiamen City were totally identical, indicating that the HEV infecting the same species of animals were similar (Figs. 2 and 3 and Additional file 1: Table S1). In our study, HEV detected in *R. losea* showed a lower percentage of identity versus in other animal species (Tables 2 and 3). When comparing the three near full-length HEV sequences obtained in our study with the rat HEV sequence reported in a previous study (GU345042) [8], the near-full-length sequence from *R. losea* in Xiamen City also showed a lower percentage of identity than the sequences from *R. norvegicus* in Guangzhou City (Table 4). The result of the phylogenetic analysis showed that two sequences from *R. norvegicus* in Guangzhou City clustered together, while the other HEV strains from *R. losea* were in another branch of the phylogenetic trees (Figs. 5, 6, 7 and 8). This might be due to the difference in HEV infection as a result of host-specificity.

The percentages of identity at the amino acid level in the partial ORF1 and ORF1-ORF2 regions were greater than the percentages of identity at the nucleotide level (Tables 2 and 3). This might be explained by the presence of degenerate codons.

A 281 bp fragment sequence of HEV from house shrews was reported in a previous study [30]. In this

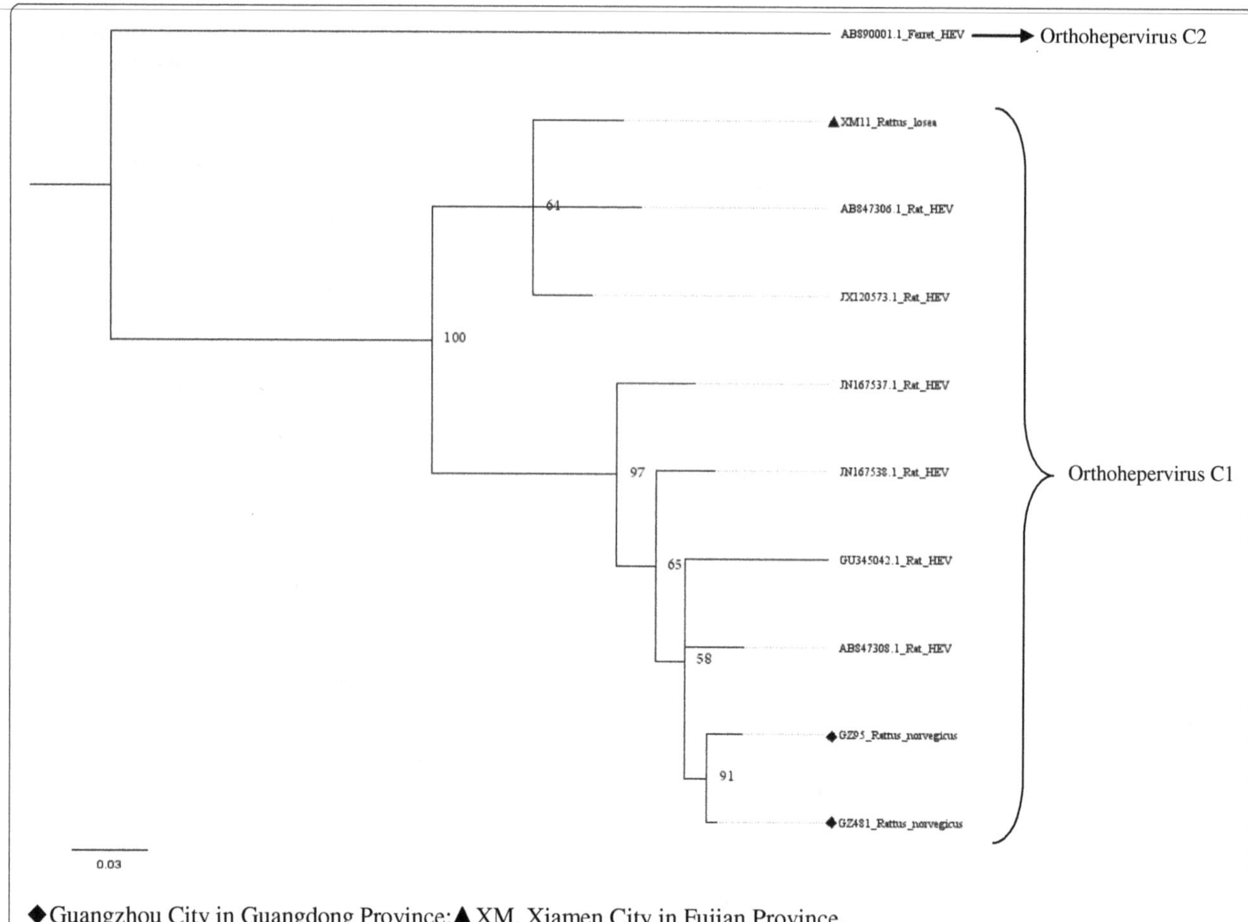

Fig. 6 Phylogenetic tree constructed based on partial ORF4 of the near full-length genomes of HEV (MrBayes, GTR + G + I nucleotide substitution model). Three rat HEV isolates and seven HEV isolates derived from rats and ferret are included for comparison. Percentages of the PP values are indicated

investigation, 778 bp to 825 bp partial nucleic acid sequences of HEV from house shrews were obtained. One partial ORF1 sequence from house shrew obtained in our study was identical to another sequence detected in *R. norvegicus*. The result of phylogenetic analysis based on partial sequences of ORF1 region showed that HEV sequence detected in house shrew clustered with rat HEV sequences instead of the HEV sequence detected from house shrew reported in previous studies (Fig. 2). This may be explained by the interspecies transmission or spillover infection of HEV between murine rodents and house shrews [30].

Conclusions

There was a high prevalence of HEV in fecal samples of *R. norvegicus*, while a high prevalence of HEV was observed in the liver samples of *R. losea*. Murine rodents and house shrews shared a high identity genome of HEV, suggesting that HEV might cross transmit among different animal species.

Methods

Samples

Between 2014 and 2017, rodents and shrews were captured near human residences with cage traps. These animals were captured in five regions of four provinces in China, as follows: Yiyang City in Hunan Province, Xiamen City in Fujian Province, Maoming City and Guangzhou City in Guangdong Province, and Malipo County in Yunan Province (Fig. 1). Inhalational anesthesia was performed with diethyl ether, and the dosage of diethyl ether was adjusted according to the heart rate, respiratory frequency, corneal reflection and extremity muscle tension of the animal. Following anesthesia, the rodents and shrews were executed via cervical dislocation by trained personnel. Liver tissue samples were collected by intraperitoneal surgery in the laboratory and stored in RNAlater (Invitrogen, California, United States). The fecal samples were soaked in phosphate-buffered saline (PBS). The liver tissue and fecal samples were stored at − 80 °C and thawed at 4 °C prior to

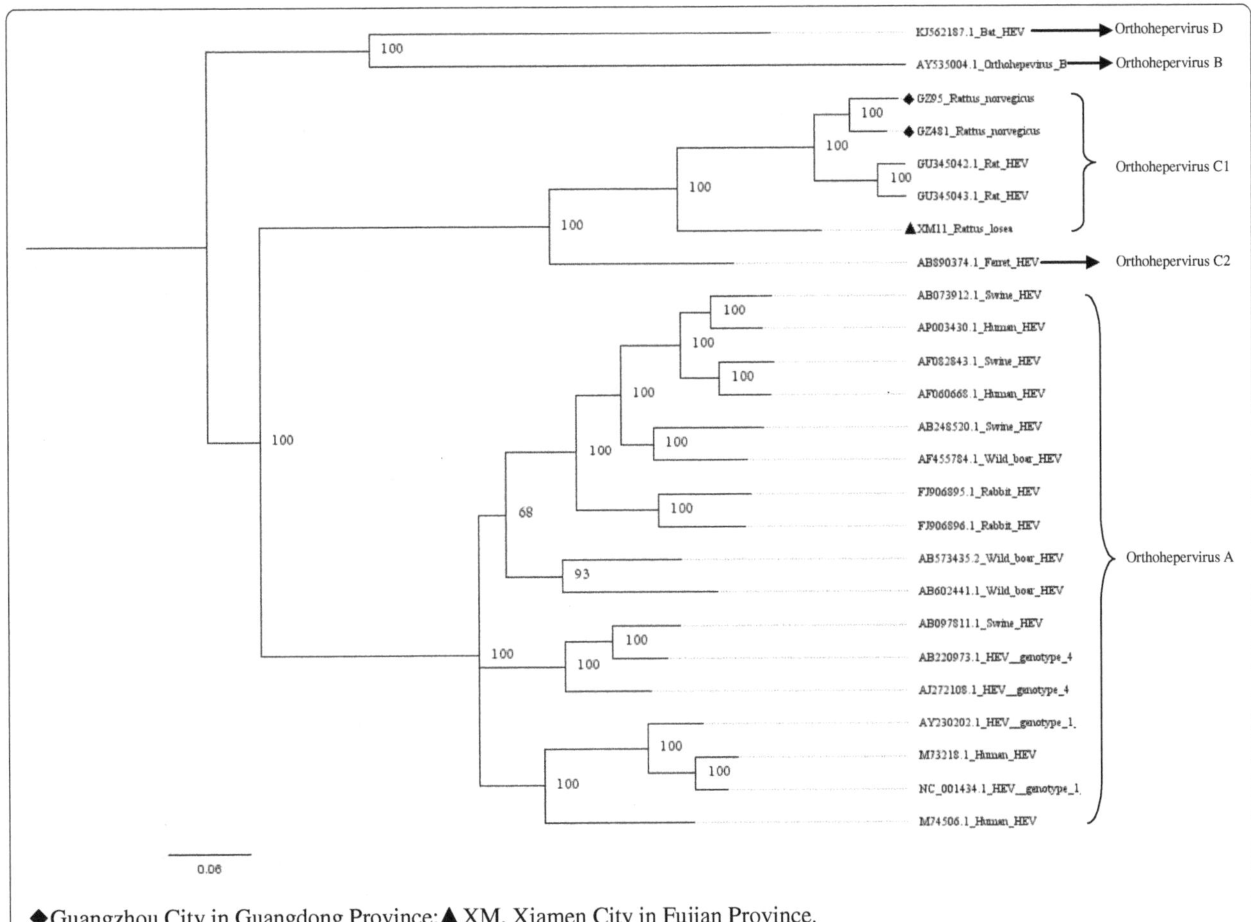

◆Guangzhou City in Guangdong Province;▲XM, Xiamen City in Fujian Province.

Fig. 7 Phylogenetic tree constructed based on partial ORF2 of the near full-length genomes of HEV (MrBayes, GTR + G + I nucleotide substitution model). Three rat HEV isolates and 22 HEV isolates derived from swine, humans, rabbits, wild boars, bat, and ferret are included for comparison. Percentages of the PP values are indicated

processing. The species of the trapped animals were identified by morphological identification and sequencing of the cytochrome B (*cytB*) gene [31].

Extraction of nucleic acid and detection of HEV

Total RNA and DNA were extracted from ~ 20 mg of liver tissue samples or 200 μL aliquot of fecal samples by using the MiniBEST Viral RNA/DNA Extraction Kit (TaKaRa, Kusatsu, Japan). We simultaneously used two polymerase chain reaction (PCR) methods for detecting HEV, one of which was a nested broad-spectrum PCR to amplify a 334 bp ORF1 fragment of all known HEV strains [8], while the other was a nested PCR to amplify the ORF1–ORF2 region of rat HEV [23]. The amplified products were separated on a 1.5% agarose gel, and the positive samples were sent to the Beijing Genomics Institute (Shenzhen, China) for sequencing.

Genome sequencing

Based on six HEV genome sequences from GenBank (GenBank accession numbers: JX120573, AB847307, GU345042, JN167538, KM516906, and LC225389), we designed eight pairs of primers to amplify the near full-length genomes of HEV (Table 5). After sequencing all of the fragments, the Lasergene SeqMan software (DNASTAR; Madison, WI, USA) was used to assemble the sequences.

Phylogenetic analysis

The selected representative ORF1 fragments and ORF1–ORF2 fragments and all of the near full-length sequences obtained in this study were aligned with HEV sequences from humans and different species of animals obtained via GenBank by using the ClustalW multiple sequence alignment program in MEGA (version 7.0; Oxford Molecular Ltd., Cambridge, UK), respectively. The phylogenetic trees were constructed based on the ORF1 fragments, ORF1–ORF2 fragments, and the near full-length sequences. We constructed the phylogenetic trees via MrBayes (version 3.2) using a GTR + G + I nucleotide substitution matrix; two million Markov chain Monte Carlo (MCMC) iterations were sampled every

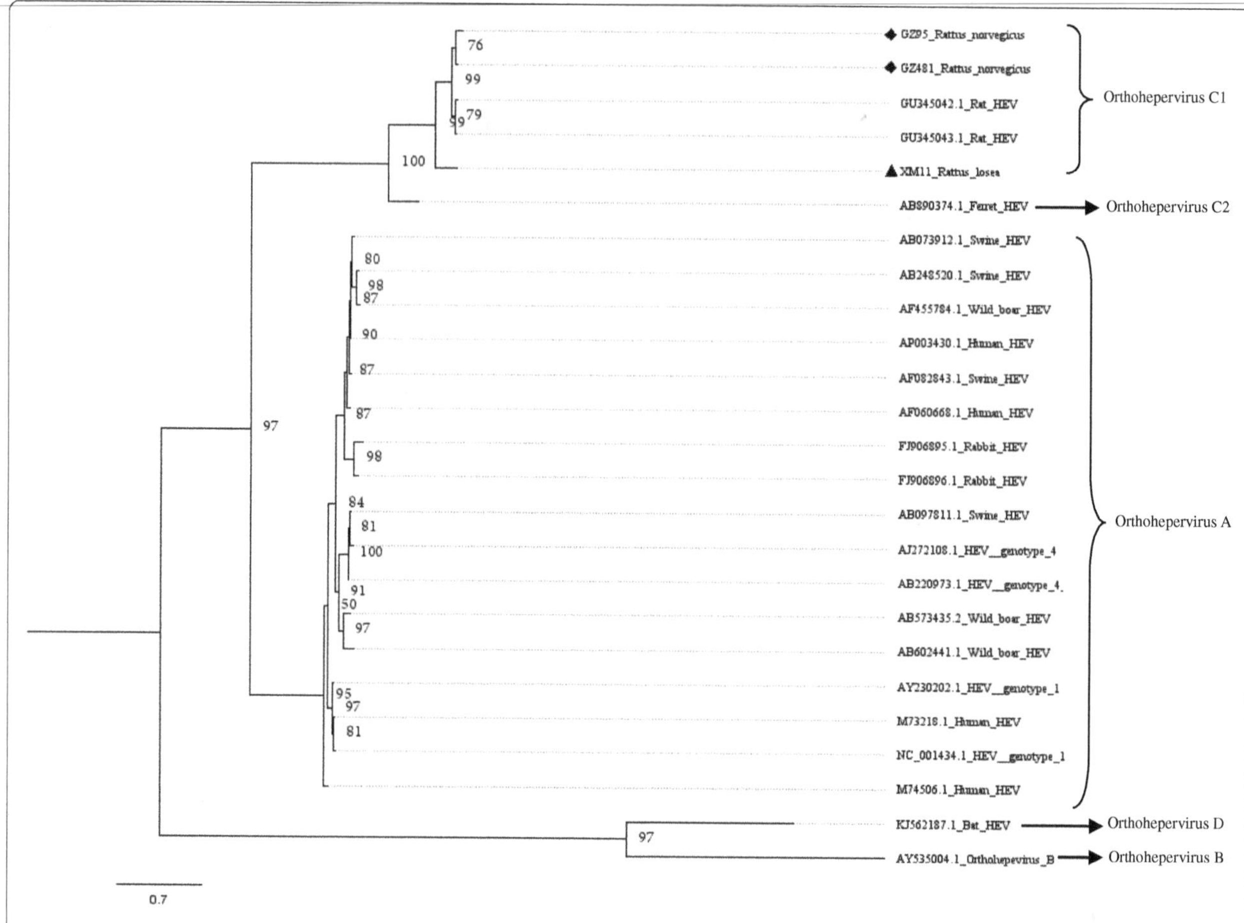

◆Guangzhou City in Guangdong Province;▲XM, Xiamen City in Fujian Province.

Fig. 8 Phylogenetic tree constructed based on ORF3 of the near full-length genomes of HEV (MrBayes, GTR + G + I nucleotide substitution model). Three rat HEV isolates and 22 HEV isolates derived from swine, humans, rabbits, wild boars, bat, and ferret are included for comparison. Percentages of the PP values are indicated

100 steps in order to obtain 20,000 trees, and burn-in was generally 25% of tree replicates [32, 33]. The phylogenetic trees were also constructed based on the neighbor-joining method in MEGA, with 1000 bootstrap replicates.

The identity of the different regions of the amino acid sequences and nucleotide sequences were estimated by use of the Sequence Identity Matrix program in BioEdit (version 7.2.5).

Statistical analysis

Data were analyzed using the Statistical Product and Service Solutions software (SPSS, version 13.0; IBM Corp., Armonk, NY, USA). The positive rates of HEV

Table 3 Nucleotide (nt) and amino acid sequence identity for the region spanning nt positions 4157–4925 or amino acid positions 1385–1636, based on strain GQ504010

Species	Nt identity (%)	Amino acid identity (%)
Rattus norvegicus (12[a])	77.7–85.3	93.3–97.2
Rattus tanezumi (8[a])	84.2–85.1	96.4–96.8
Rattus losea (2[a])	77.5–84.5	93.7–96.4
Suncus murinus (1[a])	84.9	95.6

[a]number of the sequences
GQ504010, GenBank accession number of a HEV strain isolated from *R. norvegicus* from Germany

Table 4 Nucleotide (nt) and amino acid identity for different regions of the near-full-length HEV genomes, based on strain GU345042

	Nt identity				Amino acid identity			
	Partial ORF1	Partial OFR2	ORF3	Partial OFR4	Partial ORF1	Partial OFR2	ORF3	Partial OFR4
GZ95	0.836	0.833	0.899	0.945	0.907	0.902	0.813	0.872
GZ481	0.846	0.884	0.935	0.941	0.919	0.952	0.892	0.872
XM11	0.744	0.783	0.805	0.893	0.864	0.913	0.647	0.755

GU345042, GenBank accession number of a HEV strain isolated from *R. norvegicus* from Germany

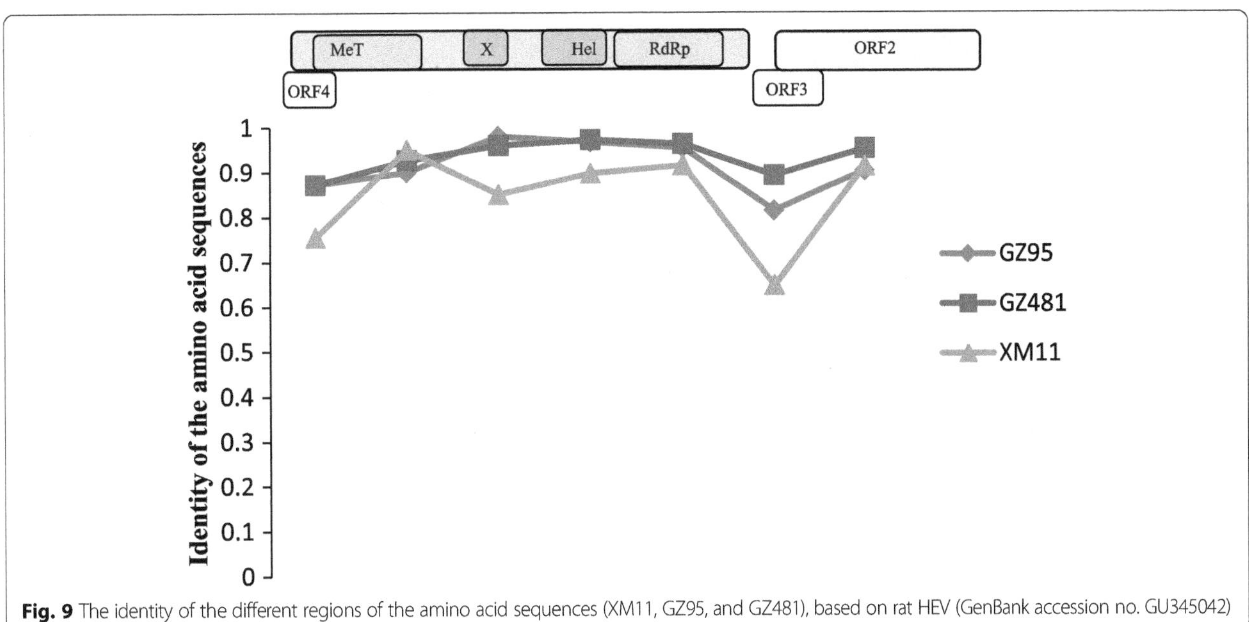

Fig. 9 The identity of the different regions of the amino acid sequences (XM11, GZ95, and GZ481), based on rat HEV (GenBank accession no. GU345042)

Table 5 HEV near-full-length PCR primers

Primer	Orientation	Sequence (5'-3')	Target fragment (base pairs)
HEV-F15	Sense (first/second round)	AGACCCATCARTATGTCG	921 (this study)
HEV-R935	Antisense (second round)	GTDGAYCKRGMCTTCTCACA	
HEV-R977	Antisense (first round)	CATRAGYCKRTCCCADAT	
HEV-F827	Sense (first/second round)	CATCTATGTGCGCAGCCTGT	1340 (this study)
HEV-R2166	Antisense (second round)	GTGGACAAACTGGGTGCGATC	
HEV-R2228	Antisense (first round)	GTGCCACAGCGTGTATTATAG	
HEV-F2084	Sense (first/second round)	GCNGTNTATGARGGRGAY	955 (this study)
HEV-R3038	Antisense (second round)	CAAAATCWATNGCNGGGATCTG	
HEV-R3053	Antisense (first round)	ATVAGVCCCTTGCTYTCAAAATC	
HEV-F2868	Sense (first/second round)	TKAARGCNCARTGGMGDG	841 (this study)
HEV-R3708	Antisense (second round)	ARBAYDGTCACCTGVTCHC	
HEV-R3740	Antisense (first round)	ATRCGRCARTGCACDGT	
HEV-F3539	Sense (first/second round)	ACCAACTTGCAGGATATAG	699 (this study)
HEV-R4237	Antisense (second round)	AAACTCGCTAAAATCATTCTCAA	
HEV-R4253	Antisense (first round)	ATTCTGGGTGCTGTCAAACTCG	
HEV-F4980	Sense (first/second round)	TCGTGCTCGTGYTTTTGCT	1029 (this study)
HEV-R6008	Antisense (second round)	CCTATRTCRCCYACMCCRTT	
HEV-R6014	Antisense (first round)	CCCTTRCCTATRTCRCCYAC	
HEV-F5552	Sense (first/second round)	GTRTCAATGTCRTTYTGG	1153 (this study)
HEV-R6704	Antisense (second round)	RTTAACAGGYCCAGYACC	
HEV-R6836	Antisense (first round)	ATWGCATCAGCMACGAGGCA	
HEV-F6300	Sense (first/second round)	CAACTGGCGGTCTGGTGATGTC	535 (this study)
HEV-R6881	Antisense (second round)	AGACACTGTCGGCTGCTGC	
HEV-R6834	Antisense (first round)	GCATCAGCCACGAGGCAGG	
HE607	Sense (first round)	CTTGGTTYAGGGCCATAGAG	880 [23]

Table 5 HEV near-full-length PCR primers *(Continued)*

Primer	Orientation	Sequence (5'-3')	Target fragment (base pairs)
HE604	Antisense (first round)	CAGCAGCGGCACGAACAGCA	
HE608	Sense (second round)	TTYAGGGCCATAGAGAAGGC	
HE606	Antisense (second round)	ACAGCAAAAGCACGAGCACG	

among different species of animals were analyzed using chi-square tests. A *p*-value of 0.05 was considered to be statistically significant.

Ethics guidelines

The protocol of this study was approved by the Ethics Committee of the Institutional Animal Care and Use Committee of Southern Medical University in Guangzhou, China.

Additional files

Additional file 1: Table S1. Identity of the representative nucleotide (nt) sequences (nt positions 4139–4393) obtained in this study. **Table S2.** Identity of the representative nucleotide (nt) sequences (nt positions 4157–4925) obtained in this study. **Figure S1.** Phylogenetic tree constructed by the neighbor-joining method based on partial nucleotide sequences of ORF1 regions (255 nt) of 19 HEV strains. Twenty five representative HEV isolates derived from rats, swine, humans, rabbits, wild boars, bat and ferret are included for comparison. Bootstrap support of branches (1000 replication) is indicated. **Figure S2.** Phylogenetic tree constructed by the neighbor-joining method based on partial nucleotide sequences of ORF1 -ORF2 regions (769 nt) of 23 HEV isolates. Five rat HEV isolates and 20 HEV isolates derived from swine, humans, rabbits, wild boars, bat and ferret are included for comparison. Bootstrap support of branches (1000 replication) is indicated. **Figure S3.** Phylogenetic tree constructed by the neighbor-joining method based on near full-length genomes of HEV. Three rat HEV isolates and 22 HEV isolates derived from swine, humans, rabbits, wild boars, bat and ferret are included for comparison. Bootstrap support of branches (1000 replication) is indicated. **Figure S4.** Phylogenetic tree constructed by the neighbor-joining method based on ORF1 of the near full-length genomes of HEV. Three rat HEV isolates and 22 HEV isolates derived from swine, humans, rabbits, wild boars, bat and ferret are included for comparison. Bootstrap support of branches (1000 replication) is indicated. **Figure S5.** Phylogenetic tree constructed by the neighbor-joining method based on ORF4 of the near full-length genomes of HEV. Three rat HEV isolates and 7 HEV isolates derived from rats and ferret are included for comparison. Bootstrap support of branches (1000 replication) is indicated. **Figure S6.** Phylogenetic tree constructed by the neighbor-joining method based on ORF2 of the near full-length genomes of HEV. Three rat HEV isolates and 22 HEV isolates derived from swine, humans, rabbits, wild boars, bat and ferret are included for comparison. Bootstrap support of branches (1000 replication) is indicated. **Figure S7.** Phylogenetic tree constructed by the neighbor-joining method based on ORF3 of the near full-length genomes of HEV. Three rat HEV isolates and 22 HEV isolates derived from swine, humans, rabbits, wild boars, bat and ferret are included for comparison. Bootstrap support of branches (1000 replication) is indicated. (DOCX 309 kb)

Abbreviation

HEV: Hepatitis E virus

Acknowledgements

We are grateful to Xueshan Zhong, Shaowei Chen, Xueyan Zheng, Shujuan Ma, Lina Jiang, Jin Ge, Ming Qiu, Shuting Huo, Xuemei Ke, Wen Zhou, Xing Li, Yun Mo, Xuejiao Chen, Yanxia Chen, Yongzhi Li, and Fangfei You for their participation in the collection of samples.

We would like to thank LetPub (www.LetPub.com) for providing linguistic assistance during the preparation of this manuscript.

Funding

This work was supported by the National Natural Science Foundation of China (grant no. 81373051). The funders had no role in the study design, data collection and analysis, decision to publish, or preparation of the manuscript.

Authors' contributions

WH and QC conceived of the project and QC obtained the funding. WH and QC contributed to the writing of the paper. WH, YW, and MZ led virus detection. WH, YX, and MC performed the statistical analysis. WH, YW, YX, MZ, and MC performed the sample collection. All of the authors have read and approved the manuscript for publication.

Consent for publication

Not applicable.

Competing interests

The authors declare that they have no competing interests.

References

1. von Wulffen M, Westholter D, Lutgehetmann M, Pischke S, Hepatitis E. Still Waters Run Deep. J Clin Trans Hepatol. 2018;6(1):40–7.
2. Khuroo MS, Khuroo MS. Hepatitis E: an emerging global disease - from discovery towards control and cure. J Viral Hepat. 2016;23(2):68–79.
3. Kamar N, Abravanel F, Lhomme S, Rostaing L, Izopet J. Hepatitis E virus: chronic infection, extra-hepatic manifestations, and treatment. Clin Res Hepatol Gastroenterol. 2015;39(1):20–7.
4. Jaiswal SP, Jain AK, Naik G, Soni N, Chitnis DS. Viral hepatitis during pregnancy. Int J Gynaecol Obstetr. 2001;72(2):103–8.
5. Li TC, Yamakawa Y, Suzuki K, Tatsumi M, Razak MA, Uchida T, Takeda N, Miyamura T. Expression and self-assembly of empty virus-like particles of hepatitis E virus. J Virol. 1997;71(10):7207–13.
6. Emerson SU, Nguyen HT, Torian U, Burke D, Engle R, Purcell RH. Release of genotype 1 hepatitis E virus from cultured hepatoma and polarized intestinal cells depends on open reading frame 3 protein and requires an intact PXXP motif. J Virol. 2010;84(18):9059–69.
7. Tanggis KT, Takahashi M, Jirintai S, Nishizawa T, Nagashima S, Nishiyama T, Kunita S, Hayama E, Tanaka T, et al. An analysis of two open reading frames (ORF3 and ORF4) of rat hepatitis E virus genome using its infectious cDNA clones with mutations in ORF3 or ORF4. Virus Res. 2018;249:16–30.
8. Johne R, Plenge-Bonig A, Hess M, Ulrich RG, Reetz J, Schielke A. Detection of a novel hepatitis E-like virus in faeces of wild rats using a nested broad-spectrum RT-PCR. J Gen Virol. 2010;91(3):750–8.
9. Mulyanto SJB, Andayani IG, Khalid TM, Ohnishi H, Jirintai S, Nagashima S, Nishizawa T, Okamoto H. Marked genomic heterogeneity of rat hepatitis E virus strains in Indonesia demonstrated on a full-length genome analysis. Virus Res. 2014;179:102–12.

10. Purdy MA, Harrison TJ, Jameel S, Meng XJ, Okamoto H, Van der Poel WHM, Smith DB, Ictv Report C. ICTV virus taxonomy profile: Hepeviridae. J Gen Virol. 2017;98(11):2645–6.
11. Pavio N, Meng XJ, Doceul V. Zoonotic origin of hepatitis E. Curr Opinion Virol. 2015;10:34–41.
12. Sridhar S, Lau SK, Woo PC, Hepatitis E. A disease of reemerging importance. J Form Med Assoc Taiwan yi zhi. 2015;114(8):681–90.
13. Smith DB, Simmonds P, Izopet J, Oliveira-Filho EF, Ulrich RG, Johne R, Koenig M, Jameel S, Harrison TJ, Meng XJ, et al. Proposed reference sequences for hepatitis E virus subtypes. J Gen Virol. 2016;97(3):537–42.
14. Syed SF, Zhao Q, Umer M, Alagawany M, Ujjan IA, Soomro F, Bangulzai N, Baloch AH, Abd El-Hack M, Zhou EM, et al. Past, present and future of hepatitis E virus infection: zoonotic perspectives. Microb Pathog. 2018;119: 103–8.
15. Kanai Y, Miyasaka S, Uyama S, Kawami S, Kato-Mori Y, Tsujikawa M, Yunoki M, Nishiyama S, Ikuta K, Hagiwara K. Hepatitis E virus in Norway rats (Rattus norvegicus) captured around a pig farm. BMC Res notes. 2012;5:4.
16. Wang B, Cai CL, Li B, Zhang W, Zhu Y, Chen WH, Zhuo F, Shi ZL, Yang XL. Detection and characterization of three zoonotic viruses in wild rodents and shrews from Shenzhen city, China. Virol Sin. 2017;32(4):290–7.
17. Dremsek P, Wenzel JJ, Johne R, Ziller M, Hofmann J, Groschup MH, Werdermann S, Mohn U, Dorn S, Motz M, et al. Seroprevalence study in forestry workers from eastern Germany using novel genotype 3- and rat hepatitis E virus-specific immunoglobulin G ELISAs. Med Microbiol Immunol. 2012;201(2):189–200.
18. Lack JB, Volk K, Van Den Bussche RA. Hepatitis E virus genotype 3 in wild rats, United States. Emerg Infect Dis. 2012;18(8):1268–73.
19. Jirintai S, Tanggis M, Suparyatmo JB, Takahashi M, Kobayashi T, Nagashima S, Nishizawa T, Okamoto H. Rat hepatitis E virus derived from wild rats (Rattus rattus) propagates efficiently in human hepatoma cell lines. Virus Res. 2014;185:92–102.
20. Li W, Guan D, Su J, Takeda N, Wakita T, Li TC, Ke CW. High prevalence of rat hepatitis E virus in wild rats in China. Vet Microbiol. 2013;165(3–4):275–80.
21. Wang B, Li W, Zhou JH, Li B, Zhang W, Yang WH, Pan H, Wang LX, Bock CT, Shi ZL, et al. Chevrier's Field Mouse (*Apodemus chevrieri*) and Pere David's Vole (*Eothenomys melanogaster*) in China Carry Orthohepeviruses that form Two Putative Novel Genotypes Within the Species Orthohepevirus C. Virol Sin. 2018;33(1):44–58.
22. Johne R, Heckel G, Plenge-Bonig A, Kindler E, Maresch C, Reetz J, Schielke A, Ulrich RG. Novel hepatitis E virus genotype in Norway rats, Germany. Emerg Infect Dis. 2010;16(9):1452–5.
23. Mulyanto DSN, Sriasih M, Takahashi M, Nagashima S, Jirintai S, Nishizawa T, Okamoto H. Frequent detection and characterization of hepatitis E virus variants in wild rats (Rattus rattus) in Indonesia. Arch Virol. 2013;158(1):87–96.
24. He J, Innis BL, Shrestha MP, Clayson ET, Scott RM, Linthicum KJ, Musser GG, Gigliotti SC, Binn LN, Kuschner RA, et al. Evidence that rodents are a reservoir of hepatitis E virus for humans in Nepal. J Clin Microbiol. 2006; 44(3):1208.
25. Van Nguyen D, Van Nguyen C, Bonsall D, Ngo TT, Carrique-Mas J, Pham AH, Bryant JE, Thwaites G, Baker S, Woolhouse M, et al. Detection and characterization of homologues of human Hepatitis viruses and Pegiviruses in rodents and bats in Vietnam. Viruses. 2018;10(3):102.
26. Simanavicius M, Juskaite K, Verbickaite A, Jasiulionis M, Tamosiunas PL, Petraityte-Burneikiene R, Zvirbliene A, Ulrich RG, Kucinskaite-Kodze I. Detection of rat hepatitis E virus, but not human pathogenic hepatitis E virus genotype 1-4 infections in wild rats from Lithuania. Vet Microbiol. 2018;221:129–33.
27. Li TC, Yoshizaki S, Ami Y, Suzaki Y, Yasuda SP, Yoshimatsu K, Arikawa J, Takeda N, Wakita T. Susceptibility of laboratory rats against genotypes 1, 3, 4, and rat hepatitis E viruses. Vet Microbiol. 2013;163(1–2):54–61.
28. Zhou Y, Zhao C, Tian Y, Xu N, Wang Y. Characteristics and functions of HEV proteins. Adv Exp Med Biol. 2016;948:17–38.
29. Shimizu K, Hamaguchi S, Ngo CC, Li TC, Ando S, Yoshimatsu K, Yasuda SP, Koma T, Isozumi R, Tsuda Y, et al. Serological evidence of infection with rodent-borne hepatitis E virus HEV-C1 or antigenically related virus in humans. J Vet Med Sci. 2016;78(11):1677–81.
30. Guan D, Li W, Su J, Fang L, Takeda N, Wakita T, Li TC, Ke C. Asian musk shrew as a reservoir of rat hepatitis E virus, China. Emerg Infect Dis. 2013; 19(8):1341–3.
31. Arai S, Bennett SN, Sumibcay L, Cook JA, Song JW, Hope A, Parmenter C, Nerurkar VR, Yates TL, Yanagihara R. Phylogenetically distinct hantaviruses in the masked shrew (Sorex cinereus) and dusky shrew (Sorex monticolus) in the United States. Am J Trop Med Hyg. 2008;78(2):348–51.
32. Drexler JF, Corman VM, Lukashev AN, van den Brand JM, Gmyl AP, Brunink S, Rasche A, Seggewibeta N, Feng H, Leijten LM, et al. Evolutionary origins of hepatitis a virus in small mammals. Proc Natl Acad Sci U S A. 2015; 112(49):15190–5.
33. Ronquist F, Teslenko M, van der Mark P, Ayres DL, Darling A, Hohna S, Larget B, Liu L, Suchard MA, Huelsenbeck JP. MrBayes 3.2: efficient Bayesian phylogenetic inference and model choice across a large model space. Syst Biol. 2012;61(3):539–42.

First retrospective studies with etiological confirmation of porcine transmissible gastroenteritis virus infection in Argentina

Pablo Enrique Piñeyro[1*], Maria Inez Lozada[2], Laura Valeria Alarcón[3], Ramon Sanguinetti[4], Javier Alejandro Cappuccio[5], Estefanía Marisol Pérez[2], Fabio Vannucci[6], Alberto Armocida[2], Darin Michael Madson[1], Carlos Juan Perfumo[2] and Maria Alejandra Quiroga[2]

Abstract

Background: In 2014, a notification of porcine transmissible gastroenteritis virus (TGEV) was made by the National Services of Animal Health of Argentina (SENASA) to the World Organization of Animal Health (OIE). The notification was based on a serological diagnosis in a small farm with a morbidity rate of 2.3% without enteric clinical signs. In order to determine if TGEV was circulating before the official report, a retrospective study on cases of neonatal diarrhea was performed. The selection criteria was a sudden increase in mortality in 1- to 21-day-old piglets with watery diarrhea that did not respond to antibiotics. Based on these criteria, three clinical cases were identified during 2010–2015.

Results: All animals that were evaluated presented histological lesions consistent with enteric viral infection. The feces and ultrathin sections of intestine that were evaluated by electron microscopy confirmed the presence of round particles of approximately 80 nm in size and characterized by finely granular electrodense nucleoids consistent with complete particles of coronavirus. The presence of the TGEV antigen was confirmed by monoclonal specific immunohistochemistry, and final confirmation of a metabolically-active virus was performed by in situ hybridization to detect a TGE mRNA encoding spike protein. All sections evaluated in this case were negative for PEDV and rotavirus A.

Conclusions: This is the first case series describing neonatal mortality with etiological confirmation of TGEV in Argentina. The clinical diagnosis of TGEV infections in endemic regions is challenging due to the epidemiological distribution and coinfection with other enteric pathogens that mask the clinical presentation.

Keywords: Porcine transmissible gastroenteritis virus, Diarrhea, Mortality, Piglets

Background

There are five coronaviruses (CoVs) known to infect swine, and the clinical disease is mainly associated with neonatal diarrhea, but respiratory and neurological signs have also been reported [1–4]. Porcine transmissible gastroenteritis virus (TGEV) and porcine epidemic diarrhea virus (PEDV) belong to the *Coronaviridae* family, *Coronavirinae* subfamily, and genus *Alphacoronavirus* [5]. A new coronavirus genetically distinct from TGEV and PEDV, porcine deltacoronavirus (PDCoV) (genus *Deltacoronavirus*), has recently been associated with enteric disease in pigs [6]. Enteric porcine coronaviruses including TGEV, PEDV and PDCoV are characterized by acute diarrhea and anorexia with rapid dissemination in naïve populations. The severities of clinical diarrhea, vomiting, and anorexia can vary based on the age of the affected pigs [7, 8]. Without adequate passive lactogenic immunity, the mortality rate in neonatal piglets can reach up to 100% [1, 9–11]. Etiological diagnosis relies mainly on molecular tools like PCR and serology, as the clinical signs and enteric lesions associated with TGEV and PEDV are indistinguishable [8, 12, 13].

* Correspondence: pablop@iastate.edu; pablop@iastae.edu
[1]Veterinary Diagnostic Laboratory, 1655 Veterinary Medicine, Iowa State University, 1850 Christensen Drive, Ames, IA 50011, USA
Full list of author information is available at the end of the article

The epidemiology of TGEV is rather complex, and infection in neonates can arise from multiple sources. In addition to swine, it has be documented that cats, dogs, and foxes can host TGEV [14]. The virus can be shed in feces for approximately 18 months, and milk shedding from infected sows can result in vertical transmission. Historically, TGEV infection has followed a seasonal pattern, becoming more prevalent during winter months perhaps due to increased viral survival in colder temperatures and with less exposure to sunlight. TGEV is susceptible to most commercial disinfectants, but resistant to digestive bile and stable at pH 3 [1]. Only one TGEV genotype has been described, however differences in pathogenicity among strains has been reported in field outbreaks, although not confirmed by an experimental study [12].

Infection with TGEV has two different clinical presentations: epidemic and endemic. In epidemics, TGEV enters a naïve herd and all pig categories are affected, particularly piglets that are 1–2 weeks old. The duration of the clinical presentation is short, approximately 3 weeks, and in small, farrow-to-finish herd, the infection can be self-limiting [1, 14]. Endemic disease scenarios, those occurring after the epidemic phase, are observed in farms with incomplete AIAO management or in breeding farms that have a continuous flow of naïve gilts. In breeding herds, the varying levels of humoral and lactogenic immunity lowers piglet mortality, but may lengthen the course of the disease [1].

Since the first description provided in the United States [15], TGEV infections have been reported all over the world. In South America, it has been reported in Colombia [16], Venezuela [17], Bolivia, and is currently seen in Brazil [18]. In Argentina, an episode of high pre-weaning mortality related to *Isospora suis* infection alone or in association with an unknown enteric virus was reported in 1998 [19]. Further studies using negative stain electron microscopy demonstrate the presence of viral particles consistent with coronavirus in feces of pre-weaning diarrheic (34.4%) and post-weaning (10%) piglets [20]. A retrospective histopathological study performed on cases of neonatal diarrhea at our laboratory during 2013 showed that 29% of neonatal diarrhea cases had lesions consistent with viral enteritis [21]. In 2014, a notification of TGEV infection was reported by the National Services of Animal Health of Argentina (SENASA) to the World Organization of Animal Health (OIE). It was detected by serology in a small farm with an apparent morbidity rate of 2.3% without clinical signs [22]. This is an unusual presentation of TGEV infection and might be related to passed infection or interspecies transmission [1].

In order to clarify the situation prior to the first official report of TGEV infection in Argentina, a retrospective study was performed on cases suspected of TGEV-like disease recorded at the Laboratory of Special Veterinary Pathology at the College of Veterinary Sciences, La Plata University. Benchmarking analyses of epidemiological behavior and clinical histories were the criteria for herd selection. Etiological diagnosis was confirmed by electron microscopy (EM), immunohistochemistry (IHC), and in situ hybridization (ISH-RNA) of archived paraffin blocks.

Results

Clinical, pathological and etiological findings

Case 1

Thirteen 1- to 7-day-old piglets with body weights ranging from 1 to 1.5 kg were submitted for pathological investigation. Pigs with clinical diarrhea were dirty, wet, had stained perinea, and showed moderate dehydration characterized by sunken eyes and diffusely pale mucosa. Their small intestinal walls were thin with unremarkable mesenteric lymphatic vessels, and were distended by gas or occasionally contained yellow watery digesta with a pH of 5–6 (Fig. 1a). Their stomachs were empty or contained floccules of undigested milk. No other macroscopic findings were observed. The histopathological evaluation showed a shortening of intestinal villi with a crypt-villous ratio of 1:2 (Fig. 1b), that villi were fused and lined by vacuolated cuboidal or attenuated epithelium, and that the lamina propria was expanded by moderate edema. Microscopic lesions were limited to the jejunum and ileum. Immunohistochemistry and ISH-RNA against TGEV showed strong staining in the epithelium of sections that presented minimal epithelial villous changes (Fig. 1c, d). Conversely, sections with the most severe epithelial damages were IHC negative. All sections evaluated in this case were negative for PEDV and rotavirus.

Case 2

A gross evaluation of five piglets between 2 and 3 days old showed marked dehydration, wet and stool-stained perinea, and poor body conditions (1.15 kg average body weight) (Fig. 2a). The small intestine contained abundant yellow-watery diarrhea with a pH of 5–6 in two pigs and alkaline pH in three pigs (Fig. 2b). A histopathology examination of multiple sections of jejunum and ileum showed mild to moderate villous shortening and fusion (Fig. 2c). The villous enterocytes showed a marked cytoplasmic vacuolization (Fig. 2d). The lamina propria was minimally infiltrated by lymphocytes and plasma cells, and was expanded by edema. The superficial villous enterocytes in the ileum showed strong hybridization signals characterized by active-replicating TGEV (Fig. 2e). No microscopic lesions were seen in the sections of colon. All sections evaluated in this case were negative for PEDV and Rotavirus A. In the ultrathin sections, rounded particles measuring approximately 80 nm in diameter were located in the cytoplasm of the intestinal epithelial cells. The particles were characterized by finely

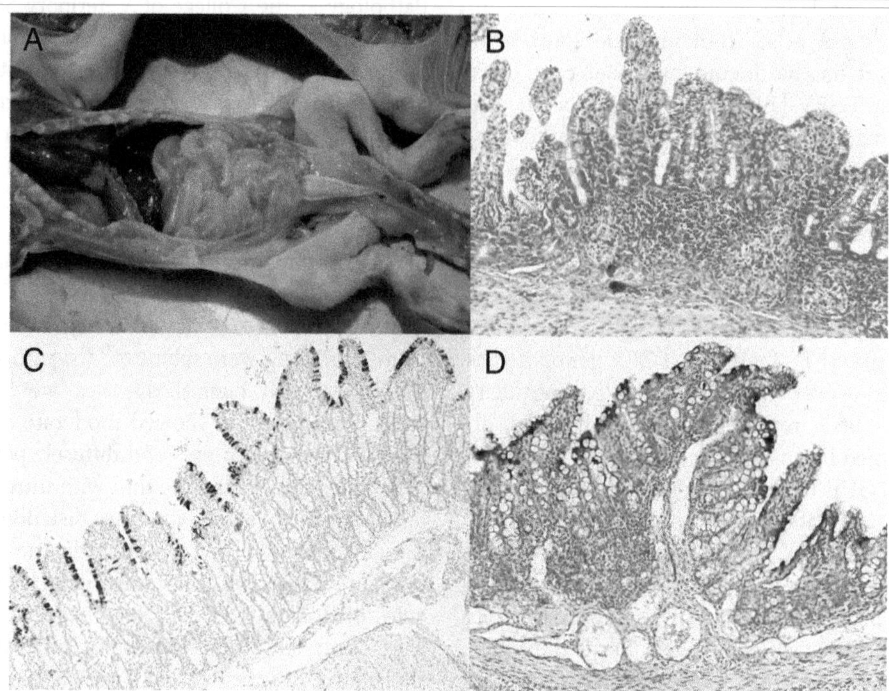

Fig. 1 Gross and histological findings in neonatal piglets affected with TGEV detected in case 1. **a** shows thin small intestinal walls and loops distended by gas containing scant yellow watery digesta. **b** shows histological changes characterized by villous shortening, fusion and moderate submucosal edema. **c** and **d** present viral detection by IHC and ISH-RNA respectively. Note the severity of viral distribution affecting the entire length of the villi

granular electrodense nucleoids with electron lucent centers compatible with complete particles of coronavirus (Fig. 2f).

Case 3

Three neonatal piglets that died naturally presented with fecal stained perinea and were markedly dehydrated. The stomachs were empty, the small intestinal wall was thin/translucent, and scant yellow watery contents were sometimes apparent. The histopathology examination of the jejunum and ileum showed moderate villous fusion and shortening, and the villi were lined by low-cuboidal to flattened/attenuated epithelium (Fig. 3a). The lamina propria was infiltrated by numerous lymphocytes and plasma cells, was expanded by edema, and exhibited lymphangiectasia (Fig. 3b). The TGEV antigen was detected by IHC in multiple sections of the jejunum and ileum (Fig. 3c). No significant lesions were observed in the section of colon. All sections evaluated in this case were negative for PEDV and Rotavirus A.

Discussion

The epidemiological and clinical presentations of outbreaks of neonatal mortality associated with enteritis and the detection of TGEV started in the gestation units. Both gilts and sows showed anorexia, diarrhea, and vomiting before enteric signs were observed in neonatal piglets. However, the prevalence of clinical signs in the breeding stock was low and no mortality was reported. When TGEV enters in a naïve herds, an epizootic form characterized by a 100% mortality of pre-weaning piglets due to diarrhea and dehydration is normally observed [1, 14]. Although in the present study, the farms had a high prevalence of diarrhea in suckling pigs, only farms A and C showed almost 100% neonatal mortality, while in farm B had approximately 20% neonatal mortality. The clinical presentation and epidemiological pattern observed in farm B resembled the TGE endemic form. Although no other etiological diagnosis was confirmed, the low mortality associated with TGEV that was observed in farm B might be the result of previous exposure to PRCV. PRCV infection can confer cross-protection against TGEV, reducing the enteric clinical signs and pre-weaning mortality [12]. Therefore, herds concomitantly infected with PRCV and TGEV develop less severe clinical signs, making the clinical differentiation from other enteric infections such as rotavirus or *E. coli* infections more challenging [1, 14]. Another potential reason for the low mortality rate due to TGE infection that was observed in farm B could be the intermittent viral exposure of the breeding stock that provided partial immunity to the neonatal piglets [14]. In all herds in this study, it was suspected that the virus entered the farms through the subclinically infected replacement animals

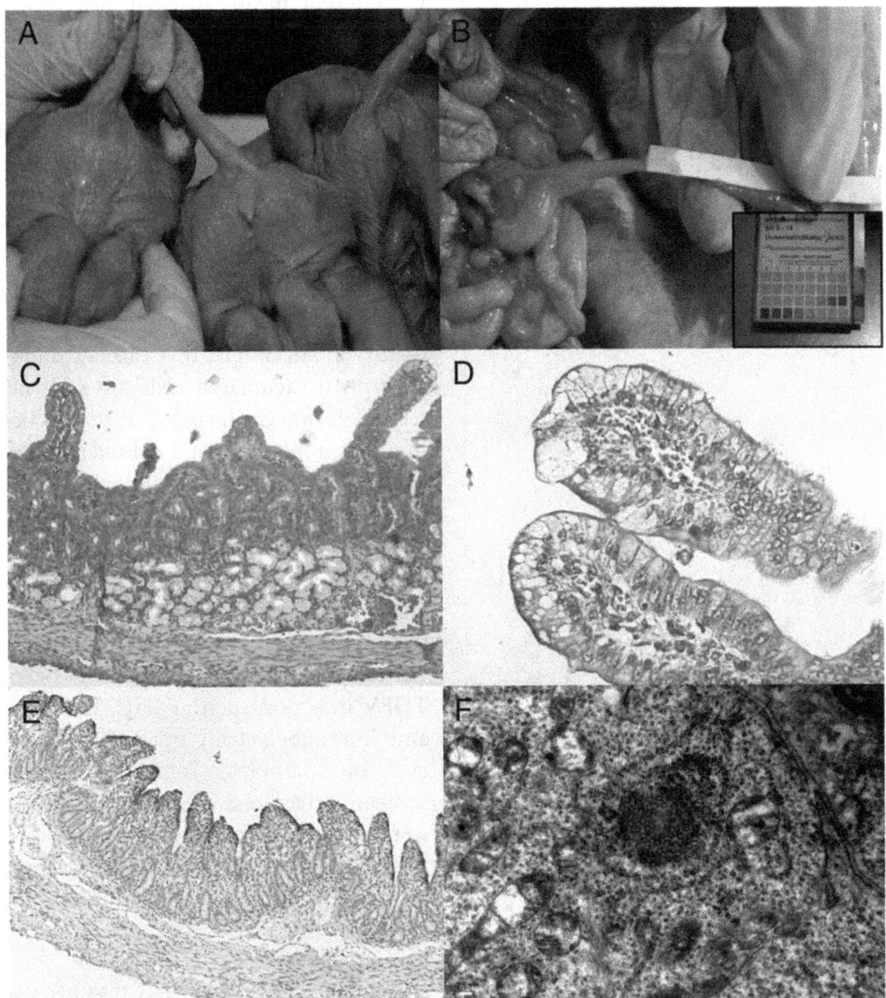

Fig. 2 Gross and histological changes and viral detection in neonatal piglets affected with TGEV in case 2. Piglets presented with stool-stained perineum (**a**). Intestinal loops have thin walled, distended by gas and contain abundant yellow-watery diarrhea with 5–6 pH (**b**). Multiple sections of small intestine showed villous atrophy, blunting and fusion (**c**) with occasional cytoplasmic vacuoles (**d**). The presence of TGE mRNA confirmed in positive ISH-RNA enterocytes through multiple section of jejunum and ileum (**e**). **f** shows intracytoplasmic particles characterized by finely granular electrodense nucleoids with electron lucent centers compatible coronavirus

that were not quarantined, or that the herds were exposed to the virus due to the proximity of a farm to a slaughterhouse (as seen in farm B).

Although the pre-weaning mortality rate in farm B was lower than that in farm A and C, the pre-weaning mortality was higher in this farm than the mortality rate seen in similar production systems due to other enteric causes such us *I suis* [19, 23], *C. perfringes* type A [24], or *C. difficile* [25]. According to a previous study, the pre-weaning mortality due to diarrhea should not exceed 20% [26]. Usually, in porcine CoVs infections, the course of the clinical disease is short and normally does not exceed 3–4 weeks [14] due to the establishment of a rapid herd immunity, as early as one week post-infection. However, in farm B the clinical presentation persisted for approximately 2 months. Potential reinfection due to poor husbandry and incomplete AIAO management are just few potential causes of viral persistence in the environment that can predispose reinfection of the farrowing units.

In clinically affected litters, most of the pigs are dirty, wet, and dehydrated, with diminished body weights. Diarrhea is watery, yellowish, and with an acidic smell from the presence of undigested milk [23]. The mortality rate is inversely correlated with the age of the piglets, reaching 100% in 2–7 day-old piglets. This predisposition is due to the slow replacement rate of villus epithelial cells (around 10 days) in neonatal piglets compared with the replacement rate of 3-week-old piglets (around 2–4 days) [14]. In this study, mortality varied from 20 to 100%.

The jejunum and ileum are the target segments of the small intestine for virus multiplication. However, TGEV infection is segmental, so multiple segments should be

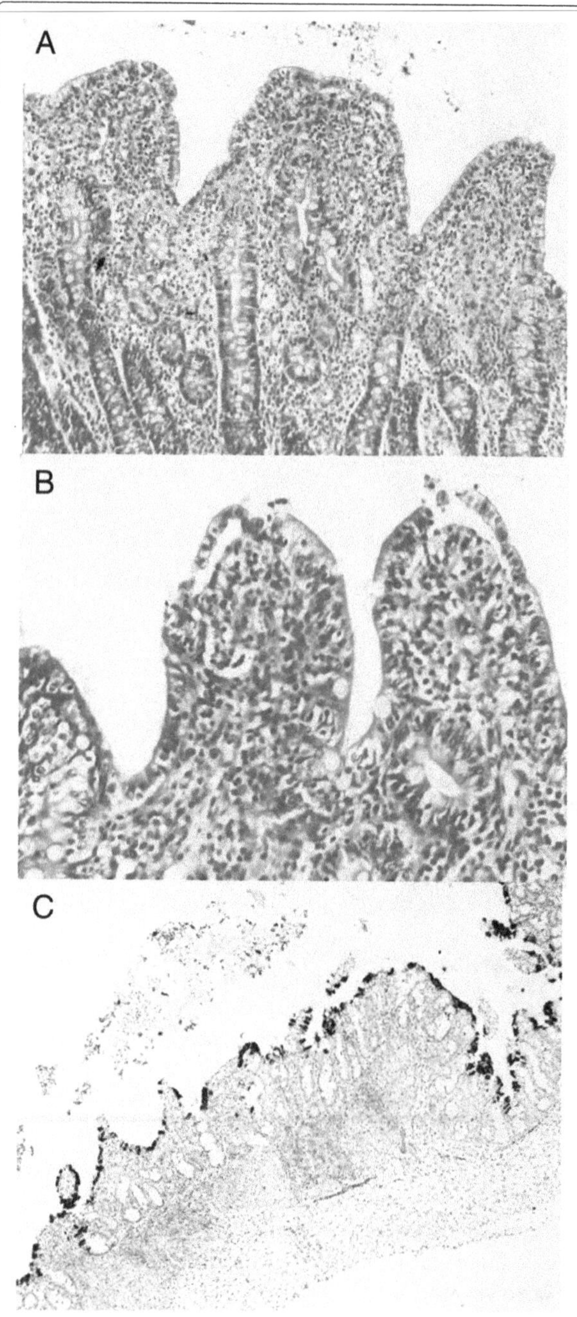

Fig. 3 Histological changes and TGE-IHC detection small intestine in neonatal piglets affected with TGEV in case 3. **a** shows moderate villous atrophy and villi lined by low-cuboidal attenuated epithelium. There is also infiltration of the lamina propria by lymphocytes and plasma cells and minimal edema (**b**). Note the strong immunoreactivity against TGEV in superficial enterocitos (**c**)

included for histopathology or etiological diagnoses in situ such as IHC or ISH-RNA. Due to the retrospective nature of this study, fixed tissue was used to confirm the presence of TGEV and the morphological changes consistent with viral infection in the intestinal mucosa. It is important to highlight that although TGEV was detected by different means in each farm, due to the segmental nature of the lesions and viral distribution, viral ARN or viral antigen was not detected in all of the intestinal segments that were evaluated. In piglets that are less than 2 weeks old, the reduction of villus length and the fusion of jejunum and proximal ileum are the main histological changes [27]. The normal villous/crypt length ratio is 7:1, however, after 24–48 h post-infection, the villous/crypt ratio can be reduced to 2:1 or 1:1 [27]. Since TGEV replicates in mature absorptive epithelial cells [28], a false negative diagnosis can be observed by IHC or ISH in specimens that display villous shortening. Few absorptive vacuolated cells are seen normally in the intestine, however, during TGEV infection, they are found in great numbers with a cuboidal shape [27]. In endemic TGEV infection, histopathological diagnosis is more difficult because only 25% of pigs display typical TGE lesions. In addition, immunofluorescence tests and IHC often fail in endemic farms due to a low number of enterocytes in the TGEV-infected because of partial protection conferred by colostral antibodies [23, 29]. Different diagnostic techniques have been used to detect TGEV infection such as IHC, ISH, electron microscopy, and immunoelectron microscopy and PCR [13]; however, histopathology remains the most useful tool for screening diagnosis [18]. In this study, although all cases were selected using clinical features and epidemiological information, the histological evaluation consistently showed lesions compatible with viral infection.

Conclusions

The application of IHC and ISH-RNA on archived paraffin blocks from cases of neonatal diarrhea with high morbidity and mortality allowed retrospective identification of TGEV infection. Diagnosis of TGEV infection in endemic regions, such as in Argentina, is complicated due to the epidemiological distribution and clinical signs that might be masked with other enteric infections. Further studies are necessary to determine the true prevalence of this pathogen and the correlation with neonatal enteric cases observed in confined production systems.

Methods

Case selection and clinical history

Case selection was based on the clinical history reported by the farm managers and referring veterinarians. The selection criteria was a sudden increase in mortality that included, significant increment of more than 2 SD from average pre-weaning mortality and last for a period of a week. In addition reported mortality should be associated with the presence of watery diarrhea in 1- to 21-day-old piglets that did not respond to antibiotics. First screening of cases was done by histopathological evaluation, and only cases presenting features of viral enteritis with no other detected pathogen were included. Based on these criteria, three clinical cases were identified from 2010 to 2015.

Case 1

A 170-sow farrow-to-finishing herd located in Buenos Aires served as the first case. The farm produced its own replacement breeding stock, however, two months before the outbreak, gilts were introduced from a breeding company. The parity of the breeding stock was distributed as 40% gilts while the rest of the reproductive stock parity varied from 2 to 6. On September 2011, approximately 10% of the pregnant sows presented acute vomiting while 30% of the pregnant sows presented acute diarrhea. During the period when the sows showed gastro-enteric clinical signs, 2- to 4-day-old piglets presented vomiting (75–80%) and diarrhea (90%), and the mortality rate of suckling pigs reached 90%. The course of the disease in both breeder stock and piglets lasted for approximately three weeks.

Case 2

Case 2 involved a one-site herd of 350 sows with its own replacement gilts and the following parity distribution: 20% gilts, 40% parity 1–2, and 27.3% distributed amongst parity 3–5. Boars were purchased from a breeding company and were incorporated into the reproductive herd without quarantine. The farm was located within a few miles of a swine slaughterhouse in Buenos Aires. In February 2012, the pregnant sows showed anorexia (14–30%) and diarrhea (1%) associated with heat returns and abortions (3.3%). In the farrowing houses, approximately 100% of the lactating sows presented with anorexia. Pre-weaning mortality associated with the presence of diarrhea varied from 16.5% at the beginning of the outbreak to 27.9% 3 to 4 weeks after the initial clinical signs. An anatomopathological evaluation showed that 93.6% of the total pre-weaning mortality was due to diarrhea.

Case 3

A one-site herd of 400 sows was the subject of Case 3. In July 2013, two boars were located close to the gestation unit. A week later, gestation sows showed anorexia (16.8%) and diarrhea (5.3%). Thereafter, in the gilts, diarrhea was evident in the nursery (3–7%) and fattener (5–23%). Two-day-old piglets showed watery diarrhea (100%) with a mortality rate of 95%. Affected piglets died from severe dehydration within two days of the onset of clinical signs. The course of the disease lasted approximately two months with an overall pre-weaning mortality of 50% during that period.

Pathological studies

At the onset of the outbreak, clinically affected suckling pigs were submitted for postmortem examination. Tissue samples from different organs, including multiple segments of small intestine (duodenum, jejunum, and ileum) and large intestine, were fixed in 10% buffered formalin, processed for routine histopathologic examination, and stained with hematoxylin and eosin.

Etiological diagnosis in tissues and feces by immunohistochemistry, in situ hybridization, and electron microscopy

Immunohistochemestry IHC was carried-out briefly to differentiate TGEV [30] from PEDV [31]. Monoclonal antibodies against TGEV (OSU: #.14E3-3C) and PEDV (OSU 6C8) at dilutions of 1:8000 were used. Antigen retrieval was performed with humid heat and revealed with peroxidase (Novocastra[a], Leica Biosystems, IL, USA). To rule out other potential viral enteritis, Rotavirus A was evaluated in all sections by IHC [32]. Rotavirus IHC was performed using a monoclonal antibody against Rotavirus A (Santa Cruz: sc-101363) at a 1:2000 dilution. Antigen retrieval was performed with Epitope Retrieval Solution 2 for 20 min, as programmed on Leica Bond III, and revealed with PowerVision Poly-HRP anti-Mouse (Leica PV 6113). All samples were tested in duplicate and sections were controlled appropriately for TGEV, PEDV and Rotavirus A with positive and negative controls (Additional file 1: Figure S1).

In situ hybridization ISH-RNA was developed through the RNAScope platform (Advanced Cell Diagnostics, Inc., CA), targeting the specific reverse complementary nucleotide sequence of the TGE viral mRNA (716–1859 region of spike gene, GenBank: KC609371.1). Therefore, positive hybridization signals represent a metabolically-active virus characterized by the TGE mRNA encoding spike protein. Unstained paraffin tissue sections were processed as previously described [33]. Briefly, tissues were deparaffinized and treated with hydrogen peroxide at room temperature for 10 min. The slides were hybridized using a hybridization buffer, and sequence amplifiers were added. The red colorimetric staining detected the TGE hybridization signal, and counterstaining occurred with hematoxylin.

Representative sections of small intestine were fixed in 2.5% glutaraldehyde and 2% paraformaldehyde in 0.1 M, pH 7.4 phosphate buffer (PBS) and post-fixed in 1% osmium tetroxide in PBS. After dehydration in an alcohol series, the fragments were embedded in epoxy resin, Quetol 812 (Nisshin EM Co., Ltd., Tokyo). Ultrathin sections were cut, double-stained with uranyl acetate-lead citrate, and observed under a JEM-1200EX (JEOL Co. Ltd., Tokyo).

Additional file

Additional file 1: Figure S1. Panel of pathogens used as control for detection of TGEV by immunohistochemistry. Row one include immunostaining of TGEV clinical cases against TGEV, PEDV, and rotavirus specific antibodies. A moderate to severe immunostaining is observe only reacting against TGEV. In row two include positive controls for each pathogen detected by immunohistochemistry. Row three present section tested negative by PCR for TGEV, PEDV, and Rotavirus that were used as negative control of the immunohistochemistry techniques. (PPTX 3545 kb)

Abbreviations

AIAO: All in all out; CoVs: Coronaviruses; EM: Electron microscopy; HRP: Horseradish peroxidase; IHC: Immunohistochemistry; ISH: In situ hybridization; mRNA: Messenger ribonucleic acid; OIE: World Organization of Animal health; PBS: Phosphate buffer saline; PCR: Polymerase chain reaction; PDCoV: Porcine deltacoronavirus; PEDV: Porcine epidemic diarrhea virus; SENASA: Argentina services of animal health; TGE: Porcine transmissible gastroenteritis; TGEV: Porcine transmissible gastroenteritis virus

Authors' contributions

PEP, CJP, MAQ performed histological examination, data analysis and conclusion and were the major contributors in writing the manuscript; DMM, Performed IHC; FV, Performed ISH; MIL, RS, JAC, EMP, AA, LVA field data collection and case identification. All authors read and approved the final manuscript.

Consent for publication

Not applicable.

Competing interests

The authors declare that they have no competing interests.

Author details

[1]Veterinary Diagnostic Laboratory, 1655 Veterinary Medicine, Iowa State University, 1850 Christensen Drive, Ames, IA 50011, USA. [2]Laboratorio de Patología Especial Veterinaria FCV-UNLP, Calle 60 y 118 S/N (1900), La Plata, Buenos Aires, Argentina. [3]HIPRA Argentina, Saenz Peña R. Pte. Av 1110, Capital Federal, Argentina. [4]DILACOT-SENASA, Av A Fleming 1653, Martinez, Buenos Aires, Argentina. [5]EEA Marcos Juaréz, INTA, CONICET, Ruta 12 km. 3 (2580), Marcos Juárez, Córdoba, Argentina. [6]Veterinary Diagnostic Laboratory, University of Minnesota, 1333 Gortner Ave, St Paul, MN, USA.

References

1. Saif LJ, Pensaert MB, Sestak K, Yeo SG, Jung K. Coronaviruses. In: Zimmerman JJ, Karriker LA, Ramirez A, Schwartz KJ, Stevenson GW, editors. Diseases of Swine. 10th ed. Ames: Willey; 2012. p. 501–23.
2. Wang L, Byrum B, Zhang Y. New variant of porcine epidemic diarrhea virus, United States, 2014. Emerg Infect Dis. 2014;20(5):917–9.
3. Quiroga MA, Cappuccio J, Pineyro P, Basso W, More G, Kienast M, Schonfeld S, Cancer JL, Arauz S, Pintos ME, et al. Hemagglutinating encephalomyelitis coronavirus infection in pigs, Argentina. Emerg Infect Dis. 2008;14(3):484–6.
4. Laude H, Van Reeth K, Pensaert M. Porcine respiratory coronavirus: molecular features and virus-host interactions. Vet Res. 1993;24(2):125–50.
5. González JM, Gomez-Puertas P, Cavanagh D, Gorbalenya AE, Enjuanes L. A comparative sequence analysis to revise the current taxonomy of the family Coronaviridae. Arch Virol. 2003;148(11):2207–35.
6. Chen Q, Gauger P, Stafne M, Thomas J, Arruda P, Burrough E, Madson D, Brodie J, Magstadt D, Derscheid R, et al. Pathogenicity and pathogenesis of a United States porcine deltacoronavirus cell culture isolate in 5-day-old neonatal piglets. Virology. 2015;482:51–9.
7. McCluskey BJ, Haley C, Rovira A, Main R, Zhang Y, Barder S. Retrospective testing and case series study of porcine delta coronavirus in U.S. swine herds. Prev Vet Med. 2016;123:185–91.
8. Song D, Moon H, Kang B. Porcine epidemic diarrhea: a review of current epidemiology and available vaccines. Clin Exp Vaccin Res. 2015;4(2):166–76.
9. Stevenson GW, Hoang H, Schwartz KJ, Burrough ER, Sun D, Madson D, Cooper VL, Pillatzki A, Gauger P, Schmitt BJ, et al. Emergence of porcine epidemic diarrhea virus in the United States: clinical signs, lesions, and viral genomic sequences. J Vet Diagn Investig. 2013;25(5):649–54.
10. Jung K, Saif LJ. Porcine epidemic diarrhea virus infection: etiology, epidemiology, pathogenesis and immunoprophylaxis. Vet J. 2015;204(2):134–43.
11. Poonsuk K, Giménez-Lirola LG, Zhang J, Arruda P, Chen Q, Correa da Silva Carrion L, Magtoto R, Pineyro P, Sarmento L, Wang C, et al. Does circulating antibody play a role in the protection of piglets against porcine epidemic diarrhea virus? PLoS One. 2016;11(4):e0153041.
12. Kim L, Hayes J, Lewis P, Parwani AV, Chang KO, Saif LJ. Molecular characterization and pathogenesis of transmissible gastroenteritis coronavirus (TGEV) and porcine respiratory coronavirus (PRCV) field isolates co-circulating in a swine herd. Arch Virol. 2000;145(6):1133–47.
13. Woods RD. Development of PCR-based techniques to identify porcine transmissible gastroenteritis coronavirus isolates. Can J Vet Res. 1997;61(3): 167–72.
14. Sestak K, Saif LJ. Porcine coronavirus. In: Trends in emerging viral infections of swine. Edited by Morilla A, Yoon K-J, Zimmerman JJ. Ames: Iowa State Press; 2002;321–30.
15. Doyle LP, Hutchings LM. A transmissible gastroenteritis in pigs. J Am Vet Med Assoc. 1946;108:257–9.
16. Piñeros R, Mogollón Galvis JD. Coronavirus en porcinos: importancia y presentación del virus de la diarrea epidémica porcina (PEDV) en Colombia. Rev Med Vet. 2015;29:73–89.
17. Marin C, Rolo M, López N, Álvarez L, Castaños H, Sifontes S. Detección de focos de gastroenteritis transmisible en Venezuela. Vet Trop. 1985;10:35–42.
18. Martins AMCRPF, Bersano JG, Ogata R, Amante G, Nastari BDB, Catroxo MHB. Diagnosis to detect porcine transmissible gastroenteritis virus (TGEV) by optical and transmission electron microscopy techniques. Int J Morphol. 2013;31:706–15.
19. Perfumo C, Venturini L, Sanguinetti H, Aguirre J, Armocida A, Petruccelli M, Moredo F. Infección por Isospora suis sola o asociada a virus entéricos como causa de alta morbimortalidad en lechones lactantes. Revta Med Vet. 1998;79:264–8.
20. Aguirre JI, Petruccelli MA, Armocida AD, Moredo FS, Risso M, Venturini L, Idiart JR, Perfumo CJ. Diarrea en lechones lactantes y posdestete de cuatro criaderos intensivos de la provincia de Buenos Aires, Argentina: identificación e índice de detección de partículas virales en materia fecal por microscopía electrónica. Analecta Vet. 2000;20(2):16–21.
21. Chavez F, Pérez E, Barrales H, Zignago F, Lozada M, Quiroga M, Machuca M, Cappuccio J, Perfumo C. Análisis de los cuadros entéricos en cerdos remitidos al Laboratorio de Patología Especial Veterinaria (2013). In Proceeding of: Memorias XII Congreso Nacional de Producción Porcina: 12–15 agosto 2014. Mardel Plata; 2014, 2014. p. 178.
22. Carné LÁ. Transmissible gastroenteritis, Argentina. In: OIE, editor. Servicio Nacional de Sanidad y Calidad Agroalimentaria (SENASA), Ministerio de Agricultura, Ganadería y Pesca: World Organization of Animal health; 2014.
23. Dewey CE, Carman S, Hazlett M, Dreumel TV, Smart NE. Endemic transmissible gastroenteritis: difficulty in diagnosis and attempted confirmation using a transmission trial. J Swine Health Prod. 1999;7(2):73–8.
24. Sanz MG, Venturini L, Assis RA, Uzal F, Risso MA, Idiart JR, Perfumo CJ. Fibrinonecrotic enteritis of piglets in a commercial farm: a postmortem study of the prevalence and the role of lesion associated agents Isospora suis and Clostridium perfringens. Pesqui Vet Bras. 2007;27:297–300.
25. Cappuccio JA, Quiroga MA, Moredo FA, Canigia LF, Machuca M, Capponi O, Bianchini A, Zielinski G, Sarradell J, Ibar M, et al. Neonatal piglets mesocolon edema and colitis due to Clostridium difficile infection: prevalence, clinical disease and pathological studies. Braz J Vet Pathol. 2009;2(1):35–40.

26. Sanz M, Sernia C, Viale G, Bustos L, Sanguinetti H, Risso M, Venturini L, Idiart J, Perfumo C. Why Should Piglets Dead at the Pre-weaning Period be Post-mortem Examined and Statistically Analysed at Weekly Intervals? In Proceeding of: 32nd Annual Meeting American Association of Swine Practitioners: February 24-27, 2001. Nashville; 2001. p. 69–74.
27. Hooper BE, Haelterman EO. Lesions of the gastrointestinal tract of pigs infected with transmissible gastroenteritis. Can J Comp Med. 1969;33(1):29–36.
28. Moeser AJ, Blikslager AT. Mechanisms of porcine diarrheal disease. J Am Vet Med Assoc. 2007;231(1):56–67.
29. Pritchard GC. Transmissible gastroenteritis in endemically infected breeding herds of pigs in East Anglia, 1981-85. Vet Rec. 1987;120(10):226–30.
30. Shoup DI, Swayne DE, Jackwood DJ, Saif LJ. Immunohistochemistry of transmissible gastroenteritis virus antigens in fixed paraffin-embedded tissues. J Vet Diagn Investig. 1996;8(2):161–7.
31. Kim O, Chae C, Kweon C-H. Monoclonal antibody-based Immunohistochemical detection of porcine epidemic diarrhea virus antigen in formalin-fixed, paraffin-embedded intestinal tissues. J Vet Diagn Investig. 1999;11(5):458–62.
32. Magar R, Larochelle R. Immunohistochemical detection of porcine rotavirus using Immunogold silver staining (IGSS). J Vet Diagn Investig. 1992;4(1):3–7.
33. Wang F, Flanagan J, Su N, Wang L-C, Bui S, Nielson A, Wu X, Vo H-T, Ma X-J, Luo Y. RNAscope: a novel in situ RNA analysis platform for formalin-fixed, paraffin-embedded tissues. J Mol Diagn. 2012;14(1):22–9.

Development and evaluation of serotype-specific recombinase polymerase amplification combined with lateral flow dipstick assays for the diagnosis of foot-and-mouth disease virus serotype A, O and Asia1

Hongmei Wang[1], Peili Hou[1], Guimin Zhao[1], Li Yu[2], Yu-wei Gao[3*] and Hongbin He[1*]

Abstract

Background: Foot-and-mouth disease (FMD) caused by foot-and-mouth disease virus (FMDV) is one of the most highly infectious diseases in livestock, and leads to huge economic losses. Early diagnosis and rapid differentiation of FMDV serotype is therefore integral to the prevention and control of FMD. In this study, a series of serotype-specific reverse transcription recombinase polymerase amplification assays combined with lateral flow dipstick (RPA-LFD) were establish to differentiate FMDV serotypes A, O or Asia 1, respectively.

Results: The serotype-specific primers and probes of RPA-LFD were designed to target conserved regions of the FMDV VP1 gene sequence, and three primer and probe sets of serotype-specific RPA-LFD were selected for amplification of FMDV serotypes A, O or Asia 1, respectively. Following incubation at 38 °C for 20 min, the RPA amplification products could be visualized by LFD. Analytical sensitivity of the RPA assay was then determined with ten-fold serial dilutions of RNA of VP1 gene and the recombinant vector respectively containing VP1 gene from FMDV serotypes A, O or Asia1, the detection limits of these assays were 3 copies of plasmid DNA or 50 copies of viral RNA per reaction. Moreover, the specificity of the assay was assessed, and there was no cross reactions with other viruses leading to bovine vesicular lesions. Furthermore, 126 clinical samples were respectively detected with RPA-LFD and real-time PCR (rPCR), there was 98.41% concordance between the two assays, and two samples were positive by RPA-LFD but negative in rPCR, these were confirmed as FMDV-positive through viral isolation in BHK-21 cells. It showed that RPA-LFD assay was more sensitive than the rPCR method in this study.

Conclusion: The development of serotype-specific RPA-LFD assay provides a rapid, sensitive, and specific method for differentiation of FMDV serotype A, O or Asia1, respectively. It is possible that the serotype-specific RPA-LFD assay may be used as a integral protocol for field detection of FMDV.

Keywords: FMDV, Serotype-specific, Recombinase polymerase amplification, Lateral flow dipstick

* Correspondence: gaoyuwei@gmail.com; hongbinhe@sdnu.edu.cn
[3]Key Laboratory of Jilin Province for Zoonosis Prevention and Control, Military Veterinary Research Institute of Academy of Military Medical Sciences, Changchun 130122, China
[1]Ruminant Diseases Research Center, Key Laboratory of Animal Resistant Biology of Shandong, College of Life Sciences, Shandong Normal University, Jinan 250014, China
Full list of author information is available at the end of the article

Background

Foot-and-mouth disease (FMD) is an acute, highly contagious disease that affects cloven-hoofed animals, often resulting in huge economical losses in terms of trade and animal productivity. Recent outbreaks of FMD in Taiwan, Japan, South Korea and the United Kingdom have directly caused the culling of millions of animals, compensated heavily by the government [1, 2]. The responsible virus, FMD virus (FMDV), was a single-stranded positive-sense RNA virus belonging to the Aphthovirus genus in the family Picornaviridae. There were seven serotypes including O, A, C, Asia 1, and South African Territories (SAT) 1, 2 and 3, which together manifest a distinct geographical distribution [3]. FMDV serotype A, C and O are widely distributed across the world while Asia 1 and SAT 1–3 mainly occur in Asia and Africa, respectively. Several outbreaks of FMD serotype Asia 1, O and A have been recorded in mainland provinces of Southern China during 1999–2013 [4–6]. Early diagnosis of FMDV is therefore essential to providing valuable epidemiological information, and initiating the appropriate prevention and control strategies.

FMDV can be detected from blood, esophageal-pharyngeal fluid, nasal fluid, saliva, and other excretions of FMDV infected animals before clinical symptoms [7, 8] start to show. Currently, there are three typical assays for FMDV diagnosis including virus isolation, antigen enzyme-linked immunosorbent assay (Ag-ELISA) and real-time RT-PCR (rRT-PCR) used in FMDV reference laboratories [9]. However, these diagnostic tests require special equipment and professionally trained personnel. Another alternative propose is to use isothermal assays for diagnosis of FMDV. To date, there are four isothermal assays to detect FMDV: reverse transcription loop-mediated isothermal amplification (RT-LAMP) [10, 11], reverse transcription recombinase polymerase amplification (RT-RPA) [12], and nucleic acid sequence based amplification [13, 14]. RT-LAMP and RT-RPA have also been used to distinguish various serotypes in clinical samples [11, 12]. However, LAMP assay needs more primers than PRA, leading to longer amplicons and difficult designs in cases of highly variable viruses.

The RPA method is probably the one promising direction capable of rapid diagnosis of many different pathogens [12, 15–18]. The amplification relies on recombinase, single stranded binding protein, and strand displacing DNA polymerase at a constant temperature. The RPA products could be analyzed with gel electrophoresis, fluorescence monitoring based on probes, or simple visualization with a lateral flow dipstick (LFD) [19–22].

In the present study, a reverse transcription serotype-specific RPA-LFD assay was established, and evaluated as a field method for diagnosis and typing of FMDV serotypes A, Asia 1 or O, respectively.

Results

Design and optimization of serotype-specific RPA primers and probe

The TwistAmp nfo reactions were performed to screen the candidate primer/probes for the RPA-LFD assay, the products were analyzed on 2% agarose gel. A6, As9, O6 sets of the primer/probes were respectively screened as serotype-specific RPA primers and probe for FMDV serotypes A, Asia-1, or O. They respectively produced 346 bp and 334 bp, 286 bp and 182 bp, 231 bp and 190 bp amplification products and their RPA-LFD test line appeared faster and darker than other sets within 5 min (Table 1 and Fig. 1).

Besides that, different reaction conditions were optimized for the serotype-specific RPA assay. Results showed that a reaction temperature at 38 °C (Fig. 2a) and a incubation time of 20 min or longer (Fig. 2b) can best promote the amplification efficiency. Thus, the amplification reaction for the serotype-specific RPA assay should be carried out at 38 °C for 20 min.

Sensitivity and specificity of the FMDV serotype-specific RPA reaction

To determine the sensitivity of RPA assay, positive vector and RNA standard of three FMDV serotypes were diluted from 3×10^6 to 3×10^0 copies/μL and 5×10^7 to 5×10^1 molecules/μL, respectively. RNA was reversely transcribed to cDNA. All reverse transcriptional cDNA and positive vectors of each dilution were respectively used as templates in the RPA reactions. All RPA reaction products were than respectively tested on LFD, and the limit of detection with the RPA-LFD was 3×10^0 dilution for positive vector (Fig. 3a) and 5×10^1 dilution for RNA standard (Fig. 3b).

The other viral pathogens with similar clinical signs, including BEV, BVDV, BEFV, VSV, IBRV, and SVDV, were used to assess the specificity of the assay. IBRV DNA and other viral cDNA were respectively detected with the RPA-LFD. There was no cross reaction with the other bovine viral pathogens with similar clinical signs (Fig. 4). The cDNA of other FMDV epidemic virus strains in China were used for detection of serotype-specific RPA-LFD. There was no cross-reactivity with different serotype strains of FMDV, so the specific primer and probe sets could differentiate the corresponding serotype virus.

Performance of FMDV serotype-specific RPA-LFD assay on clinical samples

To evaluate the diagnostic sensitivity of the FMDV RPA-LFD assay, cDNA obtained from each specimen of 126 clinical samples were detected with RPA-LFD and rPCR, respectively. RPA-LFD identified 41 negative and 85 positive samples (25 vesicular material, 14 saliva, 10

Table 1 The sequences of primers and probes designed for screening in the study

Primer/probe set Name		Sequence(5′ → 3′)	The location on accession number	Product sizes (bp)
	F2	ATGGAGCACCTGAGGCAGCACTGGACAACA	3452–3481	221
A1	P1	[FAM]CACTGGACAACACGAGCAACCCCACTGCTTA[dSpacer] TATAAAGCACCGTT CACA[C3-spacer]	3470–3519	203
	R1	[Biotin]CGTTGAGAAGGGCACAGTCGTATTGAAACA	3643–3672	
	F2	ATGGAGCACCTGAGGCAGCACTGGACAACA	3452–3481	258
A2	P1	[FAM]CACTGGACAACACGAGCAACCCCACTGCTTA[dSpacer] TATAAAGCACCGTT CACA[C3-spacer]	3470–3519	250
	R2	[Biotin]GCACGAGGAGTTCTTGGATCTCCGTGGCTC	3680–3709	
	F2	ATGGAGCACCTGAGGCAGCACTGGACAACA	3452–3481	352
A3	P1	[FAM]CACTGGACAACACGAGCAACCCCACTGCTTA[dSpacer] TATAAAGCACCGTT CACA[C3-spacer]	3470–3519	229
	R3	[Biotin]GCAGGGGCAATAATTTTCTGCTTGTGTCTG	3774–3803	
	F1	CACCTGAGGCAGCACTGGACAACACGAGCAA	3458–3488	215
A4	P1	[FAM]CACTGGACAACACGAGCAACCCCACTGCTTA[dSpacer] TATAAAGCACCGTT CACA[C3-spacer]	3470–3519	203
	R1	[Biotin]GCCGTAGTTGAAGGAGGCAGGAAGCTGTGC	3643–3672	
	F1	CACCTGAGGCAGCACTGGACAACACGAGCAA	3458–3488	252
A5	P1	[FAM] CACTGGACAACACGAGCAACCCCACTGCTTA[dSpacer] TATAAAGCACCGTT CACA[C3-spacer]	3470–3519	240
	R2	[Biotin]GCACGAGGAGTTCTTGGATCTCCGTGGCTC	3680–3709	
A6 (Optimal set)	F1	CACCTGAGGCAGCACTGGACAACACGAGCAA	3458–3488	346
	P1	[FAM]CACTGGACAACACGAGCAACCCCACTGCTTA[dSpacer] TATAAAGCACCGTT CACA[C3-spacer]	3470–3519	334
	R3	[Biotin]GCAGGGGCAATAATTTTCTGCTTGTGTCTG	3774–3803	
	F3	TGAGGCAGCACTGGACAACACGAGCAACCC	3462–3461	211
A7	P1	[FAM]CACTGGACAACACGAGCAACCCCACTGCTTA[dSpacer] TATAAAGCACCGTT CACA[C3-spacer]	3470–3519	203
	R1	[Biotin]GCCGTAGTTGAAGGAGGCAGGAAGCTGTGC	3643–3672	
	F3	TGAGGCAGCACTGGACAACACGAGCAACCC	3462–3461	248
A8	P1	[FAM]CACTGGACAACACGAGCAACCCCACTGCTTA[dSpacer] TATAAAGCACCGTT CACA[C3-spacer]	3470–3519	240
	R2	[Biotin]GCACGAGGAGTTCTTGGATCTCCGTGGCTC	3680–3709	
	F3	TGAGGCAGCACTGGACAACACGAGCAACCC	3462–3461	342
A9	P1	[FAM]CACTGGACAACACGAGCAACCCCACTGCTTA[dSpacer] TATAAAGCACCGTT CACA[C3-spacer]	3470–3519	334
	R3	[Biotin]GCAGGGGCAATAATTTTCTGCTTGTGTCTG	3774–3803	
	F1	AAAAGCAACCCATTACCCGCCTGGCACTCC	3582–3611	285
As1	P1	[FAM]ACCCGCCTGGCACTCCCTTACACCGCTCCC[dSpacer]ACCGTGTGCTTGCA ACAGT[C3-spacer]	3596–3645	271
	R2	[Biotin]GACTCTTCCCCGTAGGTTGTCTTCCCGTTG	3837–3866	
As2	F1	AAAAGCAACCCATTACCCGCCTGGCACTCC	3582–3611	145
	P1	[FAM]ACCCGCCTGGCACTCCCTTACACCGCTCCC[dSpacer]ACCGTGTGCTTGCA ACAGT[C3-spacer]	3596–3645	131
	R3	[Biotin]GGGAGTGCCAGGCGGGTAATGGGTTGCTTT	3697–3726	
	F2	AACCCAACCGCCTACCAAAAGCAACCCATT	3566–3595	153
As3	P1	[FAM]ACCCGCCTGGCACTCCCTTACACCGCTCCC[dSpacer]ACCGTGTGCTTGCA ACAGT[C3-spacer]	3596–3645	123
	R1	[Biotin]CGGTGTAAGGGAGTGCCAGGCGGGTAATGG	3689–3718	

Table 1 The sequences of primers and probes designed for screening in the study *(Continued)*

Primer/probe set Name		Sequence(5′→3′)	The location on accession number	Product sizes (bp)
	F2	AACCCAACCGCCTACCAAAAGCAACCCATT	3566–3595	301
As4	P1	[FAM]ACCCGCCTGGCACTCCCTTACACCGCTCCC[dSpacer]ACCGTGTGCTTGCA ACAGT[C3-spacer]	3596–3645	271
	R2	[Biotin]GACTCTTCCCCGTAGGTTGTCTTCCCGTTG	3837–3866	
	F2	AACCCAACCGCCTACCAAAAGCAACCCATT	3566–3595	161
As5	P1	[FAM]ACCCGCCTGGCACTCCCTTACACCGCTCCC[dSpacer]ACCGTGTGCTTGCA ACAGT[C3-spacer]	3596–3645	131
	R1	[Biotin]CGGTGTAAGGGAGTGCCAGGCGGGTAATGG	3689–3718	
	F3	CGAATCAGCAGACCCAGTTACCACCACAGT	3274–3303	445
As6	P2	[FAM]TGAAACTCACACAGCTCAAGAACACCCAAACT[dSpacer] TTGATCTTATGCAA ATC[C3-spacer]	3378–3427	341
	R1	[Biotin]CGGTGTAAGGGAGTGCCAGGCGGGTAATGG	3689–3718	
	F3	CGAATCAGCAGACCCAGTTACCACCACAGT	3274–3303	593
As7	P2	[FAM]TGAAACTCACACAGCTCAAGAACACCCAAACT[dSpacer] TTGATCTTATGCAA ATC[C3-spacer]	3378–3427	489
	R2	[Biotin]GACTCTTCCCCGTAGGTTGTCTTCCCGTTG	3837–3866	
	F3	CGAATCAGCAGACCCAGTTACCACCACAGT	3274–3303	453
As8	P2	[FAM]TGAAACTCACACAGCTCAAGAACACCCAAACT[dSpacer] TTGATCTTATGCAA ATC[C3-spacer]	3378–3427	349
	R3	[Biotin]GGGAGTGCCAGGCGGGTAATGGGTTGCTTT	3697–3726	
As9 (Optimal set)	F3	CGAATCAGCAGACCCAGTTACCACCACAGT	3274–3303	286
	P2	[FAM]TGAAACTCACACAGCTCAAGAACACCCAAACT[dSpacer] TTGATCTTATGCAA ATC[C3-spacer]	3378–3427	182
	R4	[Biotin]GAGAAGTAGTACGTCGCAGACCGAAGTAGCG	3530–3559	
	F3	CGAATCAGCAGACCCAGTTACCACCACAGT	3274–3303	445
As10	P1	[FAM]ACCCGCCTGGCACTCCCTTACACCGCTCCC[dSpacer]ACCGTGTGCTTGCA ACAGT[C3-spacer]	3378–3427	341
	R1	[Biotin]CGGTGTAAGGGAGTGCCAGGCGGGTAATGG	3689–3718	
	F3	CGAATCAGCAGACCCAGTTACCACCACAGT	3274–3303	593
As11	P1	[FAM]ACCCGCCTGGCACTCCCTTACACCGCTCCC[dSpacer]ACCGTGTGCTTGCA ACAGT[C3-spacer]	3596–3645	271
	R2	[Biotin]GACTCTTCCCCGTAGGTTGTCTTCCCGTTG	3837–3866	
	F1	CAACACCACCAACCCAACGGCGTACCATAA	3570–3599	161
O1	P1	[FAM]CGTACCATAAGGCGCCGCTTACCCGGCTTA[dSpacer] ATTGCCCTACACGG CACCA[C3-spacer]	3590–3639	141
	R1	[Biotin]GAGCCAGCACTTGGAGATCGCCTCTCACGT	3701–3730	
	F1	CAACACCACCAACCCAACGGCGTACCATAA	3570–3599	343
O2	P1	[FAM]CGTACCATAAGGCGCCGCTTACCCGGCTTA[dSpacer] ATTGCCCTACACGG CACCA[C3-spacer]	3590–3639	323
	R2	[Biotin]CAAGGACTGCTTTACAGGTGCCACTATTTT	3883–3912	
	F1	CAACACCACCAACCCAACGGCGTACCATAA	3570–3599	210
O3	P1	[FAM]CGTACCATAAGGCGCCGCTTACCCGGCTTA[dSpacer] ATTGCCCTACACGG CACCA[C3-spacer]	3590–3639	190
	R3	[Biotin]TTGATGGCACCGTAGTTGAAAGAAGTAGGC	3751–3779	
	F2	GGAGCACCTGAAGCAGCCTTGGACAACACC	3549–3578	182
O4	P1	[FAM]CGTACCATAAGGCGCCGCTTACCCGGCTTA[dSpacer] ATTGCCCTACACGG CACCA[C3-spacer]	3590–3639	141
	R1	[Biotin]GAGCCAGCACTTGGAGATCGCCTCTCACGT	3701–3730	

Table 1 The sequences of primers and probes designed for screening in the study *(Continued)*

Primer/probe set Name		Sequence(5′→3′)	The location on accession number	Product sizes (bp)
	F2	GGAGCACCTGAAGCAGCCTTGGACAACACC	3549–3578	364
O5	P1	[FAM]CGTACCATAAGGCGCCGCTTACCCGGCTTA[dSpacer] ATTGCCCTACACGG CACCA[C3-spacer]	3590–3639	323
	R2	[Biotin]CAAGGACTGCTTTACAGGTGCCACTATTTT	3883–3912	
O6 (Optimal set)	F2	GGAGCACCTGAAGCAGCCTTGGACAACACC	3549–3578	231
	P1	[FAM]CGTACCATAAGGCGCCGCTTACCCGGCTTA[dSpacer] ATTGCCCTACACGG CACCA[C3-spacer]	3590–3639	190
	R3	[Biotin]TTGATGGCACCGTAGTTGAAAGAAGTAGGC	3751–3779	
	F3	GGGGACCTTACCTGGGTGCCAAATGGAGCA	3524–3553	217
O7	P2	[FAM]CAAATGGAGCACCTGAAGCAGCCTTGGACAA[dSpacer]ACCACCAACCCAAC GGCGTAC[C3-spacer]	3542–3596	189
	R1	[Biotin]GAGCCAGCACTTGGAGATCGCCTCTCACGT	3701–3730	
	F3	GGGGACCTTACCTGGGTGCCAAATGGAGCA	3524–3553	389
O8	P2	[FAM]CAAATGGAGCACCTGAAGCAGCCTTGGACAA[dSpacer]ACCACCAACCCAAC GGCGTAC[C3-spacer]	3542–3596	371
	R2	[Biotin]CAAGGACTGCTTTACAGGTGCCACTATTTT	3883–3912	
	F3	GGGGACCTTACCTGGGTGCCAAATGGAGCA	3524–3553	256
O9	P2	[FAM]CAAATGGAGCACCTGAAGCAGCCTTGGACAA[dSpacer]ACCACCAACCCAAC GGCGTAC[C3-spacer]	3542–3596	238
	R3	[Biotin]TTGATGGCACCGTAGTTGAAAGAAGTAGGC	3751–3779	

Note: *F*:forward primer, *R* reverse primer, *P* probe, *FAM* Carboxyfluorescein, *dSpacer* A tetrahydrofuran residue, *C3-spacer* 3′-block

aerosol, 14 oesophageal-pharyngeal fluid, 9 blood and 13 nasal swabs) (Table 2). Of the 85 positive samples, 32 were serotyped as serotype A, 17 as serotype Asia 1, and 36 as serotype O. The concordance between FMDV RPA-LFD and rPCR was 98.41% (124/126).

It is worth noting that 4 aerosol specimens of FMDV serotype A were designated as positive by the RPA-LFD assay, but only 3 of them were FMDV serotype A positive detected by the rPCR, and 3 aerosol specimens of FMDV serotype O were differentiated with the RPA-LFD assay, whereas 2 of them were FMDV serotype O positive using the rPCR. BHK-21 cells were used to isolate viruses and identify the two inconsistent aerosol specimens, and CPE of cells inoculated with two inconsistent samples appeared untill the third passage. CPE-positive cell and control cell culture were respectively harvested and detected for FMDV using rPCR and RPA-LFD. As expected, CPE-positive cells were respectively identified for FMDV serotype O and A, and the control cells were FMDV negative, it showed that the sensitivity of RPA-LFD assay was higher than the rPCR in this study.

Discussion

There are a series of methods used as detection of FMDV, such as Ag-ELISA, virus separation, and rRT-PCR. However, the above methods have been either too time-consuming or require high-precision instruments to meet practical needs [19, 20]. RPA isothermal amplification techniques can amplify nucleic acids and detect the products without a requirement of special instrument or complex operations [21–24]. It is worth mentioning that human body heat can indeed incubate RPA reactions under certain limit resources [25]. Moreover, it is not required to store the lyophilized RPA reagents with a cooling chain because they can actually be stably stored at room temperature for a longer time [26].

Although it has been proved that the pan-specific real-time RT-RPA (rRT-RPA) and RPA-LFD technique can provide rapid and accurate diagnosis of FMDV [12, 18], they are difficult to distinguish the serotypes of FMDV for the rRT-RPA assay. The VP1, a surface exposed-capsid protein, take a pivotal role in the antigenicity as a major viral antigen, and plays an important role in pathogenicity of FMDV as its binding to viral receptors of host cells. Because of heterogeneity, the nucleotide sequence encoding VP1 is widely used to determine genetic relationships between different strains and to trace the provenance and transmission route of epidemic FMDV strains [27–30]. In this study, primer and probe sets specific for serotypes O, A, or Asia-1 FMDV were designed based on the alignment of the nucleotide sequences of viral VP1 gene of the above serotypes strains circulating in Asia, respectively. The primer and probe sets A6, As9, and O6 screened for RPA in this study could perform effective and accurate detection of different FMDV serotype of A/China/5/99, Asia1/AF/72,

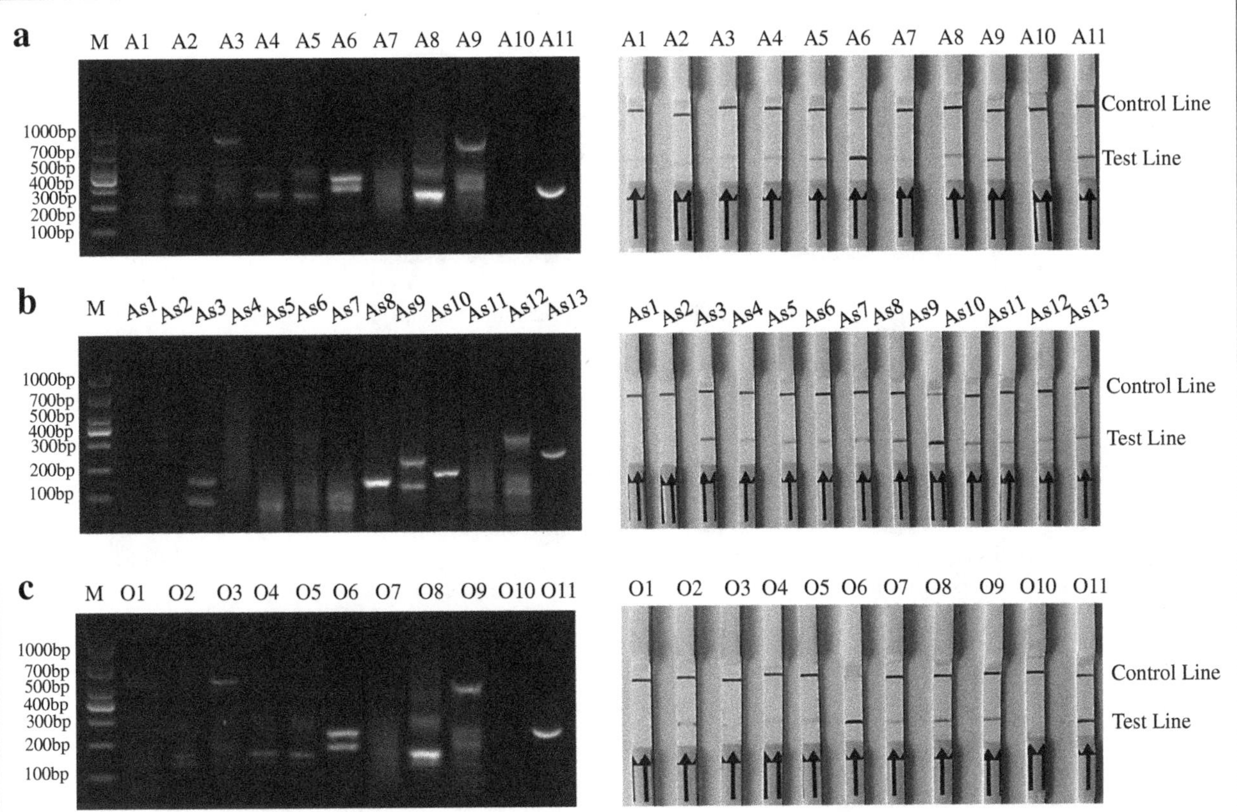

Fig. 1 Screening of the primers/probes for the FMDV serotype-specific RPA-LFD assay. **a** Agarose gel electrophoresis and LFD detection of RT-RPA products amplified with different primer and probe sets of FMDV serotype A. Lane M was DNA Marker DL1000. A1 to A9 were different primer and probe sets. A6: the optimal primer and probe set, and the estimated size of the RPA amplified fragment were 346 bp and 334 bp. A10: negative control, (DNase-free water). A11: positive control (supplied by Twist Amp nfo kit). **b** Agarose gel electrophoresis and LFD detection of RT-RPA products amplified with different primers/probe sets of FMDV serotype Asia 1. As2 to As10 were different primer and probe sets. As9: the optimal primer/probe set, and the estimated size of the RPA amplified fragment were 286 bp and 182 bp. As1: negative control (DNase-free water). As10: positive control (supplied by Twist Amp nfo kit). **c** Agarose gel electrophoresis and LFD detection of RT-RPA products amplified with different primers/probe sets of FMDV serotype O. O1 to O9 were different primer and probe sets. O6: the optimal primer/probe set, and the estimated size of the RPA amplified fragment were 231 bp and 190 bp. O10: negative control (DNase-free water). O11: positive control (supplied by Twist Amp nfo kit)

and O/HNK/CHA/05 (see Fig. 1) as intended. Furthermore, the serotype-specific RPA-LFD assay successfully detected the epidemic strains of FMDV in China (Table 2), and provided a more robust assessment method regarding the serotype specificity.

FMDV is mainly transmitted by aerosol. Viral RNA was detected in aerosol samples from FMDV suspected farm at 1–3 days before infected cattle appeared clinical signs [31]. Based on the aerogenous characteristics of the FMDV, it's proven to be a valuable technique that aerosol samples were used to detect viral RNA in infected farms [31, 32]. In our study, viruses in aerosol were also detected and their serotypes were respectively differentiated by the FMDV RPA-LFD assay. Therefore, it may be the potential integral monitoring strategies for prevention, control and eradication of FMD using this technique.

It is a crucial step for clinical detection with any molecular diagnostic assay that the nucleic acids are extracted from tissue and cell samples, the Punch-it™ kit can easily isolate nucleic acid from different samples via paper chromatography, and becomes one of the ideal tools for the extraction of DNA/RNA. Recent study showed that the DNA isolated with the Punch-it™ kit could be used in molecular assays [33]. In our study, nucleic acids of different samples, including vesicular material, blood, oesophageal-pharyngeal fluid, saliva, aerosol, and nasal swabs, were successfully extracted using the Punch-it™ kit. The method only takes 10 min to extract the nucleic acid without centrifugation, so it is relatively simple and rapid, and it is even more important that the extracted nucleic acids can directly serve as templates in RPA-LFD assays. In the present study, the extracted total RNA needed to be reverse transcribed into cDNA in the two-step RPA-LFD assay, whereas reverse transcription and RPA were performed in one reaction using the TwistAmp™ exo kit (TwistDx Limited, UK) and with the addition of reverse transcriptase in previously studied RT-RPA assays [12].

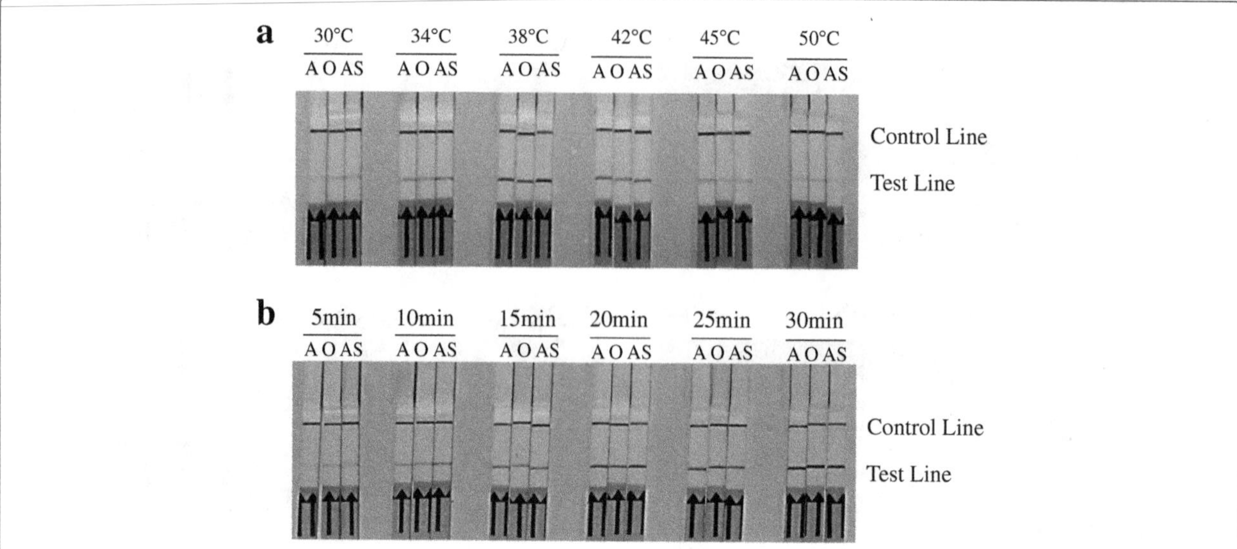

Fig. 2 Optimization of reaction temperature and time for FMDV serotype-specific RPA-LFD assays. **a** The RPA-LFD performs effectively in a wide range of constant reaction temperatures. **b** The amplified products can be visible on the LFD at 5 min or longer

We would like to further simplify the test to make it more suitable for field use in future. Moreover, sophisticated instrumentations were required in RT-RPA [12, 15, 16], whereas RPA-LFD assay in this study only needs a thermos metal bath for incubation at 38 °C and amplified products can be direct visible on the LFD without requirements of instruments. Therefore, it may be an effective way to detect clinical samples in the field.

Conclusions

In the present study, the serotype-specific FMDV RPA-LFD assay was successfully developed, and will be helpful for detection of FMDV infection during FMD outbreaks. Because RPA-LFD assay is a simple, specific, rapid, and serotype-specific method, it is possible to be a general differentiated protocol for diagnostics of FMDV, especially for detection of clinical samples in the field.

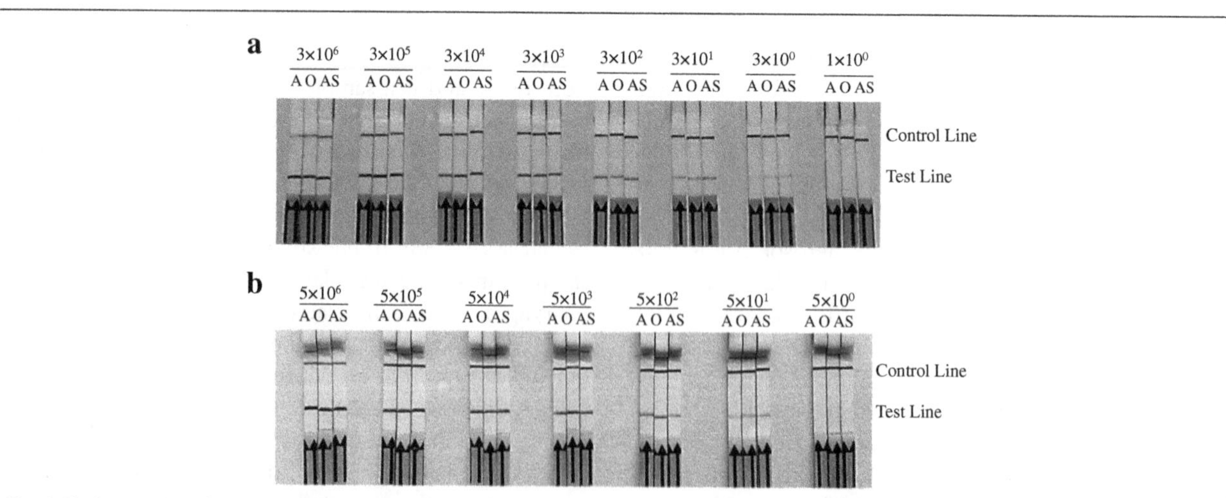

Fig. 3 The sensitivity of FMDV serotype-specific RPA-LFD assays. **a** Sensitivity of the standard plasmids. Molecular sensitivity of RPA-LFD was determined using 10-fold serially diluted 3×10^6 to 3×10^0 copies and 10^0 copy of FMDV DNA standard plasmids per reaction as template. The minimum limits for virus detection of RPA-LFD were 3×10^0 copies. **b** Sensitivity of the RNA standard. The cDNA of reverse transcription using 10-fold serially diluted 5×10^6 to 5×10^0 RNA molecular was used in molecular sensitivity of RPA-LFD. The minimum limits detection of RPA-LFD were 5×10^0 RNA. A: primers/probe set of FMDV serotype A. AS: primers/probe sets of FMDV serotype Asia 1. O: primers/probe sets of FMDV serotype O. Samples were tested in triplicate with one reaction and independently repeated 3 times

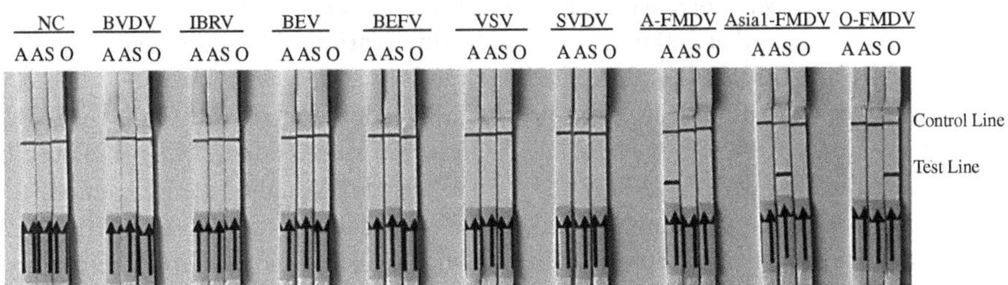

Fig. 4 The specificity of the FMDV RPA-LFD assays. Other bovine viral pathogens with similar clinic and etiologies were used to assess the specificity of the assays. There was no cross-reaction with BVDV, IBRV, BEV, BEFV, BVSV and SVDV. NC: negative control. A: primers/probe set of FMDV serotype A. AS: primers/probe set of FMDV serotype Asia 1. O: primers/probe set of FMDV serotype O. Samples were tested in triplicate with one reaction and three separate assays

Methods

Virus and clinical specimens

In this study, cDNA of three serotypes of FMDV reference strains including type O strain China/5/99, type A strain AF/72, and type Asia 1 strain HNK/CHA/05, which were provided by Harbin Veterinary Research Institute, Chinese Academy of Agricultural Sciences, were positive controls of various serotypes used to optimize primer and probe sets of RPA. The cDNA of other epidemic FMDV strains in China including type A (HuBWH/2009, Mya98, GSLX/2010, GDMM/2013), type Asia1 (ZB/58, HeB/05, YS/05, HN/06, BR/Myanmar/06, WHN/06), and type O (LY/2000, CC/03, GZ/2010, BY/2010, HKN/2011, GD/2013, GD/2015), were used to provide a more robust assessment of serotype-specific RPA. Other viral pathogens causing similar clinical vesicular signs, including bovine ephemeral fever virus (BEFV), vesicular stomatitis virus (VSV), bovine viral diarrhea virus (BVDV), bovine enterovirus (BEV), infectious bovine rhinotracheitis virus (IBRV),and swine vesicular disease virus (SVDV), were stored by the Ruminant Diseases Research Center, Shandong Normal University, and used for cross-reactivity testing. To compare the detection sensitivity between RPA-LFD reactions and rPCR, 126 clinical specimens (30 vesicular materials, 29 salivas, 14 aerosols, 16 bloods, 17 oesophageal-pharyngeal fluids, and 20 nasal swabs) were collected from suspected cases of FMD in the Chinese endemic region from 2013 to 2017.

Isolation of viral RNA/DNA and synthesis of cDNA

Viral RNA and DNA were isolated using the MiniBEST Viral RNA/DNA Extraction Kit (TaKaRa, Dalian, China) following respective instructions. The amounts of viral RNA was measured using a Biophotometer plus (Eppendorf, USA). The extracted RNA was template used to synthesized cDNA with reverse transcription using random primers in a total volume of 10 μL according to the instructions of the PrimeScript™ RT Master Mix (Takara, Dalian, China). All viral DNA and cDNA were stored at − 70 °C for further employment.

Generation of DNA/RNA molecular standard

The viral VP1 gene recombinant vectors respectively containing RPA amplified region of FMDV serotype O, A, or Asia 1 (named pET32a-A-FMDV-VP1, pET32a-AS1-FMDV-VP1 and pET32a-O-FMDV-VP1) were constructed and used for the analytical sensitivity. The Positive plasmids were measured using a Biophotometer plus (Eppendorf, USA), respectively. The quantity of copies was calculated by the formula: DNA copy number (copies/μL)

Table 2 Comparative performance of serotype-specific RT-LFD-RPA and rRT-PCR assays for detection of suspected clinical specimens and serotyping of FMDV

	LFD-RPA					Real-time (qPCR)				
Samples name	A	Asia 1	O	FMDV positive	FMDV Negative	A	Asia 1	O	FMDV positive	FMDV Negative
vesicular material	7	3	15	25	5	7	3	15	25	5
nasal swab	6	2	5	13	7	6	2	5	13	7
saliva	5	3	6	14	15	5	3	6	14	15
oesophageal pharyngeal fluid	6	4	4	14	3	6	4	4	14	3
blood	4	2	3	9	7	4	2	3	9	7
aerosol	4	3	3	10	4	3	3	2	8	6
Total	32	17	36	85	41	31	17	35	83	43

$= (M \times 6.02 \times 10^{23} \times 10^{-9})/(n \times 660)$, M: molecular weight, n:plasmid concentration (g/μL) measured at 260 nm.

RNA molecular standards were prepared as described in previous study [17] with certain modifications. The linearized recombinant vectors with SgrA I (New England Biolabs, USA) were purified with the MiniBEST DNA Fragment Purification Kit (Takara, Dalian, China), and then used as template for RNA transcription with the RiboMAX Large Scale RNA Production System-T7 (Promega, USA). Furthermore, the RNA was measured using the Quant-iTTM RiboGreen RNA Assay Kit (Thermo Fisher Scientific, Germany) according to the manufacturer's instructions. The quantity of copies was calculated by the equation: Amount (copies/μL) = [RNA concentration (g/μL)/(transcript length in nucleotides× 340)] $\times 6.02 \times 10^{23}$.

Design of serotype-specific RPA primers and probe

A multiple sequence alignment of FMDV strains of serotype A, Asia 1, or O was respectively performed to find highly conserved region of the FMDV VP1 gene. The following reference sequences of three serotypes found in GenBank database were respectively used: KT968663(A/HY/CHA/2013), FJ755082 (A/PAK/1/2006), KY322679 (A/TAI/4/2014), FJ755052 (A/IRN/51/2005), KY404935 (A/A01NL), EF149010 (Asia 1/HNK/CHA/05), EF614458 (Asia1/MOG/05), AY687334 (Asia1/IND 491/97), GU 931682 (Asia1/YS/CHA/05), AY687333 (Asia1/IND 321/01), HQ009509 (O/China/5/99), LC149720 (O/JPN/2010–362/3), JN998086 (O/GZ/CHA/2010), AF095876(O/Taipei-150). RT-RPA primers and probes specific for serotypes O, A or Asia-1 of FMDV were designed against the consensus sequence for this region and were synthesized by Sangon Biotech, respectively. RPA primers/probes were synthesized and labeled as described in previous study [18]. Oligonucleotide sequences of RPA primers and probes of specific serotype A, Asia-1, or O of FMDV are listed in Table 1 (accession numbers KT968663, EF149010 and HQ009509, respectively).

FMDV serotype-specific RPA assays

Serotype-specific primers and probes were screened as described in previous study [25] with some modifications. In brief, RPA was performed using a TwistAmp™ nfo kit (TwistDx Limited, UK). The freeze-dried enzyme pellet was dissolved with 47.5 μLof solution containing 29.5 μL rehydration buffers, 2.1 μL forward and reverse primers (10 μM), 0.6 μL probe (10 μM), 11.2 μL of sterile water, 2 μL of cDNA of FMDV reference strains, and then 2.5 μL magnesium acetate (280 mM) was added. Assays were completed in a thermos metal bath at 38 °C for 20 min. The amplified products were then put through a 2% (*w*/*v*) agarose gel electrophoresis to screen the optimal and serotype-specific primer and probe sets. The optimal reaction conditions were determined by testing various reaction temperatures and incubation times.

LFD double label with anti-FAM gold conjugates and anti-Biotin antibodies (Milenia Biotec GmbH, Germany) were used to visualize the RPA amplified products. 1 μL of RPA products were diluted with 99 μL Dipstick Assay Buffer (Milenia Biotec GmbH, Germany), and then tested by LFD. FMDV serotype-specific positives are indicated by the visualization of both a test line and control line simultaneously perceptible on the LFDs after 5 min, while the negative reactions only generate a control line. The cDNA of other FMDV epidemic virus strains in China were used for evaluation of serotype-specific RPA-LFD.

Sensitivity and specificity of the RPA-LFD assay

To determine the DNA analytical sensitivity of the RPA-LFD assay, the recombinant plasmids pET32a-A-FMDV-VP1, pET32a-As1-FMDV-VP1, and pET32a-O-FMDV-VP1 were respectively the standard DNA template of FMDV serotype A, Asia 1 and O. The RPA-LFD assays were performed with ten-fold serial dilutions of the recombinant vector ranging from 3×10^6 to 3×10^0 copies per microliter for respective serotypes. To detect the RNA analytical sensitivity, RNA standards of three FMDV serotypes were diluted from 5×10^7 to 5×10^1 molecules/μL. 10 μL RNA was used as template to synthesize cDNA in 20 μL reverse transcription reaction system using PrimeScript™ RT Master Mix (Takara, Dalian, China) in accordance with the manufacturer's instructions. 2 μL cDNA of each dilution was used as a template in the RPA reactions. DNA plasmid/RNA samples were detected with three separate assays, respectively.

The specificity of the method was assessed using other viral pathogens with similar clinical symptoms, including BEFV, VSV, BVDV, BEV, IBRV, and SVDV. The IBRV DNA was extracted and viral cDNA were reverse transcribed from isolated other viral RNA. Positive controls and negative controls for RPA were constructed using recombinant vectors and RNase free water.

Diagnosis of clinical specimens with FMDV serotype-specific RPA-LFD assays

To compare the diagnostic sensitivity between RPA-LFD and rPCR, 126 clinical specimens were collected from bovine farms suspected with infection of FMDV in China from 2013 to 2017 (details listed in Table 2). RNAs were isolated from the clinical samples using a Punch-it™ Kit (Nanohelix, Daejeon, South Korea) following manufacturers' instructions. A 1 mm punched disk, containing the nucleic acids, was added with 10 μL of reverse transcript reaction system using PrimeScript™ RT Master Mix (Takara, Dalian, China), which contained

2 μL 5 × PrimeScript RT Master Mix and 8 μL RNase Free dH_2O. Reverse transcription for each sample was completed in two tubes using a thermos metal bath at 37 °C for 15 min. 2 μL of cDNA was then used in both RPA-LFD and rPCR reactions. For the rPCR, serotype-specific primers and probes for serotypes O, A or Asia-1 FMDV were employed as previously described [23]. Reactions were performed with Premix Ex Taq™ Kit (Takara, Dalian, China) for respective serotypes.

Samples were positive in RPA-LFD, but negative in rPCR, were further tested for presence of FMDV. The virus was isolated using BHK-21 cells (provided by China Center for Type Culture Collection) as described in the previous study [18]. The cytopathic effect (CPE) was examined at 24 h, 48 h, and 72 h, respectively. If there was no CPE after 72 h, Cell cultures were passaged. CPE-positive, CPE-negative and control cell cultures were respectively harvested and examined again for FMDV using RPA-LFD and rPCR.

Abbreviations

BEFV: Bovine ephemeral fever virus; BEV: Bovine enterovirus; BVDV: Bovine viral diarrhea virus; ELISA: Immune-histopathology and enzyme-linked immunosorbent assay; FMDV: Foot-and-mouth disease virus; IBRV: Infectious bovine rhinotracheitis virus; LAMP: Loop-mediated isothermal amplification; LFD: Lateral flow dipstick; PCR: Polymerase chain reaction; RPA: Recombinase polymerase amplification; rPCR: Real time PCR; RT: Reverse transcription; SAT: South African Territories; SVDV: Swine Vesicular Disease Virus; VSV: Vesicular stomatitis virus

Acknowledgments

Yang He from Emory University in the United State revised the text of manuscript.

Funding

This study was partly supported by grants from National Natural Science Fund of China (31672556, 31872490, 31502064), Taishan Scholar and Distinguished Experts (H.HB), National Primary Research & Development Plan (2018YFD0501605-06), Primary Research & Development Plan of Shandong Province (2016GNC113006, 2018GNC113011).

Authors' contributions

W.HM and H.PL performed the experiments and drafted manuscript, Z.GM analyzed the data, Y. L isolated virus, H.HB and G.YW designed and instructed the experiments. All authors have read and approved the final manuscript.

Consent for publication

Not applicable.

Competing interests

The authors declare that they have no competing interests.

Author details

[1]Ruminant Diseases Research Center, Key Laboratory of Animal Resistant Biology of Shandong, College of Life Sciences, Shandong Normal University, Jinan 250014, China. [2]Division of Livestock Infectious Diseases, State Key Laboratory of Veterinary Biotechnology, Harbin Veterinary Research Institute, Harbin 150001, China. [3]Key Laboratory of Jilin Province for Zoonosis Prevention and Control, Military Veterinary Research Institute of Academy of Military Medical Sciences, Changchun 130122, China.

References

1. Domingo E, Escarmís C, Baranowski E, Ruiz-Jarabo CM, Carrillo E, Núñez JI, Sobrino F. Evolution of foot-and-mouth disease virus. Virus Res. 2003; 91(1):47–63.
2. Jamal SM, Belsham GJ. Foot-and-mouth disease: past present and future. Vet Res. 2013;44:116.
3. Brito BP, Rodriguez LL, Hammond JM, Pinto J, Perez AM. Review of the global distribution of foot-and-mouth disease virus from 2007 to 2014. Transbound Emerg Dis. 2017;64:316–32.
4. Zheng H, He J, Guo J, Jin Y, Yang F, Lv L, Liu X. Genetic characterization of a new pandemic Southeast Asia topotype strain of serotype O foot-and-mouth disease virus isolated in China during 2010. Virus Genes. 2012; 44(1):80–8.
5. Yang X, Zhou YS, Wang HN, Zhang Y, Wei K, Wang T. Isolation, identification and complete genome sequence analysis of a strain of foot-and-mouth disease virus serotype Asia1 from pigs in southwest of China. Virol J. 2011;8:175.
6. He CQ, Liu YX, Wang HM, Hou PL, He HB, Ding NZ. New genetic mechanism, origin and population dynamic of bovine ephemeral fever virus. Vet Microbiol. 2016;182:50–6.
7. Charleston B, Bankowski BM, Gubbins S. Chase-topping ME, Schley D, Howey R, Barnett PV, Gibson D, Juleff ND, Woolhouse MEJ. Relationship between clinical signs and transmission of an infectious disease and the implications for control. Science. 2011;332:726–9.
8. Ranjan R, Biswal JK, Subramaniam S, Singh KP, Stenfeldt C, Rodriguez LL, Pattnaik B, Arzt J. Foot-and-mouth disease virus-associated abortion and vertical transmission following acute infection in cattle under natural conditions. PLoS One. 2016;11(12):e0167163.
9. Knight-Jones TJ, Robinson L, Charleston B, Rodriguez LL, Gay CG, Sumption KJ, Vosloo W. Global foot-and-mouth disease research update and gap analysis: 4-diagnostics. Transbound Emerg Dis. 2016;63(1):42–8.
10. Ferris NP, Nordengrahn A, Hutchings GH, Reid SM, King DP, Ebert K, Paton DJ, Kristersson T, Brocchi E, Grazioli S, Merza M. Development and laboratory validation of a lateral flow device for the detection of foot-and-mouth disease virus in clinical samples. J Virol Methods. 2009;155(1):10–7.
11. Waters RA, Fowler VL, Armson B, Nelson N, Gloster J, Paton DJ, King DP. Preliminary validation of direct detection of foot-and-mouth disease virus within clinical samples using reverse transcription loop-mediated isothermal amplification coupled with a simple lateral flow device for detection. PLoS One. 2014;9(8):e105630.
12. Abd El Wahed A, El-Deeb A, El-Tholoth M, Abd El Kader H, Ahmed A, Hassan S, Hoffmann B, Haas B, Shalaby MA, Hufert FT. Weidmann M. a portable reverse transcription recombinase polymerase amplification assay for rapid detection of foot-and-mouth disease virus. PLoS One. 2013;8(8): e71642.
13. Lau LT, Reid SM, King DP, Lau AMF, Shaw AE, Ferris NP. Yu ACH. Detection of foot-and-mouth virus by nucleic acid sequence-based amplification (NASBA). Vet Microbiol. 2008;126(1–3):101–10.
14. Zheng S, Wu X, Zhang L, Xin C, Liu Y, Shi J, Peng Z, Xu S, Fu F, Yu J, Sun W, Xu S, Li J, Wang J. The occurrence of porcine circovirus 3 without clinical infection signs in Shandong Province. Transbound Emerg Dis. 2017;64(5): 1337–41.
15. Geng Y, Wang J, Liu L, Lu Y, Tan K, Chang YZ. Development of real-time recombinase polymerase amplification assay for rapid and sensitive detection of canine parvovirus 2. BMC Vet Res. 2017;13(1):311.
16. Hou P, Zhao G, He C , Wang H, , He H. Biopanning of polypeptides binding to bovine ephemeral fever virus G1 protein from phage display peptide library. BMC Vet Res 2018; 14(1): 3.

17. Wang J, Wang J, Li R, Liu L, Yuan W. Rapid and sensitive detection of canine distemper virus by real-time reverse transcription recombinase polymerase amplification. BMC Vet Res. 2017;13(1):241.
18. Wang HM, Zhao GM, Hou PL, Yu L, He CQ, He HB. Rapid detection of foot-and-mouth disease virus using reverse transcription recombinase polymerase amplification combined with a lateral flow dipstick. J Virol Methods. 2018;261:46–50.
19. Ding NZ, Qi QR, Gu XW, Zuo RJ, Liu J, Yang ZM. De novo synthesis of sphingolipids is essential for decidualization in mice. Theriogenology. 2018; 106(3):227–36.
20. Jamal SM, Belsham GJ. Development and characterization of probe-based real time quantitative RT-PCR assays for detection and serotyping of foot-and-mouth disease viruses circulating in west eurasia. PLoS One. 2015;10(8): e0135559.
21. Shalaby MA, El-Deeb A, El-Tholoth M, Hoffmann D, Czerny CP, Hufert FT, Weidmann M, Abd El Wahed A. Recombinase polymerase amplification assay for rapid detection of lumpy skin disease virus. BMC Vet Res. 2016; 12(1):244.
22. Hou P, Wang H, Zhao G, He C, He H. Rapid detection of infectious bovine Rhinotracheitis virus using recombinase polymerase amplification assays. BMC Vet Res. 2017;13(1):386.
23. Zhang F, Huang YH, Liu SZ, Zhang L, Li BT, Zhao XX, Fu Y, Liu JJ, Zhang XX. Pseudomonas reactans, a bacterial strain isolated from the intestinal flora of Blattella germanica with anti-Beauveria bassiana activity. Environ Entomol. 2013;42(3):453–9.
24. Zhao G, Wang H, Hou P, He C, Huan Y, He H. Rapid and visual detection of Mycobacterium avium subsp. paratuberculosis by recombinase polymerase amplification combined with a lateral flow dipstick. J Vet Sci. 2018;19(2): 242–50.
25. Crannell ZA, Rohrman B, Richards-Kortum R. Equipment-free incubation of recombinase polymerase amplification reactions using body heat. PLoS One. 2014;9(11):e112146.
26. Lillis L, Siverson J, Lee A, Cantera J, Parker M, Piepenburg O, Lehman DA, Boyle DS. Factors influencing recombinase polymerase amplification (RPA) assay outcomes at point of care. Mol Cell Probes. 2016;30(2):74–8.
27. Samuel AR, Knowles NJ. Foot-and-mouth disease type O viruses exhibit genetically and geographically distinct evolutionary lineages (topotypes). J Gen Virol. 2001;82:609–21.
28. Cottam EM, Thébaud G, Wadsworth J, Gloster J, Mansley L, Paton DJ, King DP, Haydon DT. Integrating genetic and epidemiological data to determine transmission pathways of foot-and-mouth disease virus. Proc R Soc B. 2008; 275:887–95.
29. Liu M, Xie S, Zhou J. Use of animal models for the imaging and quantification of angiogenesis. Exp Anim. 2018;67(1):1–6.
30. Jamal SM, Ferrari G, Ahmed S, Normann P, Curry S, Belsham GJ. Evolutionary analysis of serotype a foot-and-mouth disease viruses circulating in Pakistan and Afghanistan during 2002–2009. J Gen Virol. 2011;92:2849–64.
31. Pacheco JM, Brito B, Hartwig E, Smoliga GR, Perez A, Arzt J, Rodriguez LL. Early detection of foot-and-mouth disease virus from infected cattle using a dry filter air sampling system. Transbound Emerg Dis. 2017;64(2):564–73.
32. Nelson N, Paton DJ, Gubbins S, Colenutt C, Brown E, Hodgson S, Gonzales JL. Predicting the ability of preclinical diagnosis to improve control of farm-to-farm foot-and-mouth disease transmission in cattle. J Clin Microbiol. 2017;55(6):1671–81.
33. Kim J, Wang HY, Kim S, Park SD, Yu K, Kim HY, Uh Y, Lee H. Evaluation of the punch-it™ NA-sample kit for detecting microbial DNA in blood culture bottles using PCR-reverse blot hybridization assay. J Microbiol Methods. 2016;128:24–30.

Satellite telemetry tracks flyways of Asian Openbill storks in relation to H5N1 avian influenza spread and ecological change

Parntep Ratanakorn[1,2], Sarin Suwanpakdee[1,2], Witthawat Wiriyarat[2,3], Krairat Eiamampai[4], Kridsada Chaichoune[2,3], Anuwat Wiratsudakul[1,2], Ladawan Sariya[2] and Pilaipan Puthavathana[5,6*]

Abstract

Background: Asian Openbills, *Anastomus oscitans*, have long been known to migrate from South to Southeast Asia for breeding and nesting. In Thailand, the first outbreak of H5N1 highly pathogenic avian influenza (HPAI) infection in the Openbills coincided with the outbreak in the poultry. Therefore, the flyways of Asian Openbills was determined to study their role in the spread of H5N1 HPAI virus to poultry and wild birds, and also within their flocks.

Results: Flyways of 5 Openbills from 3 colonies were monitored using Argos satellite transmitters with positioning by Google Earth Programme between 2007 and 2013. None of the Openbills tagged with satellite telemeters moved outside of Thailand. Their home ranges or movement areas varied from 1.6 to 23,608 km^2 per month (95% utility distribution). There was no positive result of the viral infection from oral and cloacal swabs of the Openbills and wild birds living in the vicinity by viral isolation and genome detection during 2007 to 2010 whereas the specific antibody was not detected on both Openbills and wild birds by using microneutralization assay after 2008. The movement of these Openbills did not correlate with H5N1 HPAI outbreaks in domestic poultry but correlated with rice crop rotation and populations of the apple snails which are their preferred food. Viral spread within the flocks of Openbills was not detected.

Conclusions: This study showed that Openbills played no role in the spread of H5N1 HPAI virus, which was probably due to the very low prevalence of the virus during the monitoring period. This study revealed the ecological factors that control the life cycle of Asian Openbills.

Keywords: *Anastomus oscitans*, Asian Openbill, H5N1 highly pathogenic avian influenza (H5N1 HPAI), Satellite telemetry, Flyway

Background

Asian Openbills or Asian Open-billed storks, *Anastomus oscitans*, are large wading birds in the family *Ciconiidae* [1]. The birds get their name from the natural open space between the curved inner surfaces of the mandibles in adults [2]. This gap between the mandibles increases with age. Openbills are migratory birds which generally move between South Asian countries (India, Sri Lanka, Bangladesh) and Southeast Asian countries (Myanmar and Thailand) [3, 4]. Using ring bands, McClure and Kwanyen in 1973 suggested that Asian Openbills migrate from Bangladesh to Thailand for breeding and nesting [1, 3]. In Thailand, Asian Openbills gather together during the breeding season, around November to February, which are cold months of the year. These birds are colonial breeders with several nests being built in the same tree [5]. Openbills forage in marshes and paddies and feed on mollusks, frogs, crabs, aquatic animals, snakes and giant insects [5, 6]. In Thailand, their favorite food is apple snails [5, 7], including *Pomacea canaliculata* and *P. insularum* [7]. The gap between the mandibles allows the birds to hold and carry snails easily [8].

* Correspondence: pilaipan.put@mahidol.edu
[5]Center for Research and Innovation, Faculty of Medical Technology, Mahidol University, Nakhon Pathom 73170, Thailand
[6]Department of Microbiology, Faculty of Medicine Siriraj Hospital, Mahidol University, Bangkok 10700, Thailand
Full list of author information is available at the end of the article

The outbreaks of H5N1 highly pathogenic avian influenza (HPAI), occurred in Thailand in January 2004 and resulted in enormous economic losses of poultry and several human deaths. With restricted control strategy, no human case was found after July 2006 [9], while no outbreak occurred in poultry farm after 2008 [10]. At the early event of the outbreak, a lot of Asian Openbills were found dead in the rice paddy fields together with free-ranging ducks. In 2004, the prevalence of H5N1 HPAI virus infection in the Openbills of unknown health status was as high as 26.67% (95% confidential interval-CI: 13.7–39.6) as determined by virus isolation and viral genome detection. The prevalence decreased to 0.91% (95% CI: 0.1–1.7) in 2005, and 0% in 2006 and 2007 [11].

Initially, the Openbills was suspected of bringing in the virus along with their migration from Bangladesh to Thailand. Therefore, depopulation of Openbills was proposed to be an approach for avian influenza (AI) control. However, this opinion was stopped based on the information that there was no occurrence of H5N1 HPAI outbreak in South Asia at that time. According to the Office International des Epizooties (OIE) [10], the first outbreak of this virus in South Asia occurred in Bangladesh in March 2007. The new opinion speculated that the Openbills might get the H5N1 virus infection from grazing ducks and/or resident birds that shared the foraging habitats in rice paddy fields.

Our field observation revealed that the flock of Openbills returned to the old nesting places in early November every year, and by the end of February, the flocks disappeared from the nesting areas or markedly decreased in population size. No further information exists on where the birds go after February, and the role of Openbills in H5N1 AI spread has not been elucidated. This study used satellite telemetry in combination with field surveys and virological assays to monitor the flyways of Openbills and investigate the possible spread of H5N1 HPAI virus within the flocks and to other animal species living in the vicinity in Thailand.

Results

Biographic data and monitoring periods of Openbills

Biographic data of the 5 Openbills tagged with satellite transmitters are shown in Table 1. Their flyways were monitored for the duration of 1–5 years. During the cold months of the year (November–February), the Openbills gathered together for nesting, breeding and egg-laying, and the colonies became overcrowded with birds. About 50 days after hatching, the baby birds became juveniles, and the flocks left their nests to other foraging areas in February to March. The Openbills returned to their old nesting areas in the next breeding season. None of the birds flew from Thailand to other countries during the monitoring period (Table 1).

Movement areas of the Openbills

The data used in this study are available on Movebank (movebank.org, study name "Asian Openbill tagging with satellite telemetry in Thailand, 2007-2013") and are published.

in the Movebank Data Repository [12].

The Openbill ID numbers 74,793 and 74,800 nested and formed colonies at the study site in Nakhon Pathom province. Both birds were tracked foraging in several provinces in Central Thailand. The average home-range area of bird 74,793 was 1564 km^2/month (range 6.6–11,642: 95% Utility Distribution - UD); and it was 2477 km^2/month (range 2.5 to 23,608: 95% UD) for bird 74,800 (Tables 1 and 2). Unfortunately, the signals from satellite transmitters on these two Openbills lost after they flew back to their old nested area in the second season. Their habitat home ranges were relatively wide around Chao Phraya River basin where a main river of Thailand is. That area presented mainly as the rice farming and the agricultural area (Fig. 1a and b).

The Openbill ID numbers 74,794 and 74,799 from the Nakhon Sawan study site did not forage far from their origin and had average home-range areas of 76 km^2/month (range 3.4 to 752: 95% UD) and 414 km km^2/month (range 1.6 to 4481: 95% UD), respectively. The signal of bird ID 74799 lost after three years of follow-up, while the monitoring of bird ID 74794 was

Table 1 Biographic and tracking period of the Asian Openbills monitored by satellite telemetry

Bird No.	Transmitter No.	Sex	Body weight (kg)	System of satellite transmitter	Study site	Transmitter tagging date	Time at signal loss from monitoring	Tracking period
1	74793	Male	1.40	35 g, Solar PTT-100	Nakhon Pathom	02/25/2007	05/30/2008	1 year 3 months
2	74794	Female	1.50	35 g, Solar PTT-100	Nakhon Sawan	01/03/2007	09/06/2012	5 years 6 months
3	74799	Female	1.50	45 g, GPS PTT-100	Nakhon Sawan	01/03/2007	01/25/2010	3 years 1 month
4	74800	Male	1.40	45 g, GPS PTT-100	Nakhon Pathom	02/25/2007	12/18/2008	1 year 10 months
5	30123	ND	ND	35 g, Solar PTT-100	Phitsanulok	07/21/2009	09/16/2013 (Stopped tracking)	More than 4 years 2 months

Table 2 The movement areas of Asian Openbills as tracked by satellite telemetry

Bird ID	Home range area/month (km^2), 95%UD			Movement areas by provinces
	Maximum	Minimum	Average	
74793	11,642	6.6	1564	Nakhon Pathom, Ratchaburi, Nonthaburi, Phra Nakhon Si Ayutthaya, Suphan Buri, Ang Thong, Saraburi, Sing Buri, Lop Buri, Chai Nat, Nakhon Sawan and Phichit
74800	23,608	2.5	2477	Nakhon Pathom, Ratchaburi, Suphan Buri, Nakhon Sawan and Phichit
74794	752	3.4	76	Nakhon Sawan, Phichit, Phetchabun and Phitsanulok
74799	4481	1.6	414	Nakhon Sawan, Phichit, Chainat, Sing Buri and Ang Thong
30123	2833	4.6	360	Phitsanulok, Phichit, Nakhon Sawan, Phetchabun and Nakhon Pathom

terminated by the project after 5.5 years of tracking (Tables 1 and 2). Their primary habitat area was in the north-central part of Thailand, and both birds lived around the sizeable fresh marsh (Bungborapet) where rice farms were abundant (Fig. 2a and b).

The Openbill ID number 30123 from Phitsanulok province also foraged around its area of origin. The flock of this Openbill shared the same habitat with great herons in the rice paddies in Phitsanulok province. It moved within an average home range of 360 km/month (range 4.6 to 2833: 95% UD) (Tables 1 and 2). The project terminated the monitoring of this bird after 4.3 years of follow-up. Mostly, the bird moved near the site of satellite tagging (Fig. 3).

Ecology of foraging areas and foods of Asian Openbills

The spatial analysis demonstrated that the Openbills foraged in wetlands nearby water reservoirs, rivers, and agricultural areas, in particular rice paddies, in most of the time, and less frequently in dry agricultural lands (Fig. 4). Field observations revealed that the flocks of the Openbills split into small groups during non-breeding season for feeding and resting. Birds formed colonies again for breeding at their original nesting sites in the next November.

The Openbill diet included mollusks, crabs and some insects [5], but their favorite prey was the apple snails such as *P. canaliculata* and *P. insularum.* Field observations also showed that the movements of Openbills were well correlated with the rice crop rotation. Birds moved from one place to the others where new rice crops appeared, and there were plenty of rice sprouts. The apple snails came on for feeding on these young rice stems of age around 10 days, on the other hand, the snails were the favorite food of the Openbills. The land became dry during the harvesting season. In this period, the snails embedded themselves in the mud and lived silently underground. When water returned either through rainfall or irrigation canals, the land became wet and muddy, and it was time to start the new crop of rice. The snails came out of the ground to feed, and snail population attracted the Openbills again.

Surveillance for H5N1 HPAI infection in Asian Openbills

Blood, together with respiratory and cloacal swab samples, were collected from Openbills between January 2007 and 2010 to monitor the H5N1 HPAI virus infection in flocks of the tracked birds. Testing for the viral genome and isolation of H5N1 HPAI virus from 566 swab samples yielded negative results. However, MN assay of 431 healthy adult storks detected H5N1 antibody titers ≥40 in 2.2% (4 of 181) of serum samples collected in 2007, 1.7% (2 of 120) in 2008, and none in 2009 and 2010 (Table 3).

Movement of Openbills in relationship to AI infection in other wild birds

Role of Openbills in the spread of H5N1 HPAI virus to poultry and other wild birds was explored. The wild birds that lived along the flyways of the tracked flocks were trapped by mist nets. Their blood together with respiratory and cloacal swab samples was collected and investigated for H5N1 HPAI virus infection. MN assay was performed in 206 serum samples collected from 17 bird families. The result showed the MN antibody titers of ≥40 in only four birds: three Streak-eared bulbuls (family Pycnonotidae) from Nakhon Sawan province, and one Ashy wood swallow (family Artamidae) from Nakhon Pathom province in 2008 (Additional file 1). All seropositive birds were healthy adults. The viral genome and the H5N1 HPAI virus could not be detected in 851 birds that were tested.

Movement of Openbills and AI outbreaks

Positions of the tagged Openbills and their flyways were determined from the signals emitted from satellite transmitters. We sent field epidemiologists to the areas where Openbills foraged to interview the villagers for evidence of AI outbreaks. The interviews revealed no evidence of H5N1 HPAI outbreaks along the flyways of the Openbills during the monitoring period from 2007 to 2013. One AI outbreak occurred in a broiler farm for local consumption in Nakhon Sawan province in November of 2008, but that place did not overlap the habitats of our Openbills (Fig. 4).

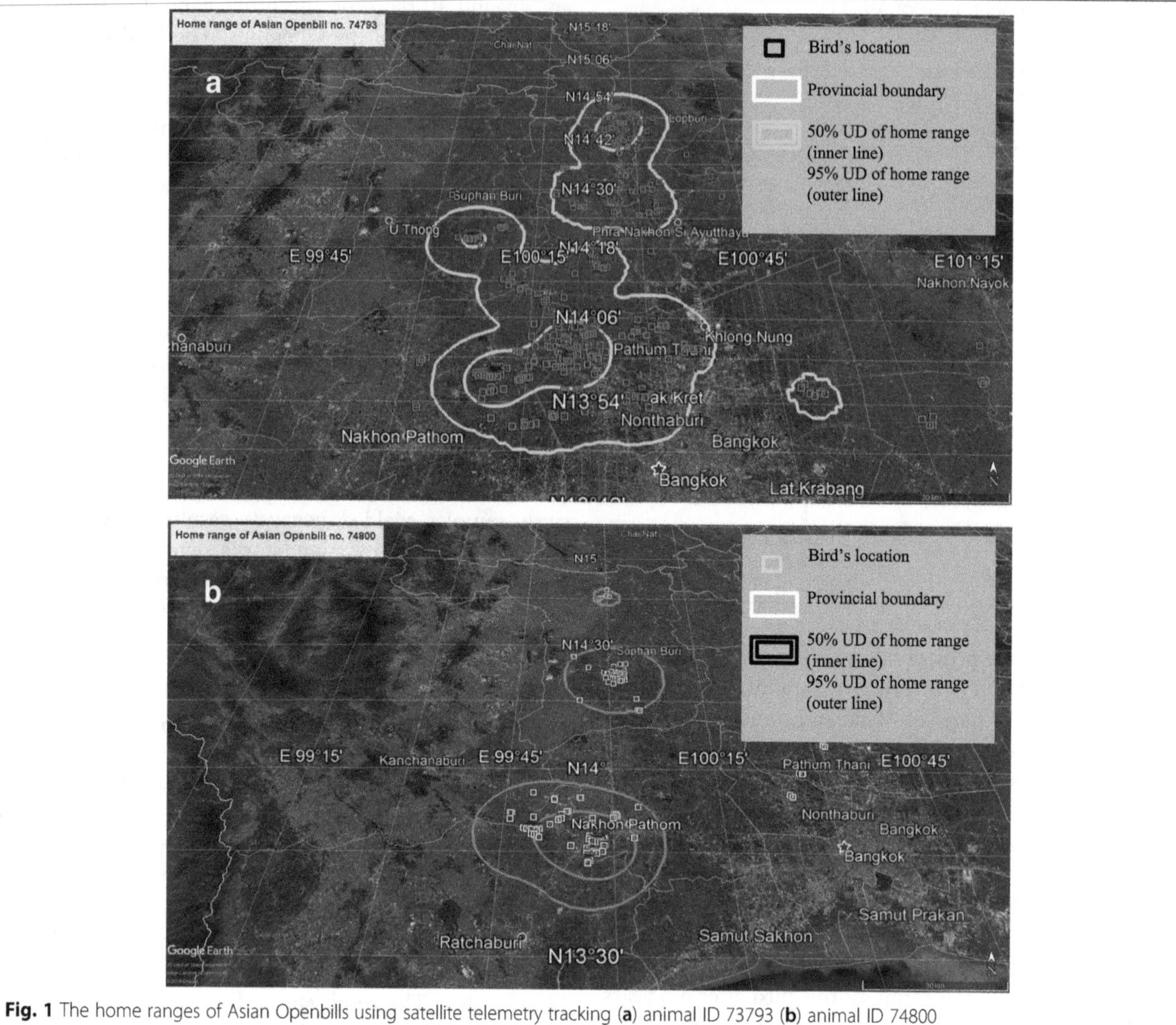

Fig. 1 The home ranges of Asian Openbills using satellite telemetry tracking (**a**) animal ID 73793 (**b**) animal ID 74800

Discussion

The present study determined the potential role of Openbills in the spread of H5N1 HPAI virus by using satellite telemetry to monitor the flyways of birds which overlapped to the locations of AI outbreaks. Using solar power PTT-100 and Argos/GPS PTT-100 transmitters, we monitored the movements of 5 Openbills for periods varying from 1 to more than 5 years between February 2007 and September 2013. During the monitoring period, the flyways or movements of these Openbills did not correlate with the locations of H5N1 HPAI outbreak. We noted that with the strict control measures of the government and private sector actors, the occurrence of AI outbreaks was rapidly declining and became rare events during our study period. Previous investigators have demonstrated an H5N1 virus infection rate of 26.7% in Openbills in 2004; subsequently, the prevalence decreased to 0.9% in 2005, and 0% in 2006 and 2007 [11]. Similarly, the H5N1 HPAI infection rate in the flocks of Openbills in this study was low as demonstrated by the prevalence of H5N1 neutralizing antibody of 2.2% in 2007, 1.7% in 2008, and 0% in 2009 and 2010. Nevertheless, the presence of H5N1 antibody in some Openbills suggested that they might have been infected and survived. Our previous work showed that Openbills in captivity were susceptible to H5N1 HPAI virus infection. All of the virus-inoculated Openbills developed clinical symptoms and died, even with a low virus inoculum of 10 TCID50 [13]. Pigeons, however, were more tolerate [14]. Another group of investigators also reported that the inoculated pigeons did not develop clinical symptoms, even with high virus inoculum of 10^6 egg infectious dose 50 [15]. Also, the prevalence of the H5N1 HPAI infection rate in a variety of wild birds was approximately 0.27% (17/6263) in the previous Thailand nationwide study from 2004 to 2007 [11], compared with 1.9% (4/206) from 2008 to 2009 as determined by MN assay for H5N1 antibody in this study (Additional

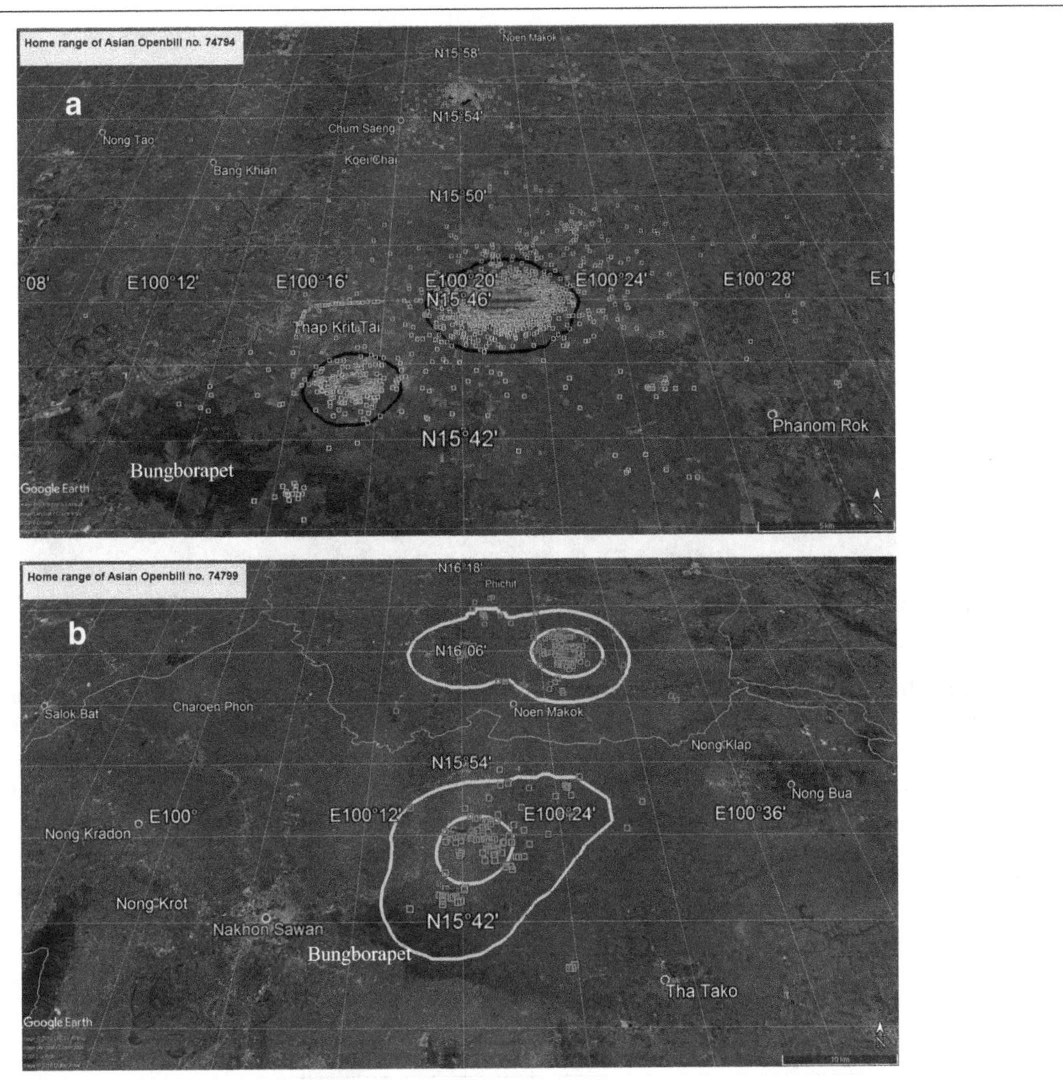

Fig. 2 The home ranges of Asian Openbills using satellite telemetry tracking (**a**) animal ID 73794 (**b**) animal ID 74799

file 1). The evidence supported that the most of Openbills would die with AI infection, the survived birds may develop the immunity and may not be a carrier of AI. Therefore, the healthy Openbills may be less likely to shed the virus to the wild birds living in the vicinity of the flyway [11, 13]. We have shown that the movement areas of Openbills were strictly related to populations of apple snails, which coupled with the rice crop rotation [5, 16]. An Asian Openbill can eat as many as 123 apple snails per day. The highest density of snails can reach 2.49 individuals/square meter [5]. This apple snail species (*P. canaliculata* Lamarck) is an alien imported from Taiwan and Japan into Thailand in 1982, for decorating and cleaning the walls of fish tanks. Snails escaped and bred very quickly, becoming overpopulated. They subsequently spread in natural waterways throughout the country. After the big flood in 1995, apple snails were found in at least 60 of 76 (number at that time) provinces of Thailand [7, 16]. In nature, the apple snails live in freshwater marshes, ditches and rice paddies (Fig. 5a). A snail can lay approximately 400–3000 eggs per clutch, and it can lay eggs throughout the wet season (Fig. 5b). During the dry season, apple snails embed themselves underground where they can survive without eating for years. These alien snails now replace most populations of native *Pila* snail species in the fields. Apple snails are an enemy that causes considerable economic losses in rice agriculture. The Openbills partially control them as predators. Eventually, the free-grazing ducks also feed on snails besides falling rice grains after harvesting season.

The primary factor that contributes to the over-population of the apple snails in Thailand is the improvement of the irrigation system and rice farming. Several decades ago, Thai farmers grew rice only once or twice a year depending on rainfall and the

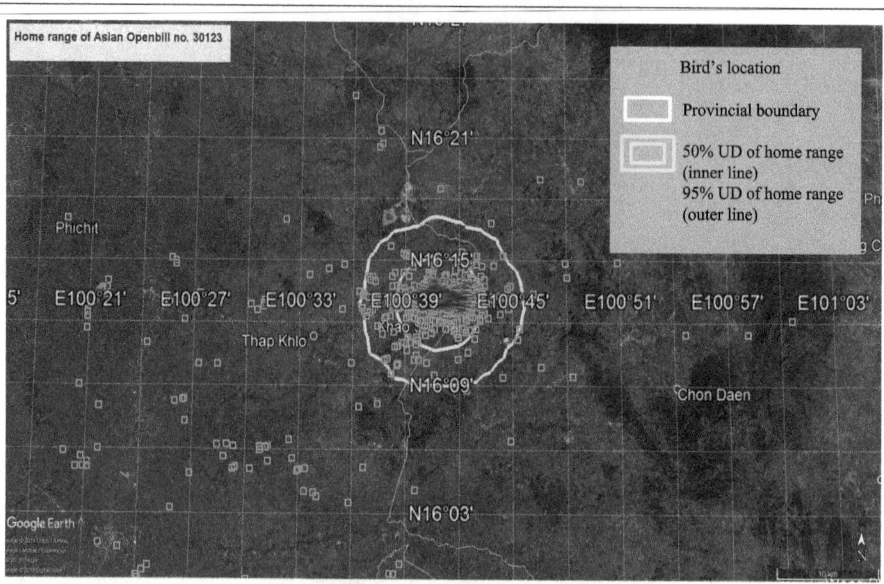

Fig. 3 The home ranges of Asian Openbill ID 30123 using satellite telemetry tracking

water supply. Improvements of the irrigation system by construction of dams, large water reservoirs, and canals, together with rice strain selection, allowed Thai farmers to grow rice all year round. Crop rotation takes place two to three times a year and generates abundant young rice stems, a large food surplus for apple snails. During our monitoring period, the movements of Openbills from their nesting places to feeding areas coincided with the beginning of a new rice crop (Additional file 1). With abundant young rice stems available all year round, the population of apple snails enlarged greatly. Consequently, with plenty of the apple snails available, the Openbills do not need to migrate back to South Asia anymore. Moreover, the Openbill is a protected animal species by Thai law, and this has facilitated their expansion in Thailand in recent years. Initially, there was only a single colony of the Openbills in Thailand, situated at Wat Pailom, Pathum Thani province. At present, there are several colonies throughout the country [17], and they have become permanent residents in Thailand. This result is opposite to our finding on

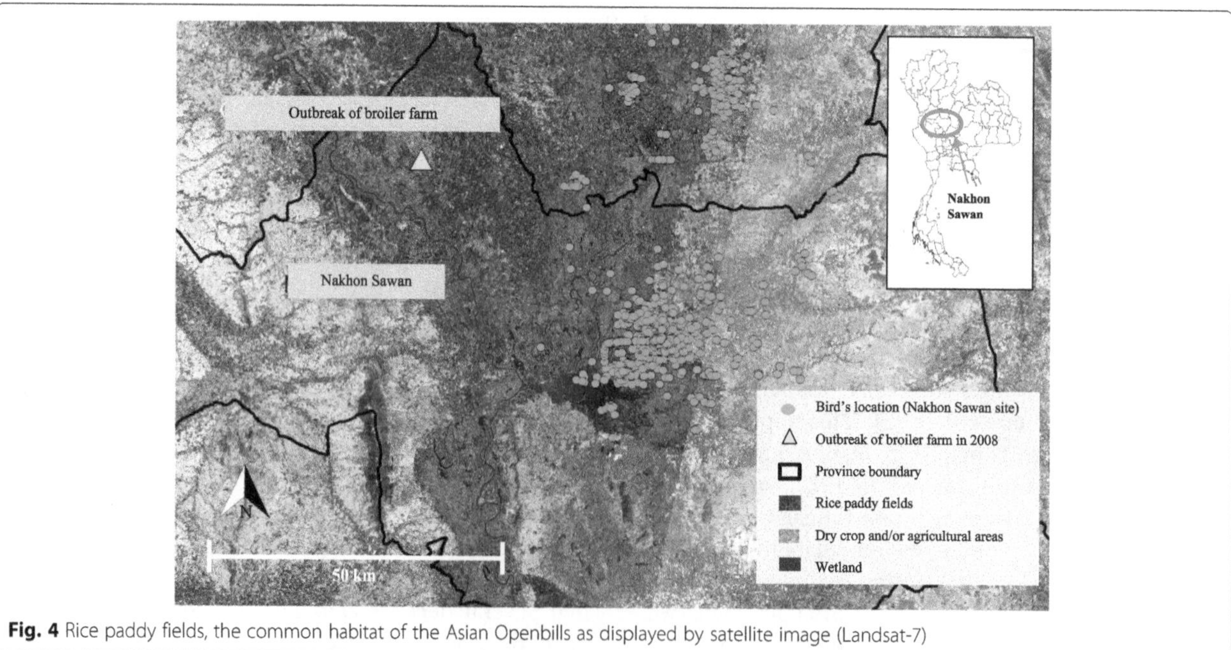

Fig. 4 Rice paddy fields, the common habitat of the Asian Openbills as displayed by satellite image (Landsat-7)

Table 3 Serosurveillance of HPAI H5N1of Asian Openbills in Thailand during 2007–2010

Month	Percentage of HPAI H5N1 antibody-positive samples			
	2007	2008	2009	2010
Jan	–	–	–	0/66 (0%)
Feb	0/48 (0%)	0/57 (0%)	–	–
Mar	0/59 (0%)	–	–	–
Apr	–	–	0/64 (0%)	–
May	–	–	–	–
Jun	–	–	–	–
Jul	–	–	–	–
Aug	–	–	–	–
Sep	0/3	–	–	–
Oct	–	–	–	–
Nov	0/1	–	–	–
Dec	4/70 (5.7%)	2/63 (3.2%)	–	–
Total antibody-positive samples per year	4/181(2.2%)	2/120 (1.7%)	0/64 (0%)	0/66 (0%)

tracking the flyways of brown-headed gulls that they are winter visitors to Thailand until the present [18]. In that study, direct linkage between the flyways of brown-headed gulls and AI spread in Thailand could not be demonstrated. Nevertheless, the gulls flew across 7 countries that faced the problem of H5N1 AI outbreaks. The role of migratory birds on AI spread has been suggested in many studies [19–21].

It is common to see Openbills sharing the same habitats with free-ranging ducks and other resident birds in marshes and rice paddies in Thailand. Improvements in rice farming have caused enormous increases in populations of Apple snails, which has changed the behavior of Openbills from being migrants to becoming permanent residents of Thailand. The Openbills have been found dead in the rice fields along with ducks. We suggest that the ducks might be transmitting the virus to Openbills, as free-ranging ducks have been shown to associate with H5N1 HPAI spread in Thailand [22].

H5N1 HPAI virus was first identified in Guangdong, China, in 1996. The emerged virus has gone through genetic evolution rapidly and resulted in at least 32 clades/subclades at present [23]. Subsequently, multiple events of gene reassortment between H5N1 virus and the other AI virus subtypes haves given rise to the H5 viruses with various N subtypes over time, i.e., H5N2, H5N5, H5N6, H5N8 and H5N9 [24]. The H5N1 viruses that spread all over Thailand since the initial epidemic waves belonged to clade 1 [25], and the virus in clade 2.3.4 was first detected in the northeastern region of Thailand between 2006 and 2007 [26]. The only AI reassortants recognized in Thailand were generated from reassortment of the internal gene segments among H5N1 virus population [27]. Even though our laboratory investigation system can pick up all subtypes of HPAI and LPAI viruses, neither H5Nx nor other AI virus had been detected in this study. Nevertheless, a recent report from the other group of Thai investigators showed that they could isolate an H5N2 LPAI virus from a cloacal specimen collected from apparently healthy free-ranging ducks in 2007 [28]. The H7N9 virus has not been detected in Thailand up to the present.

Fig. 5 Apple snails (**a**); and a clutch of snail's eggs (**b**)

Although the relationship between flyways of Openbills and H5N1 AI spread could not be established in this study, a possible risk of Openbills on acquiring H5N1 infection through sharing habitats with free-ranging ducks has been demonstrated. Many environmental factors affect the behavior of Openbills and their possible role in carrying the H5N1 virus (Fig. 6). During 5 years of our study, the life cycle demonstrating the relationships between the Openbills' flyways, apple snail populations, free-ranging ducks and wild birds and rice agriculture has been established for the first time (Fig. 6).

Conclusions

Satellite telemetry had been conducted to monitor the flyways of 5 Asian Openbills in correlation to their potential role on H5N1 AI spread to poultry and domestic birds and within their flocks. The flyway analysis showed that Openbills played no role in AI spread which was probably due to the low prevalence of H5N1 virus during the monitoring period of 6 years. The flyway data showed that the Openbills changed their behavior from being the migratory birds to be the resident birds according to the abundant surplus of the Apple snails, their favorite food and an exotic animal species of Thailand. Moreover, an improvement of irrigation system improved the efficiency of rice farming and made the young rice stem available all year round to feed the Apple snails. Instead of demonstrating the role of Openbills on AI spread, this study first revealed the influence of agricultural and ecological changes on the Openbills life cycle.

Methods

Study design

Openbills from three colonies in different geographical areas were tagged with satellite transmitters, and their flyways/movements were monitored until the transmitter signals lost or the project terminated the monitoring. Throughout the monitoring period, the location of each bird tagged was positioned from the transmitter signals emitted. Field investigators trapped the birds in their breeding colonies where the numbers of birds were at highest density. Approximately 50–60 birds per flock were trapped each time for collection of oropharyngeal and cloacal swabs and blood specimens for investigation of H5N1 AI virus infection. The wild birds living in the vicinity where Openbills foraged were also trapped and investigated for H5N1 AI virus infection, in particular, those living in the areas with poultry die-offs, and/or HPAI outbreaks.

Study sites

Three sites where Openbills nested and formed colonies were located in three provinces of Thailand: Bang Len district, Nakhon Pathom province, in Central Thailand; Bungborapet (a sizeable fresh marsh), Nakhon Sawan province, in the northern Central region; and Phitsanulok province in North Thailand; at distances of 58, 244 and 444 km from Bangkok, respectively. All study sites had good irrigation systems and are close to rivers or water reservoirs. The sites were mainly agricultural areas with paddy rice fields. Two to three rice crops are produced per year. Repeated outbreaks of H5N1 HPAI were reported in several poultry farms in these three provinces during 2004 to 2005.

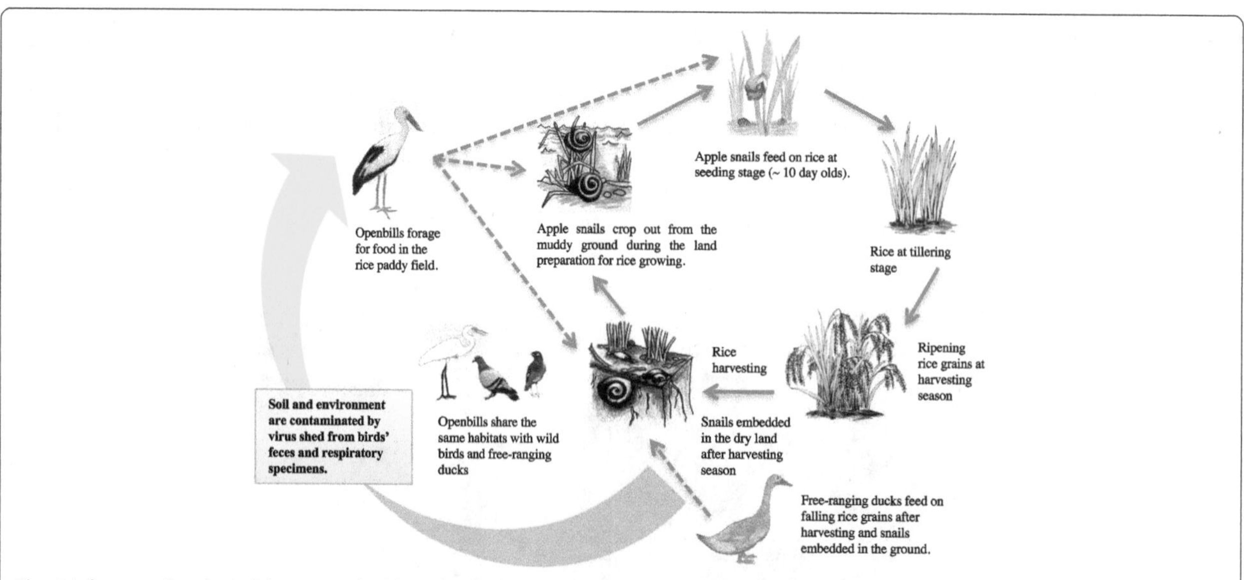

Fig. 6 Influence of ecological factors on the life cycle of Asian Openbills and their risk of getting H5N1 HPAI virus infection from free-ranging ducks and other wild birds through habitat sharing

The colony in Nakhon Pathom province was located in a mango orchard where the Openbills nested on the mango trees. This study site was surrounded by several poultry farms and various kinds of farm gardens including rice fields. The Nakhon Sawan site was close to Bungborapet where hundreds of species of resident and migratory birds shared the same habitat. The third study site, in Phitsanulok province, was surrounded by agricultural fields. This province is famous for the breeding of fighting cocks.

Tracking the flyways of Openbills using satellite telemetry

Asian Openbills were trapped from their nests by hand net during the night. Being colonial breeders, the birds built several nests in the same tree (Fig. 7a). The birds were given physical examinations for health status. Oropharyngeal and cloacal swabs were collected and kept in separate tubes of viral transport media, and 1–2 ml of blood was collected from the wing vein or jugular vein. All birds were screened at the study sites for influenza viral infection by detection of the influenza viral antigen in swab samples using immunochromatography (Bio-Chek, London, UK). Other aliquots of swab samples and blood specimens were sent for further complete investigation of H5N1 HPAI infection at the Virology Laboratory, Faculty of Veterinary Science, Mahidol University. None of the Openbills screened was infected with the H5N1 virus.

Five healthy adult Openbills negative for influenza viral antigen were selected from colonies in the three study sites for flyway monitoring by satellite telemetry. Each of these Openbills was tagged with a ring band and fitted with a 35-g-solar power PTT-100 or a 40-g solar Argos/GPS PTT-100 satellite transmitter (Microwave Telemetry Inc., Columbia, MD) on its back with a Teflon harness (Bally Ribbon Mills, Bally, PA) (Fig. 7b). The solar PTT-100 transmitters operated at a frequency of 401.650 MHz with a standard duty cycle of 10 h on and 24 h off for recharging the batteries. The solar-powered Argos/GPS PTT-100 satellite transmitter was attached to a tiny GPS receiver for locations with higher accuracy [29]. This transmitter package weighed about 0.06% of the birds' body weights. The birds were freed at the capture sites within 1 h after tagging. The activated satellite transmitters attached to the Openbills emitted ultra-high frequency signals that could be detected by special ARGOS receivers on a polar-orbit weather satellite. The data on locations of the tagged birds were retrieved from the emitted signals, recorded and relayed to a ground station in the United States every two days, and subsequently to the server of Argos CLS Company (Toulouse, France). The data on bird movements at location classes 1, 2 or 3 according to Argos was analyzed in our data manipulation laboratory, and further mapped with Google Earth Program version 7.1.5 (Google, Mountain View, CA, USA) to obtain the real-time location of birds with a precision of less than 1500-m error. If the emitted signals lost for longer than one month, it implied that the transmitter-tagged bird was sick or dead, then the flyway monitoring was terminated.

Spatial analysis

Microsoft Excel 2016 (Microsoft Corporation, Redmond, Washington, USA) was used for the data analysis. The positions of each Openbill were displayed and analyzed by ArcGIS 9.3.1 and ArcGIS 10 software (Environmental Systems Research Institute, Redlands, CA, USA). More than 10,000 records of the bird positions were cleaned and determined for their habitats and home ranges using a fixed kernel density estimator, and then displayed on Google Earth maps.

Laboratory investigation

Oropharyngeal and cloacal swabs from Openbills and other wild birds caught during 2007 to 2010 were used for detection of H5N1 HPAI viral genome by real-time reverse transcription-polymerase chain reaction (RT-PCR),

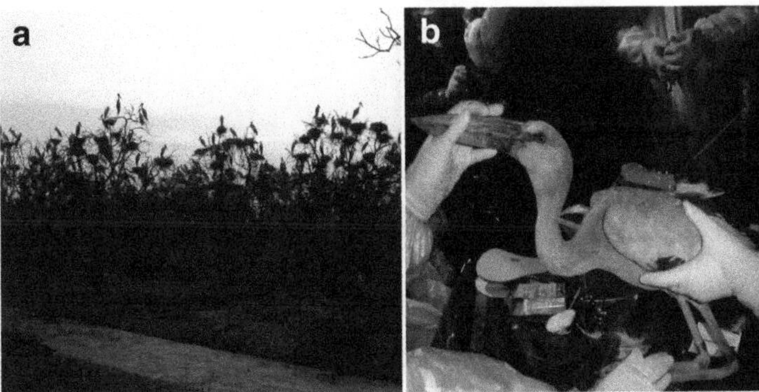

Fig. 7 Tagging of the Asian Openbills: (**a**) An overcrowded colony of Openbills with several nests on each tree; (**b**) Tagging an Openbill with a satellite transmitter and a ring band

and also for virus isolation in chick embryonated eggs and Madin Darby canine kidney (MDCK) cells; and blood specimens were used for detection of H5N1 antibody by microneutralization (MN) assays.

Viral genome detection

Real-time RT-PCR for detection of H5N1 HPAI viral genome was performed using the protocol as described by OIE (https://www.oie.int/fileadmin/Home/eng/Health_-standards/tahm/2.03.04_AI.pdf) and/or those established by the U.S. Centers for Disease Control and Prevention (CDC). Oropharyngeal and cloacal swabs from each bird were investigated in separate reaction tubes.

Viral isolation technique

Cloacal and throat swab specimens were inoculated separately in chick embryonated eggs and Madin Darby canine kidney (MDCK) cell monolayers, in duplicate. Amniotic and allantoic fluids and MDCK culture supernatant were screened for the presence of influenza virus by hemagglutination with 0.5% goose red blood cells. Two blind passages were carried out before reporting the result as negative virus isolation.

Microneutralization (MN) assay

ELISA based-MN assay for detection of neutralizing antibody to H5N1 HPAI virus was employed according to the WHO manual for avian influenza [30], and modified by our laboratory [31] in which MDCK cell monolayers were used instead of MDCK cell suspension. The assay was performed in microculture plates in duplicate using A/chicken/Thailand/ICRC-V143/07(H5N1) at a final concentration of 100 tissue culture infective dose 50 (TCID50) as the test virus. Influenza viral nucleoprotein produced in the virus-infected cells was detected by a mouse monoclonal specific antibody (Chemicon International, Inc., Tecumala, CA) as the primary antibody and horseradish peroxidase conjugated-rabbit anti-mouse Ig (Dako Cytomation, Denmark) as the secondary antibody. Antibody titer was defined as the reciprocal of highest serum dilution that causes a 50% reduction in the amount of viral nucleoprotein produced as compared to the virus controls.

Abbreviations

AI: Avian influenza; CDC: The Centers for Disease Control and Prevention; CI: Confidential interval; DLD: The Department of Livestock Development; HPAI: Highly pathogenic avian influenza; km^2: Square kilometer; MDCK: Madin Darby canine kidney; MN: Microneutralization assays; OIE: The Office International des Epizooties; OLAW: The Office of Laboratory Animal Welfare; RT-PCR: Real-time reverse transcription-polymerase chain reaction; TCID: Tissue culture infective dose; UD: Utility distribution

Acknowledgements

We are grateful to the Department of National Parks, Wildlife and Plant Conservation for collaboration in bird trapping; Dr. Adrian H. Farmer, previously from U.S. Geological.
Survey for the transfer of technology on satellite telemetry; Dr. Plern Yongyuttawichai for her help in coordinating the field investigation, and the laboratory team of the Monitoring and Surveillance Center for Zoonotic Diseases in Wildlife and Exotic Animals, Mahidol University for all sample collection and laboratory diagnosis; Assoc. Prof. Dr. Sura Pattanakiat, Faculty of Environment and Resource Studies, Mahidol University for providing the software for spatial analysis; Mr. Steve Durako and Mrs. Mekkla Thompson from Westat for coordinating the project; Dr. Sakranmanee Krajangwong and Miss Tatiyanuch Chamsai for the diagram of Openbill life cycle; and Prof. Warren Brockelman for English editing. We also thank the Center for Prevention and Disease Control (CDC) for funding.

Completing interests

The authors declare that they have no competing interests.

Funding

This study was a part of the "Avian Influenza Surveillance in Thailand-Studies at Human-Animal Interface" project supported by cooperative agreement number 1U19C1000399 from the Centers for Disease Control and Prevention (CDC). Its contents are solely the responsibility of the authors and do not necessarily represent the official views of the CDC. The funder had no role in the study design, data collection and analysis, decision to publish, or preparation of the manuscript.

Authors' contributions

Conceived and designed the study: PR, PP and WW. Performed the data collection: SS, AW and KE. Contributed to the laboratory tests: LS, KC and WW. Analyzed the data: SS and AW. Wrote and revised the manuscripts: PP and SS. All authors read and approved the final manuscript.

Consent for publication

Not applicable.

Author details

[1]Department of Clinical Science and Public Health, Faculty of Veterinary Science, Mahidol University, Nakhon Pathom 73170, Thailand. [2]The Monitoring and Surveillance Center for Zoonotic Diseases in Wildlife and Exotic Animals, Faculty of Veterinary Science, Mahidol University, Nakhon Pathom 73170, Thailand. [3]Department of Preclinic and Applied Animal Science, Faculty of Veterinary Science, Mahidol University, Nakhon Pathom 73170, Thailand. [4]Department of National Parks, Wildlife and Plant Conservation, Ministry of Natural Resources and Environment, Bangkok 10900, Thailand. [5]Center for Research and Innovation, Faculty of Medical Technology, Mahidol University, Nakhon Pathom 73170, Thailand. [6]Department of Microbiology, Faculty of Medicine Siriraj Hospital, Mahidol University, Bangkok 10700, Thailand.

References

1. Lekagul B, Round PD. A guide to birds of Thailand. Bangkok: Saha Karn Bhaet Group; 2005.
2. KL G. Scopate tomia: an adaptation for handling hard-shelled prey? Willson Bull. 1993;105:136–324.
3. McClure HE. Migration and survival of the birds of Asia. Sahamitr Karn Pim: Bangkok; 1974.
4. BirdLife International. *Anastomus oscitans*. In: The IUCN Red List of Threatened Species 2016. 2016. http://dx.doi.org/10.2305/IUCN.UK.2016-3.RLTS.T22697661A93628985.en. Accessed 20 May 2018.
5. Eiamampai K. Study on population, movement and food habitats of Asian open-billed stork (*Anastomus osticans* (Boddaert)). Department of National Parks, Wildlife and Conservation: Bangkok; 2008. (in Thai with English abstract)
6. Quasmieh S. *Anastomus oscitans*. In: Animal Diversity Web. 2013 http://animaldiversity.org/accounts/Anastomus_oscitans/. Accessed 28 May 2018.
7. Srinives S. Management for golden apple snail *(Pomacea canaliculata* Larmarck). Bangkok: Department of Agricultural Extension, Ministry of Agriculture and Cooperatives; 2004. (in Thai).
8. Thai National Parks. Asian openbill. In: Species of Thailand. Thai National Parks. 2018. https://www.thainationalparks.com/species/asian-openbill. Accessed 1 May 2018.
9. World Health Organization. Avian Influenza. In: Surveillance and outbreak alert. World Health Organization. 2017. http://www.searo.who.int/entity/emerging_diseases/topics/avian_influenza/en/. Accessed 5 Dec 2017.
10. World Health for Animal Health. Update on Highly Pathogenic Avian Influenza in Animals (Type H5 and H7). World Health for Animal Health. 2017. http://web.oie.int/wahis/reports/en_fup_0000007822_20090227_171633.pdf. Accessed 10 Dec 2017.
11. Siengsanan J, Chaichoune K, Phonaknguen R, Sariya L, Prompiram P, Kocharin W, et al. Comparison of outbreaks of H5N1 highly pathogenic avian influenza in wild birds and poultry in Thailand. J Wildl Dis. 2009;45:740–7.
12. Ratanakorn P, Suwanpakdee S, Wiriyarat W, Eiamampai K, Chaichoune K, Wiratsudakul A, et al. Asian Openbill tagging with satellite telemetry in Thailand, 2007-2013. Movebank Data Repository. 2018. https://doi.org/10.5441/001/1.1j802v05.
13. Chaichoun K, Wiriyarat W, Phonaknguen R, Sariya L, Taowan NA, Chakritbudsabong W, et al. Susceptibility of openbill storks (*Anastomius oscitans*) to highly pathogenic avian influenza virus subtype H5N1. Southeast Asian J Trop Med Public Health. 2013;44:799–809.
14. Hayashi T, Hiromoto Y, Chaichoune K, Patchimasiri T, Chakritbudsabong W, Prayoonwong N, et al. Host cytokine responses of pigeons infected with highly pathogenic Thai avian influenza viruses of subtype H5N1 isolated from wild birds. PLoS One. 2011;6:e23103.
15. Yamamoto Y, Nakamura K, Yamada M, Mase M. Persistence of avian influenza virus (H5N1) in feathers detached from bodies of infected domestic ducks. Appl Environ Microbiol. 2010;76:5496–9.
16. Eiamampai K, Chaipukdee M, Chaipukdee W, Sonsa T. Knowledge of Asian Openbill in Thailand. Bangkok: Wildlife Research Division, Wildlife Conservation Office, Department of National Parks, Wildlife and Plant Conservation; 2012. (in Thai).
17. Chanittawong W, Chaipakdee M. Migratory bird of Thailand. Bangkok: Department of National Parks, Wildlife and Plant Conservation; 2005. (in Thai)
18. Ratanakorn P, Wiratsudakul A, Wiriyarat W, Eiamampai K, Farmer AH, Webster RG, et al. Satellite tracking on the flyways of brown-headed gulls and their potential role in the spread of highly pathogenic avian influenza H5N1 virus. PLoS One. 2012;7:e49939.
19. Fourment M, Darling AE, Holmes EC. The impact of migratory flyways on the spread of avian influenza virus in North America. BMC Evol Biol. 2017;17(1):118.
20. Hu J, Xu X, Wang C, Bing G, Sun H, Pu J, et al. Isolation and characterization of H4N6 avian influenza viruses from mallard ducks in Beijing. China PLoS One. 2017;12:e0184437.
21. Onuma M, Kakogawa M, Yanagisawa M, Haga A, Okano T, Neagari Y, et.al. Characterizing the temporal patterns of avian influenza virus introduction into Japan by migratory birds. J Vet Med Sci 2017;79:943–951.
22. Songserm T, Jam-on R, Sae-Heng N, Meemak N, Hulse-Post DJ, Sturm-Ramirez KM, et al. Domestic ducks and H5N1 influenza epidemic. Thailand. Emerg Infect Dis. 2006;12:575–81.
23. Su S, Bi Y, Wong G, Gray GC, Gao GF, Li S. Epidemiology, evolution, and recent outbreaks of avian influenza virus in China. J Virol. 2015;89(17):8671–6.
24. Su S, Gu M, Liu D, Cui J, Gao GF, Zhou J, et al. Epidemiology, evolution, and pathogenesis of H7N9 influenza viruses in five epidemic waves since 2013 in China. Trends Microbiol. 2017;25(9):713–28.
25. Puthavathana P, Auewarakul P, Charoenying PC, Sangsiriwut K, Pooruk P, Boonnak K, et al. Molecular characterization of the complete genome of human influenza H5N1 virus isolates from Thailand. J Gen Virol. 2005;86(Pt 2):423–33.
26. Suwannakarn K, Amonsin A, Sasipreeyajan J, Kitikoon P, Tantilertcharoen R, Parchariyanon S, et al. Molecular evolution of H5N1 in Thailand between 2004 and 2008. Infect Genet Evol. 2009;9(5):896–902.
27. Chaichoune K, Wiriyarat W, Thitithanyanont A, Phonarknguen R, Sariya L, Suwanpakdee S, et al. Indigenous sources of 2007-2008 H5N1 avian influenza outbreaks in Thailand. J Gen Virol. 2009;90(Pt 1):216–22.
28. Taechowisan T, Dumpin K, Phutdhawong WS. Isolation of avian influenza a (H5N2) from free-grazing ducks in Thailand and antiviral effects of tea extracts. Int J Curr Res Life Sci. 2018;7(4):1810–6.
29. Microwave Telemetry Inc. Bird tracking. In: Argos Satellite Transmitters. Microwave Telemetry, Inc. 2017. http://www.microwavetelemetry.com/bird/. Accessed 7 Dec 2017.
30. World Health Organization. WHO manual on animal influenza diagnosis and surveillance. World Health Organization. 2002. http://www.who.int/csr/resources/publications/influenza/en/whocdscsrncs20025rev. pdf. Accessed 15 Dec 2013.
31. Louisirirotchanakul S, Lerdsamran H, Wiriyarat W, Sangsiriwut K, Chaichoune K, Pooruk P, et al. Erythrocyte binding preference of avian influenza H5N1 viruses. J Clin Microbiol. 2007;45:2284–6.

Permissions

All chapters in this book were first published in VR, by BioMed Central; hereby published with permission under the Creative Commons Attribution License or equivalent. Every chapter published in this book has been scrutinized by our experts. Their significance has been extensively debated. The topics covered herein carry significant findings which will fuel the growth of the discipline. They may even be implemented as practical applications or may be referred to as a beginning point for another development.

The contributors of this book come from diverse backgrounds, making this book a truly international effort. This book will bring forth new frontiers with its revolutionizing research information and detailed analysis of the nascent developments around the world.

We would like to thank all the contributing authors for lending their expertise to make the book truly unique. They have played a crucial role in the development of this book. Without their invaluable contributions this book wouldn't have been possible. They have made vital efforts to compile up to date information on the varied aspects of this subject to make this book a valuable addition to the collection of many professionals and students.

This book was conceptualized with the vision of imparting up-to-date information and advanced data in this field. To ensure the same, a matchless editorial board was set up. Every individual on the board went through rigorous rounds of assessment to prove their worth. After which they invested a large part of their time researching and compiling the most relevant data for our readers.

The editorial board has been involved in producing this book since its inception. They have spent rigorous hours researching and exploring the diverse topics which have resulted in the successful publishing of this book. They have passed on their knowledge of decades through this book. To expedite this challenging task, the publisher supported the team at every step. A small team of assistant editors was also appointed to further simplify the editing procedure and attain best results for the readers.

Apart from the editorial board, the designing team has also invested a significant amount of their time in understanding the subject and creating the most relevant covers. They scrutinized every image to scout for the most suitable representation of the subject and create an appropriate cover for the book.

The publishing team has been an ardent support to the editorial, designing and production team. Their endless efforts to recruit the best for this project, has resulted in the accomplishment of this book. They are a veteran in the field of academics and their pool of knowledge is as vast as their experience in printing. Their expertise and guidance has proved useful at every step. Their uncompromising quality standards have made this book an exceptional effort. Their encouragement from time to time has been an inspiration for everyone.

The publisher and the editorial board hope that this book will prove to be a valuable piece of knowledge for researchers, students, practitioners and scholars across the globe.

List of Contributors

Jianzhong Wang and Guixue Hu
College of Animal Science and Technology, Jilin Agricultural University, Changchun 130118, China

Na Feng
College of Animal Science and Technology, Jilin Agricultural University, Changchun 130118, China
Military Veterinary Research Institute of Academy of Military Medical Sciences, Key Laboratory of Jilin Province for Zoonosis Prevention and Control, Changchun 130122, China

Xianzhu Xia
College of Animal Science and Technology, Jilin Agricultural University, Changchun 130118, China
Military Veterinary Research Institute of Academy of Military Medical Sciences, Key Laboratory of Jilin Province for Zoonosis Prevention and Control, Changchun 130122, China
Jiangsu Co-innovation Center for Prevention and Control of Important Animal Infectious Diseases and Zoonosis, Yangzhou 225009, China

Yicong Yu, Hualei Wang, Yongkun Zhao, Songtao Yang, Yuwei Gao, Weiwei Xu and Tiecheng Wang
Military Veterinary Research Institute of Academy of Military Medical Sciences, Key Laboratory of Jilin Province for Zoonosis Prevention and Control, Changchun 130122, China

Yuxiu Liu
National Research Center for Veterinary Medicine, Luoyang, Henan 471000, China

Tiansong Li
College of Chemistry and Biology, Beihua University, Jilin 132013, China

Lei Wang
Department of Animal Science and Veterinary Medicine, Henan Institute of Science and Technology, Xinxiang 453003, China

Tao Hua, Xuehua Zhang, Bo Tang, Chen Chang, Guoyang Liu, Lei Feng, Yang Yu, Daohua Zhang and Jibo Hou
Institute of Veterinary Immunology & Engineering, Jiangsu Academy of Agricultural Sciences, Nanjing 210014, China
National Research Center of Engineering and Technology for Veterinary Biologicals, Jiangsu Academy of Agricultural Science, Nanjing 210014, China
Key lab of Food Quality and Safety of Jiangsu Province—State Key laboratory Breeding Base, Jiangsu Academy of Agricultural Science, Nanjing 210014, China
Jiangsu Co-innovation Center for Prevention and Control of Important Animal Infectious Diseases and Zoonoses, Yangzhou 225009, China

Yonghyan Kim, My Yang, Sagar M. Goyal, Maxim C-J. Cheeran and Montserrat Torremorell
Department of Veterinary Population Medicine, College of Veterinary Medicine, University of Minnesota, 1988 Fitch Ave, St. Paul, MN 55108, USA

Mahmoud Sabra
Department of Poultry Diseases, Faculty of Veterinary Medicine, South Valley University, Qena 83523, Egypt
Exotic and Emerging Avian Viral Diseases Research Unit, Southeast Poultry Research Laboratory, US National Poultry Research Center, Agricultural Research Service, USDA, 934 College Station Road, Athens, GA 30605, USA

Kiril M. Dimitrov, Poonam Sharma, Claudio L. Afonso, Dawn Williams-Coplin and Patti J. Miller
Exotic and Emerging Avian Viral Diseases Research Unit, Southeast Poultry Research Laboratory, US National Poultry Research Center, Agricultural Research Service, USDA, 934 College Station Road, Athens, GA 30605, USA

Iryna V. Goraichuk
Exotic and Emerging Avian Viral Diseases Research Unit, Southeast Poultry Research Laboratory, US National Poultry Research Center, Agricultural Research Service, USDA, 934 College Station Road, Athens, GA 30605, USA
National Scientific Center Institute of Experimental and Clinical Veterinary Medicine, 83 Pushkinskaya Street, Kharkiv 61023, Ukraine

Denys V. Muzyka
National Scientific Center Institute of Experimental and Clinical Veterinary Medicine, 83 Pushkinskaya Street, Kharkiv 61023, Ukraine

Asma Basharat and Shafqat F. Rehmani
Quality Operations Laboratory (QOL), University of Veterinary and Animal Sciences, Syed Abdul Qadir Jilani Road, Lahore 54000, Pakistan

Abdul Wajid
Quality Operations Laboratory (QOL), University of Veterinary and Animal Sciences, Syed Abdul Qadir Jilani Road, Lahore 54000, Pakistan
Institute of Biochemistry and Biotechnology, University of Veterinary and Animal Sciences, Syed Abdul Qadir Jilani Road, Lahore 54000, Pakistan

N. B. Goecke, C. K. Hjulsager, S. Rasmussen, S. E. Jorsal and L. E. Larsen
National Veterinary Institute, Technical University of Denmark, Kemitorvet, Lyngby DK-2800, Denmark
Ø. Angen
National Veterinary Institute, Technical University of Denmark, Lindholm, Kalvehave DK-4771, Denmark
Department of Veterinary and Animal Sciences, Faculty of Health and Medical Sciences, University of Copenhagen, Gronnegaardsvej 15, DK-1870 Frederiksberg, Denmark

M. Boye
National Veterinary Institute, Technical University of Denmark, Kemitorvet, Lyngby DK-2800, Denmark

F. Granberg
Department of Biomedical Sciences and Veterinary Public Health (BVF), Swedish University of Agricultural Sciences (SLU), Uppsala, Sweden

T. K. Fischer and S. E. Midgley
Statens Serum Institut (SSI), Artillerivej 5, Copenhagen S DK-2300, Denmark

L. D. Rasmussen
Statens Serum Institut (SSI), Artillerivej 5, Copenhagen S DK-2300, Denmark
National Veterinary Institute, Technical University of Denmark, Lindholm, Kalvehave DK-4771, Denmark

H. Kongsted
Pig Research Centre, Danish Agriculture and Food Council, Vinkelvej 13, DK-8620 Kjellerup, Denmark

J. P. Nielsen
Department of Veterinary and Animal Sciences, Faculty of Health and Medical Sciences, University of Copenhagen, Gronnegaardsvej 15, DK-1870 Frederiksberg, Denmark

Ignacio García-Bocanegra, Jorge Paniagua, Antonio Arenas-Montes and David Cano-Terriza
Departamento de Sanidad Animal, Facultad de Veterinaria, Universidad de Córdoba-Agrifood Excellence International Campus (ceiA3), Rabanales, 14071 Córdoba, Spain

Ana V. Gutiérrez-Guzmán, Christian Gortázar and Ursula Höfle
Instituto de Investigación en Recursos Cinegéticos IREC, (CSIC-UCLM-JCCM), Ciudad Real, Spain

Steeve Lowenski and Sylvie Lecollinet
ANSES, Laboratoire de Santé Animale de Maisons-Alfort, UMR 1161 Virologie, INRA, ANSES, ENVA, Maisons-Alfort F-94703, France

Mariana Boadella
Sabiotec, Camino de Moledores s.n., Ed. Polivalente UCLM, 13005 Ciudad Real, Spain

Shuya Mitoma, Kosuke Notsu, Yuta Sakai and Ryoji Yamaguchi
Animal Infectious Disease and Prevention, Department of Veterinary Sciences, Faculty of Agriculture, University of Miyazaki, Miyazaki, Japan

Thi Ngan Mai and Van Diep Nguyen
Animal Infectious Disease and Prevention, Department of Veterinary Sciences, Faculty of Agriculture, University of Miyazaki, Miyazaki, Japan
Faculty of Veterinary Medicine, Vietnam National University of Agriculture, Hanoi, Vietnam

Wataru Yamazaki, Tamaki Okabayashi, Junzo Norimine and Satoshi Sekiguchi
Animal Infectious Disease and Prevention, Department of Veterinary Sciences, Faculty of Agriculture, University of Miyazaki, Miyazaki, Japan
Center for Animal Disease Control, University of Miyazaki, Miyazaki, Japan

Yunyun Geng, Yan Lu, Ke Tan and Yan-Zhong Chang
College of Life Sciences, Hebei Normal University, No.20, Road E. 2nd Ring South, Yuhua District, Shijiazhuang, Hebei Province 050024, People's Republic of China

Jianchang Wang and Libing Liu
Center of Inspection and Quarantine, Hebei Entry-Exit Inspection and Quarantine Bureau, No.318 Hepingxilu Road, Shijiazhuang, Hebei Province 050051, People's Republic of China
Hebei Academy of inspection and quarantine science and technology, No.318 Hepingxilu Road, Shijiazhuang, Hebei Province 050051, People's Republic of China

Daishen Feng, Min Cui and Siyang Liu
Research Center of Avian Disease, College of Veterinary Medicine, Sichuan Agricultural University, Chengdu 611130, Sichuan, China
Institute of Preventive Veterinary Medicine, Sichuan Agricultural University, Chengdu 611130, Sichuan, China

Renyong Jia, Mingshu Wang, Shun Chen, Mafeng Liu, Xinxin Zhao, Yin Wu, Qiao Yang and Anchun Cheng
Research Center of Avian Disease, College of Veterinary Medicine, Sichuan Agricultural University, Chengdu 611130, Sichuan, China
Institute of Preventive Veterinary Medicine, Sichuan Agricultural University, Chengdu 611130, Sichuan, China
Key Laboratory of Animal Disease and Human Health of Sichuan Province, Chengdu 611130, Sichuan, China

Dekang Zhu
Research Center of Avian Disease, College of Veterinary Medicine, Sichuan Agricultural University, Chengdu 611130, Sichuan, China
Key Laboratory of Animal Disease and Human Health of Sichuan Province, Chengdu 611130, Sichuan, China

Zhongqiong Yin
Key Laboratory of Animal Disease and Human Health of Sichuan Province, Chengdu 611130, Sichuan, China

Denise A. Marston
Animal and Plant Health Agency, New Haw, Addlestone, Surrey KT15 3NB, UK

Torfinn Moldal, Turid Vikøren, Knut Madslien and Irene Ørpetveit
Norwegian Veterinary Institute, Sentrum, 0106 Oslo, Norway

Florence Cliquet
Nancy OIE/WHO/EU Laboratory for Rabies and Wildlife, French Agency for Food, Environmental and Occupational Health & Safety, CS 40009, 54220 Malzéville, France

Jeroen van der Kooij
Norwegian Zoological Society's Bat Care Centre, Rudsteinveien 67, 1480 Slattum, Norway

Maria Carolina Rocha de Medeiros Bento, Luís Manuel Morgado Tavares and Ana Isabel Simões Pereira Duarte
Centre for Interdisciplinary Research in Animal Health, Faculty of Veterinary Medicine, University of Lisbon, 1300-477 Lisbon, Portugal

Ana Luisa Marçalo and Catarina Isabel Costa Simões Eira
Department of Biology and CESAM, University of Aveiro, 3810-193 Aveiro, Portugal
Portuguese Wildlife Society, Department of Biology, Minho University, 4710-057 Braga, Portugal

Alfredo Lopez Fernandez
Department of Biology and CESAM, University of Aveiro, 3810-193 Aveiro, Portugal
Coordinadora para o Estudo dos Mamíferos Mariños, 36380 Gondomar, Pontevedra, Spain

José Vitor Vingada
Portuguese Wildlife Society, Department of Biology, Minho University, 4710-057 Braga, Portugal
Department of Biology and CESAM, Minho University, 4710-057 Braga, Portugal

Marisa Cláudia Teixeira Ferreira
Portuguese Wildlife Society, Department of Biology, Minho University, 4710-057 Braga, Portugal
Department of Biology and CBMA, Minho University, 4710-057 Braga, Portugal

Tina Naglič and Andrej Steyer
Institute of Microbiology and Immunology, Faculty of Medicine, University of Ljubljana, Zaloška cesta 4, SI-1000 Ljubljana, Slovenia

Danijela Rihtarič, Peter Hostnik and Urška Kuhar
Institute of Microbiology and Parasitology, Veterinary Faculty, University of Ljubljana, Gerbičeva ulica 60, Ljubljana, Slovenia

Nataša Toplak and Simon Koren
Omega d.o.o, Dolinškova ulica 8, Ljubljana, Slovenia

Urška Jamnikar-Ciglenečki
Institute of Food Safety, Feed and Environment, Veterinary Faculty, University of Ljubljana, Gerbičeva ulica 60, Ljubljana, Slovenia

Denis Kutnjak
National Institute of Biology, Večna pot, 111 Ljubljana, Slovenia

Olivier Péter
Central Institute of Valais Hospitals, Infectious diseases, Av Grand Champsec 86, -1950 Sion, CH, Switzerland

Nadia Rieille
Central Institute of Valais Hospitals, Infectious diseases, Av Grand Champsec 86, -1950 Sion, CH, Switzerland
Institute of Biology, Laboratory of Ecology and Evolution of parasites, University of Neuchâtel, Rue Emile-Argand 11, 2000 Neuchâtel, Neuchâtel, Switzerland

Christine Klaus
Friedrich-Loeffler-Institut, Institute of Bacterial Infections and Zoonoses, Naumburger Str. 96a, D-07743 Jena, Germany

Donata Hoffmann
Friedrich-Loeffler-Institut, Institute of Diagnostic Virology, Südufer 10, D-17493 Greifswald-Insel Riems, Germany

Maarten J. Voordouw
Institute of Biology, Laboratory of Ecology and Evolution of parasites, University of Neuchâtel, Rue Emile-Argand 11, 2000 Neuchâtel, Neuchâtel, Switzerland

Hai Quynh Do, Dinh Thau Trinh, Thi Lan Nguyen and Van Phan Le
Faculty of Veterinary Medicine, Vietnam National University of Agriculture, Hanoi, Vietnam

Thi Thu Hang Vu, Duc Duong Than and Thi Van Lo
Research and Development Laboratory, Avac Vietnam Company Limited (AVAC), Hung Yen, Vietnam

Minjoo Yeom and Daesub Song
College of Pharmacy, Korea University, Sejong, Republic of Korea

SeEun Choe and Dong-Jun An
Animal and Plant Quarantine Agency, Gyeonggi-do, Gimcheon, Gyeongsangbukdo, Republic of Korea

Evelyne Picard-Meyer, Alexandre Servat, Marine Wasniewski and Florence Cliquet
ANSES Nancy Laboratory for Rabies and Wildlife, European Union Reference Laboratory for Rabies, WHO Collaborating Centre for Research and Management in Zoonoses Control, OIE Reference Laboratory for Rabies, European Union Reference Institute for Rabies Serology, Technopôle agricole et vétérinaire de Pixérécourt, CS 40009, 54220 Malzéville, France

Matthieu Gaillard
Néomys association, Centre Ariane, 240 rue de Cumène, 54230 Neuves-Maisons, France

Christophe Borel
CPEPESC-Lorraine, Centre Ariane, 240 rue de Cumène, 54230 Neuves-Maisons, France

Fanfeng Meng, Xuan Dong, Shuang Chang, Peng Zhao and Zhizhong Cui
College of Veterinary Medicine, Shandong Agricultural University, Taian 271018, China

Tao Hu
Institute of Pathogen Biology, Taishan Medical College, Taian, Shandong, China

Jianhua Fan
Poultry Institute, Chinese Academy of Agricultural Sciences, Yangzhou, Jiangsu, China

Ivana Šimić, Ivana Lojkić, Nina Krešić and Tomislav Bedeković
Croatian Veterinary Institute, Savska cesta 143, 10000 Zagreb, Croatia

Florence Cliquet, Evelyne Picard-Meyer and Marine Wasniewski
ANSES - Nancy Laboratory for rabies and wildlife, Batiment H CS 40009, 54220 Malzeville, France
Anđela Ćukušić and Vida Zrnčić
Croatian Biospeleological Society, Demetrova 1, 10000 Zagreb, Croatia

Wenqiao He, Yuqi Wen, Yiquan Xiong, Minyi Zhang, Mingji Cheng and Qing Chen
Department of Epidemiology, School of Public Health, Guangdong Provincial Key Laboratory of Tropical Disease Research, Southern Medical University, 1838 Guangzhou North Road, Guangzhou 510515, China

Pablo Enrique Piñeyro and Darin Michael Madson
Veterinary Diagnostic Laboratory, 1655 Veterinary Medicine, Iowa State University, 1850 Christensen Drive, Ames, IA 50011, USA

Carlos Juan Perfumo, Maria Alejandra Quiroga, Alberto Armocida, Estefanía Marisol Pérez and Maria Inez Lozada
Laboratorio de Patología Especial Veterinaria FCV-UNLP, Calle 60 y 118 S/N (1900), La Plata, Buenos Aires, Argentina

Laura Valeria Alarcón
HIPRA Argentina, Saenz Peña R. Pte. Av 1110, Capital Federal, Argentina

Ramon Sanguinetti
DILACOT-SENASA, Av A Fleming 1653, Martinez, Buenos Aires, Argentina

Javier Alejandro Cappuccio
EEA Marcos Juaréz, INTA, CONICET, Ruta 12 km 3 (2580), Marcos Juárez, Córdoba, Argentina

Fabio Vannucci
Veterinary Diagnostic Laboratory, University of Minnesota, 1333 Gortner Ave, St Paul, MN, USA

Hongmei Wang, Peili Hou, Guimin Zhao and Hongbin He
Ruminant Diseases Research Center, Key Laboratory of Animal Resistant Biology of Shandong, College of Life Sciences, Shandong Normal University, Jinan 250014, China

Li Yu
Division of Livestock Infectious Diseases, State Key Laboratory of Veterinary Biotechnology, Harbin Veterinary Research Institute, Harbin 150001, China

Yu-wei Gao
Key Laboratory of Jilin Province for Zoonosis Prevention and Control, Military Veterinary Research Institute of Academy of Military Medical Sciences, Changchun 130122, China

Parntep Ratanakorn, Sarin Suwanpakdee and Anuwat Wiratsudakul
Department of Clinical Science and Public Health, Faculty of Veterinary Science, Mahidol University, Nakhon Pathom 73170, Thailand
The Monitoring and Surveillance Center for Zoonotic Diseases in Wildlife and Exotic Animals, Faculty of Veterinary Science, Mahidol University, Nakhon Pathom 73170, Thailand

Ladawan Sariya
The Monitoring and Surveillance Center for Zoonotic Diseases in Wildlife and Exotic Animals, Faculty of Veterinary Science, Mahidol University, Nakhon Pathom 73170, Thailand

Witthawat Wiriyarat and Kridsada Chaichoune
The Monitoring and Surveillance Center for Zoonotic Diseases in Wildlife and Exotic Animals, Faculty of Veterinary Science, Mahidol University, Nakhon Pathom 73170, Thailand
Department of Preclinic and Applied Animal Science, Faculty of Veterinary Science, Mahidol University, Nakhon Pathom 73170, Thailand

Krairat Eiamampai
Department of National Parks, Wildlife and Plant Conservation, Ministry of Natural Resources and Environment, Bangkok 10900, Thailand

Pilaipan Puthavathana
Center for Research and Innovation, Faculty of Medical Technology, Mahidol University, Nakhon Pathom 73170, Thailand
Department of Microbiology, Faculty of Medicine Siriraj Hospital, Mahidol University, Bangkok 10700, Thailand

Index

CPSIA information can be obtained
at www.ICGtesting.com
Printed in the USA
BVHW012147171022
649694BV00004B/43

9 781639 275199